GLACIO-FLUVIAL SEDIMENT TRANSFER

GLACIO-FLUVIAL SEDIMENT TRANSFER

An Alpine Perspective

Edited by
A. M. GURNELL and M. J. CLARK
GeoData Unit and Department of Geography
University of Southampton

A Wiley–Interscience Publication

JOHN WILEY & SONS
Chichester·New York·Brisbane·Toronto·Singapore

Library of Congress Cataloging in Publication Data:

Glacio-fluvial sediment transfer.
  'A Wiley–Interscience publication.'
  Includes index.
  1. Sedimentation and deposition. 2. Glacial
landforms. 3. Glaciology. 4. Alluvium. I. Gurnell,
A. M. (Angela M.) II. Clark, M. J.
QE571.G53   1987   551.3'03   86-13172

ISBN 0 471 90929 7

British Library Cataloguing in Publication Data:

Glacio-fluvial sediment transfer
  1. Sediments (Geology) 2. Glaciers
  I. Gurnell, A. M. II. Clark, M. J.
  551.3'1   QE576

ISBN 0 471 90929 7

Printed and bound in Great Britain.

Contributors

A. Bezinge   Chef d'Exploitation, Grande Dixence S.A., Sion, Valais, Switzerland

M. J. Clark   GeoData Unit and Department of Geography, University of Southampton, Southampton SO9 5NH, UK

C. R. Fenn   Department of Geography, Worcester College of Higher Education, Henwick Grove, Worcester WR2 6AJ, UK

B. Gomez   School of Geography, University of Oxford, Oxford OX1 3TB, UK

K. J. Gregory   Department of Geography, University of Southampton, Southampton SO9 5NH, UK

A. M. Gurnell   GeoData Unit and Department of Geography, University of Southampton, Southampton SO9 5NH, UK

H. Lang   Geographisches Institut, Hydrologie ETH, Winterthurerstr. 190, CH-8057 Zürich, Switzerland

M. M. Lemmens   Laboratoire de Géomorphologie, Université Libre de Bruxelles, Faculté des Sciences, C.P. 160, Avenue F.D. Roosevelt, 50 B-1050 Bruxelles, Belgium

R. D. Lorrain   Laboratoire de Géomorphologie, Université Libre de Bruxelles, Faculté des Sciences, C.P. 160, Avenue F.D. Roosevelt, 50 B-1050 Bruxelles, Belgium

H. Röthlisberger   Versuchsanstalt für Wasserbau, Hydrologie und Glaziologie, ETH-Zentrum, CH-8092 Zürich, Switzerland

R. J. Small   Department of Geography, University of Southampton, Southampton SO9 5NH, UK

R. A. Souchez   Laboratoire de Géomorphologie, Université Libre de Bruxelles, Faculté des Sciences, C.P. 160, Avenue F.D. Roosevelt, 50 B-1050 Bruxelles, Belgium

v

Contents

SECTION II
GLACIAL SEDIMENT TRANSFER

SECTION III
FLUVIAL SEDIMENT TRANSFER

SECTION IV
IMPLICATIONS

Preface

The transfer of ice and water in glacier basins has been the subject of detailed research for many years, but it is only recently that the associated transfer of sediment has received significant attention. The primary aim of this book is to provide an integrated review of sediment transfer in alpine glacier basins, with particular emphasis on the glacial and fluvial components of sediment transfer. In describing sediment transfer processes, consideration is given to the morphological implications of the erosion and deposition of sediment, although these are not central components of the text. It is clear from the contributions that research on sediment transfer in alpine glacier basins is being pursued with enormous enthusiasm and, although great strides are being made, there is still much to discover. We are very grateful to the contributors for making the current status of research in this field and its future prospects so clear, and for giving us great encouragement by their support for this volume.

In addition to the support of our co-authors, we have received a great deal of help which we would like to acknowledge. Judith Maizels, Keith Richards and John Gurnell read and made very helpful comments on early versions of some of the chapters. Alan Burn and his colleagues in the Cartographic Unit, University of Southampton, spent many hours producing the diagrams for the chapters which were written by authors from Southampton University, and gave invaluable general advice on presentation of material. June Ghandi and Tina Birring tirelessly typed several chapters on to a word processor so that we could edit them with ease and John Threlfall gave a great deal of help with construction of the index. To all of these people we give our very sincere thanks.

Angela Gurnell
Michael Clark
University of Southampton

SECTION I

A Background to Glacio-fluvial
Sediment Transfer

Glacio-fluvial Sediment Transfer
Edited by A. M. Gurnell and M. J. Clark
©1987 John Wiley & Sons Ltd.

Chapter 1

Introduction

A. M. GURNELL

'The alpine and high-mountain areas of the world play an extremely important and distinctive role in the hydrological processes of the planet, and in the regional hydrology of all continents. It is in the alpine regions where meteorological, glaciological, periglacial and hydrological phenomena have most intimate and complex interaction and variability on short space scales and short time scales, yet the results of these interactions have a profound effect on hydrological regions over much greater distances and much longer time scales.'

(Roots and Glen, 1982, p. V)

Glaciated alpine areas form a fascinating basis for studying environmental processes. Glaciers can be found at virtually any latitude on the globe where the mountains are high enough and the moisture supply large enough to promote permanent ice cover. Figure 1.1 provides a schematic representation of the way in which the variables of altitude, latitude and climate interact to provide regional thresholds above which glaciers may occur. Thus, in mountainous areas in maritime climatic zones and in higher latitudes, glacier snouts will frequently extend to sea level, whereas in the arid continental areas of the tropics, permanent ice and snow cover can occur only at very high altitudes. The presence or absence of glaciers in a drainage basin above the regional threshold will, of course, depend on other local factors, notably the size, orientation and slope of the basin and the way that these interact with larger scale patterns of precipitation and insolation, to control moisture inputs to the basin and losses through ablation, evaporation and runoff and thus the rate of movement and the mass balance of areas of ice or snow within the basin. The significance of these local characteristics will also vary with macro-climate as it is affected by latitude and altitude and will affect the dynamism of the glacial system.

These alpine catchments with permanent snow and ice cover are of enormous importance in local, regional and continental water resources. They are also

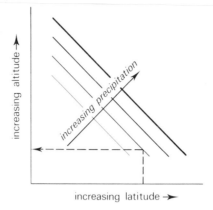

FIGURE 1.1   Schematic nomogram representing altitudinal thresholds for glacier formation in mountain areas and illustrating the interaction between latitude, altitude and climate

indicators of climatic processes and change with different glaciers responding at different rates and lags to adjustments in climate. Further, it is in these mountain zones that areas of permanent snow and ice impinge most closely upon man, acting as a source of water for agricultural and domestic use and for power generation, but also providing a range of environmental hazards which require prediction and control or alleviation.

The hydrological importance of alpine glacierized areas has long been recognized, as can be illustrated by the Proceedings of the Symposia of the International Association of Hydrological Sciences which have been devoted to the hydrology of areas of snow and ice cover, notably since the initiation of the International Hydrological Decade (1965–74) and the International Hydrological Programme which followed it. Of particular interest are the proceedings of the Cambridge Symposium in 1969 on 'The Hydrology of Glaciers' (IAHS Publication 95, 1973), the Proceedings of the Moscow Symposium in 1971 on 'Snow and Ice' (IAHS Publication 104, 1975) and the Proceedings of the Exeter Symposium in 1982 on 'Hydrological Aspects of Alpine and High Mountain Areas' (IAHS Publication 138, 1982). Information on fluctuations in the mass balance of glaciers throughout the world is regularly recorded and these data are accumulated in the World Glacier Inventory (Müller and Scherler, 1980), from which data are periodically published (e.g. Kasser, 1967, 1973; Müller, 1977).

The geological significance of glacierized areas has also been widely studied and the processes involved in the erosion, transport and deposition of sediment and the associated landforms and sedimentary structures in glacierized catchments have been the theme of several symposia (e.g. Fahey and Thompson,

1973; Jopling and McDonald, 1975; International Glaciological Society, 1979, 1981; Davidson-Arnott *et al.*, 1982).

This book considers the processes of weathering, erosion, transport and deposition of sediment in alpine glaciated catchments, emphasizing sediment transfer processes in particular, and it sets the consideration of the significance of these processes very firmly against a background of the periglacial, glacial and proglacial zones of such catchments. The glacial and proglacial zones are particularly strongly emphasized and the role of hydrological processes as a link between the glacial and proglacial zones of the catchment forms a major theme of this book. Thus the volume aims to discuss the interface between the purely hydrological (whether emphasizing snow, ice or meltwater) consideration of alpine drainage basins and the more sedimentological consideration of such basins and their deposits. The book is largely concerned with present processes, although some longer term perspectives will be provided.

One of the major problems faced in the preparation of this book was the dearth of alpine glaciated catchments in the world where integrated studies of the sediment system had been undertaken. Even the monitoring of total sediment yield from such catchments is rare and a great academic debt is owed to hydroelectric companies (particularly the Norwegian Water Resources and Electricity Board under the guidance of Dr Østrem), who have had to cope with the high sediment yields of these catchments and in doing so have collected invaluable, long-term records of sediment concentration and yield data. The shortage of catchments with integrated studies of the sediment system was the main impetus behind the production of this integrated volume. Although the book seeks to consider the glacio-fluvial sediment transfer processes of alpine areas in general, the authors contributing to the volume have all undertaken research within the same area of the Swiss Alps. Hence it has been possible to set a spatially broad consideration of different components of the sediment system of alpine catchments against a background of integrated study within the Swiss Alps and within the drainage basin of the Tsidjiore Nouve glacier in particular. Figure 1.2 describes the basic structure of the glacio-fluvial sediment system in an alpine glaciated catchment as a background to the detailed discussions of the components of the system in the book.

The book is subdivided into four parts. The first part includes review chapters on mountain sediment systems (Chapter 2), periglacial processes (Chapter 3) sediment transfer by ice and water (Chapter 4) and the hydrogeomorphology of proglacial zones within alpine catchments (Chapter 5). The second part emphasizes the ice-controlled components of the alpine sediment system, including englacial and supraglacial sediment transport and deposition (Chapter 6), subglacial processes (Chapter 7) and moraine sediment budgets (Chapter 8). Chapter 9 presents a brief overview of the glacial sediment system in alpine

SEDIMENT   WATER

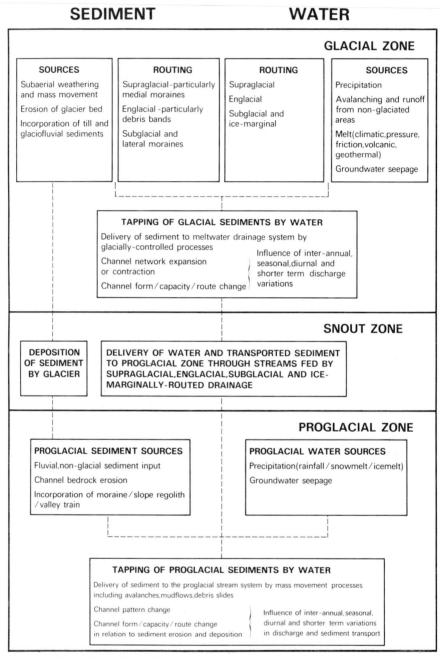

FIGURE 1.2   A description of the basic structure of the glacio-fluvial sediment system
in an alpine glaciated catchment

areas. The third part centres on the theme of glacial hydrology (Chapter 10) and the way in which meltwater transfers solutes (Chapter 11), suspended sediment (Chapter 12) and bedload (Chapter 13) from the glacial to the proglacial zone of alpine catchments. Chapter 14 considers how electrical conductivity, which provides a simple index of the solute concentration in the meltwater, can be employed to produce models of meltwater generation and routing within the glaciated zone of alpine catchments as an aid to the elucidation of the sources and routing of components of the sediment load of proglacial streams. Chapter 15 considers the difficulties of identifying components of the sediment load of proglacial streams and presents some data which allow an assessment of the relative significance of the components of the total sediment load. Finally, the fourth part considers the implications of sediment transfer in alpine glacierized catchments. Chapter 16 presents information on short-term variations in the form of proglacial streams as a response to variations in their discharge and sediment load. Chapter 17 considers the hydrology and sediment transport of proglacial streams within the applied context of the Grande Dixence hydroelectricity scheme. Finally, Chapter 18 considers the applications and implications of sediment transfer in alpine glaciated catchments.

REFERENCES

DAVIDSON-ARNOTT, R., NICKLING, W., and FAHEY, B. D. (Eds.) (1982), *Research in Glacial, Glacio-Fluvial and Glacio-Lacustrine Systems*, 6th Guelph Symposium on Geomorphology, 1980, Geo Books, Norwich, 318pp.

FAHEY, B. D., and THOMPSON, R. D. (Eds.) (1973), *Research in Polar and Alpine Geomorphology*, 3rd Guelph Symposium on Geomorphology, 1973, Geo Abstracts, Norwich, 206pp.

INTERNATIONAL GLACIOLOGICAL SOCIETY (1979), Symposium on Glacier Beds: the ice-rock interface (Ottawa, 1978), *Journal of Glaciology*, **23**, 89.

INTERNATIONAL GLACIOLOGICAL SOCIETY (1981), Symposium on Processes of Glacier Erosion and Sedimentation (Geilo, Norway, 1980). *Annals of Glaciology*, **2**.

JOPLING, A. V., and MCDONALD, B. C. (Eds.) (1975), *Glaciofluvial and Glaciolacustrine Sedimentation*, Society of Economic Palaeontologists and Mineralogists Special Publication 23, 320pp.

KASSER, P. (1967), *Fluctuations of Glaciers 1959–1965*, International Association of Scientific Hydrology and United Nations Educational, Scientific and Cultural Organization.

KASSER, P. (1973), *Fluctuations of Glaciers 1965–1970*, International Association of Hydrological Sciences and United Nations Educational, Scientific and Cultural Organization.

MÜLLER, F. (1977), *Fluctuations of Glaciers 1970–1975*, International Association of Hydrological Sciences and United Nations Educational, Scientific and Cultural Organization.

MÜLLER, F., and SCHERLER, K. (1980), Introduction to the world glacier inventory, in *World Glacier Inventory Workshop* (Proceedings of the Workshop at Reideralp, Switzerland, 17–22 September 1978), International Association of Hydrological Sciences Publication 126, pp. xiii–xx.

ROOTS, E. F., and GLEN, J. W. (1982), Preface, in GLEN J. W. (Ed.), *Hydrological Aspects of Alpine and High Mountain Areas* (Proceedings of the Exeter Symposium, July 1982), International Association of Hydrological Sciences Publication 138, pp. v–vi.

Glacio-fluvial Sediment Transfer
Edited by A. M. Gurnell and M. J. Clark
©1987 John Wiley & Sons Ltd.

Chapter 2

The Alpine Sediment System:
A Context for Glacio-fluvial Processes

M. J. CLARK

ABSTRACT

The glacio-fluvial sediment system is part of the larger and more complex alpine sediment system which incorporates slope processes as well as glacial and fluvial components. Although distinct boundaries are difficult to designate, it is clear that this sediment system can be studied meaningfully at a number of scales from the sub-continental to the individual small catchment or slope. The controlling factors on the geomorphological processes include latitude, altitude, biology, pedology and geology — all but the last having some climatic implication. These factors vary in relative and absolute importance, depending on the scale of study, and induce extremely important spatial variations which render the concept of a single alpine sediment system suspect. The same is true of temporal variations, again at a variety of scales. Indeed, the reconceptualization and recalibration of time-based models of the system or its components in recent years represent one of the most exciting aspects of the subject at the present time.

2.1 INTRODUCTION

The alpine sediment system can be thought of in two ways. Strictly, it serves as no more than a convenient conceptual framework within which the interrelated energy and material flows of the alpine zone can be studied and perhaps quantified. Alternatively, if the system can be regarded as existing in its own right outside the mind of the scientists who seek to master it, then it represents in essence the totality of materials, states and processes which combine to define the physical dynamism and variety of the mountain lands. Whilst climatologists, biologists and geologists might initially refute such a claim, on reflection they would probably see their own role in the sediment transfer system as being sufficiently powerful to accept the overall contention. On the other

9

hand, hydrologists, glaciologists and geomorphologists might initially welcome the sediment system as their own, but on reflection they too should accept that they share it with other disciplines. Above all, each of the contributing specialists should turn to the sediment system primarily for the benefit of its integrating potential: integration between processes (climatological, geological, biological, hydrological, glaciological); integration between places (supraglacial, glacial, subglacial and proglacial); and integration between times (preglacial, glacial, paraglacial and postglacial). Thus the sediment system comes to assume an importance considerably greater than might be deduced from the fact that in any given time period actual sediment movement occupies only a relatively small proportion of the total land surface.

Many other approaches seek relationships and linkages in the alpine environment, but whereas both energy flux and the ecosystem offer considerable scope for synthesis, no theme matches the breadth of the sediment transfer system. Given that the primary value of the system lies in its synthetic attributes, it would be anomalous to probe the details of any one subsystem (such as the glacio-fluvial sediment system) without first providing an alpine overview against which to evaluate the detailed treatment, and within which to identify boundaries to define the limits and neighbours of the subsystem. In effect, this approach takes the framework provided by Figure 1.2 and extends it both at the top (to the unglaciated zone above a glacier) and at the bottom (to the zone well beyond the glacier margin). It must be stressed that the following discussion is tempered by the requirement to provide a context for the glacio-fluvial sediment system, and does not presume to provide a comprehensive introduction to the alpine sediment system as a whole.

Such a view of the alpine sediment system concentrates on its formal interpretation as a conceptual framework, yet in a sense this viewpoint lacks much of the impact and power of the alternative, physical, connotation of the term. In such physical terms, the sediment system is a vast and complex machine, fuelled by tectonic and solar energy, whose role it is to shape the mountain landscape, carry its products and use them to create new sedimentary structures. Whereas the conceptual model may, even in a partial version, suggest in its elegance an element of both understanding and mastery, the physical entity impresses all researchers with a seemingly impregnable complexity and variety. Whereas the model generates respect, the physical system itself can only be regarded with awe.

However, it is inherent to systems thinking that the complexity inevitably produced by a holistic viewpoint can be assimilated by the twin strategies of generalizing the linking structures and analysing the detailed subsystems. Thus, the function of this chapter is to furnish a context for subsequent detailed discussion, with the necessary repercussion that it must stress perspective rather than precision. Any resulting deficiency will be tackled by the expedient of focusing on major components of the system in Chapter 3 (periglacial aspects),

Chapter 4 (glacial and hydrological aspects) and Chapter 5 (fluvial geomorphology) before proceeding to the detail of the glacio-fluvial system itself in the subsequent chapters. First, however, it is helpful to consider some of the characteristics of the alpine sediment system as a whole, to identify on this basis some of the major implications of current concern and to preview the methodological attributes which motivate and constrain this field of study (a topic more fully developed in Chapter 18). It will rapidly become apparent that the spatial and temporal dimensions of the sediment system are inextricably interlinked. Although this adds to the difficulty of their analysis, it also increases the justification for the attempt. Only by isolating spatial variability can a rigorous definition of temporal change (inherent in both palaeo-environmental and process-predictive tasks) be achieved, and only through a standardization of temporal phasing is it possible to identify real spatial variability.

2.2 DIMENSIONS OF THE SYSTEM: THE SPATIAL CONTEXT

2.2.1 Boundaries for the alpine system

The designation of formal boundaries for a system only becomes strictly necessary at the stage when a quantification of system attributes is to be attempted, which would be markedly premature in the case of the alpine sediment system as a whole. However, even if qualitative systematic conceptualization is the immediate aim, it is useful to be able to propose limits upon which to define both the components and the extent of the field of study. Even this very restricted target proves elusive for the alpine zone, since the term 'alpine' has overtones of environment which far transcend any strict altitudinal definition. Many attempts have been made to impose some kind of rigour on the terminology through general reviews of mountain geomorphology, for example by Löve (1970), Hewitt (1972), Ives and Barry (1974), Price (1981), Barsch and Caine (1984) and Ives (1985, 1987).

Given this degree of subjectivity, it is clearly impossible to propose a boundary for the alpine sediment system which is both comprehensive and unambiguous. Certain attributes appear to be diagnostic and will be assumed below, including high energy and characteristic landform elements (Hewitt, 1972), high elevation, steep slopes, rocky terrain and the presence of ice and snow (Barsch and Caine, 1984). Caine (1983) and Barsch and Caine (1984) proposed a dichotomizing structure for mountain geomorphology (Table 2.1) which implies that the alpine sediment system, although of considerable integrative significance, is representative of only one branch of the subject as a whole. Such overviews, of course, focus on the core content of the field rather than attempting to supply outer limits. To take the matter of system and subsystem boundaries further, it seems preferable to concentrate separately on spatial and temporal variability, on the assumption that these two dimensions will combine to yield a series of

TABLE 2.1   Research themes in mountain geomorphology (after Barsch and Caine, 1984)

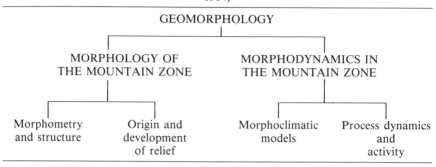

environmental gradients along which an operational definition of 'alpine' can be specified within the terms of any specific study. The failure to produce a general working definition is regrettable, but it is preferable to the designation of unrealistic or inflexible boundaries.

Perhaps the most significant role of the system boundaries in the present context is that they specify the scale of the landscape unit under consideration. Not only does this impinge directly upon the appropriate study techniques and precision, but it also relates to the process components likely to be dominant within the explanatory model. Thus, as Thorn and Loewenherz (1985) pointed out, the macro-scale alpine landscape remains dominated by past glacial and glacio-fluvial activity, leaving periglacial processes as important mainly at the micro- and perhaps the meso-scale. However, in terms of the present sediment transfer system, it may be that past glaciation is less universally dominant even at the macro-scale, at least in those areas that have not inherited major spreads of glacial debris. When it comes to present-day sediment processes, the dominant factors vary spatially at all scales depending on the catchment hydrological, geomorphological and climatic conditions. Perhaps most important in this respect are gross variations in total precipitation together with Barsch and Caine's (1984) articulation of the distinction between 'Alp type relief' (dominated by glacial erosion) and 'Rocky Mountain type relief' (displaying a lesser impact of glacial erosion and thus retaining significant summit and interfluve areas of low relief). Clearly, this book is devoted very much to the Alp-type environments.

2.2.2   Large-scale patterns of spatial variation

The two primary contemporary controls of global variability are latitude and altitude. Both act through climate mainly by inducing trends of increasing coldness, and in combination they offer a first approximation to the limits of the alpine environment. This linkage of climate and topography provides an

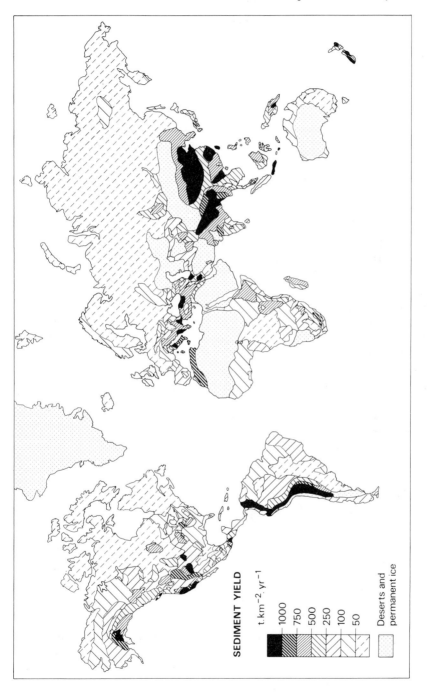

FIGURE 2.1   A revised map of global sediment yield. *Reproduced with permission from Walling and Webb (1983)*

important context for the explanation of the relationship between environment and process or morphological response, and has been referred to as topoclimate. The most obvious, but nevertheless striking, gradients which result are the loss in altitude of the snowline and the permafrost limits (Harris, 1985a) polewards along a north–south mountain range such as the Rocky Mountains. However, it is facile to suppose that such trends could be regular, even at the macro-scale. In the case of alpine permafrost, for example, regional deviations from a steady poleward decline relate both to snow depth control on ground insulation and to continental-scale air mass effects (Harris and Brown, 1982). Since individual components of the environment such as permafrost, treeline and frost cycle frequency are subject to different controlling factors, it is hardly surprising that they do not display any great correlation in spatial trend. As a result, a composite product of these environmental attributes such as sediment yield represents a major challenge for global mapping and explanation. Although there have been several attempts of increasing sophistication to define global trends in the spatial distribution of sediment yield (e.g. Fournier, 1960; Strakhov, 1967; Walling and Webb, 1983), the level of detail is necessarily low and the underlying explanation of trends remains elusive within any one climatic domain. Thus, Figure 2.1 provides a fascinating first approximation of the overall trends, and clearly distinguishes between the high sediment productivity of the alpine zone and the relatively low yields of arctic non-alpine areas. However, the internal variation within these two domains remains difficult to monitor and explain.

Nevertheless, one pattern that is immediately clear is the extent to which both latitude and altitude relate to factors other than just average temperature. For example, latitude determines solar angle, which in turn influences the variation in radiation balance (particularly in winter) and contributes to the extent to which asymmetry of environment exists on slopes of opposing aspect. Similarly, altitude not only triggers adiabatic temperature gradients, but also correlates with reduced atmospheric damping of incoming and outgoing radiation, and with the absolute relief which permits cold air drainage and ponding under suitable synoptic conditions (Harris, 1983; Karenlampi, 1972). This can lead to extremely rapid temperature fall and, particularly in northern latitudes, gives sufficiently large variations in temperature to yield an actual inversion even in the winter average temperatures. Thus, both of the primary controls of large-scale variation serve to produce not only regular global trends, but also globally differing sets of meso-scale variations. These patterns, affecting both temperature and moisture, are of sufficient magnitude to influence sediment yield, water yield and sediment transport.

Also at the macro-scale there are important climatically based longitudinal variations in environment. These act in part through precipitation shadow effects, which at the large scale tend to be manifested in increasing aridity to the lee of regional mountain masses (in the mid-latitudes generally involving a west–east reduction in precipitation). However, it must be remembered that once again meso-scale trends are superimposed, giving wind-vortex enhanced snow deposition in the lee of individual mountain barriers to the extent that it is often this snow feed factor which dominates the glacier mass budget, nival runoff and consequent sediment yield rather than any global trend. In addition, there are important longitudinal variations in temperature which are partly adiabatic and partly related to the dissipation of the ameliorating maritime effect inland. Such trends are significant in producing environmental gradients across such latitudinally elongated mountain chains as the Rockies, the Andes and the Fennoscandian cordillera, and help to explain why latitude is such a poor predictor of alpine system attributes.

Global trends in alpine environments are cogently summarized by Harris (1987) in the context of periglacial geomorphology. However, any introductory review inevitably underestimates both the breadth and depth of relationship with mountain climatology, with the result that a rigorous consideration of this important link demands reference to a volume of the substance of Barry (1981). Nevertheless, it would be a mistake to assume that the entire control of global trends rests with present-day climate, and at least two other factors need to be incorporated — geology and climatic history.

Whilst most aspects of lithology and structure can best be regarded as meso-scale factors, the global distribution of the mountainlands is essentially a product of mega-tectonics. Indeed, one connotation of the term 'alpine' refers to the young, high and dynamic mountain zones associated with contemporary converging plate margins. Not only is there a crude relationship between the age and height of mountain masses, but a link also exists between current tectonic dynamism and high rates of erosion and sediment yield, as in the Karakorum (Miller, 1984). This is partly a function of tectonic fracturing of materials, partly a result of altitude-related environmental severity and partly a reflection of the influence of slope steepness and altitude-induced precipitation enhancement on the rate of glacial and fluvial processes. Such factors must certainly be incorporated in any attempt to model global variations in the efficacy of the alpine sediment system.

Finally, it is imperative that due emphasis should be given to the role of climatic history in preconditioning the present-day sediment transfers. This topic is fully developed below in the context of temporal controls on the sediment system, but also has relevant spatial expression. For example, there is a gross difference in the magnitude of the sediment yield and sediment transport of those areas which have available extensive glacial deposits ready for reworking compared with those that are largely bereft of surficial material. Thus, whilst sediment export may serve as a crude index of denudation, it is considerably less directly related to actual erosion. Similarly, a period of intensive frost weathering (however defined) prior to the onset of glaciation will increase enormously the excavational propensity of the advancing ice. As in so many contexts, sediment yield is as much a function of sediment availability as it is of process power — a conclusion inherent in the paraglacial model discussed in Section 2.3.1.

2.2.3   Meso-scale patterns of spatial variation

Many of the controls of meso-scale variation have already been introduced, but it has been noted above that the controlling variables change with scale and thus attention must particularly be paid to the dominating effect of topography. This acts both through meso- and micro-climate, and through soil mechanics influences — both involving indirect associations with vegetation. In both the glacial and the extra-glacial zones, few factors are more important than snow distribution, which controls water storage, water release and ground temperature insulation and is subject to great variability in distribution at all scales. Within the mountain zone, snow depth relates to a complex of topographic and vegetative variables, the interactions between which produce a mosaic of varied snow terrain units of some inter-seasonal consistency. During the thaw period, a separate set of radiative (e.g. Williams *et al.*, 1972; Isard, 1983) and advective controls influence the phasing of the melting of the snow store mosaic, as was effectively demonstrated in a northern rather than high-altitude context by Kuusisto (1984). The resulting balance between snow input and output conditions the alpine snowmelt hydrograph strongly influences the activation of spatially distinct sediment sources and dominates the glacier mass budget.

It is also particularly important to emphasize again that moisture variability is just as important as temperature variability in affecting weathering, sediment release and sediment transport on the slopes above and beyond the glaciers. Given that water storage and release are also the fundamental variables of the glacial and fluvial systems, it is apparent why the spatial pattern of efficacy of the alpine sediment system discourages simple modelling such as might be possible if control lay with the more systematically variable patterns of temperature. Further, a number of the slope mass movement processes responsible for feeding sediment into rivers or glaciers are triggered by threshold conditions of such complexity that climatic change may well be unnecessary to explain some variations in sediment transport, and known climatic trends may still have relatively low power to predict sediment transfer response.

Although variations in snow retention and soil or surface moisture combine to provide a powerful explanation of the zonation of alpine sediment processes, a further primary factor requiring discussion is that of relative location on the slope profile. Whether this can be regarded as a causal factor rather than a product of the slope system is debatable, but its importance transcends semantics. Various frameworks have been proposed for handling variations in spatial location, and these are considered further as components of the sediment system in Chapter 3.

A number of case studies are now available which indicate sediment yield from glaciated basins (see Chapters 8 and 9), but so variable are the terrain, climatic and historical attributes of these studies that it is difficult to use them to isolate the underlying influence of any one variable on sediment production or transfer. This topic is also explored further in Chapter 3, but at a more general level it is possible to identify the paucity of comparable data sets and analyses as one of the most serious deficiencies in the study of the alpine sediment system. The mere proliferation of field or laboratory investigations will not in itself solve this problem. The need is for the imposition of certain sampling and descriptive standards, or for study boundaries to be extended across a number of catchments so as to permit some variables to be held pseudo-constant, thus allowing a clearer analysis of the influence of other factors. Clearly, there is a divergence here between academic priorities and research capabilities, and it may be that a solution must await the development of more automated data collection or of improved techniques of remote sensing or regional extrapolation. Certainly, although there is still an enormous challenge in monitoring, explaining and predicting within-catchment processes, the next fundamental stage in modelling the alpine sediment system lies at the multi-catchment scale. In the absence of such studies, this volume serves an important purpose in its attempt to codify existing alpine investigations within a common framework of comparison. In a sense, regional compilations such as that for the Canadian Rocky Mountains by Slaymaker and McPherson (1972) represent an earlier move in the same direction.

2.2.4   Micro-scale patterns of spatial variation

At the micro-scale of the single glacier, slope or river reach there is a considerable and growing understanding of sediment processes, although it is inherent to this scale that interrelationships between major components of the system are particularly difficult to identify. Indeed, it may be that there are limitations on the study of the complete sediment transfer system at anything below the meso-scale. This being so, we might regard the component studies which occupy much of this volume as supplying potential modules for future incorporation into a holistic model. The recent heavy concentration on investigating subsystems derives both from the analytical simplicity of this approach and from its suitability for accessing via the present generation of environmental monitoring instruments. However, there are limits to the advantages of reductionism, and signs that a more integrative priority may soon be in the ascendancy. As mentioned above, this does not represent a withdrawal from micro-studies, but rather a realization that such studies have the potential to build and calibrate larger scale hypotheses.

In moving from the meso- to the micro-scale, we witness some significant changes in the controlling variables, with local terrain factors (including vegetation, microlithology and soil) becoming increasingly important, and periglacial processes playing an ever more important role alongside glacial and fluvial components. Increasing influence is also attributable to variations in slope angle and aspect, acting through radiation balance, precipitation storage and slope drainage. Inevitably, this increases the degree of variability encompassed by a study, because generalizations become progressively less acceptable. Hence the production of a representative investigative design is rendered extremely difficult. Paradoxically, this decreases the extent to which micro-scale results can readily be incorporated into meso-scale models, and as a consequence the rigour of investigative (particularly sampling) design becomes crucial to the value and validity of the project. This important implication of the present scale of investigation is considered fully in Chapter 18.

2.3   DIMENSIONS OF THE SYSTEM: THE TEMPORAL CONTEXT

2.3.1   Large-scale temporal patterns

Alpine sediment processes vary in relation to time just as markedly as they do in relation to spatial location, and a further similarity is that the nature of this variation depends upon scale. It was pointed out by Barsch and Caine (1984) that part of this variability is predictable in terms of external influences such as diurnal or seasonal temperature regime, but that part seems to be infrequent and essentially random in nature. At the largest scale, it is accepted that many mountain areas, particularly in mid-latitude and mid-altitude locations, have undergone a time-bound oscillation between glacial, periglacial and non-glacial conditions — with associated contrasts in the attributes of the sediment system. However, whilst some aspects of this phasing are reasonably well documented and understood, the synthesis of such sequences into a coherent model of systematic change has only fairly recently been attempted by Church and Ryder (1972). In their recognition of a series of important changes in the state of the sediment system, they codified a number of previous observations and ideas within a single powerful terminology, and thus focused attention rightfully on the importance of always considering sediment dynamics against a background of relative chronology.

  If we concentrate on landscape change rather than on the absolute state of the system at any one time, then it is clear that the change from glacial to non-glacial conditions will represent a series of significant stresses as the various components of the system adjust at different rates to the altered and altering environment. Inevitably, there will ensue marked changes in the rates and patterns of sediment production and transfer, reflecting the susceptibility of the system during a phase of disequilibrium. It is not necessary to postulate that a single predictable sequence of states will be followed, but such a model does provide a convenient working basis for subsequent elaboration and evaluation.

  Although the term 'paraglaciation' had been previously introduced (Ryder, 1971a,b), it was first fully defined by Church and Ryder (1972) to cover:

'. . . nonglacial processes that are directly conditioned by glaciation. It refers both to proglacial processes, and to those occurring around and within the margins of a former glacier that are the direct result of the earlier presence of the ice. It is specifically contrasted with the term "periglacial", which does not imply the necessity of glacial events occurring.'

(Church and Ryder, 1972, p. 3059)

The underlying assumption of this concept is that glaciation represents a fundamental change in terrestrial erosional environment such that major quantities of sediment are produced in the form of glacial drift. This sediment may have reached stability with the depositional environment of the ice margin, but it is unstable with respect to the fluvial environment which succeeds the glacier spatially and temporally. Thus the proglacial and postglacial rivers evacuate the glacial sediment at a rate which is far in excess of the 'normal' material supply to be expected in the nonglacial environment. In effect, for

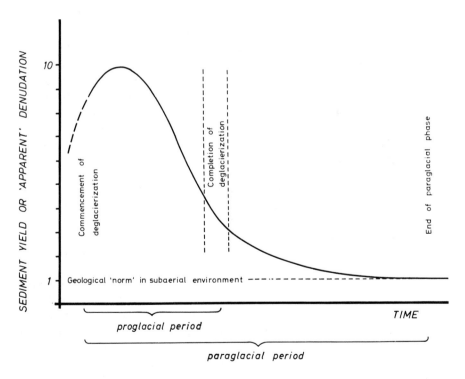

FIGURE 2.2   Schematic diagram of sedimentation during a paraglacial phase. *Originally published in Church and Ryder,* Bulletin of the Geological Society of America, **83** *3059–3077 (1972)*

periods of up to several thousands of years, sediment yield is unrelated to the concurrent primary production of debris by extraglacial weathering processes — an imbalance reflected in the widespread creation of alluvial fans, cones, plains and deltas in the proglacial zone, and the subsequent incision and erosion of these forms. The end-product of this sequence of change from a glacial to a nonglacial sediment-yield equilibrium is depicted in Figure 2.2.

Although both the concept of paraglacial sedimentation and the summary sequence of associated process changes are speculative, they represent an extremely valuable attempt to integrate spatial and temporal glacio-fluvial characteristics, with important implications for the estimation of long-term sediment transport and the inference of denudational trends from such patterns of transport (see Chapter 18). In the present context, however, it is particularly pertinent to highlight the hydrological consequences which are implicit in the paraglacial model but which receive only cursory discussion by Church and Ryder (1972). In its original formulation, the paraglacial sequence is controlled largely by variations in sediment availability. Abundant supplies of glacial drift available in the deglacial period facilitate high fluvial sediment loads, but in due course sediment exhaustion brings the system back to a more 'normal' pattern of sediment yield conditioned by concurrent rates of debris production by primary weathering. However, equal emphasis should be given to variations in discharge through the glacial, deglacial and postglacial periods, since these too influence the temporal pattern of sediment transport. The consequences in terms of proglacial fluvial geomorphology have been considered by Maizels (1979, 1983a,b), but purely hydrological causes and effects have not been fully investigated and warrant further attention.

In combination, the sequences of discharge and sediment availability go far to explaining the variability of sediment transport on both local and regional scales. The sequence can be likened to the hysteresis apparent in the annual pattern of proglacial sediment flux, and this comparison suggests that there may exist fruitful grounds for relating the behaviour of proglacial fluvial systems on long and short timescales — a possibility which indicates that contemporary field studies may have potential as analogues of the mechanism of the paraglacial sedimentation sequence. If this is the case, then it perhaps offers a route towards more sophisticated extrapolation of present day sediment yields or denudation rates than has hitherto been achieved.

Church and Ryder's (1972) model provides an invaluable hypothetical framework within which to consider variations in the alpine sediment system. Clearly, however, it is based on a blend of deductions from first principles together with an amalgam of spatially distinct observations fused into a single model through the ergodic theorem (see Section 18.2.3). As a consequence, it is perhaps best to regard the model as an hypothesis of great potential power, but one which now requires careful evaluation.

Real-time testing of the model at a site is impossible in view of the historical time-scale involved. Although much can be achieved simply by incorporating advances in process understanding since the publication of the model, confirmation and refinement would seem best sought through carefully regulated spatial comparisons. In practice, this is difficult to achieve, and a compromise is necessary. One possible approach is to assemble a literature-based tabulation of published studies, and to use this to demonstrate sediment system variations attributable to chronological phasing within a time-bound model. This has the potential for accessing major contrasts in phasing, but a glance at Tables 12.1, 12.2 and 12.3 is sufficient to demonstrate that standardization of measurement techniques and of other catchment characteristics is so difficult that a rigorous

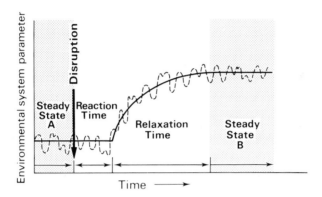

FIGURE 2.3   A rate law model of the response of a geomorphic system (such as sediment transfer) to disruption. *Reproduced with permission from Graf (1977)*

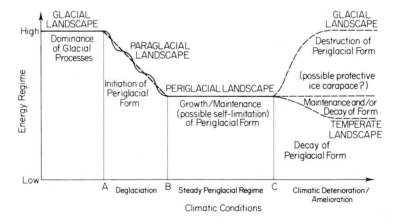

FIGURE 2.4   The evolution of a periglacial landscape with respect to changing environmental conditions. *Reproduced with permission from Thorn and Loewenherz (1985)*

evaluation of the phasing model is impractical. Alternatively, there is much potential in carefully standardized inter-catchment comparative studies within a single region. Such an approach could minimize technique and regional environmental attribute variation, leaving phasing as a more easily isolatable variable. However, the phasing variations concerned within a single region at a given time are those which are dependent on catchment characteristics (particularly those relating to snow reception and retention), so that the model tested is actually one of the meso-scale rather than the macro-scale. The test thus depends on a scale shift as well as the spatial shift implicit in the ergodic principle. Clearly, there is no simple means of testing the Church and Ryder model, though the strength and scope of the hypothesis renders such evaluation a high research priority for the alpine sediment system.

Two further aspects of the macro-scale demand attention before turning to other dimensions of the system. Firstly, it must be acknowledged that all of the process characteristics highlighted by Church and Ryder (1972), those associated with the disequilibrium of paraglaciation have attracted the greatest interest, perhaps because of their manifest dynamism and their relative neglect in the past. However, the rate law concept introduced by Graf (1977) as a negative exponential function with which to model the relaxation time of a geomorphic system (Figure 2.3), has been used explicitly by Thorn and Loewenherz (1985) to suggest that a similarly sensitive and active phase may be that associated with the initial or repeated refrigeration of the sediment system (Figure 2.4). This possibility has long been recognized in terms of the enhanced efficacy of individual processes such as the entrainment of pre-weathered debris by advancing ice, and as such is further discussed in Chapter 4. However, just as the notion of paraglaciation synthesizes many previous ideas into a single and more powerful unified concept, so that of refrigeration codifies a number of fragmented observations and suggests that the period of glacial onset might be worthy of greater attention in terms of sediment yield and hydrological change than it has previously been accorded.

Secondly, the process-based focus of much recent work has highlighted attempts to designate suites of process attributes pertaining at a site and between given times—an essentially inductive real-time investigative design. It is, therefore, pertinent to acknowledge a more fundamental but more traditional morphological interest in the conditions inherited by any area from its historical antecedants. Because many of the controls of the sediment system vary only slowly, they easily become out of phase with rapidly changing environments and their associated processes. Thus current attributes such as a mantle of glacial deposits, a mass of buried ice or frozen ground or an over-steepened rock slope may be inherited from the past but have a powerful influence on the sediment system of the present. This notion is, of course, inherent in the instability of the paraglacial and refrigeration phases, but it is helpful to be aware that similar underlying factors have also long been incorporated into other modes of explanation.

2.3.2 Meso-scale temporal variations

Meso-scale variations may reflect local, regional or global inputs, but they have the advantage that they are amenable to study at the catchment scale and within a period spanning historical and instrumental records. As a consequence, they are perhaps better understood than macro-scale patterns, and it is no surprise that the meso-scale is often used to model the macro-scale (for example, with a short-term glacial advance period serving as a model for paraglaciation). The justification for such temporal extrapolation rests on the assumption that the controls remain the same at all scales—an unproved, and indeed largely uninvestigated, assertion. Further, short-term changes are used to evaluate and calibrate dating and sequencing techniques. Once again, it is not necessary to challenge the principle of uniformitarianism in order to demonstrate that such projections require considerable caution. These reservations form one component in the evaluation of the modern analogue concept, which is discussed in Section 18.2.3.

In the present context, two aspects of the meso-scale pattern invite particular attention. The first concerns the role and interpretation of the sediment stores which are a product of reduced energy in the sediment transfer system. Such stores occur at a range of spatial and temporal scales, and yield significant landform assemblages such as talus slopes, protalus ramparts, solifluction lobes, moraines, channel bars or lake deltas. Whether they are regarded as sediment outputs or punctuated sediment transfers depends on system boundaries, and in many ways the dynamism and chronology of the system can be defined through the frequency, intensity and duration of the activation of these sedimentary forms. However, their overall significance lies as much in the sediment budget as in morphological evolution (cf. Walling and Webb, 1983). The interpretation of sediment yield or transport rates often rests on a measure of sediment output from a designated subsystem, and thus assumes that the magnitude of the stores within the subsystem remains constant. Whilst this may be valid for the micro-scale, it is less reliable at the meso-scale, and requires catchment-specific verification. It also varies between fine-sediment, coarse-sediment and geochemical fluxes. As in other branches of hydrology and geomorphology, there is a growing recognition that low-frequency but repeated medium magnitude events have been hitherto underestimated as means of accessing sediments which are unavailable during the intervening periods. Glacier surges, glacial outbursts and abnormally powerful or widespread avalanches and slushflows are examples of this type of event.

At a still smaller scale, there is growing interest in the variation in sediment transfer associated with advancing, static and retreating glaciers—a variation which is relevant in both spatial and temporal contexts (e.g. Maizels, 1979). The increasing frequency of meso-scale outbursts and the possibility of rising glacio-fluvial sediment loads associated with periods of advance are a case in point, and are considered at several points in Chapters 12 and 15, where it is concluded that further data are required before the significance of such relationships can be assessed. Clearly, such variations have important implications, and invite explanation. Are they a product of increased process intensity, or merely a reflection of access to previously inaccessible sediment sources? Are all components of the system enhanced simultaneously, or do the non-sympathetic changes in snow retention, temperature, ice flow and runoff produce a complex of changes which make it difficult to generalize the overall behaviour of the sediment system? Is it the nature of the system dynamism that is significant (i.e. the attributes of advance and retreat), or merely the fact that the sediment system responds to the instability related to systems change, and uses sediment flux to set in motion a series of morphological responses? If it is difficult to answer questions such as these, is it safe to use glacier advance and retreat to model longer term refrigeration and paraglacial phases?

2.3.3   Micro-scale temporal variations

Changes of between-day and seasonal duration dominate hydrological and climatic data from the alpine zone and have a significant influence on glacial processes and on many of the more rapidly responding periglacial processes. So great is the short-term variability that the recognition of meso- and macro-scale trends is often rendered difficult. Thus, even when the focus of study is elsewhere, the micro-scale cannot be ignored since it represents noise to be removed from the data set — a problem exemplified in Chapter 16. The failure to take account of this background variability must be regarded as a significant weakness in the investigative design of many long-term studies, which overlook the need to specify a sampling framework standardized with respect to short-term oscillations. Sampling design is even more critically important when the study aim is the specification, explanation or prediction of micro-scale temporal patterns. Sampling intervals should reflect the rate of change of the variable concerned. Thus, rockwall retreat can be modelled from annual measurements, but ground temperature may require weekly or daily data and many aspects of the hydrological subsystem are masked by anything substantially less than continuous records. The influence of sampling design on the nature of the patterns revealed by the resulting data is so great that this methodological problem is accorded detailed treatment in several of the subsequent chapters. As an interim generalization, however, it can be suggested that studies based on a sampling interval that is greater than a small fraction of the average phase length of the smallest event deemed significant are likely to delude their authors as much as they mislead their readers (not least because both will remain oblivious to the potential significance of events too short to be recorded). Given the logistic difficulties of achieving high-frequency sampling in the alpine zone, this standard has to be conceded to be an ideal to which relatively few projects can aspire.

Not surprisingly, despite its possible pitfalls, the attraction of direct instrumentation is such that many recent studies have centred on the micro-scale, although the results of such research have frequently been used to predict or reconstruct longer term trends. The dangers of such an approach are manifest, although its power may warrant occasional cautious acceptance. However, blatant extrapolation of trends whose representativeness and stability are unverified should be avoided, or should be accompanied by confidence limits to serve as a credibility index. In the context of the alpine sediment system, it is perhaps the field of estimating total catchment denudation trends that most often demonstrates a need for such precautions.

These essentially quantitative reservations should be coupled with qualitative vigilance if micro-scale studies are used to model meso- or macro-scale systems. Whilst the types of response involved may be similar (e.g. glacier advance and retreat, stage rise and fall, sediment concentration increase and decrease), it by no means follows that the controls of such response or the links between responses will be the same at all scales. Some properties, such as kinematic waves, appear to have counterparts at many scales. Others, such as the control of solute concentration by residence time of the water, may be effective over periods of tens of hours rather than tens of days, and would have little practical meaning over tens of centuries. The challenge of assessing the relative significance of processes of varying intensity and magnitude, and of assigning explanations to these various levels, is as great in the alpine sediment system as in most other branches of geoscience.

2.3.4 Models of time

Whilst a great deal of attention has been given by geomorphologists to the problem of conceiving a general model to incorporate both time-bound and time-independent changes, this type of discussion has mainly been focused on traditional segments of the landscape such as the slope or channel system. The literature is extensive, but can be accessed through comparative perspectives by Thornes and Brunsden (1977), Thorn (1982) and Gregory (1985). However, some advances in this field are specifically pertinent to the conceptualization of the alpine sediment system (e.g. Maizels, 1983a; Thorn and Loewenherz, 1985). For example, it is potentially rewarding to note that the three timescales discussed above could be reviewed in the context of Schumm and Lichty's (1965) distinction between cyclic, graded and steady time, and many of the connotations of that model apply to the alpine sediment system — including its designation of the changes of variables between dependent and independent status at

TABLE 2.2   The status of drainage basin variables during time spans of decreasing duration (*Reproduced with permission of the American Journal of Science from Schumm and Lichty, 1965*). Cyclic time equates to the macro-temporal and steady time to the micro-temporal

| No. Drainage basin variables | Status of variables during designated time spans | | |
	Cyclic	Graded	Steady
1 Time	Independent	Not relevant	Not relevant
2 Initial relief	Independent	Not relevant	Not relevant
3 Geology	Independent	Independent	Independent
4 Climate	Independent	Independent	Independent
5 Vegetation (type and density)	Dependent	Independent	Independent
6 Relief or volume of system above base level	Dependent	Independent	Independent
7 Hydrology (runoff and sediment yield per unit area within the system)	Dependent	Independent	Independent
8 Drainage network morphology	Dependent	Dependent	Independent
9 Hillslope morphology	Dependent	Dependent	Independent
10 Hydrology (discharge of water and sediment from the system)	Dependent	Dependent	Dependent

different timescales (see Table 2.2). Some redefinition would be appropriate for the alpine zone, but the concept remains helpful.

Similarly illuminating is Thornes and Brunsden's (1977) recognition of three types of serial dependence: 'trend' (a long-term pattern from which shorter term variations have been removed), 'periodicity' (regular fluctuations of repeating form, controlled by attributes intrinsic or extrinsic to the system) and 'persistence' (pattern relating either to a controlling physical factor or to data handling techniques). Whilst these three types of dependence certainly do not equate in any simple fashion with the timescales defined above, they do vary in relative importance at different scales, and thus have relevance in the search for an appropriate description and explanation of an observed time series. In much the same way, the notion of thresholds (Schumm, 1979) has been widely employed in interpreting the behaviour of systems which are subject to timebound input variations (such as those relating to fluctuations triggered by climatic history) or to internal changes in the state of controlling variables (such as progressive weathering or loading of sediments), both common occurrences in the alpine zone.

Finally, two extensions of these concepts can be proposed as having particular relevance to the alpine sediment system. First, the timing of the response of a system element to disruption is subject to a complex set of internal and external controls, but is modelled effectively by the application of the rate law (Graf, 1977). This distinguishes between the reaction time during which the system absorbs the impact of a disruption and the relaxation time during which adjustment of the system takes place to transform it to a new steady state (Figure 2.3). An application of this formulation to the Church and Ryder (1972) paraglacial model by Thorn and Loewenherz (1985) has already been discussed above. Given the repeated external and internal changes to the sediment system over the Pleistocene, there would appear to be considerable scope for further application of the threshold and rate law concepts.

Second, the very notion that a controlled change from one steady state to another takes place as a predictable response to given alterations to a system has recently been challenged, for example by Thornes (1983, 1986), who argues that analyses of long-term behaviour of landforms should be soundly based in theory rather than being inferentially derived from historical or extrapolative studies. Starting with the notion of process domains (regions characterized by a given process assemblage, or by a location on a series of environmental gradients), he distinguishes between stable and unstable systems in terms of the extent of their tendency to remain constant or return to a previous state after imposition of a disruption. Moving from domain to domain, the relative dominance of different processes changes, and if the boundary between domains is sharp then a small change in the controlling variables may result in a rapid and extensive change in behaviour.

This concept has spatial expression in the frequent terrain changes witnessed in alpine areas, and temporal manifestation in the way in which relatively small climatic fluctuations during the Pleistocene have triggered substantial and widespread icesheet growth phases. Perhaps even more interesting in the present context is the suggestion that a given set of systems variables may achieve any one of a number of equally viable equilibria, rather than being constrained to a unique mechanistically defined end product. This approach, encapsulated by Thornes in the concepts of dynamical (as opposed to dynamic) equilibrium and chaotic behaviour, frees natural science from many of the apparent limitations of systems analysis, and opens the way to relating many ostensibly varied systems states within a single model. It has yet to be applied in detail to the alpine sediment or glacio-fluvial sediment systems, but appears to have attributes of great advantage to both.

2.4 ARCTIC AND ALPINE SEDIMENT SYSTEMS—
LINKS AND CONTRASTS

Some attention has already been paid in Section 2.2.1 to the problem of defining the alpine zone and the reasons for attempting to do so. However, specific comparison of the arctic and alpine zones is of value because the justifiable interest in ensuring that similarities and links are exploited (e.g. Ives and Barry, 1974) can otherwise lead to the facile assumption that contrasts are insignificant. Although this latter viewpoint is rarely articulated, it is implicit in many studies which draw on the proglacial and periglacial literature without reference to the detailed nature of the environment concerned—another example of the limited extent to which the real significance of spatial variation has been recognized.

In addition to the largely climatic points already made in Section 2.2.2, it must be stressed that there are important differences between polar and alpine permafrost (Harris, 1985b; Hegginbottom, 1984), and between cold-based and warm-based glaciers (summarized by Paterson, 1981), both of which induce major contrasts between the sediment system driven by high latitude and that of high altitude areas. Further, the varied mountain topography will inevitably exercise considerable influence on sediment processes. For example, the interfluve and talus-foot slope zones recognized by Caine (1974) are characterized by slower sediment transport and a greater degree of sediment comminution than is found on the other steeper alpine slope components. However, the relative proportions of the alpine zone represented by each of these components varies markedly from region to region. For example, in the Colorado Front Range of the Rocky Mountain system where the model was developed, interfluve plateau remnants are markedly better preserved than is the case in much of the European Alps. In general, the similarity between arctic and alpine processes is greater for low-angle slopes than if steep alpine slopes are compared with low-angle arctic areas, since this standardizes the gravity component—a conclusion which is surprising only in the extent to which it is frequently overlooked.

Similar semantic and systematic problems are met in the attempt to distinguish between glacial and periglacial geomorphological processes. Whilst the end members of this spectrum are easily identified, there is a central set of processes and forms, often loosely (and perhaps erroneously) ascribed to nival dominance of the environment, that defies simple classification. This problem is admirably exemplified by the contention surrounding the concept of nivation, a term introduced by Matthes (1900) that has gained widespread acceptance as a collective noun for the assemblage of processes that are intensified by a late-lying snowpatch, and for the associated landforms (Thorn, 1987). Thorn concluded that '. . . nivation is a term that no longer withstands close inspection,' and that 'the emotional stress of abandoning the term would in large part be alleviated by the clarity gained.' More important in the present context, he found little support for the notion of a continuum of processes and forms, linked through both evolution and morphological similarity, from the snowpatch in its hollow to the cirque glacier in its cirque. In this and several other contexts, there would seem to be a strong case for distinguishing more carefully between glacial and non-glacial processes, for recognizing the important role played by non-glacial processes and for according them considerably more detailed inspection (as opposed to a preconception-bound glance) than has been common in the past.

Finally, it is clear that considerable differences exist between arctic and alpine hydrology, and that these have considerable impact on the respective sediment systems. Such contrasts impinge on all aspects of hydrology, including precipitation totals and distribution, precipitation efficiency, storage and the release of water during diurnal, synoptic or seasonal warming phases, and the mechanical, chemical and thermal processes which render sediment available for evacuation. The essentially alpine hydrological processes of this book can be compared with the arctic hydrology summary by Clark (1987).

2.5 CONCLUSIONS

Having considered the broad framework of spatial and temporal dimensions within which the glacio-fluvial sediment system has to be handled, and having recognized some of the distinctions between the arctic and alpine variants of this system, it is now possible to consider more specifically the process inputs to this system. Chapter 3 thus provides an introduction to the systems and non-systems models which have been proposed, and considers the periglacial processes which are pertinent but which are not addressed in detail elsewhere in this volume. The glacial and hydrological components are then introduced in Chapter 4, together with their links to fluvial geomorphology in Chapter 5. These introductory reviews provide a general setting for the detailed research reports on glacio-fluvial sediment transfer which follow. On this basis, it is then possible in Chapter 18 to identify some of the underlying methodological issues which relate to many aspects of the study of glacio-fluvial sediment processes, and which provide an evaluative perspective on the field as a whole.

REFERENCES

BARRY, R. G. (1981), *Mountain Weather and Climate*, Methuen, London.

BARSCH, D., and CAINE, N. (1984), The nature of mountain geomorphology, *Mountain Research and Development*, **4**, 287–298.

CAINE, N. (1974), The geomorphic processes of the alpine environment, in IVES, J. D., and BARRY, R. G. (Eds.), *Arctic and Alpine Environments*, Methuen, London, pp. 721–748.

CAINE, N. (1983), *The Mountains of Northeastern Tasmania. A Study of Alpine Geomorphology*, Balkema, Rotterdam.

CHURCH, M., and RYDER, J. M. (1972). Paraglacial sedimentation: a consideration of fluvial processes conditioned by glaciation. *Bulletin of the Geological Society of America*, **83**, 3059–3071.

CLARK, M. J. (1987), Periglacial hydrology, in CLARK, M. J. (Ed.), *Advances in Periglacial Geomorphology*. Wiley, Chichester (in press).

FOURNIER, F. (1960). *Climat et Érosion*, P.U.F., Paris.

GRAF, W. L. (1977), The rate law in fluvial geomorphology, *American Journal of Science*, **277**, 178–191.

GREGORY, K. J. (Ed.) (1985), *The Nature of Physical Geography*, Edward Arnold, London.

HARRIS, S. A. (1983), Cold air drainage west of Fort Nelson, British Columbia, *Arctic*, **35**, 539–541.

HARRIS, S. A. (1985a), Distribution and zonation of permafrost along the eastern ranges of the Cordillera of North America, *Biuletyn Peryglacjalny*, **30**, in press.

HARRIS, S. A. (1985b), *The Permafrost Environment*. Croom Helm, Beckenham, Kent, 288pp.

HARRIS, S. A. (1987), The alpine periglacial zone, in CLARK, M. J. (Ed.), *Advances in Periglacial Geomorphology*, in press.

HARRIS, S. A., and BROWN, R. J. E. (1982), Permafrost distribution along the Rocky Mountains in Alberta, in FRENCH, H. M. (Ed.), *Proceedings of the 4th Canadian Permafrost Conference, Calgary, Alberta*, National Research Council of Canada, Ottawa, pp. 59–67.

HEGGINBOTTOM, J. A. (1984), The mapping of permafrost, *Canadian Geographer*, **28**, 78–83.

HEWITT, K. (1972), The mountain environment and geomorphic processes, in SLAYMAKER, H. O., and McPHERSON, H. J. (Eds.), *Mountain Geomorphology: Geomorphological Processes in the Canadian Cordillera*, B.C. Geographical Series, No. 14, Tantalus Research, Vancouver.

ISARD, S. A. (1983), Estimating potential direct insolation to alpine terrain, *Arctic and Alpine Research*, **15**, 77–89.

IVES, J. D. (1985), Mountain environments, *Progress in Physical Geography*, **9**, 425–433.

IVES, J. D. (1987), The mountain lands, in CLARK, M. J., GREGORY, K. J., and GURNELL, A. M. (Eds.), *Horizons in Physical Geography*, Macmillan, Basingstoke and London (in press).

IVES, J. D., and BARRY, R. G. (Eds.) (1974), *Arctic and Alpine Environments*, Methuen, London.

KÄRENLAMPI, L. (1972), Comparison between the microclimates of the Kevo ecosystem study sites and the Kevo meteorological station, *Reports from the Kevo Subarctic Research Station*, **9**, 50–65.

KUUSISTO, E. (1974), Snow accumulation and snowmelt in Finland, *Publications of the Water Research Institute*, National Board of Waters, Finland, **55**, 149 pp.

LÖVE, D. (1970), Subarctic and subalpine: where and what?, *Arctic and Alpine Research*, **2**, 63–73.

MAIZELS, J. K. (1979), Proglacial aggradation and changes in braided channel patterns during a period of glacier advance: an alpine example, *Geografiska Annaler*, **61A**, 87–101.

MAIZELS, J. K. (1983a), Proglacial channel systems: change and thresholds for change over long, intermediate and short timescales, in COLLINSON, J., and LEWIN, J. (Eds.), *Modern and Ancient Fluvial Systems*, International Association of Sedimentologists, Special Publication, Blackwell Scientific Publications, Oxford, 251–266.

MAIZELS, J. K. (1983b), Palaeovelocity and palaeodischarge determination for coarse gravel deposits, in GREGORY, K. J. (Ed.), *Background to Palaeohydrology*, Wiley, Chichester, 101–140.

MATTHES, F. E. (1900), *Glacial Sculpture of the Bighorn Mountains, Wyoming*, United States Geological Survey 21st Annual Report, 1899–1900, pp. 167–190.

MILLER, K. J. (Ed.) (1984), *The International Karakoram Project*, 2 vols., Cambridge University Press, Cambridge.

PATERSON, W. S. B. (1981), *The Physics of Glaciers*, 2nd ed., Pergamon Press, Oxford.

PRICE, L. W. (1981), *Mountains and Man*, University of California Press, Berkeley.

RYDER, J. M. (1971a), The stratigraphy and morphology of paraglacial alluvial fans in south-central British Columbia, *Canadian Journal of Earth Sciences*, **8**, 279–298.

RYDER, J. M. (1971b), Some aspects of the morphometry of paraglacial alluvial fans in south-central British Columbia, *Canadian Journal of Earth Sciences*, **8**, 1252–1264.

SCHUMM, S. A. (1979), Geomorphic thresholds: the concept and its applications, *Transactions Institute of British Geographers*, **NS4**, 485–515.

SCHUMM, S. A., and LICHTY, R. W. (1965), Time, space and causality in geomorphology, *American Journal of Science*, **263**, 110–119.

SLAYMAKER, H. O., and McPHERSON, H. J. (Eds.) (1972), *Mountain Geomorphology: Geomorphological Processes in the Canadian Cordillera*, B.C. Geographical Series, No. 14, Tantalus Research, Vancouver.

STRAKHOV, N. M. (1967), *Principles of Lithogenesis*, Vol. 1, Oliver and Boyd, Edinburgh.

THORN, C. E. (Ed.) (1982), *Space and Time in Geomorphology*, Allen and Unwin, London.

THORN, C. E. (1987), Nivation: a geomorphic chimera, in CLARK, M. J. (Ed.), *Advances in Periglacial Geomorphology*, Wiley, in press.

THORN, C. E., and LOEWENHERZ, D. S. (1985), Interaction between spacial and temporal trends in alpine periglacial landform evolution: implications for palaeocharacter, Lecture presented to First International Geomorphology Conference, Manchester.

THORNES, J. B. (1983), Evolutionary geomorphology, *Geography*, **68**, 225–235.

THORNES, J. B. (1987), Environmental systems: patterns, processes and evolution, in CLARK, M. J., GREGORY, K. J., and GURNELL, A. M. (Eds.), *Horizons in Physical Geography*, Macmillan, London, in press.

THORNES, J. B., and BRUNSDEN, D. (1977), *Geomorphology and Time*, Methuen, London.

WALLING, D. E., and WEBB, B. W. (1983), Patterns of sediment yield, in Gregory, K. J. (Ed.), *Background to Palaeohydrology*, Wiley, Chichester, 69–100.

WILLIAMS, L. D., BARRY, R. G., and ANDREWS, J. T. (1972), Application of computed global radiation for areas of high relief, *Journal of Applied Meteorology*, **11**, 526–533.

Glacio-fluvial Sediment Transfer
Edited by A. M. Gurnell and M. J. Clark
© 1987 John Wiley & Sons Ltd.

Chapter 3

Geocryological Inputs to the Alpine Sediment System

M. J. CLARK

ABSTRACT

The alpine sediment system provides a broad cryogenic framework within which
it is possible to assess both the role and linkage of the glacio-fluvial sediment
transfer subsystem. Models of alpine sediment transfer have often been little more
than structured inventories of process, but a growing interest in process integration
has produced several contributions which have systems characteristics. Whilst a
comprehensive review of non-glacial sediment processes is inappropriate, it is
helpful to establish the significance of geocryological processes in three contexts —
alpine permafrost, the role of snow and the slope processes themselves. Even such
a brief and partial discussion reveals numerous influences on the glacio-fluvial
system, and at the same time indicates that current research is achieving progress
both through a growing consensus and through a new willingness to face contention
and accept change in traditional views.

3.1 THE NATURE OF THE ALPINE SEDIMENT TRANSFER SYSTEM

Whilst the glacio-fluvial sediment system is often dominated by the glacial
component, it is misleading to assume that nonglacial processes and materials
play a trivial role. Figure 1.2 represents a simple model of the system which
demonstrates two contexts within which nonglacial inputs of both water and
sediment are significant in all mountain catchments except those which are
currently wholly glacierized. First, the slopes above the glacier are an important
source of water and sediment yield to the ice, subsequently to become part of
the short- or long-term glacier store. The proportion of the water and sediment
derived from the extra-glacial zone varies greatly, depending on topography,
terrain characteristics and percentage of area glacierized, but can be substantial.
Second, further important additions of water and sediment can take place
directly into proglacial rivers from the slopes above. These mix with the glacial

discharge and render the accurate definition of water and sediment source difficult, unless monitoring is undertaken close to the snout so as to distinguish between glacial and nonglacial components. This in turn makes accurate functional definition of 'proglacial' difficult, and reduces the extent to which studies categorized as proglacial can be compared in any meaningful way.

Thus, whether the focus is on the alpine sediment transfer system as a whole or on the glacio-fluvial subsystem, it is important to recognize the significance of nonglacial inputs. However, since the emphasis of this volume is deliberately on the glacial and fluvial components, the nonglacial aspects are introduced simply as a background perspective in this chapter. Whilst integrated approaches to mountain geomorphology have developed strongly over the last decade (e.g. Barsch and Caine, 1984; Ives, 1987), much of the literature on nonglacial alpine processes has been associated with the study of slopes or periglaciation, the latter being a particularly significant focal point but one which raises increasingly problematic issues. This is another case where definition of a term has attracted considerable academic attention over a long period (e.g. Hamelin, 1964). The concept of periglaciation was introduced to characterize the zone adjacent to present ice, but has subsequently been applied to the morphological features of that zone, to similar features in zones not currently adjacent to ice, to the processes responsible for such features (whether or not juxtaposed to ice) and to the overall environmental system dominated by (but not exclusively related to) these processes. Given such a marked absence of semantic stability, it is not surprising that in recent years there has been a tendency to abandon the term in favour of a more robust nomenclature.

Although consensus is lacking, 'geocryological' appears to be gaining favour as a general term encompassing all the cold-region processes, but avoiding the spatial (i.e. near the glaciers) and temporal (i.e. before or after the glaciers) connotations of 'periglacial.' Geocryology assumes that low temperature is a significant characteristic of the environment, but does not demand that frost, permafrost or ice should be ubiquitous. The term thus includes the cold-region phase of azonal processes such as fluvial, aeolian, soil mechanic and rock mechanic processes. In short, geocryology subsumes periglaciation and extends beyond it to include related geological, geomorphological and hydrological attributes which render it useful in defining boundaries for a framework within which to assemble a model of the nonglacial components of the high-altitude and high-latitude environments. However, the structure of that framework is open to considerable variation, and warrants specific attention before the individual components are considered.

3.2 FRAMEWORKS FOR AN ALPINE SEDIMENT TRANSFER MODEL

Many different models of alpine sediment transfer have been introduced either explicitly or implicitly, and the variety of possible structures is considerable.

With the benefit of hindsight, it can be suggested that two broad approaches have been adopted, the first concentrating on inventory and hierarchy (i.e. stressing components) and the second on linkage and inter-relationship (i.e. stressing structure). The suggested distinction between the two is instructive but in no way rigorous, and should not be regarded as an exclusive classification. Further, given the history of geomorphology over the last three decades, it would not be surprising to find that the list reflected a chronology which involved trends from morphology to process, from univariate to multivariate, from discipline-focused to interdisciplinary, from mesoscale to microscale and back and from pure to applied. However, whilst echoes of these trends are discernable, there are some interesting departures.

3.2.1 Structured inventories of sediment processes

Nothing illustrates more clearly the capacity of science to surprise than Rapp's (1960) monumental study of the Kärkevagge Valley in northern Sweden — certainly qualifying as mountain geomorphology, although less 'alpine' in aspect than Switzerland. This paper still has complete currency a quarter century after its publication, and has not since been bettered in its completeness (spatial, temporal or constituent) and rigour. It is ironic, then, that this integrated multicomponent quantitative study — by any standards way ahead of its time — should be quoted as a conceptual foundation rather than culmination. Yet in many senses Rapp's contribution was in his pioneering demonstration of the components of the system, together with a refreshingly open-minded indication of their relative significance, and he shared with his contemporaries a lesser interest in the structure of inter-relationship which linked the elements.

Kärkevagge is a glaciated mountain trough 5 km in length and 2 km wide situated in the mountains of northern Sweden at latitude 68° N. Rapp attempted a comprehensive quantitative inventory of sediment transfer processes with the exception of surface wash, which he regarded as insignificant in this area. His results are summarized in Table 3.1, which indicates the two different concepts of measurement used — one focusing on the amount of material subject to

TABLE 3.1 Annual values of slope denudation in Kärkevagge, Sweden (after Rapp, 1960)

Process	Volume (m³)	Density	Tons	Average movement	Average gradient	Ton metres vertical
Rockfalls	50	2.6	130	90–225	45°	19 565
Avalanches	88	2.6	229	100–200	30°	21 850
Earthslides	580	1.8	1036	0.5–600	30°	96 375
Talus creep	300 000	1.8		0.01	30°	2 700
Solifluction	650 000	1.8		0.02	15°	5 300
Solutes	150	2.6	390	700	30°	136 500

transport (tons) and the other combining this with an estimate of the average vertical distance across which this movement took place (ton metres vertical). The two indices suggest fundamentally different rankings for the processes involved. In terms of simple amount of material in motion, mass movement is clearly dominant, but perhaps the most striking outcome of the study was its suggestion that by incorporating transport distance the role of solutes became dramatically enhanced, in fact being quantitatively almost equal to all the other processes combined. Given the considerable detail on each of the component processes, the close attention to micro- and meso-temporal distribution of events and the innovative views on the ranking of sediment transport processes on mountain slopes, it is not surprising that this study has had a substantial influence on both mountain and periglacial geomorphology.

Perhaps as a result of the conceptual and technical turmoil of the 1960s, or of the resulting concentration on analytical studies, it was a decade or so before the next major steps were taken in the formulation of a synthetic approach to the alpine sediment system.

Systems-based advances are discussed in Section 3.2.2, but in parallel with these it is possible to identify a continuation of structured tabular thinking. For example, a work of lasting significance was that of Ives and Barry (1974), which represented the influence of the Institute of Arctic and Alpine Research (University of Colorado) at its zenith, and projected INSTAAR's philosophy of blending interdisciplinary integrated study with long-term rigorous monitoring—championing the earlier approach of Rapp.

Slaymaker's (1974) review of alpine hydrology in Ives and Barry's (1974) book demonstrates both the achievements and the continuing immaturity of the subject, but is most interesting in the present context for its structured inventory of alpine hydrological processes (Table 3.2). This tabulates the processes against terrain type (defining boundary conditions) and timescale, a covert recognition of the importance of spatial and temporal variability. The approach is overtly mesoscale, so that detail is tantalizingly lacking, but the categorization incorporates many of the distinctions still designated as significant in the present volume. In the case of mountain streams, for example, Slaymaker suggests glacial erosion as the major control over periods of thousands of years, fluvial downcutting and slope debris supply over hundreds of years, discharge regime and sediment availability over months to decades and flow conditions or individual hydrograph response over periods of less than a month. However, although relationship may be implicit it is certainly not explicit, and this is clearly a classification rather than a system. This should not be taken to indicate a constraint on the concept, since Caine (1971) had already demonstrated in an alpine context that it was entirely possible to present the same ideas in tabular and systems frameworks.

The extent to which functional relationships can be built into a non-systems presentation was further indicated by Stäblein (1984) in a study of geomorphic

TABLE 3.2 Summary of major hydrological processes operating in mesoscale alpine environments (*reproduced with permission from Slaymaker, 1974*)

Alpine environment	> 1000 yr	10–1000 yr	0.1–10 yr	< 0.1 yr (1 month)
Glaciers	Water storage term in hydrological cycle	Glacier retreat and advance	Ablation Accumulation	Physics of glacier motion
Snowpacks			Snow metamorphism Firn ice	Melt Storage Accumulation
Alpine lakes	Glacier scour Lake formation Sedimentation Meadow forming	Sedimentation	Temp. strata Draining event Ice forming and breakup Sedimentation	Seiches Level change Sediment movement Evaporation
Mountain streams	Glacial erosion	Downcutting Supply (creep)	Discharge Sediment (talus)	Runoff concentration Response to precipitation/ evaporation
Morainic mounds	Glacial deposition	Moraine degradation	Chemical weathering Frost action Veg. change	Infiltration Interflow
Alp/subalp meadows, valley bottoms	Glacial lake sedimentation	Dissection by streams Flood plains Soil forms	Periodic inundation Jökulhlaups Veg. change	Precipitation Infiltration Interflow Baseflow
Alp/subalp meadows, adret	Postglacial soil development	Slopes degrade Channel dissection	Mass wasting Veg. change	Precipitation Infiltration Interflow
Alp/subalp treed adret	Postglacial soil development Soil creep Gullying	Slopes degrade Channel dissection	Frost action Mass wasting	Interception Precipitation Infiltration Interflow
Alp/subalp treed ubac and ridge top	Postglacial soil development Earth/mud slide/flow	Slopes degrade	Frost action Mass wasting	Overland flow Precipitation Interception Infiltration Interflow
Alpine	Glacial	Slopes degrade	Frost action	Overland flow
Barren	Erosion Postglacial falls/talus		Mass wasting	Precipitation

TABLE 3.3   Geoecological/geomorphological catena for the arctic/alpine zone
(Reproduced with permission from Stäblein, 1984)

Geoecological/ geomorphological zone	Characteristics
1. Zone of peaks and plateaus	Belt of peaks, plateaus and saddle sites with frost and solution weathering, slope wash, cryoturbation and deflation forms
2. Zone of upper slope sites	An upper slope belt where frost weathering, rockfall and debris creep occur
3. Zone of middle slope sites	A mid-slope belt with slope wash, solifluction, nivation and slope dissection
4. Zone of lower slope and terrace sites	A lower slope belt often with terraces having cryoturbation, frost heave and wind deflation and accumulation
5. Zone of coastal sites	Coastal sites with marine terraces, deltaic deposition, frost weathering and cliff formation by marine and thermal erosion

variance in the mountains of Greenland. He sees the primary controlling factor as being altitude acting through permafrost distribution and the geomorphological transfers of energy. No claim is made that a simple regional altitudinal zonation exists:

'Cryogenic processes of the periglacial morphodynamic system depend more on local climatic parameters that are conditioned by geomorphological, edaphic and orographic factors than on altitudinal limits.'

(Stäblein, 1984, p. 330)

Nevertheless, Stäblein does invoke at local scale an underlying catena model which is summarized in Figure 3.1 and in Table 3.3. The significance of this 'geomorphodynamic catena' is that it retains the basic tabular form but very clearly indicates transfers and links, and thus moves substantially towards the systems mode. Surprisingly for the mid-1980s, the catena is entirely two-dimensional. It represents downslope transfers, but involves no link with the glacier upvalley (or in the past), nor does it depict the role of the slope-foot channel (whether fluvial or glacial) either in terms of sediment evacuation or as a local base level control. Perhaps it is this dimensional limitation which most obviously categorizes Stäblein's model, and others like it, as structured inventories rather than as truly functional systems.

This distinction was reinforced by Barsch and Caine (1984), who blended systematic and tabular thinking to the point where the attempt to differentiate the two appears meaningless. Ostensibly, the main statement on alpine sediment

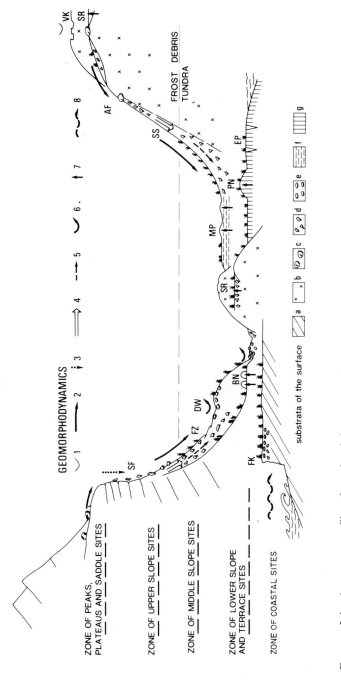

FIGURE 3.1 A catena profile of arctic–alpine mountain geomorphodynamics with relative altitudinal belts. *Reproduced with permission from Stäblein (1984)*

TABLE 3.4   Summary of a four-way classification of mountain sediment transport process systems (after Barsch and Caine, 1984)

Process system	Characteristics
1. The glacial system	Controlled by elevation and climate. Water movement in solid phase transports debris from rock walls and ice/rock erosion. High transporting power as far as glacier margin where deposition/fluvial transport occurs. Small change of glacier mass gives large change in eroding power
2. The coarse debris system	Also characterizes high mountain terrain. Transfers detritus from cliffs to the talus etc., below, and includes slope failure. In maritime zones slopes feed the glacial system, but in continental areas the downvalley extension is as rock glaciers. Climatic change producing glaciers allows periodic removal of the accumulated debris
3. The fine sediment system	An open system transporting sediment from aeolian dust and weathering, much being yielded to the fluvial system. Includes slope mass wasting, accelerated soil erosion and reservoir sedimentation. May involve interfluve wasting and erosion. Closely related to hydrologic variables
4. The geochemical system	High water quality in alpine areas often suggests only slight chemical activity, but recent work supports the quantitative importance of solution weathering. Mass involved in solute transport is small, but transport efficiency and large scale of water flux renders transport significant

transfer takes the form of a four-fold classification summarized in Table 3.4. However, in this case the classificatory criteria are not terrain-based, as they were with Slaymaker (1974) and Stäblein (1984), but process-based. The study is particularly important in its recognition of the scales of temporal and spatial variability of sediment transfer, and also in its pioneering attempt to compare rates of geomorphic activity attributable to different processes—a topic which is further considered below. The glacial, coarse sediment, fine sediment and geochemical components that are recognized can be traced back to the systems thinking of Caine (1971, 1974). The sediment flux is complete in space and time, and binds the components together in a fashion which is systems oriented in all but mode of presentation.

3.2.2   Systematic frameworks for sediment processes

The main attribute of systems thinking that is pertinent in the present context is its focus on functional and morphological linkage between components. In this sense it has already been acknowledged that the diagnostic criteria of tabular

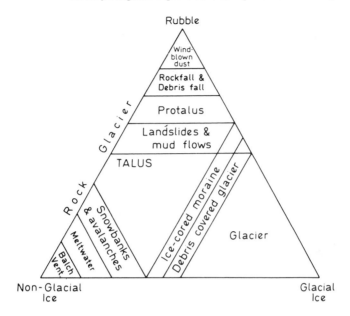

FIGURE 3.2   Classification of rock glaciers based on origins of materials. *Reproduced with permission from Johnson (1973)*

classifications may fulfil a similar role, but the important step is taken when relationship begins to dominate a conceptualization. Thus a strong measure of linkage is apparent in Johnson's (1973) three-dimensional organization of the factors that influence rock glaciers and associated alpine debris forms (Figure 3.2). A genetic element is implicit, but this is demonstrably not a fully systematic presentation.

Although both slopes and channels were subject to intense re-conceptualization in systems terms during the 1960s, there is little evidence that this theoretical and general exercise had any substantial impact on the alpine zone until Caine's (1971) development of a conceptual model for alpine slope processes, and his later (Caine, 1974) more detailed and broad-ranging exposition of these ideas. The fundamental model of the alpine sediment system as depicted in Figure 3.3 is built from familiar components, but its structure acknowledges more directly than had most previous studies the extent and strength of its internal linkage. The qualitative representation of the relative magnitudes of the waste transfers involved is also significant as a basis for comparison — although it models an equilibrium state rather than the paraglacial disequilibrium which leads to many current alpine areas having much greater fluvial transfers. On the basis of this model, Caine proposed a standard five-element alpine slope profile (Figure 3.4) which can be selectively applied to both Rocky Mountain

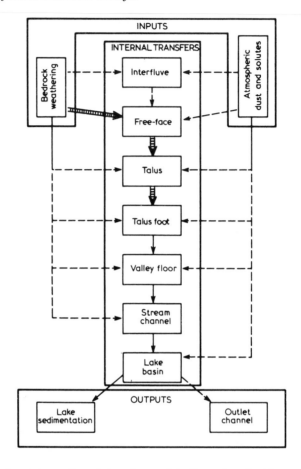

FIGURE 3.3   An alpine sediment transfer model. *Reproduced with permission from Caine (1974)*

type and Alp type mountain landscapes. This profile is simpler than the geomorphodynamic catena proposed by Stäblein (1984), but by being linked conceptually to the alpine debris transfer system it achieves greater coherence.

Perhaps equally important is Caine's (1974) demonstration of the general parentage of his proposed alpine system. Thus, the hillslope waste budget and stream channel sediment budget systems that he uses both reveal clear 1960s systems ancestry; although changes of detail might be suggested in the light of more recent work, these two formulations represent a valuable recognition that the alpine system is but a variant on the global sediment system, with the imposition of glacial inputs, snowmelt and permafrost as perhaps the most important points of distinction. Caine's influence on subsequent thinking has

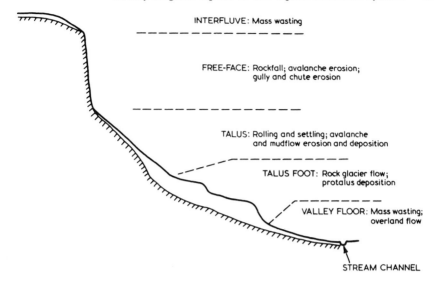

FIGURE 3.4 A hypothetical five-element alpine slope profile. *Reproduced with permission from Caine (1974)*

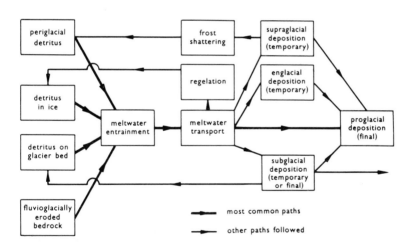

FIGURE 3.5 Flow diagram for the fluvioglacial sediment system. *Reproduced with permission from Sugden and John (1976)*

been justifiably great, and continues to provide enlightened perspective through such syntheses as his study of the mountains of Tasmania (1983) and his joint review of mountain geomorphology (Barsch and Caine, 1984).

Understandably, most aspects of alpine geomorphology were reconsidered during the 1970s in the light of the changes of emphasis inherent in systems thinking. Thus Sugden and John (1976) reflect contemporary interests in their depiction of both the glacial and glacio-fluvial sediment systems in this mode (Figure 3.5), with the immediate benefit of clarifying relationships and highlighting some of the feedbacks through which quasi-equilibrium may be approached. Nevertheless, the reconceptualization is qualitative, perhaps because the greater difficulty of mountain slope and glacial study renders comprehensive quantitative study (as opposed to single feature, single site study) rare. Ironically, therefore, it was left to Barsch and Caine (1984) to rework three classic studies (Jäckli, 1956; Rapp, 1960; Caine, 1976) in quantitative systems terms to provide a first approximation of the rankings of the major components involved. A detailed review would be inappropriate here, but the broad indication that coarse sediment flux (accounting for 10–60% of geomorphic work) and solute transport (accounting for 12–50% of geomorphic work) might be of comparable magnitude is thought provoking.

The above consideration of frameworks for the study of alpine sediment transfer concentrates on descriptive and functional attributes, and thus in many cases leads to a spatial allocation of the resulting components. It would be a mistake to underestimate the importance of the equivalent temporal organization of the system. However, this has already been reviewed in Chapter 2, where the equivalent of a structured inventory approach is introduced through concepts such as Schumm and Lichty's (1965) recognition of three timescales, whilst systems linkage is reflected in Church and Ryder's (1972) paraglacial model, Graf's (1977) rate law concept and Thorn and Loewenherz's (1985) synthesis of paraglaciation and a rate law framework.

3.3   GEOCRYOLOGICAL ATTRIBUTES OF THE ALPINE SEDIMENT SYSTEM

It has been stressed in Chapter 2 that the general organization and trend derived from the operation of frameworks such as those discussed in Section 3.2 is, in reality, subject to considerable local variation. Part of this relates to uniquely geocryological factors, of which the distribution of mountain permafrost and the role of snow would appear to warrant particular attention in an introductory perspective.

3.3.1   Mountain permafrost

It is apparent that the significance of mountain permafrost in nonarctic areas was underestimated and understudied until relatively recently. For example, although the first tentative mapping of permafrost distribution in the Canadian Rocky mountains was attempted by Brown (1967), detailed instrumentation was not forthcoming for more than a decade (Harris and Brown 1978, 1982). This increasing pace and sophistication of study is apparent in other alpine regions.

Whilst Harris (1985a,b) summarizes the global distribution of mountain permafrost, it is more appropriate for present purposes to review just a representative sample of contributions before turning to the sediment transfer implications of permafrost in the high mountain zone. Thus the discussion below concentrates on two mountain regions — the Rocky Mountains of North America and the European Alps.

In terms of the Canadian Rocky Mountains, attention must focus first on the studies by Harris and Brown (1978, 1982) using instrumented boreholes at latitudes of 50–55° N to indicate a consistent distribution pattern, with the lower limit of continuous permafrost lying at altitudes of between 2180 and 2575 m, whilst discontinuous permafrost extended as much as 100 m lower. Significantly, a simple north–south gradient in the altitude of the permafrost lower limit was not observed. Instead, the pattern appeared related to depth of winter snowfall. Thick snow around the area of glaciers and ice caps insulates the ground and yields an anomalously high altitude for the lower limit of permafrost, whilst areas of restricted winter snowfall have permafrost at lower altitudes in the same latitude. Such results are complicated by very considerable microscale variation (for example, related to slope aspect), but the establishment of underlying trends, however speculative, represents a major advance in the provision of working hypotheses for the understanding of mountain permafrost.

On this basis, work has developed both on the climatic relationships and distribution of permafrost (Harris 1981a,b, 1985b, 1987) and the morphological indices of permafrost (Harris, 1981c,d), making this suite of projects one of the most productive available sources of information. For example, it has allowed Harris (1985a, 1987) to generalize the Canadian Rocky Mountain distribution still further, recognizing in particular a reduction in the overall altitudinal range of discontinuous permafrost. Whilst sporadic permafrost may occur over a range of 1000 m, discontinuous permafrost characteristically occupies an elevation zone of no more than 70 m before continuous permafrost takes over. This may be a function of the low soil moisture content of the alpine areas compared with the arctic, but remains speculative.

Another area of concentrated research activity has been the Indian Peaks region adjacent to the Mountain Research Station of the Institute of Arctic and Alpine Research, University of Colorado. The first patterns were tentatively established for Niwot Ridge by Ives and Fahey (1971) and Ives (1973, 1974) on the basis of sampled thermister strings. Early indications were that the effective lower limit of continuous permafrost was at an altitude of 3750 m, whilst that for discontinuous permafrost was at 3500 m for these sites at latitude 40° N — a broader altitudinal range than that suggested by Harris and Brown (1982) for the Canadian Rockies at 50° N, although detailed comparison is difficult without a standard definition of the distinction between discontinuous and sporadic permafrost. Greenstein (1983), in an attempt to extend this work, inferred permafrost distributions using freezing and thawing indices based on a 30-year climatic record extrapolated to other sites through a series of assumed lapse rates. Despite acknowledged problems, this approach was felt to warrant use because the study area had relatively low snowfall (less than the 50 cm maximum recommended by Harris, 1981a,b, 1982), thus obviating one of the primary sources of discrepancy. The results of the exercise suggested that continuous permafrost might extend down to altitudes of 3600 m on south-facing slopes and 3550 m on north-facing slopes, whilst discontinuous permafrost reaches 3300 and 3200 m, respectively. The significance of aspect, snowdepth and presence of water or ice as controls on mountain permafrost distribution was confirmed.

Detailed studies such as these form the basis for Péwé's (1983) synthesis of American literature to provide a broad indication of the scale and pattern of alpine permafrost distribution in the USA. He relates permafrost occurrence to mean annual air temperatures below 0 °C, but notes that permafrost may be out of equilibrium with present temperatures, since the 0 °C isotherm was about 1000 m lower during the Pleistocene than at the present time. The lower limit of permafrost occurrence lies at about 4500 m altitude at latitude 20° N, and falls at about 80 m per degree of latitude to reach about 2000 m at latitude 50° N — generalizations that are broadly in agreement with the Canadian and Indian Peaks data presented above. The ubiquitous control of snowdepth is again apparent — deep snowfall restricts permafrost occurrence to higher altitudes. Similarly, vegetation may insulate the ground against low winter temperatures and encourage snow retention, so that most alpine permafrost lies above the treeline.

Studies of permafrost in the European Alps are less extensive. A number of contributions have noted individual occurrences of alpine permafrost. For example, Barsch *et al.* (1979) and Haeberli *et al.* (1979) described discontinuous permafrost up to 50 m thick associated with the Gruben active rock glacier (altitude 2800–2900 m) close to the Grubengletscher, Switzerland, but suggested that it is probably in equilibrium with Little Ice Age temperatures about 0.5–1.0 °C colder than present. Significantly, these studies indicate the presence of excess ice in the frozen silty sand which comprises the rock glacier, but discount the possibility of this relating to buried glacier ice. True permafrost ground ice is thus suggested. As was the case in the development of North American studies (reviewed above), such specific investigations (e.g. Barsch, 1971) have contributed to a growing general awareness of the scale, distribution and implications of alpine permafrost (e.g. Barsch, 1977a,b, 1978; Haeberli, 1978, 1983).

The last-mentioned study is particularly important in proposing that the vertical distance between the lower boundary of the permafrost in the Swiss Alps today and the contemporary equilibrium line on the glaciers can be related to mean annual air temperature at the equilibrium line. This temperature relationship is strongly linked to the continentality of the climate. Thus in the 'humid margin' conditions which characterize much of the Alps, precipitation exceeds 2000 mm. The glacier equilibrium line descends below 2600 m altitude and has a mean annual air temperature of -1.0 to $-2.5\,°C$. The glaciers are thus temperate throughout, the equilibrium line is close to the lower limit of discontinuous permafrost, and permafrost is restricted to rock outcrops such as summits, crests and rock walls. Frozen ground sedimentary features such as active rock glaciers and push moraines are absent, and continuous permafrost is found only above about 3500 m altitude. By contrast, in the less familiar 'dry interior' of the Alps with annual precipitation of 1000 mm or less, the equilibrium line lies at 3000 m or higher, where equilibrium line temperature is -5 to $-7\,°C$. Since the lower limit of discontinuous permafrost is at 2400 m, there is a considerable vertical zone within which glaciers are probably frozen to their beds, and much unconsolidated sediment is permanently frozen and thus subject to frozen ground sediment transport processes. This model is yet to be tested by other authors, but offers a potentially extremely important general explanation of gross differences in glacier behaviour and sediment transport mode within the Swiss Alps. Although some of the spatial controls are very different, the broad argument has many instructive similarities with the general model of Scandinavian alpine permafrost developed by King (1983).

It is clear that there is a growing but still fragmentary knowledge of midtemperate high-altitude permafrost, but in the present context it is essential to establish that the phenomenon is of sufficient significance to the sediment transfer system to warrant attention. At a general level, Péwé (1983) justifies his study in terms of the relevance of alpine permafrost to the increasing pace of oil and mineral exploration in the mountain zone, and the increased construction of communication and recreation facilities — none of which relates specifically to sediment transfer. Haeberli *et al.* (1979) established clearly that present day rock glacier motion is permafrost-related, and the same would apply to any surficial creep or flow process within the permafrost zone. Although not explicitly discussed in these terms, this kind of assumption is implicit in Gamper's (1983) study of solifluction lobe movement in the Swiss Alps, which is relatively slow at present compared with the considerably higher rates achieved over the frozen ground at several times during the Holocene. As is noted above, even in the 'dry interior' of the Swiss Alps, Haeberli (1983) reported creep of frozen sediments, supersaturated with ice, to be widespread at a rate about one order of magnitude less than the corresponding rate of glacier flow. Further, Haeberli's proposed relationship between permafrost, glaciers and precipitation can be used to model palaeoprecipitation (Kerschner, 1983), and thus has potential in calibrating modern analogues of sediment movement and hydrological inputs.

In addition to influencing both mass movement and glacier behaviour, alpine permafrost also has considerable significance for surface wash processes, particularly during snowmelt. Little direct work on this topic has been forthcoming in the midtemperate mountain zone, and the likely effects can perhaps best be inferred from arctic counterparts. However, it is implicit in studies of mountain slopewash processes (e.g. Thorn, 1979) that sediment transport is closely constrained by thaw penetration into the ground. Overall, it can be concluded that alpine permafrost is a major geocryological attribute of the alpine sediment system, and that it shows a number of important links with glacio-fluvial sediment transfer processes. The topic is, as yet, under-researched, and has great potential as a research focus for the future.

3.3.2   The role of snow in mountain environments

It is deceptively easy, particularly in a book concerning glacio-fluvial sediment transfer, to overestimate the importance of ice and underestimate that of snow. Although snow accounts for only about 5% of global precipitation (Hoinkes, 1967), it is more extensive than ice both spatially and temporally, and it occurs in a wider range of topographic and climatic environments. Whilst nival environments are fully acknowledged as producers of avalanches, their other attributes are less often stressed in general studies, and it was left to just a few authors (e.g. Caine, 1976; Thorn, 1978) to suggest a more balanced view which gives proper recognition to the range and magnitude of snow-based inputs to the alpine system. Thorn, for example, proposed that significant geomorphic process thresholds relate to snowfall, snowpack surface, snowpack mobility (creep and avalanches), snowpack duration, snowdepth and meltwater — but accepted that field verification for several of these is inadequate. Nevertheless, even if details were to be questioned or re-evaluated, the importance of the study is firmly established by its strong emphasis on the critical role of moisture in cold environments, and by its breadth of snow-related components.

Snowfall is of immediate significance simply as a direct input of precipitation, but its geomorphological interest and hydrological implications are greatly enhanced by the extent to which redistribution takes place, giving marked variability in local and regional patterns of snow reception and retention. Regionally, this may place control of the location of centres of ice accumulation with topographic lee effects (such as the eastern slope of the Colorado Front Range) rather than with simple altitudinal or latitudinal gradients. Locally, redistribution opens the possibility of abrasion by snow at low temperatures, and introduces drifting as a significant water distribution factor (Tabler, 1975; Daly, 1984). The amount of snow retention at a site depends on local topographic indices together with ground cover attributes, amongst which vegetation is highly significant (Thorn, 1978; Clark *et al.*, 1985), and the resulting snow surface then acts to influence radiative energy budget (Moore and Owens, 1984), surficial sediment transport (Rapp, 1960; Caine, 1969), and over-snow meltwater flow early in the melt season (Woo, 1979). The snowpack also has a major function in insulating the ground from low winter temperatures, thus influencing the occurrence and intensity of alpine permafrost formation (Section 3.3.1).

The snow itself is subject to downslope creep, and may thus be a direct sediment transporting agent in areas of steep slope with debris yield on to the snow surface. Far more familiar, of course, are the transportational effects of snow avalanches and slush avalanches, both of which have an important influence through the distance of transport rather than through the volume of sediment carried. So extensive is the avalanche literature that any attempt to summarize it within a paragraph would be both redundant and trivializing. Apart from the geomorphological and hazard aspects, avalanches represent an important mountain hydrological agent since they transport large water-equivalent volumes and influence snowmelt rates by carrying snow to lower altitudes, decreasing its albedo and altering its density. The result may be either an increase or attenuation in the snowmelt peak (Martinec and de Quervain, 1971). The implications for glacio-fluvial sediment transfer in the snowmelt-fed rivers are perhaps greater than those for the broader alpine sediment transfer system within the avalanche zone itself, particularly if the avalanche takes place over ground protected by snow cover.

The more conventional hydrological role of snow lies in its function as a water storage and release mechanism (Glen, 1982). Storage is related both to retention during the winter and to subsequent duration of snow cover before melt. Release takes the form of snowmelt, the energy conditions for which and forecasting of which represent specialist topics in their own right (Obled and Harder, 1979; Kuusisto, 1984). The attributes of the snowpack, especially in the form of late-lying snow patches, and the activity of the water released by its melting, have long been regarded by geomorphologists as representing an important process for weathering and sediment removal in the mountain zone. This process assemblage was codified by Matthes (1900) under the title of 'nivation,' a vague and tenuous concept with an emotive appeal great enough to ensure its uncritical acceptance by generations of geomorphologists. The process still has many adherents, but has been critically evaluated by Thorn (1976, 1987), and found to be lacking in substance and verification. A period of re-assessment is necessary before the topic can be rigorously redefined.

Snowmelt runoff has unique characteristics which place it midway between glacial and temperate hydrological regimes. The annual rise to flow peak can be extremely rapid, and hydrograph recession is also much more dramatic than is the case in glacier-fed catchments. The magnitude of these flow variations increases in arctic latitudes, especially in lowland areas. Conversely, high altitude may elongate both the rising and falling limbs of the hydrograph, since seasonal warming spreads sequentially up the mountain side rather than occurring simultaneously across the whole catchment. Nevertheless, the snowpack on the slopes functions as an extremely important temporary store for meltwater, and may delay runoff from the slope by several days (FitzGibbon and Dunne, 1981).

Avalanche redistribution of snow is also significant in this context, altering the proportion of the stored winter precipitation that is available for release in each altitudinal belt, and varying the width–depth ratio of the snow stores so as to control duration of melting. In arctic sites, ice-jamming may delay the onset of flow in the channel, and greatly increase the rate of the flood rise once the jam is released. Nival basins also experience jamming, but often on a lesser scale — flow being initiated over-snow or as slushflow early in the season (Woo 1979, 1982; Woo and Sauriol, 1980). Within the alpine zone, snow-jamming is often associated with the run-out zones of avalanches. The avalanche thus delivers sediment to the channel zone, and increases stream power per unit time by storing water until the jam is released, both influences being significant for glacio-fluvial sediment transport.

Sediment transport in the channel is an integral part of the processes discussed in Chapters 10–15, but snowmelt sediment flux on the slopes is less well documented. Both rates and processes are under-researched, but available studies suggest that the mechanism is effective and that it relates closely to ground temperature during the thaw period (Kane *et al.*, 1978; Thorn, 1979; Lewkowicz and French, 1982; Lewkowicz, 1983). The efficacy of slope snowmelt sediment transport thus varies depending on the occurrence of continuous permafrost, discontinuous permafrost or seasonal ground freezing — all of which are themselves linked to depth of snowcover in the alpine zone (Section 3.3.1). Finally, it should be noted that neither stream power nor discharge regime has exclusive control of sediment transport, since sediment availability is also crucial. Many aspects of availability have already been discussed in the previous chapter, but in addition the potential role of bed and bank freezing or ice-armouring should not be under-estimated.

Despite the brevity of this review of alpine nival processes, it is clear that snow has major importance as an influence on micro-climate, ground temperature, water storage, redistribution and release, river hydrograph and power and sediment availability and transport. Further, as a fundamental control on the temporal and spatial distribution of moisture, it is basic to the nature and efficacy of many of the geomorphological processes active on alpine slopes, to which attention must now turn.

3.3.3   Alpine sediment transfer processes

An exhaustive review of sediment transfer processes would be out of place in the present general introduction to alpine sediment transfer. However, brief mention of the major components of the weathering and sediment transport systems provides a basis for commenting on their overall significance in terms of denudation, and serves to indicate some of the relevant thresholds which account for the apparent step function attributes which characterize mountain sediment processes. At the same time, it is instructive to note the current research status of the fields involved, since some parts of the system are the subject of considerable contention, the resolution of which may necessitate a re-evaluation of sediment flux components in the near future.

Nowhere is this more applicable than in the case of weathering, which provides the main input to the extra-glacial sediment system except in those areas where a paraglacial disequilibrium yields a disproportionate supply of glacially derived sediment to the post-glacial sediment transfer processes (Section 2.3.1). Traditional views on mechanical 'weathering in cold regions were often qualitative, and assumed ice segregation within rock fissures to be of great importance in triggering weathering. Subsequent work has followed two main paths. Laboratory studies have been extensively used to probe weathering mechanisms (e.g. McGreevy, 1981; Lautridou and Ozouf, 1982; Fahey and Dagesse, 1984; Lautridou, 1987), and have identified a number of characteristics of both material and process which might explain variations in weathering rate and intensity. Others have regarded laboratory studies with some suspicion, considering the degree of duplication of field conditions to be inadequate, and have stressed the need for a considerable increase in field studies before definitive results will be achieved (e.g. Thorn 1979, 1980; Hall, 1980; Thorn and Hall, 1980). Whilst no final resolution of these differences of approach is in sight, it is interesting to note the possibility of a compromise position, attempting to combine the results of field and laboratory studies (McGreevy and Whalley 1982, 1985). These papers suggest that the experimental approach might exaggerate natural weathering effects by incorporating unrealistic levels of saturation, although the temperature extremes often claimed on field grounds to be necessary (e.g. $-5\,°C$ as the upper limit for frost shattering) may be too restrictive.

The other main controversy in the context of weathering and sediment release concerns the role of chemical processes and solute load. Despite Rapp's (1960) demonstration of the overall significance of solutes, the traditional view that chemical processes are ineffective at low temperatures (and thus high altitudes) remains widespread. This overlooks the proven acidity of much snow meltwater, the established efficacy of chemical activity at temperatures little above freezing point and the mounting evidence of a number of disparate but supportive recent studies (e.g. Lewis and Grant, 1980; Stednick, 1981; Eyles *et al.*, 1982; Collins, 1983; Dixon *et al.*, 1984; Singh and Kalra, 1984). The prospect that chemical processes might come to be regarded as of increasing significance both in alpine weathering and sediment transport is interesting, but Thorn (1987) offers a salutory warning against excessive expectations:

'In general, temperate alpine environments exhibit chemical weathering processes largely similar to other temperate environments (e.g. Dixon *et al.*, 1984). Yet again, this should not be interpreted as replacing one set of over-generalised assumptions (i.e. the absence of chemical process in periglacial areas) with another (i.e. suggested ubiquitous and intense chemical weathering). Rather, it should signal to us that what appeared to be a closed topic is actually an open one.'

(Thorn, 1987)

Mass movement processes are considerably less contentious at present, although this does not imply that there remains no scope for refinement. An increasing awareness of mechanism is currently restrained by residual complexities of terminology and by a paucity of unambiguous studies of the range and causes of variability in nature and rate of process. This is perhaps surprising, given the emphasis placed on scale-dependent variation in sediment yield of the Canadian Rocky Mountains by Slaymaker (1972). Nevertheless, major reviews such as that for periglacial slope processes (Lewkowicz, 1987) and the alpine counterpart by White (1981) exhibit an encouraging stability and consensus at the general level, which provides a strong framework within which detailed controversies can be handled without constraining progress in the subject as a whole.

The combination of rock weathering, evacuation of available debris and sediment flux within and beyond the catchment provides a basis for assessing overall catchment denudation. This is of considerable geomorphological interest, but raises an extreme challenge for investigative design. It is difficult to apportion sediment losses to individual processes or sediment sources within the contemporary catchment, and equally difficult to distinguish between past and present yields. Thus the aims of providing either a current sediment budget or a long-term reconstruction of palaeo-sedimentary environments remain elusive, although the target has attracted many researchers. The techniques used and the difficulties raised are discussed more fully in Section 18.2.

3.4   THE ROLE OF GEOCRYOLOGICAL PROCESSES IN THE GLACIO-FLUVIAL SEDIMENT SYSTEM: AN ASSESSMENT

Whilst many of the individual process components of the alpine sediment transfer system can thus be specified with some confidence, integration of these elements continues to present a substantial challenge. Moreover, most individual specifications are more site-specific than is desirable, so that scales of variability are ill-defined. This is apparent at all scales, from the global to the local, and includes the imperfect designation of the thresholds which divide one system state from another, and which characterize several important general contributions to the literature. Thus, Hewitt (1972) drew attention to the importance of regolith stability (related to slope steepness) as a threshold which is very sensitive to changes triggered by climatic, topographic and process inputs. Similarly sensitive thresholds could relate to tree line, vegetation line, regolith line and snow or ice stability. It has already been noted that Thorn (1978) regards snowfall, snowpack surface, snowpack mobility, snowpack duration and snowdepth as all introducing variables which act as thresholds by conditioning abrupt boundaries where one process is eliminated and/or another is added. Although not discussed in such terms, Haeberli's (1983) glacier/permafrost relationship also has threshold-like characteristics. To date, the main outcome of this type of process differentiation has been the recognition of numerous process-diagnostic 'geomorphodynamic' slope zones in alpine areas (Section 3.2), but the scope for more detailed applications represents a major future research priority.

Although this chapter has identified numerous contexts within which the present research status is deficient, the overall dynamism of the alpine sediment system and its inextricable relationship with the glacio-fluvial sediment system are undeniable. Cryogenic processes are fundamental to the supply of water and sediment from the slopes above glaciers, and directly into proglacial rivers, and in both locations represent components which have not always received adequate attention in glacial and glacio-fluvial studies. In order to examine the potential for a greater integration between these elements, it is appropriate to review the glacial and fluvial aspects of sediment transfer (Chapter 4) and their implications for the hydrogeomorphology of proglacial zones (Chapter 5) before turning to detailed studies. However, the integrative aims of this volume will be addressed by returning in Chapter 18 to an attempt to take an overall view of the way in which glacio-fluvial sediment processes relate to the broader issues of the alpine sediment system as a whole.

REFERENCES

BARSCH, D. (1971), Rock glaciers and ice cored moraines, *Geografiska Annaler*, **53A**, 203–206.

BARSCH, D. (1977a), Ein Permafrostprofil aus Graubünden, Schweizer Alpen, *Zeitschrift für Geomorphologie*, **NF21**, 79–86.

BARSCH, D. (1977b), Nature and importance of mass wasting by rock glaciers in alpine permafrost environments, *Earth Surface Processes*, **2**, 231–245.

BARSCH, D. (1978), Rock glaciers as indicators for discontinuous alpine permafrost. An example from the Swiss Alps, *Proceedings of the Third International Permafrost Conference, July 10–13, 1978, Edmonton, Alberta, Canada*, Vol. 1, National Research Council of Canada, Ottawa, pp. 349–352.

BARSCH, D., FIERZ, H., and HAEBERLI, W. (1979), Shallow core drilling and bore-hole measurements in the permafrost of an active rock glacier near the Grubengletscher, Wallis, Swiss Alps, *Arctic and Alpine Research*, **11**, 215–228.

BARSCH, D., and CAINE, N. (1984), The nature of mountain geomorphology, *Mountain Research and Development*, **4**, 287–298.

BEZINGE, A., PERRETEN, J.-P., and SCHAFER, F. (1973), Phénomènes du lac glaciaire du Gorner, in *Symposium on the Hydrology of Glaciers (Proceedings of the Cambridge Symposium, 7–13 September 1969)*, International Association of Scientific Hydrology Publication **95**, pp. 65–78.

BROWN, R. J. E. (1967), *Permafrost Map of Canada*, National Research Council NRC 9767 and Geological Survey of Canada Map 1246A.

CAINE, T. N. (1969), A model for alpine talus slope development by slush avalanching, *Journal of Geology*, **77**, 92–100.

CAINE, N. (1971), A conceptual model for alpine slope processes, *Arctic and Alpine Research*, **3**, 319–329.

CAINE, N. (1974), The geomorphic processes of the alpine environment, in IVES, J. D., and BARRY, R. G. (Eds.), *Arctic and Alpine Environments*, Methuen, London, 721–748.

CAINE, T. N. (1976), The influence of snow and increased snowfall on contemporary geomorphic processes in alpine areas, in STEINHOFF, H. W., and IVES, J. D. (Eds.), *Ecological Impacts of Snowpack Augmentation in the San Juan Mountains, Colorado*, Colorado State University, Fort Collins, pp. 145–200.

CAINE, N. (1976), A uniform measure of subaerial erosion, *Bulletin of the Geological Society of America*, **87**, 137–140.

CAINE, N. (1983), *The Mountains of Northeastern Tasmania. A Study of Alpine Geomorphology*, Balkema, Rotterdam.

CHURCH, M., and RYDER, J. M. (1972), Paraglacial sedimentation: a consideration of fluvial processes conditioned by glaciation, *Bulletin of the Geological Society of America*, **83**, 3059–3067.

CLARK, M. J., GURNELL, A. M., MILTON, E. J., SEPPÄLÄ, M., and KYÖSTILÄ, M. (1985), Remotely sensed vegetation classification as a snow depth indicator for hydrological analysis in sub-arctic Finland, *Fennia*, **163**, 195–225.

COLLINS, D. N. (1983), Solute yield from a glacierized high mountain basin, in *Dissolved Loads of Rivers and Surface Water Quantity/Quality Relationships (Proceedings of the Hamburg Symposium, August 1983)*, International Association of Hydrological Sciences Publication 141, pp. 41–49.

DALY, C. (1984), Snow distribution patterns in the alpine krummholz zone, *Progress in Physical Geography*, **8**, 157–175.

DIXON, J. C., THORN, C. E., and DARMODY, R. G. (1984), Chemical weathering processes on the Vantage Peak nunatak, Juneau Icefield, southern Alaska, *Physical Geography*, **5**, 111–131.

EYLES, N., SASSERVILLE, D. R., SLATT, R. M., and ROGERSON, R. J. (1982), Geochemical denudation rates and solute transport mechanisms in a maritime temperate glacier basin, *Canadian Journal of Earth Sciences*, **19**, 1570–1581.

FAHEY, B. D., and DAGESSE, D. F. (1984), An experimental study of the effect of humidity and temperature variations on the granular disintegration of argillaceous carbonate rocks in cold climates, *Arctic and Alpine Research*, **16**, 291–298.

FITZGIBBON, J. E., and DUNNE, T. (1981), Land surface and lake storage during snowmelt runoff in a subarctic drainage system, *Arctic and Alpine Research*, **13**, 277–285.

GAMPER, M. W. (1983), Controls and rates of movement of solifluction lobes in the Eastern Swiss Alps, in *Proceedings of the Fourth International Permafrost Conference*, National Academy Press, Washington, DC, pp. 328–333.

GLEN, J. W. (Ed.) (1982), *Hydrological Aspects of Alpine and High Mountain Areas*, International Association of Hydrological Sciences Publication 138.

GRAF, W. L. (1977), The rate law in fluvial geomorphology, *American Journal of Science*, **277**, 178–191.

GREENSTEIN, L. A. (1983), An investigation of midlatitude alpine permafrost on Niwot Ridge, Colorado Rocky Mountains, USA, in *Proceedings of the Fourth International Permafrost Conference*, National Academy Press, Washington, DC, pp. 380–383.

HAEBERLI, W. (1978), Special aspects of high mountain permafrost methodology and zonation in the Alps, in *Proceedings of the Third International Permafrost Conference, July 10–13, 1978, Edmonton, Alberta, Canada*, Vol. 1, National Research Council of Canada, Ottawa, pp. 378–384.

HAEBERLI, W. (1983), Permafrost–glacier relationships in the Swiss Alps — today and in the past, in *Proceedings of the Fourth International Permafrost Conference*, National Academy Press, Washington, DC, pp. 415–420.

HAEBERLI, W., KING, L., and FLOTRON, A. (1979), Surface movement and lichen-cover studies at the active rock glacier near the Grubengletscher, Wallis, Swiss Alps, *Arctic and Alpine Research*, **11**, 421–441.

HALL, K. (1980), Freeze–thaw activity at a nivation site in northern Norway, *Arctic and Alpine Research*, **12**, 183–194.

HAMELIN, L. E. (1964), La famille du mot 'périglaciaire,' *Biuletyn Peryglacjalny* **14**, 133–152.

HARRIS, S. A. (1981a), Climatic relationships of permafrost zones in areas of low winter snow cover, *Arctic*, **34**, 64–70.

HARRIS, S. A. (1981b), Climatic relationships of permafrost zones in areas of low winter snow cover, *Biuletyn Peryglacjalny*, **28**, 227–240.

HARRIS, S. A. (1981c), Distribution of active glaciers and rock glaciers compared to the distribution of permafrost landforms, based on freezing and thawing indices, *Canadian Journal of Earth Sciences*, **18**, 376–381.

HARRIS, S. A. (1981d), Distribution of zonal permafrost landforms with freezing and thawing indices, *Erdkunde*, **35**, 81–90.

HARRIS, S. A. (1985a), Distribution and zonation of permafrost along the eastern ranges of the Cordillera of North America, *Biuletyn Peryglacjalny*, **30**, in press.

HARRIS, S. A. (1985b), *The Permafrost Environment*, Croom Helm, Beckenham, Kent, 288 pp.

HARRIS, S. A. (1987), The alpine periglacial zone, in CLARK, M. J. (Ed.), *Advances in Periglacial Geomorphology*, Wiley, Chichester, in press.

HARRIS, S. A., and BROWN, R. J. E. (1978), Plateau Mountain: a case study of alpine permafrost in the Canadian Rocky Mountains, in *Proceedings of the Third International Permafrost Conference, July 10–13, 1978, Edmonton, Alberta, Canada*, Vol. 1, National Research Council of Canada, Ottawa, pp. 385–391.

HARRIS, S. A., and BROWN R. J. E. (1982), Permafrost distribution along the Rocky Mountains in Alberta, in FRENCH, H. M. (Ed.), *Proceedings of the Fourth Canadian Permafrost Conference, Calgary, Alberta*, National Research Council of Canada, Ottawa, pp. 59–67.

HEWITT, K. (1972), The mountain environment and geomorphic processes, in SLAYMAKER, H. O., and MCPHERSON, H. J. (Eds.), *Mountain Geomorphology: Geomorphological Processes in the Canadian Cordillera*, B. C. Geographical Series, No. 14, Tantalus Research, Vancouver.

HOINKES, H. (1967), Glaciology in the International Hydrological Decade, *IUGG General Assemby, Bern, IASH Commission on Snow and Ice, Reports and Discussions*, Publication 79, pp. 7–16.

IVES, J. D. (1973), Permafrost and its relationship to other environmental parameters in a mid-latitude, high-altitude setting, Front Range, Colorado Rocky Mountains, in *Permafrost — The North American Contribution to the Second International Permafrost Conference, Yakutsk*, National Academy of Sciences, Washington, DC, pp. 13–28.

IVES, J. D. (1974), Permafrost, in IVES, J. D., and BARRY, R. G. (Eds.), *Arctic and Alpine Environments*, Methuen, London, pp. 159–194.

IVES, J. D., and BARRY, R. G. (Eds.) (1974), *Arctic and Alpine Environments*, Methuen, London.

IVES, J. D. (1987), The Mountain Lands, in CLARK, M. J., GREGORY, K. J., and GURNELL, A. M. (Eds.), *Horizons in Physical Geography*, MacMillan, Basingstoke, in press.

IVES, J. D., and FAHEY, B. D. (1971), Permafrost occurrence in the Front Range, Colorado Rocky Mountains, USA, *Journal of Glaciology*, **10**, 105–111.

JÄCKLI, H. (1956), Gegenwartsgeologie des bundnerischen Rheingebietes—ein Beitrag zur exogen Dynamik alpiner Gebirgslandschaften, *Beitrage zur Geologie des Schweiz, Geotechnische Serie*, No. 36.

JOHNSON, J. P. (1973), Some problems in the study of rock glaciers, in FAHEY, B. D., and THOMPSON, R. D. (Eds.), *Research in Polar and Alpine Geomorphology*, Geo Books, Norwich, pp. 84–94.

KANE, D. L., FOX, J. D., SEIFERT, R. D., and TAYLOR, G. S. (1978), Snowmelt infiltration and movement in frozen soils, in *Proceedings of the Third International Permafrost Conference, July 10–13, 1978, Edmonton, Alberta, Canada*, Vol. 1, National Research Council of Canada, Ottawa, pp. 201–206.

KERSCHNER, H. (1983), Lateglacial palaeotemperatures and palaeoprecipitation as derived from permafrost:glacier relationships in the Tyrolean Alps, Austria, in *Proceedings of the Fourth International Permafrost Conference*, National Academy Press, Washington, DC, pp. 589–593.

KING, L. (1983), High mountain permafrost in Scandinavia, in *Proceedings of the Fourth International Permafrost Conference*, National Academy Press, Washington, DC, pp. 612–617.

KUUSISTO, E. (1984), *Snow Accumulation and Snowmelt in Finland*, Publications of the Water Research Institute, National Board of Waters, Finland, 55, 149 pp.

LAUTRIDOU, J. P. (1987), Recent advances in cryogenic weathering, in CLARK, M. J. (Ed.), *Recent Advances in Periglacial Geomorphology*, Wiley, Chichester (in press).

LAUTRIDOU, J. P., and OZOUF, J. C. (1982), Experimental frost shattering. 15 years of research at the Centre de Geomorphologie du C.N.R.S., *Progress in Physical Geography*, **6**, 217–232.

LEWIS, W. M., JR., and GRANT, M. C. (1980), Relationships between snow cover and winter losses of dissolved substances from a mountain watershed, *Arctic and Alpine Research*, **12**, 11–17.

LEWKOWICZ, A. G. (1983), Erosion by overland flow, central Banks Island, western Canadian arctic, in *Proceedings of the Fourth International Permafrost Conference*, National Academy Press, Washington, DC, pp. 701–706.

LEWKOWICZ, A. G. (1987), Slope processes, in CLARK, M. J. (Ed.), *Advances in Periglacial Geomorphology*, Wiley, Chichester (in press).

LEWKOWICZ, A. G., and FRENCH, H. M. (1982), The hydrology of small runoff plots in an area of continuous permafrost, in FRENCH, H. M. (Ed.), *Proceedings, Fourth Canadian Permafrost Conference, Calgary, Alberta, 1981*, National Research Council of Canada, Ottawa, pp. 151–162.

MARTINEC, J., and DE QUERVAIN, M. R. (1971), The effect of snow displacement by avalanches on snowmelt and runoff. IUGG General Assembly, Moscow, in *Symposium on Interdisciplinary Studies of Snow and Ice in Mountain Regions*, International Association of Scientific Hydrology Publication 104, pp. 364–377.

MATTHES, F. E. (1900), *Glacial Sculpture of the Bighorn Mountains, Wyoming*, United States Geological Survey, 21st Annual Report, 1899–1900, pp. 167–190.

MCGREEVY, J. P. (1981), Perspectives on frost shattering, *Progress in Physical Geography*, **5**, 56–75.

MCGREEVY, J. P., and WHALLEY, W. B. (1982), The geomorphic significance of rock temperature variations in cold environments: a discussion, *Arctic and Alpine Research*, **14**, 157–162.

McGreevy, J. P., and Whalley, W. B. (1985), Rock moisture content and frost weathering under natural and experimental conditions: a comparative discussion, *Arctic and Alpine Research*, **17**, 337–346.

Moore, R. D., and Owens, I. F. (1984), Controls on advective snowmelt in a maritime alpine basin, *Journal of Climate and Applied Meteorology,* **23**, 135–142.

Obled, Ch., and Harder, H. (1979), A review of snowmelt in the mountain environment, in Colbeck, S. C., and Ray, M. (Eds.), *Proceedings, Modeling of Snowcover Runoff*, US Army Cold Regions Research and Engineering Laboratory, Hanover, 179–204.

Østrem, G. (1975a), Sediment transport in glacial meltwater streams, in Jopling, A. V., and MacDonald, B. C. (Eds.), *Glaciofluvial and Glaciolacustrine Sedimentation*, Society of Economic Palaeontologists and Mineralogists Special Publication 23, pp. 101–122.

Péwé, T. L. (1983), Alpine permafrost in the contiguous United States: a review, *Arctic and Alpine Research*, **15**, 145–156.

Rapp, A. (1960), Recent development of mountain slopes in Karkevagge and surroundings, northern Scandinavia, *Geografiska Annaler*, **42**, 71–200.

Schumm, S. A., and Lichty, R. W. (1965), Time, space and causality in geomorphology, *American Journal of Science*, **263**, 110–119.

Singh, T., and Kalra, Y. P. (1984), Predicting solute yields in the natural waters of a subalpine system in Alberta, Canada, *Arctic and Alpine Research*, **16**, 217–224.

Slaymaker, H. O. (1972), Sediment yield and sediment control in the Canadian Cordillera, in Slaymaker, H. O., and McPherson, H. J. (Eds.), *Mountain Geomorphology: Geomorphological Processes in the Canadian Cordillera*, B. C. Geographical Series, No. 14, Tantalus Research, Vancouver, pp. 235–272.

Slaymaker, H. O. (1974), Alpine hydrology, in Ives, J. D., and Barry, R.G. (Eds.), *Arctic and Alpine Environments*, Methuen, London, pp. 134–155.

Stäblein, G. (1984), Geomorphic altitudinal zonation in the arctic–alpine mountains of Greenland, *Mountain Research and Development*, **4**, 319–331.

Stednick, J. D. (1981), Hydrochemical balance of an alpine watershed in southeast Alaska, *Arctic and Alpine Research*, **13**, 431–438.

Sugden, D. E., and John, B. S. (1976), *Glaciers and Landscape*, Edward Arnold, London.

Tabler, R. D. (1975), Predicting profiles of snowdrifts in topographic catchments, *Proceedings of the Western Snow Conference*, **43**, 87–97.

Thorn, C. E. (1976), Quantitative evaluation of nivation in the Colorado Front Range, *Bulletin of the Geological Society of America*, **87**, 1169–1178.

Thorn, C. E. (1978), The geomorphic role of snow, *Annals of the Association of American Geographers*, **68**, 414–425.

Thorn, C. E. (1979), Ground temperatures and surficial transport in colluvium during snowpatch meltout: Colorado Front Range, *Arctic and Alpine Research*, **11**, 41–52.

Thorn, C. E. (1980), Alpine bedrock temperatures: an empirical study, *Arctic and Alpine Research*, **12**, 73–86.

Thorn, C. E. (1987), Nivation: a geomorphic chimera, in Clark, M. J. (Ed.), *Advances in Periglacial Geomorphology*, Wiley, Chichester, in press.

Thorn, C. E., and Hall, K. (1980), Nivation: an arctic–alpine comparison and reappraisal, *Journal of Glaciology*, **25**, 109–124.

Thorn, C. E., and Loewenherz, D. S. (1985), Interaction between spatial and temporal trends in alpine periglacial landform evolution: implications for palaeocharacter, Lecture presented to First International Geomorphology Conference, Manchester.

White, S. E. (1981), Alpine mass movement forms (noncatastrophic): classification, description, and significance, *Arctic and Alpine Research*, **13**, 127–137.

Woo, M. K. (1979), Breakup of streams in the Canadian high arctic, *Proceedings of the 36th Eastern Snow Conference*, 95–107.

Woo, M. K. (1982), Snow hydrology of the high arctic, *Proceedings of the Western Snow Conference*, 63–74.

Woo, M. K., and Sauriol, J. (1980), Channel development in snow-filled valleys, Resolute, N.W.T., Canada, *Geografiska Annaler*, **62A**, 37–56.

Glacio-fluvial Sediment Transfer
Edited by A. M. Gurnell and M. J. Clark
©1987 John Wiley & Sons Ltd.

Chapter 4

Sediment Transfer Processes in Alpine Glacier Basins

C. R. FENN

ABSTRACT

This chapter sets out to provide an overview of the sediment production and routing processes operating in glacierized alpine drainage basins. Information relating to sediment source areas, production processes and transport pathways is summarized in tabular and diagrammatic form, and the transfer system is expressed in balance terms. The interdependency between *in situ* physical and chemical weathering processes and moving ice, water and debris mass erosive processes is highlighted, as is the spatial and temporal variability of sediment transfer. The outflow stream is considered to be the sole means of conveying sediment out of the basin. The basic characteristics of solute, suspended and bed load transport in proglacial streams are outlined, and are incorporated into a flow-based model of sediment transport in glacial outflow stream systems.

4.1 INTRODUCTION

Research interests in the sediment transfer system of alpine glacier basins typically relate to questions concerning:

(a) Process mechanics:
    e.g. what mechanisms are operating?; what are the controls on the efficiency of these processes?
(b) Sediment source-areas and pathways:
    e.g. where are sediments produced?; how is sediment moved through the basin?
(c) Sediment budgets and balances:
    e.g. how much sediment is produced?; how much sediment is stored?; how much sediment is exported?

Such questions are not easily answered. The presence of a mass of (usually) impenetrable overlying ice renders the subglacial environment 'uniquely inaccessible to direct or remote study' (Boulton, 1982, p. 2). Essential information on the characteristics of basal ice, subglacial surfaces and interface processes has been gleaned from studies undertaken in natural subglacial cavities and in man-made tunnels and boreholes [see Vivian (1980) for a useful review]. Observations from such sites are invaluable but, inevitably, insufficient: they provide a view from all too small a window. Direct field study has, in consequence, to be complemented by inferences drawn from the characteristics of debris presently in transit, and from the morphology and sedimentology of residual landforms (e.g. Boulton, 1978; Lister, 1981; Small and Gomez, 1981); by laboratory modelling of process mechanics (e.g. Budd *et al.*, 1979; Riley, 1982); and by theoretical modelling of process systems (e.g. Boulton, 1974, 1979a; Hallet, 1979, 1981). The proceedings of the 1978 Ottawa and 1980 Geilo International Glaciological Society Symposia (on, respectively, 'Glacier Beds: the Ice–Rock Interface' and 'Processes of Glacier Erosion and Sedimentation') illustrate the progress which has been made in integrating results from these various approaches, and from the pronounced interdisciplinary focus which has characterized recent research into the processes of glacier erosion and sedimentation.

This chapter aims to provide an overview of the sediment production and routing processes operating in alpine glacier basins. It is strictly intended to be a study of glacial sediment transfer processes. It is not intended to be a study of glacial geology, rheology or geomorphology: such topics have been covered recently in Eyles (1983). Nor is it intended to re-review details relating to particular erosive processes: such matters have been dealt with at length in individual research papers (e.g. Boulton, 1974; Hallet, 1979) and in general texts (e.g. Sugden and John, 1976; Derbyshire *et al.*, 1979; Embleton and Thornes, 1979). Rather, this chapter adopts a 'whole-system' approach to the problem of sediment transfer in glacierized basins, attempting to isolate, summarize and collate essential information in order to provide a framework against which the detailed studies presented in the glacial and hydrological sections of this book may be set.

4.2   THE SEDIMENT TRANSFER SYSTEM OF GLACIERIZED ALPINE BASINS

4.2.1   Specification of the system

Glacier basin sediment transfer systems are readily represented schematically (Figure 4.1). In essence, sediments may be derived from weathering processes and from glacial and fluvial erosion processes from three main sources (subaerial,

Sediment Sources

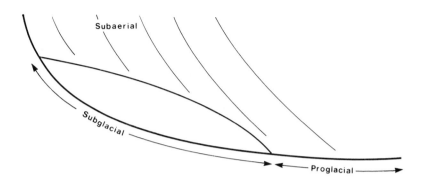

Sediment Production Processes

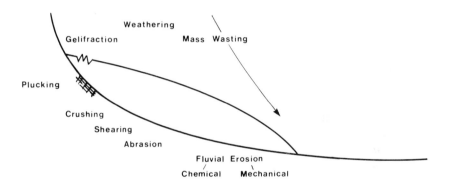

Sediment Transport Pathways

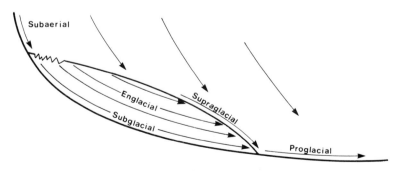

FIGURE 4.1 Sediment sources, production processes and transport paths

TABLE 4.1   Glacial erosion in alpine basins: a summary of mechanisms and rate controls

PLUCKING/QUARRYING

Block removal in association with failure along physical discontinuities
Adhesion between ice and block exceeds adhesion between block and parent material

Mechanisms:
1. Plucking out — block excavated by ice deforming around it
2. Wedging out — block gives under pressure from overriding rock particles
3. Freezing on — block lifted by freezing to base of ice

Favourable conditions:
In discontinuous rock mass (DRM) zones (Weertman, 1961; Addison, 1981)
Where high pore water pressures in subglacial rocks/sediments promote shear failure (Moran *et al.*, 1980; Addison, 1981)
In the lee of bedrock highs, in association with stress unloading (Eyles and Menzies, 1983)
In zones of frost penetration and heaving (Eyles and Menzies, 1983)
Where a reduction of pressure in temperature ice leads to freezing at the ice–rock interface:
  e.g. (1) in subglacial cavities which open and close under water pressure fluctuations (Röthlisberger and Iken, 1981)
  e.g. (2) where a 'cold patch' exists in the lee of bed bumps [Robin's (1976) 'heat pump effect']
Where the glacier margin freezes to its bed during cold seasons (Anderson *et al.*, 1982)
Where a transition from a zone of cold, non-sliding ice to one of basal sliding promotes failure (Hutter and Olunloyo, 1981)
Where basal stresses are sufficient to tear off parts of the bed (Hutter and Olunloyo, 1981)
Where subglacial sediments are ice-cemented, and behave like bedrock elements (Boulton, 1979a)

CRUSHING AND SHEARING

Dislodgment/disintegration of boundary elements due to force exerted by overriding ice–debris matrix

Mechanisms:
1. Loading on — failure due to overburden pressure on roof of obstacle
2. Butting up — shear failure due to pressure on upstream face of obstacle

Favourable conditions:
In discontinuous rock mass (DRM) zones (Morland and Morris, 1977; Addison, 1981)
In intact rock mass (IRM) zones subject to large pressure fluctuations (Gordon, 1981)
Where cold patches enhance basal stresses (Robin, 1976)

(continued)

TABLE 4.1   *(continued)*

Where entrained debris enhances basal stresses (McCall, 1960; Carson, 1971; Hagen *et al.*, 1983)

Under thick ice masses, up to a threshold at which sticking occurs (Boulton, 1974)

ABRASION

Wear of subglacial surface achieved when the force exerted by particles dragged across the bed during basal sliding exceeds the indentation hardness and fracture toughness of the bed (Riley, 1982)

Mechanisms [after Rabinowicz (1976), Riley (1982)]:
  1. Abrasive wear — indentation and grinding due to contact between a hard 'tool' and a softer bed
  2. Brittle fracture — (i)   surface fracture: cracking due to the loading and unloading of a brittle surface during basal sliding (axial cracking on loading, lateral cracking on unloading);

  3. Corrasion — channel boundary worn by bed load (cf. abrasion by ice); evorsion possible in eddy zones

Favourable conditions:
  General requirements: basal ice debris, sliding of basal ice, debris moving toward bed (Sugden and John, 1976)
  Basal sliding a *sine qua non* (Small, 1982)
  Where forward sliding velocities are high (Hallet, 1979)
  Up to a critical ice thickness, beyond which lodgement occurs (Boulton, 1974)
  Where upstream plucking provides tools for abrasion (Boulton, 1974)
  Downstream of sites where a debris-rich basal layer is produced by freezing of water (Souchez and Tison, 1981)
  Where basal melting and longitudinal extension bring particles into contact with the bed (Hallet, 1979)
  Where basal ice contains only scattered rock fragments (dense concentrations of debris impede sliding and promote lodgement) (Hallet, 1979)
  Where wear debris are moved away from the wearing interface, e.g. by subglacial waters (Riley, 1982)
  Seasonal and spatial variability in wear rates attendant upon meltwater activity (Riley, 1982)
  On topographic highs (troughs missed out, leading to reduction in bed irregularities) (Hallet, 1979)
  Where bed obstructions produce lateral segregation of the BIL, leading to streaming and differential abrasion (Boulton, 1975, 1979a)

Under Boulton's (1974, 1979a) theoretical model:
  Rate of abrasion directly dependent on the effective normal pressure, and on the concentration, size and shape of particles in traction with the bed, and inversely dependent on bed material hardness

Under Hallet's (1979, 1981) theoretical model:
  Rate of abrasion dependent on the flux of particles in contact with the bed, the effective force with which the particles press against the bed, and the relative hardness of the bed and the abrading particle

TABLE 4.2   Fluvial erosion in alpine basins: a summary of mechanisms and rate controls

---

FLUVIAL EROSION

Erosion of materials by the action of flowing water (as sheet-wash or as channel flow)

Mechanisms:
1. Wash — dislodgement and entrainment of fine particles by sheet or channel flow
2. Scour — downgrading of channel by removal of bed materials as bed load (hydraulic lift important)
3. Corrasion — channel boundary worn by bed load (cf. abrasion by ice); evorsion possible in eddy zones
4. Abrasion — bed load worn by chipping, cracking, splitting, crushing and grinding during transport
5. Cavitation — channel boundary broken down by pressure waves generated on collapse of airless bubbles formed within the flow
6. Bank collapse — slumping/sliding/undercutting of bank materials
7. Network extension/change — erosion/entrainment of debris previously untapped by the fluvial system
8. Leaching — flushing of solutes from weathered materials

Favourable conditions:
Where subglacial tills provide a reservoir of readily erodible sediments (Church and Ryder, 1972; Collins, 1979c)
Where proglacial moraines provide a reservoir of readily erodible sediments (Church and Ryder, 1972; Hammer and Smith, 1983; Richards, 1984)
Under aggressive, high-velocity turbulent flow conditions (Church and Gilbert, 1975)
Platey pebbles particularly transportable under turbulent flow conditions (Bradley *et al.*, 1972)
Where drainage systems incorporate a dense network of small, unstable tributary channels (Collins, 1979c)
Where drainage channels are prone to shifts in position into zones of unworked sediments (Collins, 1979c; Hammer and Smith, 1983)
Where high magnitude drainage events or outbursts tap previously unexploited areas of the glacier bed and/or the proglacial zone (Collins, 1979c; Beecroft, 1983; Richards, 1984)
Where bank slumping is favoured [e.g. (1) in sand/gravel proglacial channels (Church and Gilbert, 1975; Hammer and Smith, 1983); e.g. (2) in deformable subglacial conduits, which may collapse during high ice pressure/low water pressure conditions (Collins, 1979c)
Cavitation favoured by rapid changes in conduit capacity generating molecular instabilities and pressure changes within the flow (Pitty, 1971; Theakstone, 1980)
Chipping of angular edges an effective process in high-energy stream environments (Whalley, 1979)
Particles stored as outwash deposits between transport events may experience weathering which enhances breakdown on re-entrainment (Richards, 1982)
Freezing and thawing of deposits over winter/spring may produce a temporary reservoir of fine sediments available for transport in the early stages of the melt season (Hammer and Smith, 1983)
Where rain-generated runoff events wash sediments from non-glacial portions of the basin (Richards, 1984)

TABLE 4.3 Weathering in alpine basins: a summary of mechanisms and rate controls

CHEMICAL WEATHERING

Decomposition of minerals into ionic constituents. Water plays a greater or lesser part in all processes. The various actions invariably operate concurrently, and often interdependently

Mechanisms:
1. Solution — dissociation and evacuation of soluble ions from minerals (water acts as a solvent)
2. Hydrolysis — hydrogen ions dissociated from water replace mineral cations, leading to expansion and decomposition of mineral structures. The displaced mineral ions may combine with the free hydroxides to form solutions (water acts as a reactant)
3. Carbonation — soluble products (and insoluble residues) produced by reaction between a mineral and the carbonate and bicarbonate ions contributed to water by the absorption of atmospheric CO_2
4. Hydration — absorption of water into mineral lattices, causing volumetric changes promoting disintegration and/or decomposition (water acts as a catalyst)
5. Oxidation — combination of a mineral with dissolved oxygen to form oxides or hydroxides, weakening the original mineral structure
6. Reduction — addition of an electron to the elements of a mineral under anaerobic conditions, resulting in a change in susceptibility to decomposition
7. Cation exchange — one-to-one interchange of cations between a water film rich in one cation and a mineral rich in another cation.

Favourable conditions:
   Where dilute meltwaters first encounter reactive rock debris (Collins, 1981; Raiswell and Thomas, 1984)
   Where freshly ground particles are abundant (Collins, 1979a, 1981; Eyles *et al.*, 1982; Raiswell, 1984)
   Where progressive solution is not inhibited by the formation of insoluble residues
   Where flushing/leaching of ionically saturated solutions enables subsequent chemical activity to occur (Raiswell and Thomas, 1984)
   Where contact times between meltwaters and debris are sufficient to enable efficient chemical activity (Raiswell and Thomas, 1984)
   Where access to atmospheric CO_2 favours chemical weathering and dissolution (Thomas and Raiswell, 1984)
   Where equilibration of meltwaters with atmospheric CO_2 produces H^+-bearing acid (Raiswell, 1984)
   Where weathering of sulphide minerals in bedrock produces H^+-bearing acid (Raiswell, 1984)
   Chemical weathering determined by rates of H^+ supply, rates of mineral reaction and meltwater–rock contact times (all of which are spatially and temporally variable) (Raiswell, 1984)
   Weathering and dissolution likely to be high in subglacial environments, and low in supraglacial and englacial environments, as controlled by the contact time, contact area and reactivity characteristics of meltwater–debris interactions (Raiswell, 1984)

(continued)

TABLE 4.3   *(continued)*

Carbonation the single most important process in subglacial environments (Raiswell, 1984; Souchez and Lemmens, see Chapter 11)

Oxidation favoured by well oxygenated waters (Brunsden, 1979)

Reduction favoured by cool, wet conditions (Brunsden, 1979)

Selective leaching: monovalent ions most mobile, bivalent next, trivalent least mobile (Pitty, 1971)

Cation-exchange capacity maximized where grinding produces clay-sized particles (Collins, 1977)

PHYSICAL WEATHERING

Mechanical disintegration of rock into finer grade debris via the application of gravitational, expansion or hydraulic forces

Mechanisms:
  1. Pressure release — removal of confining pressures causing fracturing parallel to the freed surface
  2. Gelifraction — fracturing due to pressures exerted by freezing of water in joints and fissures
  3. Hydration shattering — fracturing of rock associated with expansion on absorption of water into the mineral lattice
  4. Impact shattering — fracturing/shattering by impact during mass movements (rock falls, debris slides, avalanches)

Favourable conditions:
  In discontinuous rock mass (DRM) zones (Rapp, 1975)
  On unstable, oversteepened ice-margin slopes (Small, 1982)
  Rapid freezing favours gelifraction (produces closed system pressures by sealing the fissure) (Pissart, 1970)
  Gelifraction favoured by freezing from the crack downward (forming a closed system) (Battle, 1960)
  In areas where sediments are alternately taken into and out of transport according to flow fluctuations (Bradley, 1970)
  Rockfall an important source of debris, both volumetrically and in terms of providing tools for glacial processes (Boulton, 1979b; Mills, 1979; Small, 1982)

subglacial and proglacial) and may be routed by glacial, fluvial and mass movement transport processes along five main paths (subaerial, supraglacial, englacial, subglacial and proglacial). The various sediment production and transport processes involved may be represented, respectively, by the following expressions:

$$\Sigma S_p = G_p + F_p + W_p \tag{4.1}$$

and

$$\Sigma St = G_t + F_t + M_t \tag{4.2}$$

where p and t refer, respectively, to sediment production processes and sediment transport processes, and where ΣS denotes total sediment volumes and G, F, W and M denote, respectively, glacial, fluvial, weathering and mass movement processes.

The term G_p includes abrasion, quarrying and fracture/crushing/shearing of subglacial surfaces. F_p includes water-related mechanical and chemical erosion of subglacial and proglacial surfaces. W_p includes the various processes involved in the physical disintegration and chemical decomposition of rock materials. Tables 4.1, 4.2 and 4.3 list the major mechanisms involved within each of these process categories, and identify conditions under which the various mechanisms are reportedly most productive/effective.

The term G_t includes materials transported along the surface (via moraines), in the body (particularly via debris-charged 'septa' or debris layers) and at the base [in the 'sole' or basal ice layer (BIL)] of the glacier. Following Boulton (1975, 1978), a distinction may be drawn between the 'passive' englacial pathway and the 'active' basal transport zone (the degree of comminution and further erosion associated with the former is negligible, with the latter substantial). The term F_t includes the transport of sediments as dissolved, wash, suspended and bed load in subglacial and proglacial sheet and channel flow. M_t includes mass movements of the fall, slide and flow type (rockfalls, debris slides, debris flows, slush avalanches and snow avalanches) on the rock faces and scree slopes flanking the basin both above and beyond the glacier. Table 4.4 generalizes the basic routing characteristics of each pathway in terms of the volume of material involved, the frequency and rate of movement and the extent of comminution and further erosion that occurs.

The foregoing representation simplifies an extremely complex erosion–transport–sedimentation system in which (*nem. con.*) the various sediment production and routing processes operate interdependently rather than independently, differentially rather than uniformly and intermittently rather than continuously. Moreover, the system invariably incorporates lags and sediment stores: not all of the sediments produced during a given time span are necessarily removed concurrently, and not all of the materials removed from the basin in a given period are necessarily produced contemporaneously. Expressed as a simple balance equation taking storage into account, the transfers involved reduce to

$$\Delta S_S = G_p + F_p + W_p - G_t - F_t - M_t \tag{4.3}$$

where the term S_S describes sediments 'stored' in the basin (i.e. materials produced, but not removed, in past periods). The term t in the above expression refers to sediments transported from the basin. The relationship assumes that the basin is effectively closed to imports of sediment from outside—a condition which may not be met in a zone of ice difluence or divide transgression.

FIGURE 4.4 Sediment routing in alpine glacier basins: travel paths and transit characteristics

Route		Type of routing	Rate of routing		Degree of modification	Degree of further erosion	Volume
Glacial:							
Supraglacial	High-level transport	Continuous	Slow	Glacier flow control	Low	Nil	High
Englacial	High-level transport / Low-level transport	Continuous	Slow	Glacier flow control	Medium	Nil	Low
Subglacial	Low-level transport	Continuous	Slow	Glacier flow control	High	High	High
Fluvial:							
Supraglacial		Continuous (episodic)	Rapid	Stream flow control	Low	Nil	Low
Subglacial		Continuous (episodic)	Rapid	Stream flow control	Medium/high	High	High
Proglacial		Continuous (episodic)	Rapid	Stream flow control	Medium/high	High	High
Marginal		Continuous (episodic)	Rapid	Stream flow control	Medium/high	Low–high	Low–high
Periglacial		Continuous/episodic	Slow–rapid	Gravity control	High	Low–high	Low–high

Glacial and mass movement mechanisms may export sediments directly from some alpine basins. This may happen, for example, if the marginal boundary of the basin is marked by moraine ridges which may be overridden during phases of ice expansion, or if avalanches or rock slides are able to run down the main axis of the basin. However, in most partially glacierized alpine basins the morphology of the catchment is such that sediments are exported as bed, suspended, wash and solute load in the proglacial streams. In balance terms, glacial and mass movement outputs may accordingly be treated as inputs to the proglacial sediment store, with all sediment removal from the basin being effected by fluvial transport processes. Further, given the protracted transit times of the G_t and M_t pathways (whether involving slow, continuous transport or rapid, but episodic transport) the term S_S may be taken to include sediments 'stored' on subaerial rockfaces and scree slopes, on supraglacial surfaces (as moraine), in englacial and subglacial ice-debris layers, in subglacial tills and in proglacial valley trains. Equation (4.3) accordingly reduces at basin level to

$$\Delta S_S = G_p + F_p + W_p - F_t \tag{4.4}$$

This indicates that changes in the basin sediment store are controlled by the balance between the volume of sediments produced by active glacial, fluvial and weathering processes and the volume of sediments removed by active fluvial processes. Restated, noting that $F_t = SQ$, the total sediment discharge from the basin, equation (4.4) becomes

$$F_t = SQ = G_p + F_p + W_p \pm \Delta S_S \tag{4.5}$$

This indicates that the volume of material removed from glacier basins (as stream sediment load) is determined by the quantity of sediments delivered by active glacial, fluvial and weathering processes and by the volume of sediment released from or lost to sediment stores in the basin.

The remainder of this section (4.2) examines the production and storage characteristics of the glacier basin sediment system [i.e. it focuses upon issues relating to the right-hand side of equation (4.5)]. Section 4.3 then examines the characteristics of sediment transport in glacial outflow streams [i.e. it focuses on issues relating to the left-hand side of equation (4.5)].

4.2.2 Characteristics of the sediment system

Let us consider some of the central aspects of the sediment system outlined above.

4.2.2.1   Efficiency of sediment production

The efficiency of the erosive processes operating in glacial environments has long been a subject of debate. On the one hand, there exists persuasive morphological and sedimentological testimony to the former power of alpine glaciers, and evidence in the high specific sediment and solute loads presently exported from highly debris-mantled alpine basins to indicate that present-day alpine glaciers are at least acting as efficient debris transport systems if not as efficient debris production systems (Small, 1982). On the other hand, geomechanical evaluations of the relative strengths of ice and rock matrices have consistently cast doubt upon the ability of moving ice to erode intact rock (McCall, 1960; Glen and Lewis, 1961; Carson 1971; Morland and Morris, 1977; Addison, 1981). Similar reservations have been placed on the geomechanical effectiveness of gelifraction in field conditions (Ives, 1973). The situation, at first sight, bears some analogy to the aerodynamic objections to the flight of the bumble bee! The apparent dilemma is eased, however, when the problem is tackled from a whole-system perspective, admitting the interdependency of processes, differential erosive efficiency and spatial (and temporal) selectivity in sediment production.

4.2.2.2   Interdependency in process operation

Proof that processes operate in an interactive rather than in an isolated fashion is plentiful. That gelifracted or plucked debris must be supplied to the basal ice transport zone before abrasion (and possibly crushing and shearing) may occur is a well documented tenet (Carson, 1971; Boulton, 1974). That preglacial weathering enhances the ability of a glacier to erode its bed is increasingly accepted (Feininger, 1971; Whalley, 1979). That the various reactions of chemical weathering are often accompanied by rock volume increases which favour mechanical disaggregation is also well known (Feininger, 1971; Brunsden, 1979; White, 1976). However, process interdependency is perhaps best evidenced in the multifarious role of water beyond its direct action in mechanical and chemical fluvial erosion: viz. (a) in physical and chemical weathering processes (e.g. in *in situ* crystallization, hydration and hydrolysis actions, and in triggering mass movement events) and (b) in glacial erosion processes (e.g. in promoting freezing-on in cavities subject to water pressure fluctuations; in controlling the permeability and shear strength of ground materials, whether via pore-water or ground-ice status; in maintaining an active ice–rock interface by washing debris into stream channels; in promoting basal sliding, and thence abrasion; by effecting regelation entrainment of debris into the BIL, and thence promoting abrasion).

It is possible to draw a distinction between those primary production processes which are able to break down rock under their own actions alone (e.g. plucking, cavitation, gelifraction, hydrolysis) and those secondary sediment production processes which rely on another process to generate the circumstances necessary for them to achieve erosion (e.g. abrasion, shearing, corrasion). Those processes falling into the former category may be regarded as pre-eminent in a qualitative sense, since the processes in the latter set are inoperative/ineffective in their absence. The relative importance of the various processes in terms of their quantitative contribution to the overall sediment budget of the basin is not so clear, however (see Section 4.2.2.4). Overall, it would appear appropriate to envisage moving ice, water and debris processes operating conjointly within a basin erosional system. The net effect of interactions of the type described above is generally reinforcing, in that efficiency levels are enhanced by the provision of optimal conditions. It follows that sediments are likely to be produced differentially according to variations in the efficiency of processes as well as variations in the resistance of bed materials.

4.2.2.3 *Selectivity in subglacial sediment production*

At a general level, the efficiency of sediment production may be viewed as being dependent on the strength–stress relationship between interface materials and processes. It has long been accepted that a glacier may efficiently exploit inherent rock mass instability, and that erosion is, in consequence, likely to be spatially uneven, with sediments being produced preferentially from 'soft' sources [i.e. discontinuous rock mass (DRM) zones, whether structural or process-prepared] (Carson, 1971; Addison, 1981). It is also accepted that there are likely to be particular points or zones underneath a glacier at which the forces exerted on the bed are greater than they are elsewhere. Recent research at subglacial sites has shown just how effective a glacier may be when circumstances combine favourably. The findings of Robin (1976) and Hagen *et al.* (1983) are particularly interesting in this respect. Robin (1976) showed that basal shear stresses may increase by around 10 bar in cold patches within temperate ice zones. Hagen *et al.* (1983) observed substantial variations in basal ice pressure under Bondhusbreen, Norway, noting (a) that unsteadiness during basal sliding produced average basal pressures 3–4 bar higher than hydrostatic pressure; (b) that large variations in basal pressures accompanied the growth and closure of subglacial cavities produced by water pressure fluctuations; (c) that huge rises in basal pressures were produced by boulders passing through the basal transport zone (pressure rises of up to 90 bar were recorded); (d) that basal pressures varied spatially with differences of up to 30 bar being measured. It would appear, therefore, that effective ice pressures may be considerably enhanced by dynamic effects associated with glacier sliding, entrained debris and thermal behaviour to levels well beyond the theoretical maxima assumed in steady-state geomechanical tests. It follows that the forces available for erosion may, at least locally, be much higher than would be expected from traditional temperate ice theory.

It seems clear that the absolute and relative efficiencies of subglacial erosive processes vary according to spatially variable bed conditions (e.g. structure, lithology, morphology) and spatially and temporally variable 'process' conditions (e.g. flow type, flow rate, basal ice temperature, subglacial hydrology). Some of the conditions which are likely to favour locally enhanced erosion are summarized in Table 4.1. Zonal variability in erosion has also been modelled. Boulton (1972), for example, attached erosional and depositional attributes to zones defined according to thermal regime; Clayton and Moran (1974) advanced a glacial form-process model based on thermal, groundwater and flow regimes; while Andrews' (1972) concept of glacier power embodies a flow-based distinction in erosive potential, with effective power (cf. erosive ability) increasing as a proportion of total power (a maximum condition) as basal sliding increases as a proportion of total glacier flow.

4.2.2.4   Sediment budgets and balance

It is possible in generalized, qualitative terms, to identify the major controls on the quantity of sediment exported from a glacier basin. Equation (4.5), for example, may be likened to the water balance equation of a drainage basin. However, it is not so easy to quantify process-specific, source-specific or time-specific contributions to sediment outputs (there are parallels here with the difficulties involved in deconvoluting riverflow hydrographs). A number of studies have, however, provided indications of the importance of contributions from particular processes and/or origins. We can cite examples relating to particular issues, as follows.

(a) On the relative importance of particular mechanisms. Boulton (1974, 1979a,b) argued that plucking is likely to be quantitatively more important than abrasion in circumstances where supraglacial debris inputs are limited (as might be expected on ice-caps); the volume of sediments produced by plucking is unlikely, however, to exceed that resulting from abrasion in alpine valley glacier situations, where valley walls may be expected to deliver substantial quantities of debris on to and into the glacier (see also Röthlisberger, 1979 and Vivian, 1979).

(b) On the importance of a particular group of processes. Recent studies of the cationic load of proglacial streams indicate that current rates of cationic denudation in glacial catchments are noticeably higher than the continental mean rate of 390 mequiv. $m^{-2} yr^{-1}$; rates of 454, 947 and 960 mequiv. $m^{-2} yr^{-1}$ have been quoted for, Gornergletscher (Collins, 1983), Berendon (Eyles *et al.*, 1982) and South Cascade (Reynolds and Johnson, 1972), respectively. Whether the action of flowing water directly produces the breakdown of rock (i.e. erosion *sensu stricto*) or simply removes material produced by other (e.g. glacial, cryergic or, as in the above case, weathering) processes, its influence is evidently manifest. Hagen *et al.* (1983), for example, noted that around 90% of the material passing through a cross-section monitored under Bondhusbreen, Norway, was carried in meltwaters rather than by the glacier itself.

(c) On contributions from particular origins. Sediment delivery in glaciated basins seems to be highly variable according to geological, glaciological, geomorphological and hydrological circumstances. Mills (1979), for example, reported that most of the Nisqually river's suspended sediment load appears to be entrained by the time it emerges from the Nisqually glacier, with over two thirds of the total sediment load being derived subglacially. The Storbreen catchment in Norway, in contrast, is reportedly characterized by a discontinuous process of sediment transfer from subglacial sediment sources to proglacial sediment stores during melt-dominated periods, and by evacuation of sediments from the proglacial zone during rainfall-controlled high flow events (Richards, 1984). Perhaps the most comprehensive sediment source area results produced to date are those of Hammer and Smith (1983). Their studies in the Hilda glacier basin, in Canada, enabled them to estimate the proportion of bedload and suspended load delivered to the stream from supraglacial, subglacial and proglacial sources: 46% of the bedload was deemed to be of supraglacial origin, 27% of subglacial origin and 27% of proglacial origin; only 6% of the suspended load was of supraglacial origin, but with subglacial and proglacial sources each yielding 47% of the total load.

(d) On past and present production contributions to present sediment output. A number of studies have derived estimates of rates of denudation in glacial catchments on the basis of the sediment output in proglacial streams and/or from marginal moraines (e.g. Østrem, 1975; Small *et al.*, 1984). Exactly what proportion of the material presently transported out of a glacier basin in proglacial streams is the result of currently active erosion processes is a matter of some uncertainty. Present sediment outputs strictly represent present sediment removal and not necessarily present sediment production. Views on the relationship between present sediment inputs and outputs vary widely. Metcalf (1979), for example, took the annual suspended sediment load to be an order of magnitude approximation to the annual mass of subglacial abrasion as a general premise in a sediment budget study of the Nisqually glacier basin, USA. Boulton (1979c, in discussion of Collins) commented that long-term net increases or net losses of sediments in subglacial stores are unlikely to be large in most cases. Conversely, however, Church (1972) reported that the sediment yield in glacier basins in Baffin Island bears no relationship to present primary rates of sediment production, a condition which has been taken by Church and Ryder (1972) to characterize 'paraglacial' environments. While sediment budget and balance characteristics undoubtedly differ from basin to basin, it is probably safe to assume the existence of some storage of sediments in all basins: the magnitude–duration characteristics of debris stores may vary from basin to basin, but there is always likely to be an element of storage present.

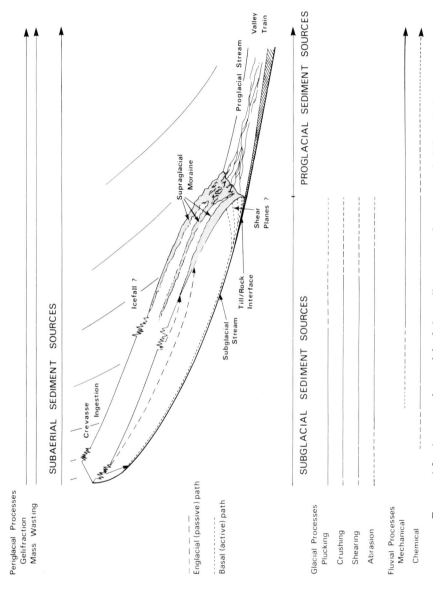

FIGURE 4.2   A composite model of the sediment transfer system of alpine glacier basins

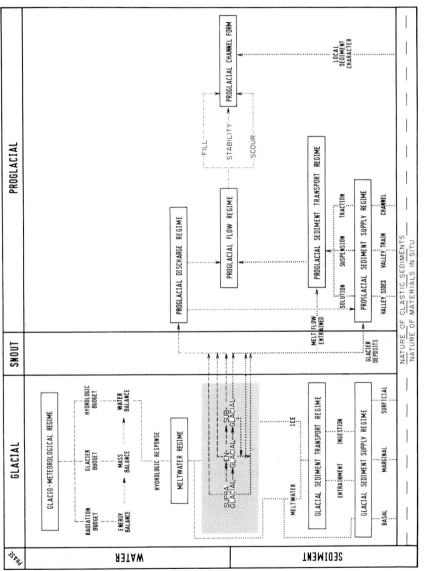

FIGURE 4.3 A model of the runoff and sediment system of alpine glacier basins

4.2.3   A model of the debris transfer system of alpine glacier basins

The main elements of the sediment transfer system described above are shown in diagrammatic form in Figure 4.2. The model attempts to illustrate (i) the sources from which debris is derived, (ii) the processes involved in acquiring material from these sources and (iii) the pathways along which debris is transported.

As noted in Section 4.2.1, there are plausible grounds for accepting (as a simplifying assumption) that the outflow stream acts as the sole means of conveying sediment out of the basin. Having addressed the input, store and transfer elements of the glacial sediment system in this section (4.2), the discussion now turns to focus upon the 'output' phase of the system.

4.3   SEDIMENT TRANSPORT IN GLACIER OUTFLOW STREAMS

This section examines the characteristics of fluvial sediment transport in alpine glacier basins. It sets out to summarize the general characteristics of the runoff and related sediment system of alpine glacier basins, and then to outline the characteristics of outflow streamflow and sediment transport regimes.

4.3.1   The runoff and sediment system of alpine glacier basins

Figure 4.3 combines the basic elements of the runoff and sediment production-routing system of glacier basins into a model of the glacier basin runoff and sediment system. The model summarizes:

   (i) the hydrometeorological processes involved in the production of runoff in the glacial phase of the basin;
  (ii) the sources and processes involved in the production of sediments in the glacial phase of the basin;
 (iii) the intraglacial hydrological network through which runoff is routed from the glacial to the proglacial phase of the basin;
 (iv) the intraglacial routes along which sediment is transported to the glacier snout;
  (v) the integration of water and sediment routes at or near the glacier snout;
 (vi) the sources and processes involved in the production of sediments in the proglacial phase of the basin;
(vii) the processes involved in the determination of proglacial flow regime and channel form.

4.3.2   Characteristics of proglacial streamflow and sediment transport

Proglacial streamflow is derived from rainfall, snowmelt, icemelt and groundwater contributions, plus releases of waters held in such subaerial, supraglacial, englacial, subglacial and ground water stores as may exist. Runoff is transferred along supraglacial, englacial, subglacial and proglacial routes (incorporating veins, pipes, channels and sheets) at a rate determined by the volume of flow input, the structure of the route taken and the nature of the hydraulic conditions experienced in transit (Björnsson, 1982). The routing process typically leads to the reduction and attenuation of surface-specific input hydrographs, and the separation of basin-integrated outputs into a quickflow component and a delayed flow component (Collins, 1977).

Outflow hydrographs are accordingly composed of a peaked diurnal quickflow component superimposed on a flatter delayed flow component. While the volume of delayed flow in the hydrograph typically exceeds the volume of quickflow, the absolute and relative volumes of the two components vary over the ablation season according to changes in input conditions (as affected, primarily, by climate) and routing conditions (as affected, for example, by expansion/contraction of the volume, capacity and efficiency of drainage channels), so that outflow hydrographs commonly vary in size and shape from day to day over an ablation season (Elliston, 1973). The discharge record of a proglacial stream over the ablation season is, therefore, normally composed of a chain of repeating outflow hydrographs that vary in size and shape over the ablation season. Runoff events generated by processes other than snow and ice ablation (i.e. rainfall-induced floods or outburst floods) modify the cyclical pattern to a greater or lesser degree: direct rainfall-runoff contributions, for example, may variously augment meltwater runoff volumes, raise baseflow levels, accentuate the peakedness of the diurnal hydrograph or introduce secondary peaks into the flow record.

The dissolved solids (solutes) carried in proglacial waters are derived from water–sediment contact zones above, beneath and in front of the glacier. While chemical enrichment may occur wherever waters pass over or through subglacial and proglacial surfaces, certain preferential conditions invariably exist. Chemical weathering processes are preferentially active where extended contact between water and materials is possible, and solute uptake in meltwaters is maximized where supplies of readily mobilized ions exist. Subglacial locations where glacier grinding has broken down the mineral lattices of materials held in subglacial tills and in the basal ice layer of the glacier, and where waters are delayed or restricted in transit (whether held in short- or long-term storage in subglacial cavities or within subglacial tills, or simply delayed as a result of conduit incapacity or blockage, or excess water pressure) are, therefore, especially favourable sites (Collins, 1977, 1981; Lemmens and Roger, 1978). Chemical enrichment on proglacial surfaces is typically less than that on subglacial surfaces (Lemmens and Roger, 1978; Gurnell and Fenn, 1985).

In terms of Collins' (1977) englacial (Q_e)–subglacial (Q_s) flow model, most solutes are contributed to the proglacial stream by the subglacial flows. Simple mixing models of this type are commonly invoked to explain the inverse relationship between stream discharge and solute concentration (or electrical conductivity) typically found in glacial catchments (Rainwater and Guy, 1961; Behrens *et al.*, 1971; Collins, 1979a; Collins and Young, 1981). Lemmens and Roger (1978), working in the Tsidjiore Nouve basin, Switzerland, have proposed, however, that the inverse relationship reflects a variable solute enrichment mechanism, with low solute uptake occurring during high flow conditions as a result of more rapid flow velocities and less intimate water-boundary contact. Whichever of the two models (the mixing model or the variable enrichment model) is taken, it is clear that streamflow and solute levels are inversely related (see Chapters 11 and 14 for further discussion).

The solute concentration record of a proglacial stream during an ablation season will, therefore, consist of a series of repeating cycles that are inversely related to (although not necessarily inversely synchronous with) those of the streamflow discharge series. The actual size and shape of each daily solute concentration 'trough,' and the timing of it, will be partly dependent on the nature of flow factors (e.g. quickflow — delayed flow volumes), and partly dependent on the nature of solute supply factors (sources, quantity).

Bottom and suspended sediments arrive at a particular point in a proglacial stream system via supraglacial, subglacial and proglacial channels that acquire sediments from direct input from glacial and periglacial processes and from a series of in-channel processes which result in the stream accessing and tapping sediment supplies. Such accessing of sediment supplies may be achieved through a number of ways, e.g. (1) the extension of stream networks in response to increases in the size of runoff contributing areas (notably across subglacial surfaces); (2) changes in stream course in response to changes in the source of water inputs, or to restrictions or blockages in existing stream courses; (3) changes in channel pattern, in association with changes in flow (an increase in discharge may, for example, result in a change from a single thread to a multiple thread channel pattern); (4) changes in channel capacity, resulting from scouring of channel beds and banks. It is likely that most sediments are fed into the streamflow system from small, unstable tributary channels, which shift course, contacting new sediment sources in the process, more readily and more frequently than do the main channels of the fluvial system (Collins, 1979a, 1981).

In general, where sediment supplies are available, the quantity of sediments transported in outflow streams varies directly (although not necessarily synchronously) with flow discharge. However, sediments are invariably differentially released into the streamflow system according to the temporally discontinuous, spatially non-uniform action of the sediment production processes described above. Moreover, available sediment supplies are commonly flushed out by 'first-flood' conditions, after which the exhaustion of sediments capable of being transported by flows of equivalent magnitude results in lower sediment loads being carried (Kjeldsen *et al.*, 1975; Fenn *et al.*, 1985). Thereafter, similar or higher sediment loads are associated only with higher flows, or with increases in sediment supply (as produced, for example, by greater sediment inputs to the existing channel system or by changes in the structure of that system). It follows that on both diurnal and seasonal scales, the sediment load associated with a flow of a given level is typically considerably greater on the rising limb of the flow hydrograph than on the falling limb. Diurnal and seasonal hysteresis effects accordingly characterize the relationship between proglacial stream discharge and suspended sediment concentration and bedload transport. While the loops generated are most frequently clockwise (Østrem, 1975; Collins, 1979c; Hammer and Smith, 1983) sediment supplies to the proglacial system are often so erratic as to result in a multiplicity of sediment peaks (ranging in magnitude, frequency and duration from major sediment transport events to more transient, although not necessarily minor, flushes) accompanying a single flow cycle. Sediment transport fluctuations may therefore vary considerably in their timing, their magnitude and their number relative to discharge fluctuations.

The sediment transport record of a proglacial stream during an ablation season typically consists, therefore, of a sequence of peaks of irregular magnitude and timing superimposed on a more general trend related to fluctuations in stream discharge. The relative significance of these two effects (an irregular, stochastic component resulting from sediment supply variations, and a more regular, more deterministic component resulting from streamflow variations) varies from basin to basin, and over time in a given basin, according to the nature of runoff and sediment production-routing conditions. A more detailed account of the characteristics and controls of the proglacial suspended sediment and bedload records may be found in Chapters 12 and 13, respectively.

4.3.3 A flow-based model of sediment transport in proglacial streams

Figure 4.4 presents hydrograph-based models of proglacial streamflow, solute and sediment transport dynamics. The concepts embraced follow those of Martinec (1976), Kjeldsen *et al.* (1975) and Collins (1977, 1979c,d), and relate closely to the discussion in Section 4.3.2. The model attempts to draw together the following:

(i) Time–space changes in diurnal input hydrograph form. Diurnal input hydrographs resulting from supraglacial ablation are depicted as varying in magnitude and amplitude through space (both properties always increasing towards the glacier terminus) and through time (magnitude varying broadly with seasonal trends in climatic and contributing area conditions; amplitude always increasing through the ablation season as drainage conduits increase in capacity, connectivity and efficiency).

(ii) Temporal changes in the form of diurnal outflow hydrographs. Diurnal outflow hydrographs are depicted as being composed of a peaked

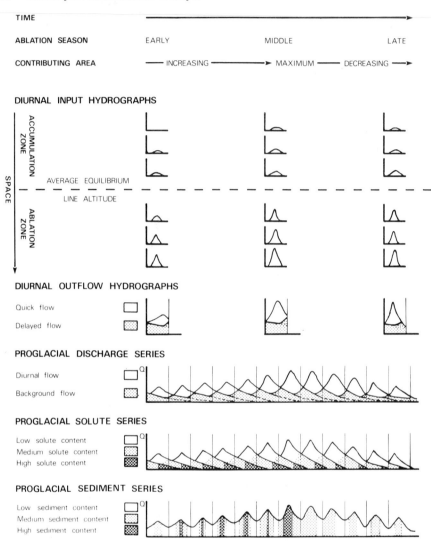

FIGURE 4.4   Hydrograph-based models of the nature of proglacial discharge, solute and
sediment series

quickflow component superimposed on a more invariant background flow
component composed of flows delayed in travelling through the glacier.
(iii) The resulting quickflow and delayed flow components of the proglacial
discharge time series.
(iv) The associated solute concentration (electrical conductivity) characteristics
of the streamflow record. Solute concentrations are shown as being

dependent on the duration of meltwater–material contact [i.e. delayed (subglacial) flow related].

(v) The pattern of sediment transport associated with the streamflow record. Sediment transport is depicted as being dependent on flow competence and sediment supply.

The models are clearly normative. Departures from the responses indicated may result from the operation of other water/solute/sediment supply and routing mechanism (only the simplest are incorporated in the model shown). The value of such models lies, however, in the identification of dissimilarities as well as similarities with 'real' records (e.g. in the timing, the number, the size and the shape of fluctuations).

4.4   CONCLUSIONS

This chapter has attempted to provide a broad overview of the nature and functioning of the debris transfer system of glacierized alpine basins. Tabular and diagrammatic summaries have been given in order that the details presented in the following chapters may be seen against a broad background, and set into a broad context.

REFERENCES

ADDISON, K. (1981), The contribution of discontinuous rock-mass failure to glacier erosion, *Annals of Glaciology*, **2**, 3–10.

ANDERSON, R. S., HALLETT, B., WALDER, J., and AUBRY, B. F. (1982), Observations in a cavity beneath Grinnel Glacier, *Earth Surface Processes and Landforms*, **7**, 63–70.

ANDREWS, J. T. (1972), Glacier power, mass balances, velocities and erosion potential, *Zeitschrift für Geomorphologie*, **13**, 1–17.

BATTLE, W. R. B. (1960), Temperature observations in bergschrunds and their relationship to frost shattering, in LEWIS, W. V. (Ed.), *Norwegian Cirque Glaciers*, Royal Geographical Society Research Series 4, pp. 83–95.

BEECROFT, I. R. (1983), Sediment transport during an outburst from Glacier de Tsidjiore Nouve, Switzerland, 16th–19th June, 1981, *Journal of Glaciology*, **29**, 185–190.

BEHRENS, H. BERGMAN, N. MOSER, H. RAUERT, W. STICHLER, W. AMBACH, W. EISNER, H., and PESSL, K. (1971), Study of the discharge of alpine glaciers by means of environmental isotopes and dye-tracers, *Zeitschrift für Gletcherkunde und Glazialgeologie*, **7**, 79–102.

BJÖRNSSON, H. (1982), Drainage basins on Vatnajökull mapped by radio echo soundings, *Nordic Hydrology*, **13**, 213–232.

BOULTON, G. S. (1972), The role of thermal regime in glacial sedimentation, in PRICE, R. J. and SUGDEN, D. E. (Eds.), *Polar Geomorphology*, Institute of British Geographers Special Publication 4, pp. 1–19.

BOULTON, G. S. (1974), Processes and patterns of glacial erosion, in COATES, D. R. (Ed.), *Glacial Geology*, State University of New York, Binghampton, pp. 41–87.

BOULTON, G. S., WRIGHT, A. E., and MOSELY, F. (Eds.) (1975), Processes and patterns of subglacial sedimentation: a theoretical approach, in *Ice Ages: Ancient and Modern*, Seal House Press, Liverpool, pp. 7-42.

BOULTON, G. S. (1978), Boulder shapes and gain-size distribution of debris as indicators of transport paths through a glacier and till genesis, *Sedimentology*, **25**, 773-799.

BOULTON, G. S. (1979a), Processes of glacier erosion on different substrata, *Journal of Glaciology*, **23**, 15-38.

BOULTON, G. S. (1979b), General discussion, *Journal of Glaciology*, **23**, 385.

BOULTON, G. S. (1979c), Comment following Collins (1979c), *Journal of Glaciology*, **23**, 256.

BOULTON, G. S. (1982), Subglacial processes and the development of glacial bedforms, in DAVIDSON-ARNOTT, R., NICKLING, W., and FAHEY, B. D. (Eds.), *Research in Glacial, Glacio-Fluvial and Glacio-Lacustrine Systems*, Geo Books, Norwich, pp. 1-31.

BRADLEY, W. C. (1970), Effect of weathering on abrasion of granitic gravel, Colorado River, (Texas), *Geological Society of America Bulletin*, **81**, 61-80.

BRADLEY, W. C., FAHNESTOCK, R. K., and ROWEKAMP, E. T. (1972), Coarse sediment transport by flood flows on Knik River, Alaska, *Geological Society of America Bulletin*, **83**, 1261-1284.

BRUNSDEN, D. (1979), Weathering, in EMBLETON, C., and THORNES, J. B. (Eds.), *Process in Geomorphology*, Arnold, London, pp. 73-129.

BUDD, W. F., KEAGE, P. L., and BLUNDY, M. A. (1979), Empirical studies of ice sliding, *Journal of Glaciology*, **23**, 157-169.

CARSON, M. A. (1971), *The Mechanics of Erosion*, Monographs in Spatial and Environmental Systems Analysis, Pion, London, 174pp.

CHURCH, M. (1972), Baffin island sandurs: a study of Arctic fluvial processes, *Bulletin of the Geological Survey of Canada*, **216**, 208pp.

CHURCH, M., and GILBERT, R. (1975), Proglacial fluvial and lacustrine environments, in JOPLING, A. V. and MCDONALD, B. C. (Eds.), *Glaciofluvial and Glaciolacustrine Sedimentation*, Society of Economic Palaeontologists and Mineralogists Special Publication 23, pp. 22-100.

CHURCH, M., and RYDER, J. M. (1972), Paraglacial sedimentation: a consideration of fluvial processes conditioned by glaciation, *Geological Society of America Bulletin*, **83**, 3059-3071.

CLAYTON, L., and MORAN, S. R. (1974), A glacial process form model, in COATES, D. R. (Ed.), *Glacial Geomorphology*, State University of New York, Binghampton, pp. 88-119.

COLLINS, D. N. (1977), Hydrology of an alpine glacier as indicated by the chemical composition of meltwater, *Zeitschrift für Gletscherkunde und Glazialgeologie*, **13**, 219-238.

COLLINS, D. N. (1979a), Hydrochemistry of meltwaters draining from an alpine glacier, *Arctic and Alpine Research*, **11**, 307-324.

COLLINS, D. N. (1979b), Quantitative determination of the subglacial hydrology of two Alpine glaciers, *Journal of Glaciology*, **23**, 347-362.

COLLINS, D. N. (1979c), Sediment concentration in meltwaters as an indicator of erosion processes beneath an alpine glacier, *Journal of Glaciology*, **23**, 247-257.

COLLINS, D. N. (1979d), Meltwater characteristics as indicators of the hydrology of Alpine glaciers, unpublished PhD Thesis, Nottingham University.

COLLINS, D. N. (1981), Seasonal variation of solute concentration in meltwaters draining from an alpine glacier, *Annals of Glaciology*, **2**, 11-16.

COLLINS, D. N. (1983), Solute yield from a glacierised high mountain basin, in WEBB, B. W. (Ed.), *Dissolved Loads of Rivers and Surface Water Quantity/Quality Relationships (Proceedings of the Hamburg Symposium, 15-17 August, 1983)*, International Association of Hydrological Sciences Publication 141, pp. 41-50.

COLLINS, D. N., and YOUNG, G. J. (1981), Meltwater hydrology and hydrochemistry in snow and ice-covered mountain catchments, *Nordic Hydrology*, **12**, 319–334.

DERBYSHIRE, E., GREGORY, K. J., and HAILS, J. R. (1979), *Geomorphological Processes*, Studies in Physical Geography, Dawson, Folkestone, 312pp.

ELLISTON, G. R. (1973), Water movement through the Gornergletscher, in *Symposium on the Hydrology of Glaciers (Proceedings of the Cambridge Symposium, 9–13 September 1969)*, International Association of Scientific Hydrology Publication 95, pp. 79–84.

EMBLETON, C., and THORNES, J. B. (1979), *Process in Geomorphology*, Arnold, London, 436pp.

EYLES, N., (Ed.) (1983), *Glacial Geology: An Introduction for Engineers and Earth Scientists*, Pergamon Press, Oxford, 409pp.

EYLES, N. SASSEVILLE, D. R., SLATT, R. M., and ROGERSON, R. J. (1982), Geochemical denudation rates and solute transport mechanisms in a maritime temperate glacier basin, *Canadian Journal of Earth Sciences*, **19**, 1570–1581.

EYLES, N., and MENZIES, J. (1983), The subglacial landsystem, in EYLES, N. (Ed.), *Glacial Geology: An Introduction for Engineers and Earth Scientists*, Pergamon Press, Oxford, pp. 19–70.

FEININGER, T. (1971), Chemical weathering and glacial erosion of crystalline rocks and the origin of till, *United States Geological Survey Professional Paper*, 750C, C65–C81.

FENN, C. R., GURNELL, A. M., and BEECROFT, I. (1985), An evaluation of the use of suspended sediment rating curves for the prediction of suspended sediment concentration in a proglacial stream, *Geografiska Annaler*, **67A**, 71–82.

GLEN, J. W., and LEWIS, W. V. (1961), Measurements of side-slip at Austerdalsbreen, 1959, *Journal of Glaciology*, **3**, 1109–1122.

GORDON, J. E. (1981), Ice-scoured topography and its relationship to bedrock structure and ice movement in parts of northern Scotland and West Greenland, *Geografiska Annaler*, **63A**, 55–65.

GURNELL, A. M., and FENN, C. R. (1985), Spatial and temporal variations in electrical conductivity in a proglacial stream system, *Journal of Glaciology*, **31**, 108–114.

HAGEN, J. O., WOLD, B., WIESTOL, O. ØSTREM, G., and SOLLID, J. L., (1983), Subglacial processes at Bandhusbreen, Norway: preliminary results, *Annals of Glaciology*, **4**, 91–98.

HALLET, B. (1979), A theoretical model of glacial abrasion, *Journal of Glaciology*, **23**, 39–50.

HALLET, B. (1981), Glacial abrasion and sliding: their dependence on the debris concentration in basal ice, *Annals of Glaciology*, **2**, 23–28.

HAMMER, K. M., and SMITH, N. D. (1983), Sediment production and transport in a proglacial stream: Hilda glacier, Alberta, Canada, *Boreas*, **12**, 91–106.

HUTTER, K., and OLUNLOYO, V. O. S. (1981), Basal stress concentrations due to abrupt changes in boundary conditions: a cause for high till concentration at the bottom of a glacier, *Annals of Glaciology*, **2**, 29–33.

IVES, J. D. (1973), Arctic and Alpine geomorphology—a review of current outlook and notable gaps in knowledge, in FAHEY, B. D., and THOMPSON, R. D. (Eds.), *Research in Polar and Alpine Geomorphology, 3rd Guelph Symposium on Geomorphology*, Geo Books, Norwich, pp. 1–10.

KJELDSEN, O. ØSTREM, G., and OLSEN, H. C. (1975), *Materialtransportundersøkelser: Norske Bre-Elver 1974*, Rapport Nr. 3–75, Varsdragsdirektorat, Hydrologisk Avdeling, Oslo.

LEMMENS, M., and ROGER, M. (1978), Influence of ion exchange on dissolved load of Alpine meltwaters, *Earth Surface Processes*, **3**, 179–188.

LISTER, H. (1981), Particle size, shape and load in a cold and a temperate valley glacier, *Annals of Glaciology*, **2**, 39–44.

MARTINEC, J. (1976), Snow and ice, in RODDA, J. C. (Ed.), *Facets of Hydrology*, Wiley, Chichester, pp. 85–118.

MCCALL, J. G. (1960), The flow characteristics of a cirque, glacier and their effect on glacial structure and cirque formation, in LEWIS, W. V. (Ed.), *Norwegian Cirque Glaciers*, Royal Geographical Society Research Series 4, pp. 39–62.

METCALF, R. C. (1979), Energy dissipation during subglacial abrasion of Nisqually glacier, Washington, *Journal of Glaciology*, **3**, 233–246.

MILLS, H. M. (1979), Some implications of sediment studies for glacial erosion on Mount Rainier, Washington, *Northwest Science*, **53**, 190–199.

MORAN, S. R., CLAYTON, L., HOOKE, R. L., FENTON, M. M., and ANDRIASHEK, L. D. (1980), Glacier-bed landforms of the prairie region of North America, *Journal of Glaciology*, **25**, 457–476.

MORLAND, L. W., and MORRIS, E. M. (1977), Stress in an elastic bedrock hump due to glacier flow, *Journal of Glaciology*, **18**, 67–75.

ØSTREM, G. (1975), Sediment transport in glacial meltwater streams, in JOPLING, A. V., and MCDONALD, B. C. (Eds.), *Glaciofluvial and Glaciolacustrine Sedimentation*, Society of Economic Palaeonotologists and Mineralogists Special Publication 23, pp. 101–122.

PISSART, A. (1970), Les phénomènes physiques essentiels liés au gel, les structures périglaciaires quî en resultent et leur signification climatique, *Annals Societé Géologie Belgique*, **93**, 7–49.

PITTY, A. F. (1971), *Introduction to Geomorphology*, Methuen, London, 526pp.

POTTS, A. S. (1970), Frost action in rocks: some experimental data, *Institute of British Geographers Transactions*, **49**, 109–124.

RABINOWICZ, E. (1976), Wear, *Materials Science and Engineering*, **25**, 23–28.

RAINWATER, F. H., and GUY, H. P. (1961), Some observations on the hydrochemistry and sedimentation of the Chamberlain Glacier Area, Alaska, *United States Geological Survey Professional Paper*, 414-C, 14pp.

RAISWELL, R. (1984), Chemical models of solute acquisition in glacial meltwaters, *Journal of Glaciology*, **30**, 49–57.

RAISWELL, R., and THOMAS, A. G. (1984), Solute acquisition in glacial meltwaters. I. Fjallsjokull (South East Iceland): bulk meltwaters with closed system characteristics, *Journal of Glaciology*, **30**, 35–43.

RAPP, A. (1975), Studies of mass wasting in the Arctic and in the tropics, in YATSU, E., WARD, A. J., and ADAMS, F. (Eds.), *Mass Wasting, 4th Guelph Symposium on Geomorphology*, Geo Abstracts, Norwich, pp. 79–104.

REYNOLDS, R. C., JR., and JOHNSON, N. M. (1972), Chemical weathering in the temperate glacial environment of the Northern Cascade mts., *Geochimica and Cosmochimica Acta*, **36**, 537–554.

RICHARDS, K. S. (1982), *Rivers: Form and Process in Alluvial Channels*, Methuen, London, 358pp.

RICHARDS, K. S. (1984), Some observations on suspended sediment dynamics in Storbregrova, Jotenheimen, *Earth Surface Processes and Landforms*, **9**, 101–111.

RILEY, N. W. (1982), Rock wear by sliding ice, unpublished PhD Thesis, University of Newcastle upon Tyne, 145pp.

ROBIN, G. DE Q. (1976), Is the basal ice of a temperate glacier at the pressure melting point?, *Journal of Glaciology*, **16**, 183–196.

RÖTHLISBERGER, H. (1979), General discussion, *Journal of Glaciology*, **23**, 385.

RÖTHLISBERGER, H., and IKEN, A. (1981), Plucking as an effect of water pressure variations at the glacier bed, *Annals of Glaciology*, **2**, 57–62.

SMALL, R. J. (1982), Glaciers—do they really erode?, *Geography*, **6**, 9–14.

SMALL, R. J., and GOMEZ, B. (1981), The nature and origin of debris layers within Glacier de Tsidjiore Nouve, Valais, Switzerland, *Annals of Glaciology*, **2**, 109–113.

SMALL, R. J., BEECROFT, I. R., and STIRLING, D. M. (1984), Rates of deposition on lateral moraine embankments, Glacier de Tsidjiore Nouve, Valais, Switzerland, *Journal of Glaciology*, **30**, 275–281.

SOUCHEZ, R. A., and TISON, J. L. (1981), Basal freezing of squeezed water: its influence on glacier erosion, *Annals of Glaciology*, **2**, 63–66.

SUGDEN, D. E., and JOHN, B. S. (1976), *Glaciers and Landscape: a Geomorphological Approach*, Edward Arnold, London, 376pp.

THEAKSTONE, W. H. (1980), Glacial geomorphology, *Progress in Physical Geography*, **4**, 241–253.

THOMAS, A. G., and RAISWELL, R. (1984), Solute acquisition in glacial meltwaters. II. Argentière (French Alps) bulk meltwaters with open system characteristics, *Journal of Glaciology*, **30**, 44–48.

VIVIAN, R. A. (1979), General discussion, *Journal of Glaciology*, **23**, 386.

VIVIAN, R. (1980), The nature of the ice–rock interface: the results of investigation on 20 000 m of the rock bed of temperate glaciers, *Journal of Glaciology*, **25**, 267–278.

WEERTMAN, J. (1961), Mechanism for the formation of inner moraines found near the edge of cold ice caps and ice sheets, *Journal of Glaciology*, **3**, 965–978.

WHALLEY, W. B. (1979), Quartz silt production and sand grain surface textures from fluvial and glacial environments, in O'HARE, A. M. F. (Ed.), *Scanning Electron Microscopy*, SEM Inc., Illinois, pp. 547–554.

WHITE, S. E. (1976), Is frost action really only hydration shattering?, *Arctic and Alpine Research*, **8**, 1–6.

Glacio-fluvial Sediment Transfer
Edited by A. M. Gurnell and M. J. Clark
©1987 John Wiley & Sons Ltd.

Chapter 5

The hydrogeomorphology of alpine proglacial areas

K. J. Gregory

ABSTRACT

Fluvial processes in alpine proglacial areas reflect an unusual combination of discharge variations, which are variable at several temporal scales, of sediment characteristics which are often not supply-limited and of local characteristics where slope is especially variable. Against the background of these controls three broad domains can be identified; namely, sandur with braided channel patterns, bedrock domains and areas dominated by palaeoforms. Analysis of recent changes has not been able to utilize the general palaeohydrologic approaches which apply in other areas, but interpretations based on the use of sedimentary characteristics to develop palaeohydraulic approaches offer considerable potential. Information from contemporary fluvial systems and derived from analyses of recent changes should be a useful input to the enhanced design of river channel management strategies in proglacial areas.

5.1 INTRODUCTION

Alpine proglacial areas are characterized by an unusual combination of very active river channel morphology with a striking legacy of fluvial features dating from the recent past. The distinctiveness of these environments can be demonstrated in other ways. For example, in his review of river processes in different climates, Sundborg (1978) selectively takes as his examples arctic, humid mid-latitude, dry climates, and tropical climates—a fragmented approach necessitated because there is as yet no integrated view of the way in which fluvial systems operate in all areas of the world. The multi-focus nature of the field is also apparent in the fact that over the last two decades fluvial geomorphology has come to be associated with landforms and with the chronology of landform development, whereas the distinct concept of hydrogeomorphology has been used to connote the study of fluvial processes and their interrelationships with associated forms (Gregory, 1979).

Since 1970, emphasis has been placed on achieving an understanding of recent changes in the fluvial system. This has led to important conceptual developments, although such system changes may not easily be predicted. The general problem, of which proglacial areas are a special case, is inherent in Burkham's (1981) argument that uncertainty still characterizes attempts to predict changes of the fluvial system. This assertion is elaborated by Schumm (1985) in his recognition of seven sources of uncertainty: scale (leading to partial coverage, temporal and spatial), location (since change is not registered simultaneously at all locations), convergence (different processes producing similar results), divergence (similar processes producing different results), singularity (individual departures from trend triggered by internal or external variables), sensitivity (unpredictable systems performance when close to a threshold) and complex behaviour. It follows that any consideration of the hydrogeomorphology of alpine proglacial areas must therefore be tentative, not only because research has not yet proceeded to the stage at which the distinctive characteristics of specific areas of the world are clearly identified, but also because some measure of uncertainty must prevail when one attempts to explain recent sequences of development.

To provide a reconnaissance of alpine proglacial hydrogeomorphology which concentrates on present fluvial systems as a basis for understanding recent changes, it is appropriate to review four topics. Firstly, to scrutinize the relationships between fluvial process and the morphological characteristics which together provide a distinctive environment (5.2); secondly, to indicate the distinctiveness of the fluvial landform assemblages (or domains) which occur in alpine proglacial areas (5.3); thirdly, to outline the characteristics of the palaeohydrology of such areas (5.4); and finally, to note issues which are relevant for the management of such proglacial areas (5.5). This sequence allows consideration of the controls of present fluvial processes (5.2) as a basis for understanding present spatial patterns (5.3), and for suggesting how major changes have occurred (5.4). However, it must be emphasized that the environment being considered is somewhat paradoxical because on the one hand it is very distinctive but on the other hand, as shown in Chapter 2, it is not completely distinguishable from the broader fluvial environments of mountain areas as a whole, a problem exemplified in reviews by Price (1981) and in the context of mountain geomorphology by Slaymaker and McPherson (1972). There are also many similarities with arctic fluvial environments.

5.2   FLUVIAL FORM AND PROCESS

Interaction between the fluvial processes operating in alpine proglacial areas and the morphological characteristics of those areas is basic to hydro-geomorphology. Stream channel features can be visualized as a function of three major groups of controls, namely discharge, sediment and local characteristics. These three controls combine in a particular way in proglacial areas to present a distinctive fluvial environment.

5.2.1 Discharge

Discharge is acknowledged to be a major control upon the fluvial landforms in any area, and it is the range of peak discharges which generally has the greatest significance (Pickup and Warner, 1976). However, in alpine areas a number of features of the discharge regime give substantial variability. River regimes show a broadly seasonal pattern with the major contrast between low flows or no flow during the winter months, followed by higher flows during the ablation season. During the ablation period river discharge reflects not only diurnal variations but also variations in radiation balance and air temperature. Thus, as an example, the glaciers in the Mount Tomur area of China have been shown (Keog Ersi *et al.*, 1982) to produce runoff derived mainly from surface ablation of the accumulation area and by internal ablation, but runoff varied according to the radiation balance on particular days and to values of air temperature. When patterns of discharge are compared from one proglacial area to another, it is not only variations in ablation during the meltwater season that differentiate peak discharges, but in addition the presence of other terrain types within the glacier basin can be significant. Thus, Young (1982) made a primary distinction between glacier and non-glacier areas and subdivided the latter into rock, moraines and snow patches, and firn zones, and indicated the way in which the summer hydrograph can reflect variations in such terrain composition of the glacier basin.

In addition to these variations which affect the detail of the annual hydrograph, there are occasional glacier outburst floods or jökulhlaups. These occur when an englacial or ice marginal lake is suddenly released and a very substantial flood wave affects the proglacial area. Estimates have been made of the magnitude of glacier outburst floods from many glaciers including Lake Donjek, Canada (Clarke and Matthews, 1981), which suggested that peak discharges from Lake Donjek can lie between 677 and 5968 m^3 s^{-1}, with values in the range 3968–5968 m^3 s^{-1} being considered most probable. Just as the cross-sectional area or channel capacity of a river channel is related to the discharges that have passed through it, so the cross-sectional area of the outlet from a glacier-dammed lake can be proportional to the amount of water that has passed through that outlet. Such a relationship was demonstrated for the Snow river in south central Alaska by Chapman (1981). It is, therefore, difficult to model the production of peak flows in proglacial areas, although two recent developments have enabled modelling strategies to be enhanced. First, the sources of water generating flood peaks are now more completely understood. For example, by using ^{18}O–^{16}O ratios of precipitation and runoff, it was possible to separate fluvial and glacial runoff components from an alpine glacier during one single storm (De Griend and Arwert, 1983). They demonstrated, especially during the rising limb of the hydrograph, that the bulk of discharge consisted of pre-storm meltwater stored in the glacier body and that the glacial meltwater could be distinguished from water generated by summer precipitation. Other approaches to such flow separation modelling are discussed later in this volume. A further way in which modelling of peak flows may be enhanced is by utilizing information from satellite imagery. Thus, Ferguson (1985) has shown that the only practical means available to forecast spring and summer runoff in remote high mountain basins such as the upper Indus in Pakistan is based on pre-season snow cover as recorded by satellite images.

In addition to these variations, there are two further reasons for variability in the pattern of peak discharges in alpine proglacial areas. Jams may cause temporary impoundments which, when they fail, give rise to flood waves and, therefore, to very high discharges which may be transmitted down the channels and across the surfaces of alpine proglacial areas. Such jams may occur when a glacier impounds stream flow, when temporary ice accumulations cause short-term impoundments, or as a result of rock falls or landslides. Ice jams are particularly common during the spring ice breakup along river courses, and may lead to extreme water velocities and to rapidly rising water levels, as demonstrated by Beltaos and Krishnappan (1982). The incidence of such dams has been reviewed by Hewitt (1982), who suggested that the southern Alaska/Yukon ranges and Karakoram, Himalaya, are the two areas where particularly large and dangerous examples have been noted. In the Karakoram there are 30 glaciers which may form substantial dams on the Upper Indus and the Yarkand river systems. The large flood waves that feature in proglacial areas after the occasional failure of such a natural dam can be very significant in relation to fluvial landforms because the size of the discharges and the associated velocities can induce a very high rate of erosion and of sediment transport. Not only do the flood waves possess a very high erosional energy and very often also considerable depth, but also they are associated with processes which extend beyond the usual channels and cut into the lag deposits or into recent terraces, and very often this erosion in turn induces landslides on the terraces and the valley sides. Substantial channel widening, deepening and even marked changes of course have therefore been reported after the passage of such flood waves (Hewitt, 1982).

A further reason for variation in the incidence of large discharges is that, in contrast to the pattern of discharges in many other areas, in the alpine proglacial zone the peak discharges may not increase substantially downstream. In some cases as peak discharges are routed through the proglacial environment a particularly significant feature is the fact that influent seepage may take place so that there are transmission losses of water draining away from a glacier. Therefore, the very high discharges that occur in the zone close to the margin of a glacier may be reduced downstream because of transmission loss of water into the sandur gravels. This feature of icefed systems has been noted by Maizels (1983b) and has been shown to be particularly important in relation to estimation of palaeohydrological conditions.

Whereas in many parts of the world it is possible to suggest a relationship between river channel dimensions and the pattern of peak discharges, in alpine proglacial areas the variability of discharge regime is such that a simple pattern of fluvial landforms is unlikely to occur. These features of the discharge regime and of sediment availability, which are further discussed in the next section, are important ingredients of paraglacial situations as noted in Chapter 2.

5.2.2. Sediment availability

Materials available for transport by river discharge also influence fluvial landforms. The dominant characteristic of materials in alpine proglacial areas is that they are not usually supply-limited. This means that sediment transport equations should apply to reaches of gravel bed channels where the course of the channel may change easily. A further feature is that there is usually a great range of sediment sizes which are not systematically sorted and which are stochastically supplied to the fluvial process system. This ample sediment supply arises from several potential sources. For example, in the Canadian Rocky Mountains, it has been shown that the Hilda Glacier yields two principle types of sediment: firstly ablation till, and secondly basal lodgement till rich in fines and derived mainly by subglacial erosion. Recent recession of this glacier has provided abundant material for erosion and transport by the meltwater stream and measurements over two summer periods showed that bedload and suspended load occur in approximately equal proportions. Supraglacial debris provided only about 25% of the fluvial sediment but nearly 50% of the bedload (Hammer and Smith, 1983).

In the meltstream of Storbreen in the Jotunheimen, Richards (1984) demonstrated that fine sediment is transferred from the subglacial to the proglacial environment during the low flows of melt-dominated periods, but is then removed from the catchment during high flows occasioned by rainfall-controlled events. In some proglacial areas the suspended sediment load can be very high indeed, and Ferguson (1984) has shown that the Hunza river basin (13 200 km^2) in the Western Karakoram Mountains carries 39% of the suspended sediment load of the upper part of the Indus basin, and this area has one of the highest denudation rates in the world.

In addition to the sediment which is derived from glacier sources, there are additional sources of sediment derived from storage between the slopes and the stream channel, whilst avalanches, rock slides and debris falls further supplement the sources available in the proglacial zone. In torrential drainage basins in the eastern Alps of Europe, it has been shown that there is a range of sediment sources including landslides, gulley erosion and more general soil erosion

(Kronfellner-Kraus, 1982). Occasionally, where deposits are stored in alpine proglacial valleys, such deposits may fail and temporarily give very high values of sediment yield. In the Southern Alps of New Zealand there are accumulations of sediments in the sub-alpine environment which are similar to sieve deposits described from south eastern California (Ackroyd and Blakely, 1984). In one particular catchment of the Torlesse Stream, the catastrophic failure of such a deposit resulted in a sediment yield equivalent to a third of the total recorded over a period of 8 years and the bedload transport rates represented at least a four-fold increase over calculated average transport rates.

In proglacial areas an ample supply of material is complemented by variations in a downstream direction. Downstream reduction in the median grain size is responsible for variations in channel behaviour of channels on the valley train of Sunwapta River in the Jasper National Park of Alberta (Rice, 1982). As one proceeds downstream in the proglacial zone it is sometimes possible that an armour coat develops which limits the availability of sediment for fluvial transport. This can then mean that the actual bedload transport rates are less than the values which are predicted from sediment transport equations. In the region of Switzerland on which the present volume focuses, the significance of the development of an armoured surface has been demonstrated along the Borgne d'Arolla, where Gomez (1983) showed that armouring affects the supply of transportable material. Although armouring can reduce the sediment available for transport, studies undertaken in Japan have shown that once the armoured bed is destroyed then the scour of deposits can take place so that there may be a sudden increase in the availability of sediment to the fluvial system (Sawada *et al.*, 1983).

5.2.3 Local characteristics

Local catchment characteristics have to be considered in addition to the significance of discharge and of sediment availability. Three aspects of local characteristics merit consideration. First is the fact that longitudinal slope, although variable, often tends to be high. This may reflect variations in valley form conditioned by bedrock outcrops, and the size of valley cross profiles and long profile are inversely interrelated. Secondly, proglacial zones tend to be largely free of vegetation cover unless recently colonized, so that the vegetative resistance, which is frequently a major roughness component in other environments, tends to be minimal. Thirdly, there are significant influences due to human activity such as gravel exploitation, flow abstraction or the incidence of purges.

5.2.4 Form and process adjustment

In the light of the above features of discharge, sediment and local characteristics, two major conclusions can be reached about fluvial form and process adjustment in alpine proglacial areas. Firstly, these areas are characteristically high-energy environments. The energy available in a fluvial environment can be indicated by the stream power (Gregory, 1982), and the excess in any situation can be visualized in relation to critical power (Bull, 1979). Stream power (ω) can be envisaged as the product of density of stream flow (ϱ), discharge (Q) and slope (S) in the form $\omega = \varrho\, QS$. High availability of sediment for transport can mean that ϱ values are relatively high, discharge varies considerably and includes high values of Q, and over some reaches the slope (S) tends to be high so that for certain time periods stream power values are very high.

A second feature is that landscape sensitivity is also high. This can be illustrated by the transient form ratio expressed by Brunsden and Thornes (1979) as the forces promoting change divided by the barriers to change. Because stream power tends to be high during certain parts of the year, so the forces leading to change are also high.

5.3   FLUVIAL LANDFORM DOMAINS

Distinctive form/process relationships in alpine proglacial areas give rise to distinctive combinations of features which may be thought of as domains (Thornes, 1983), and three major domains may be distinguished: firstly, sandurs, with braided channel patterns; secondly bedrock domains; and thirdly, areas dominated by palaeoforms and recently produced fluvial landforms.

5.3.1   Channel patterns

The characteristics of sandur plains in proglacial areas are reviewed in Chapter 16. However, although the characteristics of the several kinds of braided channel pattern and the bars which are integral parts of those patterns are considered elsewhere, it is necessary here to consider channel size and also the threshold between single and multi-thread channel patterns.

TABLE 5.1   Relative dimensions of stream channels in proglacial and humid temperate areas

	Drainage area (km^2)		
	1	10	50
Average channel capacity (m^2) from regression equations:			
A. Lowland UK (Gregory and Ovenden, 1979)	0.46	1.44	3.48
B. Upland UK (Gregory and Ovenden, 1979)	0.75	3.61	11.16
C. Val d'Herens, Switzerland (Gurnell, 1983)	1.49	9.76	36.51
Ratios:			
C:A	3.2	6.8	10.5
C:B	2.0	2.7	3.3

It is very difficult to compare the capacities of channels in alpine proglacial areas with channel capacities in other areas. However, a preliminary attempt is made in Table 5.1 by comparing data for lowland and upland UK basins (Gregory and Ovenden, 1979) with data from the Val d'Herens, Switzerland (Gurnell, 1983). The data demonstrate that for standard sizes of drainage area the channel capacities in proglacial areas are at least two and possibly as much as six times greater than the capacities for comparable catchment areas in humid temperate basins of the UK. Such a comparison has to be treated with great caution because not only are the problems of the determination of channel capacity particularly difficult, but also the data from the UK are based on eleven widely separated areas, whereas the data from Switzerland are based on measurements made along one proglacial stream. However, the ratios indicated in Table 5.1 indicate that there is a considerable difference in channel dimensions. In the case of channels in proglacial areas the natural drainage area is, of course, a value which includes the area of glacier surface. The difference in dimensions between Swiss and UK channels could reflect the controls of channel dimensions (Section 5.2).

In alpine proglacial areas, river channel patterns are frequently multi-thread in character and it is therefore important to know the extent to which such channel patterns are close to the threshold at which multi-thread channels change to single-thread channels. This is not easy to analyse, however, because sediment sizes are extremely varied in proglacial streams. Rust (1972), for example, noted the very considerable range of sediment sizes which characterize proximal and distal variations in sediments and also contrast active and stable zones in proglacial areas. However, it has generally been appreciated that flow shear stress is an important influence on the incidence of braided multi-thread or meandering single-thread channels. Thus, Begin (1981) provided a discharge slope diagram indicating the relative shear stress, but concluded that the change from one pattern to another is a gradual one and that geomorphic thresholds as defined by relative shear stress are fuzzy. The discriminant function proposed by Leopold and Wolman (1957) related slope (S) to estimated bankfull discharge (Q) in the form

$$S = 0.013 \ Q^{-0.44}$$

In his review of this threshold between meandering and braiding, Ferguson (1984) proposed that the threshold gradient for braiding depends on channel materials in addition to discharge and slope. In a reappraisal of the meandering/braided river threshold, Carson (1984a) also suggested that the search for a pattern threshold based on discharge and slope seems to be a futile exercise. He concluded that such functions merely state that gravel bed streams are more likely to be braided than channels in finer sediment. His conclusions are based on data from New Zealand rivers, including some proglacial ones. These indicate

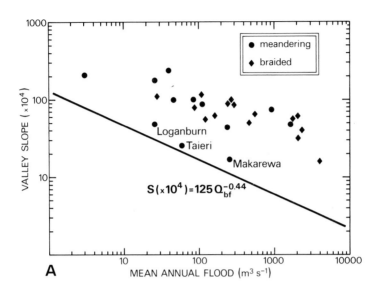

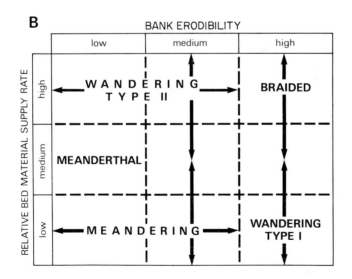

FIGURE 5.1   Variability of channel planform. (A) Carson (1984a) plotted data from New Zealand channels in relation to the discriminant function used by Leopold and Wolman (1957). (B) shows a classification of channel patterns of gravel bed rivers as proposed by Carson (1984b)

that the prerequisite for braiding is local shoaling of the thalweg (giving high relative width), and this is less dependent on a threshold hydraulic condition such as stream power than it is on a threshold state of bed material transport, although Carson noted that the two are often related. He also concluded that any hydraulic threshold will vary significantly with bed and bank material, so that it is meaningless to seek a general threshold for all streams unless the features of sediment are considered (Figure 5.1A).

In the case of the channel pattern of gravel bed rivers on the Canterbury Plains studied by Carson (1984b), it was suggested that these channel patterns take the form of multiple flow dissection of the active flood plain rather than of multiple bars featuring in a single wide channel. The braided channels on the Canterbury Plains therefore resemble 'wandering' rivers rather than the classic patterns of proglacial sandurs. The Bella Coola river draining the coast mountains and the Fraser Plateau of British Columbia also has a morphology which Church (1983) has described as that of a wandering gravel river. The Bella Coola river flows over alluvial cobble gravels, and sediment entrained from the river banks and delivered from tributaries is stored in the channel principally in sedimentation zones where the river is laterally unstable. Church (1983) showed that these zones are linked by stable, cobble-paved transport reaches. Along the Bella Coola river there is an irregular pattern of channel instability, although there is a single dominant channel evident everywhere supplemented by a sequence of braided or anastomosed reaches connected by relatively stable single-thread reaches.

In alpine proglacial areas, therefore, three types of channel planform may occur, namely, single-thread meandering, wandering, or classic braided. Carson (1984b) suggested that the extent to which wandering and braiding are each associated with different sets of processes is still not clear. A simple distinction could be that braiding is a truly depositional pattern controlled by mid-channel bars, whereas wandering is an erosional pattern formed by the dissection of the active flood plain by multiple flows. Nevertheless, the investigation of specific examples in proglacial situations suggests that the natural situation is more complicated than this simple division would suggest. Based on the channel patterns in New Zealand, Carson (1984) has proposed a simplified classification of channel patterns of gravel bed rivers in relation to the bed material supply rate and bank erodibility (Figure 5.1B). In this classification, a distinction is made between the wandering type 1 behaviour (which is irregular, involves very rapid bend migration and produces very wide point bars which are often dissected), and wandering type 2 behaviour (which is characterized by persistent avulsion of flow out of the main channel). Although the use of this scheme is difficult because the bed material supply rate and bank erodibility are factors which are not readily measurable, nevertheless it provides a scheme which may be suitable for the interpretation of channel patterns and which incorporates the spatial variety which exists.

5.3.2 Bedrock domains

In areas where bedrock outcrops and there are small amounts of superficial deposits, either glacial or fluvial, a distinctive type of morphological domain exists. Such areas are frequent where there is a high longitudinal slope, because the river then tends to follow a single-thread path and there are frequently sections of gorge which have been produced during the course of recent glacier fluctuations. In such bedrock-dominated areas there may be features whch have been recently produced, but which are now relict elements in the current proglacial area. A good example of such features is very large potholes or moulins. Such features are apparently water worn, they can be as much as 10–15 m in depth, they contain very large rounded boulders at their base and they often include evidence of having developed in a series of stages. Such a system of potholes or moulins is exemplified in southwest Switzerland immediately down-valley from the snout of the Ferpecle glacier. In that area, there are several systems which appear to have developed at a series of stages. The existence of such moulins or glacier mills devolves upon three possible explanations. The traditional explanation is that meltwater falling into crevasses in glaciers was responsible for scouring at the base of the crevasse so that a moulin or glacier mill would be produced at the bottom of the crevasse. A second explanation involves the notion that fluvial erosion can occur along subglacial stream courses. Along the line of such subglacial stream courses with a very high energy fluvial transport system there would be frequent scouring and erosion of potholes. This suggestion accords with observations of subglacial hydrology (e.g. Hooke *et al.*, 1985) and it also accords with the fact that potholes in areas like that adjacent to the Ferpecle glacier are arranged in stages. A third explanation, offered by Gjessing (1967), invokes plastic scour, particularly on the lee side of rock barriers, being accomplished by a viscous mixture of water, ice and rock fragments.

It is possible that in proglacial areas each of the three types of pothole may exist and therefore that the three hypotheses are not mutually exclusive but rather can be complementary, depending on the local environment and the fluvial and glacial processes which obtain.

5.3.3 Areas dominated by palaeoforms

A third type of domain encountered in alpine proglacial areas is one which is largely dominated by features produced prior to the contemporary process response system, although this covers a great range. At one extreme are changes during recent decades that have been associated either with fluctuations in the extent of alpine glaciers or associated with the mechanism of fluvial processes themselves. In the first case, the recession of glaciers in the European Alps has left a series of recessional moraines which present major features of the landscape in alpine proglacial areas. Between and associated with these moraines, subsequent and contemporary fluvial erosion have merely diversified the pattern of landscape that already existed. Secondly, however, there have been changes in the incidence of particular fluvial features. Thus the Bella Coola river in British Columbia studied by Church (1983) seems, according to a sequence of maps beginning in the late 19th century, to have become more stable and to have had a downstream progression of lateral instability from near 25 km in 1900 to near the mouth today. Church (1983) suggested that this effect may have occurred either by the introduction of unusual volumes of sediment into the main channel by the erosion of 18th and 19th century moraines of alpine glaciers which are now becoming exhausted, or alternatively as a result of the recent progradation of the alluvial fan of the Nusatsum river constricting the main channel and blocking the transfer of sediment downstream. Similarly, in his analysis of the channel patterns of the Canterbury Plains in New Zealand, Carson (1984b) has indicated that it is probable that many of the type 1 wandering channels were formally meandering, and that some of the braided channels were formerly type 2 wandering (Figure 5.1B).

At the other extreme it is possible to identify those areas in alpine regions in which a major component of the landscape is provided by features produced during late glacial and postglacial oscillations of glaciers. Thus, in the Doralen-Rondane area of Norway it has been shown that there is a complex pattern of alluvial cones and terraces containing dead ice depressions, and that this pattern of features was produced largely by supraglacial sediments melting on top of an ice mass which was also correlated with terraces downvalley (Gehrenkemper and Treter, 1983). In some areas there are sequences of terraces which have been produced during stages of recession of the margins of glaciers. Thus, in the Val d'Herens a series of terraces downstream from Les Hauderes has been interpreted as a system of braiding terraces (Small, 1973).

5.4 PALAEOMORPHOLOGY AND PALAEOHYDROLOGY

Changes of hydrology and associated morphological characteristics have occurred recently in alpine proglacial areas as well as on many occasions during the Quaternary. The fluvial system exhibits change particularly in relation to the threshold conditions which separate different types of channel patterns. One major recent reason for changes of the fluvial system can be found in human activity because such activity can influence the character of discharge, sediment and local characteristic controls on the fluvial system.

Thus, for example, Gurnell (1983) has demonstrated how the effects of the inception of the Grande Dixence hydroelectric power scheme in south-western Switzerland led to greatly reduced discharges downstream of water intakes, which in turn were responsible for changes in channel geometry. Although such changes can be reflected in substantial decreases of channel cross-section, they

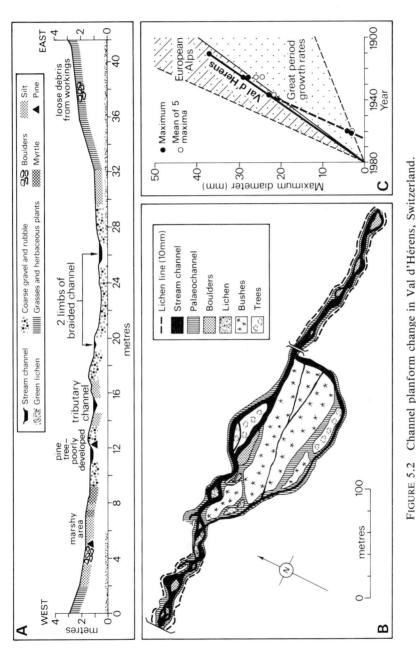

FIGURE 5.2   Channel planform change in Val d'Hérens, Switzerland.
A levelled cross-section in (A) shows a series of palaeochannels abandoned after water abstraction by Grande Dixence; (B) illustrates the planform in a segment of the Ferpècle valley; (C) provides the lichen *Rhizocarpon geographicum* growth curve, which can be used to suggest ages for boulders in abandoned channels and in fragments of former flood plains

can also lead to different type of channel pattern. Some sections surveyed across the valley floor of the Borgne D'Arolla could not easily be reconciled with the hypothesis of a contemporary channel much smaller than the original channel (Figure 5.2). This was because the reduction in discharge has been responsible for a change in channel pattern which has led to a shift across a threshold. Therefore, in some cases former braided channel patterns have been succeeded either by a wandering channel, incised into the original proglacial deposits, or by a single-thread meandering one. This has been demonstrated in an example in south-western Switzerland by reference to lichenometry. It was possible to measure the maximum diameter of lichen thalli on the residual boulders across the now stabilized floodplain. The size of these lichen thalli was related to a provisional growth curve which was constructed using historical information relating to moraines in the local area of Switzerland (Figure 5.2c).

In alpine proglacial areas, it has not been possible to utilize palaeohydrologic approaches based on regional curves related to mean annual climatic statistics or on general equations derived from temperate latitudes (Gregory, 1983). Approaches such as that indicated by Ethridge and Schumm (1978) which allow the estimation of palaeochannel and palaeohydraulic values are difficult to apply to gravel bed situations of proglacial zones and there are a number of reasons for this. First, as indicated in Section 5.2, variations in discharge, in sediment characteristics and in local characteristics are very substantial and can take place over short distances, and in addition there can be episodic variations due, for example, to particularly high discharges. Related to this is the fact that there are substantial downstream variations, so that for example in south-western Switzerland the effects of a reduced discharge on channel geometry can sometimes increase downstream rather than declining, as is the case in other areas (Gurnell, 1983).

A major control on former discharges and fluvial processes is the substantial variation in glacier extent, and the related changes in ice mass as well as in snow cover and associated snow melt, which can compound the extent to which peak discharges vary over short periods. In addition, over the time scale of the late Quaternary and certainly during the late Pleistocene, areas that were not completely covered by glaciers were subject to much higher discharges than at the present time, and general estimates have been deduced by Cogley (1973). Because of these variations, it is difficult to extend conventional techniques of palaeohydrologic investigation to alpine zones. Thus the approach developed during the International Geological Correlation Programme 158, the palaeohydrology of the temperate zone in the last 15 000 years (Starkel, 1983), is not easily applied to proglacial zones.

Earlier work tended to utilize the stratigraphy and chronology of deposits as a basis for correlation and environmental reconstruction, but such approaches have been succeeded and amplified by attempts to reconstruct the palaeohydraulics of former flows. Although in some cases analogues can be

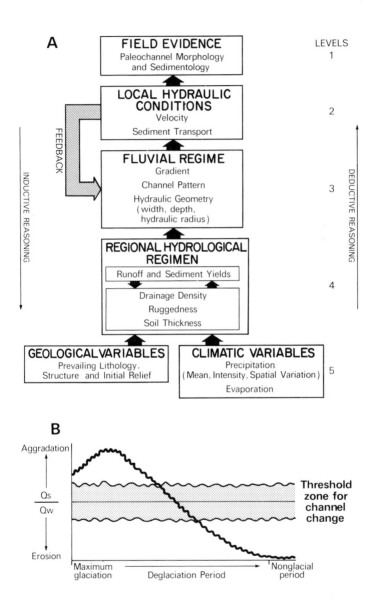

FIGURE 5.3 Palaeoenvironmental reconstruction. (A) shows how 'think back' reasoning can be undertaken based on Baker (1983) and (B) suggests the sequence of change important for channel change in proglacial areas as suggested by Maizels (1983b)

useful, as demonstrated by Bryant (1983) comparing temperate areas with contemporary arctic situations, such an analogue approach is not so easily utilized in alpine areas because of the topographical variety that exists—a constraint that is further explored in Section 17.3.1.

In general terms, it has been suggested by Baker (1983) that although relict fluvial features are a tempting basis for palaeohydrologic estimation, the current state of the art 'in the reconstruction of late Pleistocene fluvial palaeohydraulics and palaeohydrology reveals both promise and problems.' He suggests that the problems derive from the connection between palaeoclimatic or palaeohydrologic events and discernible field evidence through the chain of interlinked processes and responses (Figure 5.3A). The promise of these approaches has been reviewed generally by Maizels (1983a), who suggested that palaeovelocity determination could be based on tractive force approaches or on the use of velocity equations and friction factors. By such palaeovelocity determination it is then possible to obtain estimates for former discharges. However, in his review of more than 100 equations available for palaeohydrologic estimation, Williams (1984a) includes very few examples of equations that may be applied in the context of proglacial environments. One group of equations that are quoted by Williams (1984a) is that developed by Clague (1975) in relation to studies in Canada and Alaska. These equations (Table 5.2) largely utilize the extent of glacierized area.

TABLE 5.2   Palaeohydrologic equations developed for Canada and Alaska (after Clague, 1975)

$Q = 2.1 A_g^{0.99}$	Proglacial rivers in Canada and Alaska
$Q = 75 V_g^{0.67}$	Jökulhlaups
$Q_y = 0.26 A_g^{0.98}$	Proglacial rivers in Canada and Alaska
$Q = 0.43 A_g^{1.06}$	Proglacial rivers in Canada and Alaska

Q = instantaneous water discharge ($m^3 s^{-1}$)
Q_y = mean discharge of the year of maximum discharge ($m^3 s^{-1}$)
A_g = glacierized area (km^2)
V_g = Volume of water catastrophically drained from ice-dammed lakes, ($m^3 s^{-1} \times 10^6$)

Other equations are necessary to estimate palaeohydraulics, but Costa (1983) stressed the importance of the reasons for discrepancies between actual discharges and those reconstructed using palaeohydraulic equations. He cites the possibility that floods may have been able to move boulders larger than those available to be moved; that slope area/discharge calculations are the basis for overestimates because high water marks may have been set prior to erosion of the channel; that roughness coefficients may be underestimated; and that there may be macro-turbulent effects during fast deep flows. Whereas many approaches have focused on the critical or threshold flow necessary to move sediment of a certain size, Williams (1983) has suggested that variations in particle size and of threshold shear stress are such that the determination of threshold flow conditions is very difficult. He therefore determines palaeoflows from empirical relationships devised to indicate minimum unit stream power, bed shear stress and mean flow velocity. The approach used by Williams (1983) allows equations developed for contemporary fluvial environments in Sweden to be employed to estimate the minimum palaeoflows that could have transported the boulders of two ancient fluvial deposits. His approach is probably the one which is most applicable to the environment of the alpine proglacial areas.

Although the nature of palaeohydrologic and palaeomorphologic change in alpine proglacial areas is very difficult to model, the essence of change typically involves the development of terrace systems. This is in effect an extension of the example quoted above, where a multi-thread channel system is replaced either by a wandering channel or by a single-thread meandering one. However, the development of such terrace systems is often difficult to interpret for two groups of reasons: firstly, because of the complicating short-term factors mentioned above, but secondly, as indicated by Maizels (1983b), it is only the final stages of evolution that are usually preserved. Thus, in a proglacial environment we have the contemporary system and the one into which it is incised but in many cases remnants of former systems are either very sparse or non-existent. However, Maizels concludes that in areas experiencing long-term deglaciation the general tendency is that the earliest channel systems were characteristically braided with a low sinuousity and high width/depth ratios, and were bedload channels typical of flow through proximal proglacial meltwater systems. Subsequently, channel systems appear to have developed to include a number of deeper and more sinuous channels, and these were subsequently replaced by single-thread channels which have lower width/depth ratios, higher sinuosities and characteristically were suspended load channels similar to the meltwater system of non-glacial humid alluvial basins.

In such changes it is not simply the relationship between the sediment discharge and the water discharge that is significant, but distance from the ice margin may also be important. Using information from five different areas, Maizels (1983b) has suggested a model which demonstrates a possible relation between the balance of water and sediment supply around a threshold zone in relation to channel change during the period of deglaciation. Recent research has demonstrated that channel changes which have been dramatic in the alpine proglacial areas have occurred because of the fact that the fluvial system in such areas is close to threshold conditions. Although such threshold conditions vary substantially from one area to another according to local geomorphic, hydrologic, climatologic, vegetation and base level conditions (Maizels, 1983b), nevertheless it is because channel systems are transitional in such areas that they are sensitive to small changes. The model suggested by Maizels (1983b) is indicated in a simple form in Figure 5.3B.

5.5　CONCLUSIONS AND APPLICATIONS

The fluvial system of alpine proglacial areas is therefore characterized by a number of distinctive attributes of which two, namely high stream power values and sensitivity in relation to threshold conditions, are particularly evident. In such areas, ample evidence of recent fluvial changes can be developed by the techniques of palaeohydrology and palaeohydraulics to furnish useful data on late Quaternary change. In addition, such techniques could usefully provide an input to the future of alpine proglacial systems and particularly to the management of river courses in such areas. In environments elsewhere, it has been advocated by a number of researchers (Brookes *et al.*, 1983; Brookes, 1985) that management strategies can be more effective overall if they work 'with' the river rather than 'against' it. Although there are problems of applying such a philosophy to high-energy gravel bed proglacial river channels, nevertheless there is considerable scope for modification of existing procedures to retain some of the characteristics of the alpine proglacial river environment when the river channel is managed and controlled.

REFERENCES

ACKROYD, P., and BLAKELY, R. J. (1984), En masse debris transport in a mountain stream, *Earth Surface Processes and Landforms* **9**, 307–320.

BAKER, V. C. (1983), Late Pleistocene fluvial systems, in WRIGHT, H. E., JR. (Ed.), *Late Quaternary Environments of the United States, Vol. 1, The Late Pleistocene*, Longman, London, 115–129.

BEGIN, Z. B. (1981), The relationship between flow shear-stress and stream pattern, *Journal of Hydrology*, **52**, 307–319.

BELTAOS, S., and KRISHNAPPAN, B. G. (1982), Surges from ice jam releases: a case study, *Canadian Journal of Civil Engineering*, **9**, 276–284.

BROOKES, A. (1985), River channelization: traditional engineering methods, physical consequences and alternative practices, *Progress in Physical Geography*, **9**, 44–75.

BROOKES, A., GREGORY, K. J., and DAWSON, F. H. (1983), An assessment of river channelization in England and Wales, *The Science of the Total Environment*, **27**, 97–111.

BRUNSDEN, D., and THORNES, J. B. (1979), Landscape sensitivity and change, *Transactions of the Institute of British Geographers*, NS4, 463–484.

BRYANT, I. D. (1983), The utilization of Arctic river analogue studies in the interpretation of periglacial river sediments from southern Britain, in GREGORY, K. J. (Ed.) *Background to Palaeohydrology*, Wiley, Chichester, pp. 413–431.

BULL, W. B. (1979), Threshold of critical power in streams, *Geological Society of America Bulletin*, **90**, 453–464.

BURKHAM, D. E. (1981), Uncertainties resulting from changes in river form, *Proceedings American Society of Civil Engineers, Journal Hydraulics Division*, **107**, 593–610.

CARSON, M. A. (1984a), The meandering braided river threshold: a reappraisal, *Journal of Hydrology*, **73**, 315–334.

CARSON, M. A. (1984b), Observations of the meandering braided river transition, the Canterbury Plains, New Zealand: Part One, *New Zealand Geographer*, **40**, 12–17.

CARSON, M. A. (1984c), Observations of the meandering braided river transition, the Canterbury Plains, New Zealand: Part Two, *New Zealand Geographer*, **40**, 89–99.

CHAPMAN, D. L. (1981), *Jökulhlaups in Snow River in South Central Alaska; a Compilation of Recorded and Inferred Hydrographs and a Forecast Procedure*, United States National Weather Service, Anchorage, 55pp.

CHITALE, S. V. (1973), Theories and relationships of river channel patterns, *Journal of Hydrology*, **19**, 285–308.

CHURCH, M. (1983), Pattern of instability in a wandering gravel bed channel, *International Association of Sedimentologists Special Publication*, **6**, 169–180.

CLAGUE, J. J. (1975), Sedimentology and palaeohydrology of late Wisconsin outwash, Rocky Mountain Trench, South Eastern British Columbia, in JOPLING, A. V., and MACDONALD, B. C. (Eds.), *Glaciofluvial and Glaciolacustrine Sedimentation*, Society of Economic Palaeontologists and Mineralogists Special Publication **23**, pp. 223–237.

CLARKE, G. K. C., and MATTHEWS, W. H. (1981), Estimates of the magnitude of glacier outburst floods from lake Donjek, Yukon Territory, Canada, *Canadian Journal of Earth Science*, **18**, 1452–1463.

COGLEY, J. G. (1973), Meanders. On runoff at the time of deglaciation, *Area*, **5**, 33–37.

COSTA, J. E. (1983), Palaeohydraulic reconstruction of flash flood peaks from boulder deposits in the Colorado Front Range, *Geological Society of America Bulletin*, **94**, 986–1004.

DE GRIEND, O. A. A. V., and ARWERT, J. A. (1983), The mechanism of runoff generations from an alpine glacier during a storm traced by $^{18}O/^{16}O$, *Journal of Hydrology*, **62**, 263–278.

ETHRIDGE, F. G., and SCHUMM, S. A. (1978), Reconstructing palaeochannel morphologic and flow characteristics: methodology, limitations and assessment, in MIALL, A. D. (Ed.), *Fluvial Sedimentology*, Canadian Society of Petroleum Geologists, Memoir 5, pp. 703–721.

FERGUSON, R. I. (1984), The threshold between meandering and braiding, in SMITH, K. V. H. (Ed.), *Channel and Channel Control Structures*, Springer Verlag, pp. 6–15 to 6–19.

FERGUSON, R. I. (1985), Runoff from glacierized mountains: a model for annual variation and its forecasting, *Water Resources Research*, **21**, 702–708.

GEHRENKEMPER, J., and TRETER, U. (1983), Untersuchungen zur deglaziation und talentwicklung im Doralen/Rondane (Norwegen), in SCHROEDER-LANZ, H. (Ed.), *Late and Postglacial Oscillations of Glaciers: Glacial and Periglacial Forms*, Balkema, pp. 171–185.

GJESSING, J. (1967), Potholes in connection with plastic scouring forms, *Geografiska Annaler*, **49A**, 178–187.

GOMEZ, B. (1983), Temporal variations in bedload transport rates: the effect of progressive bed armouring, *Earth Surface Processes and Landforms*, **8**, 41–54.

GREGORY, K. J. (1979), Hydrogeomorphology: how applied should we become? *Progress in Physical Geography*, **3**, 84–100.

GREGORY, K. J. (1982), River power, in ADLAM, B. H., FENN, C. R., and MORRIS, L. (Eds.), *Papers in Earth Studies*, Geo Books, pp. 1–20.

GREGORY, K. J. (Ed.) (1983), *Background to Palaeohydrology*, Wiley, Chichester, 486pp.

GREGORY, K. J., and OVENDEN, J. C. (1979), Drainage network volumes and precipitation in Britain, *Transactions of the Institute of British Geographers*, **NS4**, 1–11.

GURNELL, A. M. (1983), Downstream channel adjustments in response to water abstraction for hydroelectric power generation from Alpine glacial meltwater streams, *Geographical Journal*, **149**, 342–354.

HAMMER, K. M., and SMITH, N. D. (1983), Sediment production and transport in a proglacial stream: Hilda Glacier, Alberta, Canada, *Boreas*, **12**, 91–106.

HEWITT, K. (1982), Natural dams and outburst floods of the Karakoram Himalaya, in J. W. GLEN (Ed.), *Hydrological Aspects of Alpine and High Mountain Areas (Proceedings of the Exeter Symposium, July 1982)*, International Association of Hydrological Sciences Publication 138, pp. 51–59.

HOOKE, R. LE B., WOLD, B., and HAGEN, J. O. (1985), Subglacial hydrology and sediment transport at Bondhusbreen, southwest Norway, *Geological Society of America Bulletin*, **96**, 388–397.

KEOG ERSI, ZHU SHOUSEN and HUANG HINGMIN (1982), Characteristics of glacial hydrology in the Mount Tomur area of China, in GLEN, J. W. (Ed.), *Hydrological Aspects of Alpine and High Mountain Areas (Proceedings of the Exeter Symposium, July 1982)*, International Association of Hydrological Sciences Publication 138, pp. 271–283.

KRONFELLNER-KRAUS, G. (1982), Estimation of extreme sediment transport from terrestrial drainage basins in the East Alps, in WALLING, D. E. (Ed.), *Recent Developments in the Explanation and Prediction of Erosion and Sediment Yield (Proceedings of the Exeter Symposium, July 1982)*, International Association of Hydrological Sciences Publication 137, pp. 269–273.

LEOPOLD, L. B., and WOLMAN, M. G. (1957), River channel patterns: braided, meandering and straight, *United States Geological Survey Professional Paper*, 282-A.

MAIZELS, J. K. (1983a), Palaeovelocity and palaeodischarge determination for coarse gravel deposits, in GREGORY, K. J. (Ed.), *Background to Palaeohydrology*, Wiley, Chichester, pp. 101–140.

MAIZELS, J. K. (1983b), Proglacial channel systems: channel and thresholds for change over long, intermediate and short timescales in COLLINSON, and LEWIN, J. (Eds.) *Modern and Ancient Fluvial Systems*, International Association of Sedimentologists Special Publication 6, Blackwell, Oxford, pp. 251–266.

PICKUP, G., and WARNER, R. F., (1976), Effects of hydrologic regime on magnitude and frequency of dominant discharge, *Journal of Hydrology*, **29**, 51–75.

PRICE, L. (1981), *Mountains and Man: a Study of Process and Environment*, University of California Press, Berkeley.

RICE, R. J. (1982), The hydraulic geometry of the lower portions of the Sunwapta River valley train, Jasper National Park, Alberta, in ARNOTT, R. D., NICKLING, W., and FAHEY, B. D. (Eds.), *Research in Glacial, Glacio-Fluvial and Glacio-Lacustrine Systems*, Geo Books, pp. 151–174.

RICHARDS, K. S. (1984), Some observations on suspended sediment dynamics of Starbregrova, Jotunheimen, *Earth Surface Processes and Landforms*, **9**, 101–112.

RUST, B. (1972), Pebble orientation in fluvial sediments, *Journal of Sedimentary Petrology*, **42**, 384–388.

SAWADA, T., ASHIDA, K., and TAKAHASHI, T. (1983), Relationship between channel pattern and sediment transport in a steep gravel bed river, *Zeitschrift für Geomorphologie*, Supplement Band 46, 55–66.

SCHUMM, S. A. (1985), Explanation and extrapolation in geomorphology: Seven reasons for geologic uncertainty, *Transactions Japanese Geomorphological Union*, **6**, 1–18.

SLAYMAKER, H. D., and MCPHERSON, H. J. (Eds.) (1972), *Mountain Geomorphology*, Tantalus Research: Vancouver, Canada.

SMALL, R. J. (1973), Braiding terraces in the Val d'Herens, Switzerland, *Geography*, **58**, 129–135.

STARKEL, L. (1983), The reflection of hydrologic changes in the fluvial environment of the temperate zone during the last 15 000 years, in GREGORY, K. J. (Ed.), *Background to Palaeohydrology*, Wiley, Chichester, pp. 213–236.

SUNDBORG, A. (1978), River processes in different climates, *Journal of Geography, Tokyo Geographical Society*, **87**, 115–128.

THORNES, J. B. (1983), Evolutionary geomorphology, *Geography*, **68**, 225–235.

WILLIAMS, G. P. (1983), Palaeohydrological methods and some examples from Swedish fluvial environments. I. Cobble and boulder deposits, *Geografiska Annaler*, **65A**, 227–243.

WILLIAMS, G. P. (1984a), Palaeohydrologic equations for rivers, in COSTA, J. C., and FLEISCHER, P. J. (Eds.), *Developments and Applications of Geomorphology*, Springer-Verlag, pp. 343–367.

WILLIAMS, G. P. (1984b), Palaeohydrological methods and some examples from Swedish fluvial environments. II. River meanders, *Geografiska Annaler*, **66A**, 89–102.

YOUNG, G. J. (1982), Hydrological relationship in a glacierized mountain basin, in GLEN, J. W. (Ed.), *Hydrological Aspects of Alpine and High Mountain Areas (Proceedings of the Exeter Symposium, July 1982)*, International Association of Hydrological Sciences Publication 128, pp. 51–59.

SECTION II

Glacial Sediment Transfer

Glacio-fluvial Sediment Transfer
Edited by A. M. Gurnell and M. J. Clark
©1987 John Wiley & Sons Ltd.

Chapter 6

Englacial and Supraglacial Sediment: Transport and Deposition

R. J. SMALL

ABSTRACT

Following a brief review of glacial sediment transport, by way of supraglacial, englacial and subglacial pathways, the principal characteristics of glacial sediment, in terms of clast shape and roundness and particle size distribution related to modes of origin, are critically examined. The occurrence of sediment within Alpine glaciers, either in dispersed form or as longitudinal debris septa and transverse sediment concentrations, is discussed, together with possible modes of sediment incorporation (sedimentary layering in the accumulation zone, incorporation via open crevasses, transference from the glacier base by upward-turning flow lines or thrust faulting and the formation of anticlinal ice structures by longitudinal or transverse glacier compression). Case studies of englacial sediment within three Swiss glaciers (Glacier de Tsidjiore Nouve, Bas Glacier d'Arolla and Haut Glacier d'Arolla) are presented. Finally, a general consideration of the role of englacial debris in the formation of supraglacial moraines (medial and lateral) is followed by a discussion of moraine form and genesis in relation to the three Swiss glaciers, and a brief examination of the problem of moraine classification.

6.1 THE GLACIER AS A TRANSPORT SYSTEM (Figure 6.1)

Sediment transported by glaciers follows three basic routes: *supraglacial*, *englacial* and *subglacial*. However, in terms of source, this sediment can be only supraglacial or subglacial; englacial sediment, which may be located at any level within the ice, is the result of either incorporation from above or from below. In most Alpine glaciers the debris-rich basal layer, containing sediment released mainly by quarrying, crushing and abrasion, is typically thin (usually <1 m thick), and contrasts with the sediment-enriched lower layers of sub-Polar glaciers (the product of numerous 'freezings-on' of basal meltwater) which often attain considerable thicknesses. It is probable that transfer of sediment upwards into the body of the ice can occur only in particular situations, and that the

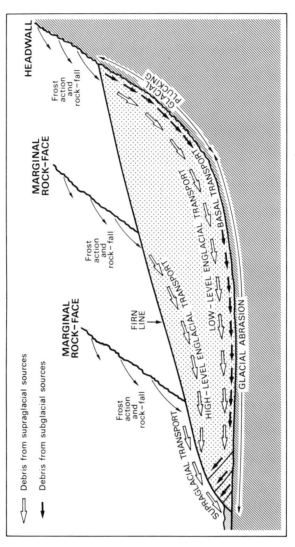

FIGURE 6.1   The glacial transport system

greater proportion of the englacial sediment load of many Alpine glaciers is of supraglacial derivation.

6.1.1 The supraglacial transport path

On glaciers where exposed rock faces and walls occur immediately adjacent to the ice margins, debris released by freeze–thaw weathering and rock falls accumulates in large quantities on the ice surface. This debris gives rise, via the differential ablation of debris-covered and bare ice surfaces, to supraglacial lateral moraines. These moraines are well developed below the equilibrium line, and appear to be directly nourished by falls of debris from above. The sediment cover remains on the glacier surface (except for some minor 'recycling' associated with the 'filling' of marginal crevasses and the subsequent melting-out of this 'shallow' englacial layer). The debris associated with supraglacial lateral moraines undergoes relatively little modification in clast shape or reduction by comminution; this is essentially a passive mode of transport (Boulton, 1978).

6.1.2 The englacial transport path

Englacial sediment is mainly incorporated within the ice in the firn zone, by a variety of mechanisms (see below). Since the debris is of supraglacial origin, the greatest quantities of englacial sediment are found (i) along the glacier margins, (ii) in the lee of nunataks and rock exposures and (iii) along the glacier 'centreline' where two or more contributory ice streams exist. Englacial sediment of supraglacial origin is by no means contained only within the upper layers of the glacier. This will be the case only where the rock exposure providing the sediment is close to the equilibrium line, resulting in a 'high level' englacial sediment path. Where the glacier occupies a basin with an extended headwall, at the base of which sediment accumulates and is subsequently incorporated in large quantities, there is commonly a marked concentration of englacial sediment in the lower layers of the ice, giving rise to a 'low-level' englacial sediment path. However, although this is proximate to the basal zone of traction, the two are quite distinct, and the contained debris can be clearly differentiated in terms of sedimentological characteristics.

It will be realized, of course, that in reality the postulation of high- and low-level englacial sediment paths is to a large extent arbitrary, since supraglacial sediment can be incorporated along the length of the valley wall from the headwall to the equilibrium line, giving a marginal zone of debris concentration extending throughout the depth of the ice. Whatever its exact position within the ice, englacial sediment—like that on the ice surface—is unlikely to experience much modification. Since it frequently occurs at a relatively low concentration, grain-to-grain contacts are uncommon and thus attrition does not occur. (The only exception might be where the debris is concentrated into layers or bands, although even here investigation of sedimentological characteristics indicates only minor modification.) Thus englacial transport, like supraglacial transport, is a 'passive' mode of transport. The original characteristics of the sediment (particle shape and size) are retained (by contrast with debris in the basal zone of transport, which is subjected to modification by rolling, sliding, crushing and abrasion).

Within the glacier transport system, transfers of sediment from one transport path to another are an important feature. Thus, englacial sediment following a low-level path may, as a result of bottom melting or the convergence of flow-lines at the ice–rock interface, become incorporated in the basal sediments, in which case their characteristics will quickly become altered. On a much larger scale, englacial sediment will be revealed at the glacier surface by ablation below the equilibrium line. This will contribute to the general surface covering of debris towards the snout, in addition to adding to the debris resting on medial and lateral supraglacial moraines. It is also possible that there may be transference upwards of subglacial sediment to an englacial position (and, subsequently, as a result of surface ablation to a supraglacial position). This is most likely to occur close to the snout where under compressing flow flow-lines turn upwards or thrust-faulting is developed.

6.2   SEDIMENTOLOGICAL CHARACTERISTICS OF GLACIAL DEBRIS

Supraglacial debris is typically coarse, heterogeneous, angular and poorly sorted, reflecting its origin by such processes as frost shattering of well jointed or stratified rocks. Englacial sediment, revealed at the ice surface by ablation, usually displays identical characteristics. Basal sediment, by contrast, may be initially angular (reflecting its origin in the detachment of joint-bounded rock masses by the plucking process or the inclusion of angular supraglacial clasts which have penetrated the ice–headwall junction), but will quickly become modified in shape. Boulton (1978), in a valuable discussion of glacier transport paths, pointed out that boulders at the ice bed are, as a result of friction between themselves and associated sediment and the underlying rock, retarded with respect to the sliding glacier ice. 'They crush and striate bedrock during their movement and are themselves progressively worn down by contact with the bed and other particles in traction. Striated abrasional facets are very rapidly produced on their surfaces. If they are already equidimensional with high rollability, new areas of the boulder are successively subjected to abrasion and the boulder tends to become spheroidal. If they are plate or blade shaped their "rollability" is low and they tend to slide, although irregularities on the bed will tend to ensure that they are from time to time rotated about a vertical axis. . . . These interactions tend to produce boulders with several sets of superimposed striae on their surfaces.' A study by Boulton of roundness and sphericity of supraglacially derived clasts in 'high-level transport' and boulders from the 'zone of traction' of Breidamerkurjökull in Iceland and Søre Buchananisen in Spitzbergen has revealed (i) that sphericity is very variable and roundness values low in the supraglacial material and (ii) that there is slightly enhanced sphericity and a clear increase in roundness for the subglacial boulders.

Boulton has also convincingly demonstrated that the particle size distributions of supraglacial/englacial sediment and debris in the basal zone of traction are significantly different where glacial erosion of igneous bedrock is involved. Studies of sediments from Glacier d'Argentière, Breidamerkurjökull and Søre Buchananisen show clearly that the tractional debris is depleted in the coarse fraction and enriched in the fine fraction compared with the debris undergoing high-level transport; the 'crossover' occurs at $0.5-2\phi$. Boulton infers that 'the grain size distribution typical of high-level transport may be transformed into the distribution typical of the basal zone of traction by the addition of fines, which in the examples used . . . have a mean mode between 4 and 6ϕ.' The source of these fines is the comminution occurring between large particles and the glacier bed, which forms a unique type of 'crushing mill.' The essential distinguishing features of particle size graphs for basal (as opposed to 'high-level') sediment are (i) larger standard deviation, (ii) smaller mean size and (iii) a decreasing skewness, because of the large quantities of fine sand-, silt- and clay-size rock flour generated by glacial abrasion. Boulton was able to argue that grain size distributions of the type described provide a powerful discriminatory tool. 'The grain-size contrasts demonstrated make me confident to have identified layers a few tens of centimetres above the glacier bed which I believe to have been *supraglacially* derived, but which have never been in contact with the bed in the zone of traction.'

6.3 THE DISTRIBUTION OF ENGLACIAL SEDIMENT WITHIN ALPINE GLACIERS

Englacial sediment is either widely dispersed throughout the glacier at a low concentration (visual inspection of ice exposures, in crevasses and at the snout, suggests that coarser particles, such as stones and boulders, constitute less than 5% by volume and in many instances less than 1%) or occurs as distinct sediment concentrations or even debris bands and layers. Where the debris is dispersed, individual fragments are usually angular or sub-angular, a form consistent with an origin from supraglacial weathering or rock fall. There is sometimes evidence to suggest that the clasts are more frequent in the upper than the lower horizons of the ice. For example, in 1951, Grande Dixence S.A. inserted 40 electrothermal probes into the Glacier de Tsidjiore Nouve (close to the base of the Pigne d'Arolla ice fall). Of these probes 22 reached the glacier bed, at a depth of 160–183 m; 18 failed as a result of contact with englacial clasts, and of these

50% were terminated at depths between 48 and 95 m, and none at a depth greater than 100 m.

Englacial debris concentrations assume a variety of forms, including the following:

1. *Longitudinal debris septa* (Sharp, 1949) occur at the margins of glaciers, as well as along 'centrelines' of compound glaciers formed by the coalescence of smaller tributary glaciers. On the Glacier de Tsidjiore Nouve and Bas Glacier d'Arolla strongly developed folia (either near vertical or dipping steeply towards the glacier centre) form a marginal zone 10–20 m or more in width, in which there are both debris bands and relatively high concentrations of discrete particles. Boulton and Eyles (1979) suggest that this lateral debris septum consists of debris derived supraglacially from flanking valley walls and entrained along foliation planes parallel to the glacier bed and for some distance above it. However, they suggest that 'use of the word lateral may hide the real disposition of this debris in the ice. In outlet valley glaciers, this debris septum may primarily be lateral, but in a valley glacier with an exposed headwall it forms an almost continuous stratum of dispersed supraglacially derived debris some distance above the bed and parallel to it. For this we propose the term bed-parallel debris septum.' Where two or more valley glaciers coalesce, the bed-parallel septa from the individual glaciers merge to produce *medial debris septa.*

2. *Transverse debris concentrations* may be associated with ogive structures, formed on temperate glaciers at the base of prominent ice falls (King and Lewis, 1961; Fisher, 1962). The development of transversely orientated ice-cored debris mounds and ridges (consisting of englacial sediment which has been revealed by ablational lowering of the ice surface) can be noted on glaciers with a well developed ogive pattern (for example, Glacier de Tsidjiore Nouve). Down-glacier the individual mounds usually merge to give a medial moraine of 'beaded' form. One assumption to account for this surface pattern of debris is that greater quantities of debris exist within the dark-ice layers of the ogive suite. These dark bands have been attributed to the passage of ice down the ice fall in summer (when freeze–thaw weathering of adjacent rocks and incorporation of the resultant debris via open crevasses are feasible). The intervening white bands (supposedly of winter origin on the ice fall) would by contrast contain little debris, since crevasses would be closed and both rock and ice surfaces cushioned by a thick snow layer. Eyles (in correspondence in *Journal of Glaciology*, 1976) has stated, however, that in Austerdalsbreen, Norway, 'substantial contributions to moraine debris made by a well developed ogive suite is lacking; debris of the dirty summer ogive bands is diffuse only. An immature beading of the moraine in harmony with summer ogive troughs (where diffuse surficial debris lowers the differential ablation ratio below 1) can be explained by mass movement of moraine sediments over the flanks of the ice core into such areas.' This seems a likely explanation of beading for other glaciers, although on the Glacier de Tsidjiore Nouve other factors are also important (see p. 126).

3. In some instances englacial debris is concentrated into debris-rich layers or debris bands, varying in width from 5 to 30 cm or more and frequently containing 80–100% sediment by volume (in other words, there is little or no interstitial ice). Many lateral or medial debris septa comprise both quantities of dispersed sediment and numerous individual debris bands. In the case of lateral debris septa the bands are usually orientated approximately parallel to the ice margin (longitudinal debris bands), whereas in medial debris septa they may be orientated either across the glacier (transverse debris bands) or parallel to glacier flow (longitudinal debris bands). For a further description of the morphology and sedimentary characteristics of debris bands, see the discussion on pp. 119–127.

Debris layers and bands within glaciers may be formed in a variety of ways:

(i) Some may consist of sedimentary layering derived from the glacier accumulation zone. Debris may fall on to the ablated surface of the firn, and is preserved by the subsequent winter's snow. Over a period of years a layered structure can be built up, and transferred down-glacier by flow, to be exposed again at the lower end of the glacier tongue as a series of dirt bands, consisting of relatively small amounts of fine sediment or layers of coarser debris from supraglacial rock falls (McCall, 1960). In his study of the small cirque glacier Vesl-Skautbreen, McCall demonstrated that although the sedimentary layers initially dip down-glacier, they subsequently undergo 'rotation' and at the glacier snout dip *up-glacier* at angles of as much as 70° from the horizontal (Grove, 1960). Along the lateral margins of the glacier, ablation surfaces and associated sediments dip steeply towards the centreline of the glacier, approximately in conformity with the slope of the rock bed.

(ii) On many present-day Alpine glaciers supraglacial debris can be observed sliding into crevasses. On the ablation zone this sediment is subsequently re-exposed at the ice surface, sometimes as a result of squeezing out where the crevasse becomes closed or, more frequently, by ablational lowering of the glacier surface. A series of crevasse fillings will often form a shallow sub-surface zone of debris, up to a maximum of 30 m in depth. However, where debris is incorporated within crevasses on the accumulation zone (at the base of the headwall or valley sidewalls, where bergschund-type crevasses and marginal crevasses occur), subsequent burial by firn and (in turn) glacier ice will result in the crevasse fillings forming a debris 'layer' at some depth within the glacier. For example, crevasse fillings from the headwall will be at maximum depth within the ice at the equilibrium line, but will then 'rise' towards the glacier surface on the ablation zone, where they will be re-exposed as strongly localized debris concentrations close to the terminus.

(iii) Debris bands may in some circumstances result from the upward transference of sediment from the glacier base. For example, Boulton (1978) has observed that when glaciers flow over large obstacles on their bed, there is a tendency for flanking ice to flow more rapidly into a lee-side position than does ice moving over the crest of the obstacle. As a result flow-lines diverge from the bed in that position. 'This tendency would be enhanced if the obstacle were larger and strong transverse compression were produced by convergent flow of two glacier septa. Thus, on the lee of a spur in a glacier headwall, or down-glacier of a glacier confluence or nunatak, basal flowlines can be expected to transport material from the basal transport zone up into high level transport.' This is the explanation of rounded and striated boulders sometimes found on the lower parts of medial moraines derived from glacier confluences or nunataks.

In other instances, the upward movement of basal sediment is associated with the development of secondary structures within the glacier, related principally to ice compression. Close to the snouts of valley glaciers, where the tendency for compressive flow is emphasized by the presence of dead-ice masses or large morainic accumulations, thrust-faulting appears to raise significant quantities of subglacial sediment to the glacier surface. The mechanism appears to be most characteristic of polar and sub-polar glaciers (see, for example, the 'shear moraines' close to the margins of Sørbreen, Ny Friesland, Vestspitzbergen, discussed by Boulton, 1967). However, well developed thrust planes, occupied by debris with subglacial characteristics, can be observed at the snouts of many Alpine glaciers (for example, Glacier de Mont Miné, Valais, Switzerland).

On a much larger scale the formation of large folds and/or faults in the zone of intense compressive stress at the base of a large ice fall may lead to the upward transport of large quantities of basal sediment. Posamentier (1978) postulated that the increase in ice velocity as the glacier passes through an ice fall will lead to extending flow and the initiation of folia parallel to the bedrock floor (see also Ragan, 1969). With a major decrease in velocity at the ice fall base, related to a sudden diminution of gradient, compressive flow will result in folding and faulting of the folia. In Posamentier's view, two types of ice structure may be formed: isoclinal folding, with or without associated axial plane reverse faulting, or reverse faulting with associated drag folding. In either case there is the probability that 'dirt-rich, bottom-most tectonic folia' will be lifted towards the glacier surface, where they can be exposed as a result of ice-surface ablation. This could account for the differential development of surface debris on dark and white ogives, although, as pointed out above, the occurrence of greater concentrations of englacial sediment in dark ice is not accepted by all authorities. Another possibility, not specifically stated by Posamentier, is that debris may be raised along the fault-lines themselves, thus producing discrete bands of sediment orientated transversely at the glacier surface.

The possible effects of 'lateral ice compression' (the process whereby compressive stresses may lead to the formation of longitudinal ice folding) were considered by Shaw (1980). This phenomenon results either from constriction of the glacier within a valley that narrows down-glacier, or from the formation of large lateral moraine embankments which act as semi-permanent obstacles to lateral ice spreading. The likely effects of lateral ice compression on glacier structures are considered more fully in the discussion of medial moraines of the Glacier de Tsidjiore Nouve (pp. 136–142).

6.4 CASE STUDIES OF ENGLACIAL SEDIMENT FROM THE VAL D'HÉRENS, SWITZERLAND

6.4.1 Englacial sediments of Glacier de Tsidjiore Nouve

The lower part of the Glacier de Tsidjiore Nouve consists of an ice-tongue extending about 2 km north-eastwards from the base of the ice fall flanking the northern face of the Pigne d'Arolla (Figure 6.2). Towards the snout the glacier surface is virtually covered with a layer of supraglacial debris (up to 0.5 m in thickness) consisting of angular blocks and smaller fragments of gneiss of the Arolla Series and highly fissile schists of the Schistes Lustrés Series of the Dent Blanche Nappe. Up-glacier, the debris forms two well defined medial moraines, each approximately 1 km in length (Small and Clark, 1974). That to the south is relatively broad and high (maximum width exceeding 100 m and maximum height up to 30 m), and that to the north is narrower and lower (maximum width 60 m and maximum height less than 10 m). At their upper limits (approximately 400 m from the base of the Pigne d'Arolla ice fall), both moraines appear to 'grow out' of the glacier surface, and indeed are clearly nourished by englacial debris which is exposed by summer ablation (as measured, in the order of 3–4 m per annum). At first the debris is revealed in patches which, by way of differential ablation, give rise to a series of transverse ice-cored mounds and ridges; as the surface debris increases in quantity down-glacier these mounds become more extensive and merge to give continuous medial moraine ridges. The latter are at first of 'beaded form' (p. 116) but further down-glacier become increasingly regular in outline. There appears to be some interrelationship between the debris mounds and the ogive suite of the Glacier de Tsidjiore Nouve. However, Eyles (correspondence in *Journal of Glaciology*, 1976) has argued that this does *not* reflect the greater concentration of englacial debris within dark ogives (p. 116).

The nature of the englacial debris septa which give rise to the Tsidjiore Nouve moraines has been studied in some detail (Small and Clark, 1974; Small and Gomez, 1981). As shown on p. 117, the septa are probably best developed within the upper layers of the ice, indicating that the contained sediments are derived not from the glacier headwall (in fact, the firn zone of the Glacier de Tsidjiore Nouve terminates in an ice-covered ridge), but from supraglacial rock faces approximately half-way between the upper glacier limit and the equilibrium line (which follows the summit of the Pigne d'Arolla ice fall at ca. 3000 m). The detailed form of the debris concentrations within the septa was revealed when, in July 1979, all the mounds and ridges, in a zone extending 360 m up-glacier from the southern medial moraine ridge, were wholly or partly cleared of

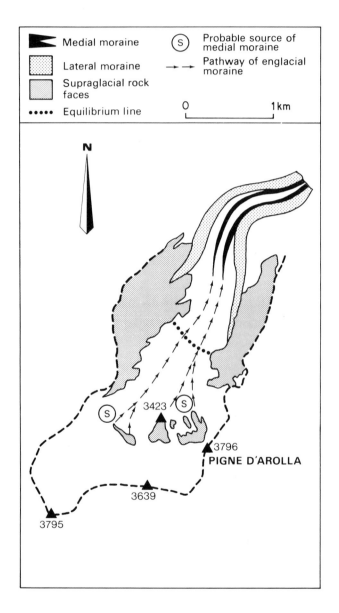

FIGURE 6.2　Map of Glacier de Tsidjiore Nouve. *Reproduced with permission from Small, Clark and Cawse (1979)*

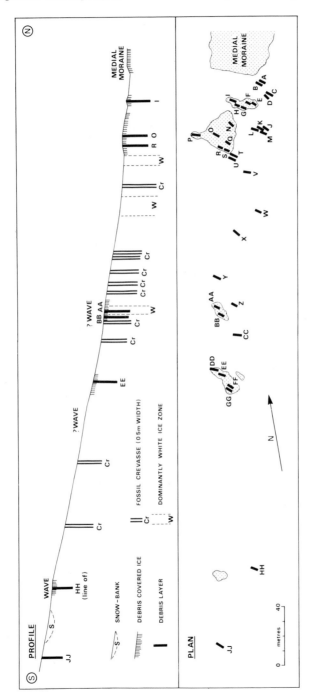

FIGURE 6.3 Debris bands of Glacier de Tsidjiore Nouve, in profile and plan. *Reproduced with permission from Small and Gomez (1981)*

FIGURE 6.4   Transverse englacial debris band, Glacier de Tsidjiore Nouve

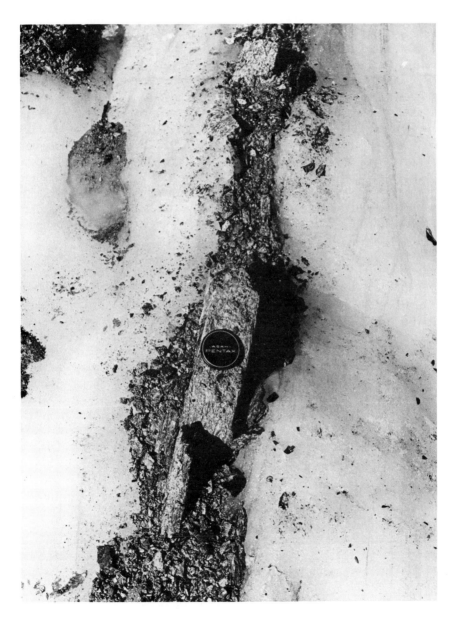

FIGURE 6.5   Sediment within transverse englacial debris band, Glacier de Tsidjiore Nouve

overlying debris. This clearance revealed the presence of 35 englacial debris bands, as shown in Figure 6.3. From the plan view it will be seen that the bands (which are up to 5 m in length) increase in density down-glacier. Only 12 bands outcrop in the 260 m below the 'emergence' of the first band, but 23 bands are concentrated in the final 100 m before the head of the medial moraine ridge. All the debris bands are orientated 'transversely' (median bearing 305°, by comparison with the glacier centreline bearing of 10°). Owing to the irregularity of the margins and the absence of good sections (there are no longitudinal crevasses), the true dips of the bands could not be determined precisely. Nevertheless, many could be seen to dip at 70–80° up-glacier, and some appeared to be near vertical (Figure 6.4). There is some variation in thickness, both within and between bands. Band D is particularly rich in debris, and reaches a maximum thickness of 30 cm; many other bands lie within the range 10–20 cm. All the bands appear to be overwhelmingly composed of angular debris, and sometimes contain slab-like clasts with major axes arranged vertically and/or parallel to the margins of the bands, within an admixture of gravel and coarse-medium sands (Figure 6.5). Only layers O and BP are different, consisting of thinner seams composed largely of fine sand and silt.

The long profile of the glacier surface along the centreline of the debris bands (Figure 6.3) depicts both debris bands and other structural features (notably narrow bands of compact white ice, evidently metamorphosed snow-filled crevasses formed initially a considerable distance up-glacier, either on or above the Pigne d'Arolla ice fall). Three subdued waveforms are indicated — an expected phenomenon in this zone at the base of the ice fall. If the Posamentier hypotheses were correct, debris emergence would be concentrated on the crests of these waves (the product of ice-folding or faulting). In reality, although englacial sediment does emerge in large quantities from debris bands exposed at the crest of the lowest wave, the two higher waves are almost debris free. There is, therefore, a *prima facie* case to be made against the hypothesis that the Tsidjiore Nouve debris bands consist of sub-glacial sediments raised from the ice bed, either as a result of thrust-faulting or the upward folding of debris-rich basal folia. Additionally, the limited lateral extent of the bands (up to 5 m) is inconsistent with a process of massive shearing through a layer of ice greater than 150 m in thickness. [Boulton (1967) has demonstrated that thrust planes containing subglacial debris near the snout of Sørbreen extend over distances of 150 m or more.]

An examination of sediment samples taken from 24 of the 35 debris bands revealed on the Glacier de Tsidjiore Nouve also tends to counter any suggestion of a predominantly subglacial origin. For purposes of comparison samples were also taken of four shear planes (the origins of which are in little doubt) at the snout of the nearby Bas Glacier d'Arolla. Figure 6.6 shows mean weight percentages at 0.5ϕ intervals of *all* debris bands and *all* shear planes. The relative coarseness of the debris band samples and the greater proportion of fines in

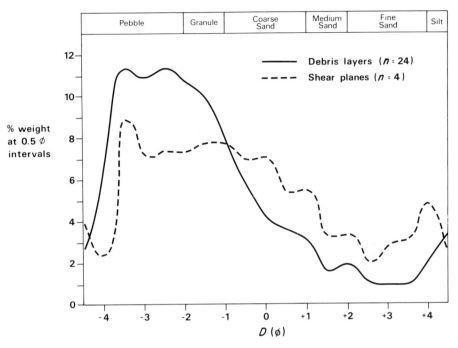

FIGURE 6.6   Particle size graph, debris bands, Glacier de Tsidjiore Nouve. *Reproduced with permission from Small and Gomez (1981)*

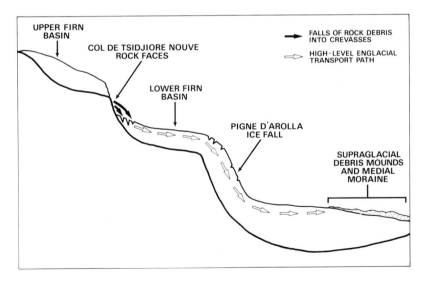

FIGURE 6.7   Sediment transport paths through Glacier de Tsidjiore Nouve

the shear plane samples are immediately apparent (the mean grain size of sediment in the former is 1.22ϕ and that in the latter 0.3ϕ). In reality, this analysis represents an under-assessment of the true coarseness of the debris bands, since in the collection of the samples in the field all large clasts had to be left (since many were of small-boulder size) and in the laboratory analysis particles in excess of 32 mm were eliminated. In the individual debris bands O and BB there was, however, a significantly greater amount of sediment in the sand–silt range (the mean grain size of O was 1.32ϕ), giving a particle size distribution more similar to those of the shear plane samples.

In general character, the particle size graphs for the Tsidjiore Nouve debris bands showed a close resemblance to those of supraglacial sediments on Breidamerkurjökull, Søre Buchananisen and Glacier d'Argentière (Boulton, 1978), whilst the shear plane sediments, as expected, are more typically 'subglacial.' There is little doubt that the Tsidjiore Nouve sediment is predominantly of supraglacial origin; indeed, likely sources of the debris are found in the rock outcrops which separate the upper and lower firn basins of the glacier (Figure 6.7). The likely route of the sediment, and the possible source of any subglacial sediment (possibly represented by bands O and BP), is shown diagrammatically in the figure.

The mechanism of debris incorporation remains an open question. As viewed in the field, the debris bands closely resemble infilled crevasses, although some misgivings must be felt about the near verticality of the bands, preserved despite 'passage' over a distance of some 2 km. An alternative view is that the debris in the bands is the product of episodic rock falls on to the glacier surface in the firn zone. These surface layers were then incorporated in the manner described on p. 117, 'rotated' by glacier flow and revealed in a steeply uptilted form at the base of the Pigne d'Arolla ice fall where intensive compressive stress is released by near vertical flow lines rising towards the glacier surface.

Along the northern and southern margins of the Glacier de Tsidjiore Nouve the upper parts of the bed-parallel debris septum (p. 116) are exposed at the ice surface, and contribute by way of ablation to the supraglacial debris layer and the formation of prominent, sharp-crested supraglacial lateral moraines. Within marginal transverse crevasses numerous individual debris bands, developed parallel to the folia, can be observed dipping inwards towards the glacier centreline. Some debris bands are near vertical, and most have an angle of dip in excess of 70°. In plan, the bands are either orientated parallel to the glacier margin or meet it at an acute angle. In superficial appearance, many of the debris bands resemble the *transverse* bands nourishing the medial moraines of the Glacier de Tsidjiore Nouve (Small, 1983). Sediment samples from six of the lateral moraine debris bands exposed in July 1981 were subjected to particle size analysis, and the results are shown in Figure 8.9. When compared with the graph of the 24 medial moraine englacial bands, that for the lateral moraine englacial debris again displays a predominantly supraglacial character, although there is some lack of material in the -3ϕ to -1ϕ range. It therefore seems likely that a proportion of the sediment within the upper parts of the bed-parallel debris septum of the Glacier de Tsidjiore Nouve consists of debris released by weathering of marginal supraglacial rock-faces above the firn line, and incorporated along the glacier edges as a result of burial by firn or falling into transverse crevasses and the gap between the ice and the valley wall. However, it must be added that much of the marginal supraglacial debris of the lower part of the Glacier de Tsidjiore Nouve is derived not from the melting out of englacial sediments, but from direct fall of clasts on to the ice surface from the rock faces, lying below the firn line, on either side of the Pigne d'Arolla ice fall. The sedimentological characteristics of the surface debris layer (an admixture of supraglacial debris from above and below the firn line) are shown in Figure 8.7.

6.4.2   Englacial sediments of Bas Glacier d'Arolla

In certain respects there is a close resemblance between the pattern of englacial debris within the Bas Glacier and the Glacier de Tsidjiore Nouve. The bed-parallel debris septum is well developed along the western edge of the glacier, and is exposed in marginal crevasses and, most spectacularly, in the large ice cave at the base of the Mont Collon ice fall. The Bas Glacier also supports two medial moraines although, in contrast to the Glacier de Tsidjiore Nouve, these are relatively small ridges that 'emerge' within 200 m of the glacier snout (Small *et al.*, 1979).

The eastern medial moraine begins as a patchy cover of angular boulders and fragments which merge down-glacier to give a continuous debris layer. The resulting differential ablation of the glacier surface then results in the development of a prominent ice-cored ridge. At the head of the moraine no debris bands were identified, although some individual rock fragments were seen 'emerging' from the glacier surface. The longitudinal debris septum nourishing the moraine thus seems to consist of a 'zone' within which clasts are more or less uniformly dispersed. The 'late emergence' of debris is suggestive either of a low-level englacial sediment path (p. 113) or of upward transfer of subglacial sediment in the zone of compression close to the glacier snout. However, the extreme angularity of all the individual clasts observed, and the complete absence of rounded or sub-rounded, polished and striated boulders, effectively counters the latter possibility.

The western moraine also begins as a line of patchy superficial debris, extending down-glacier for 75 m, beyond which point the moraine ridge grows rapidly. In this instance, however, the moraine is nourished by well-defined transverse englacial debris bands. In all 28 individual bands were identified, varying in lateral extent from 1 to 11 m. Spacing of the bands was irregular, but tended to increase down-glacier, 20 of the bands occurring within 30 m of the commencement of the ridge proper (Figure 6.8). In general structure the

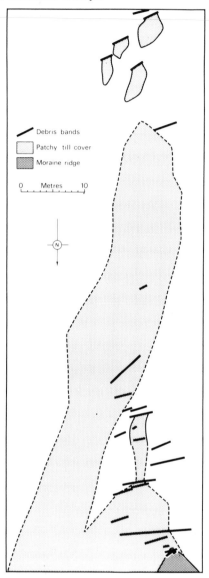

FIGURE 6.8   Map of debris bands, western medial moraine, Bas Glacier d'Arolla

debris bands of the Bas Glacier resembled those of the Glacier de Tsidjiore Nouve, although large, sharply angular fragments were more common and relatively fine debris (in the -2ϕ to $+4\phi$ range) was almost totally absent. Again the debris bands appear to constitute a longitudinal septum developed close to the glacier base (that is, in very low-level transport). The most likely sources

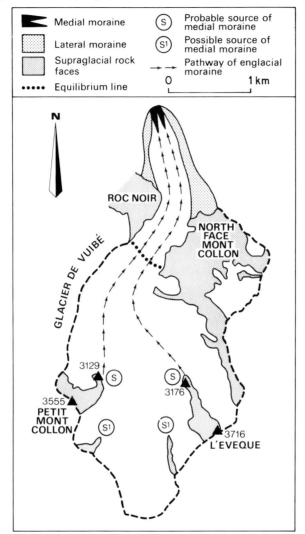

FIGURE 6.9   Map of Bas Glacier d'Arolla. *Reproduced with permission from Small, Clark and Cawse (1979)*

of the debris, and that of the easternmost medial moraine, are supraglacial rock outcrops at the head of the firn basin of the Bas Glacier. The possible routing of the sediment is shown in Figure 6.9.

6.4.3   Englacial sediments of Haut Glacier d'Arolla

The Haut Glacier differs from the Bas Glacier and the Glacier de Tsidjiore Nouve

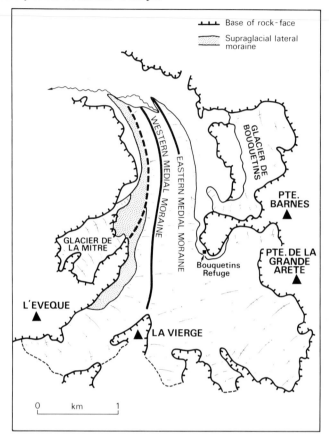

FIGURE 6.10   Map of Haut Glacier d'Arolla. *Reproduced with permission from Gomez and Small (1985)*

in its possession of a relatively gentle overall gradient and the absence from it of a major ice fall. It possesses a compound firn basin within which there are three distinct elements: a relatively large accumulation area delimited by the peak of La Vierge in the south-west and the Pointe de la Grande Arête in the east; a smaller area in the form of a glacier tongue between L'Eveque and La Vierge; and a still smaller area in the re-entrant to the south-west of Pointe Barnes (Figure 6.10). The moraine pattern of the glacier, as described by Small *et al.* (1979), consists of two well developed medial moraines, plus a 'lateral–medial' moraine complex along the western margin beyond the point at which the Haut Glacier is joined by a small tributary glacier, the Glacier de la Mitre. Additional observations on these moraines, and their relationships to englacial debris septa, were discussed by Gomez and Small (1985).

(i) The eastern moraine is developed from large quantities of englacial debris

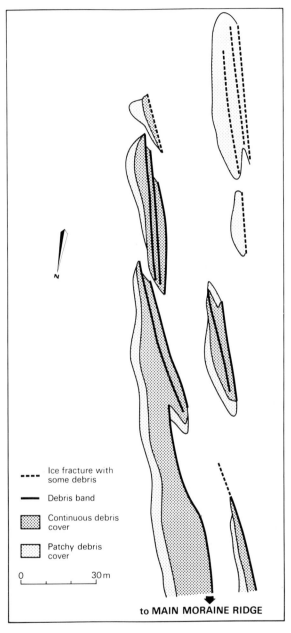

FIGURE 6.11   Map of debris bands, eastern medial moraine, Haut Glacier d'Arolla.
Reproduced with permission from Gomez and Small (1985)

which 'emerge' at a point approximately 1 km from the glacier snout to form a series of relatively small, initially discrete sub-parallel ridges running approximately parallel to the centreline of the glacier. However, the individual ridges merge within a distance of 200–300 m to form one large moraine ridge. Field examination has revealed that each ridge in the zone of emergence is related to one or more longitudinal debris bands, which are characteristically 20–40 m in length and form overall a crude *en echelon* pattern (Figure 6.11). The bands are less than 0.5 m in thickness, and are disposed either vertically or dip at a very steep angle (in excess of 80°) towards the centreline of the Haut Glacier. The constituent debris lies entirely in the coarser size range (-5ϕ to -2ϕ), is loosely packed and is invariably angular or very angular (Figure 6.12). The lithology of the included fragments (schists and gneisses, comparable to those exposed on the western face of the Pointe Barnes–Pointe de la Grande Arête ridge) suggests (a) a supraglacial origin in the eastern accumulation zone of the Haut Glacier and (b) a low-level transport path through the glacier.

The debris bands are interpreted provisionally as steeply dipping 'tabular' bodies of englacial sediment which are being exposed at the glacier surface by substantial annual ablation (of the order of 5–10 m). They are evidently passive, and do not involve the upward transference of debris either along shear planes or flow lines—certainly the extreme angularity of the contained fragments is not consistent with the morphology and characteristics of subglacially derived and transported clasts. The most likely mechanism to account for the bands involves a series of rock falls in the accumulation zone and incorporation within the glacier by burial beneath layers of firn. The actual spacing of the debris bands indicates that the falls were not themselves annual, but occurred at intervals of several (perhaps 10–20) years. As the sediment accumulations were subsequently transported down-glacier, they were increasingly transformed into a series of comparatively narrow bands running parallel to the direction of glacier flow and dipping steeply inwards towards the glacier centreline, approximately in conformity with the shape of the glacier bed.

(ii) The western moraine commences much farther up-glacier (2.6 km from the snout), on the equilibrium line of the Haut Glacier close to the base of La Vierge. Immediately down-glacier the moraine consists of two or three distinct but small ridges which extend for about 1.4 km but then merge to form one major ridge extending to the glacier snout. In the field the moraine can be seen to be nourished in part by direct falls of supraglacial debris from the rapidly weathered schists on the flanks of La Vierge, but more detailed examination reveals that englacial sources also make a contribution. The latter occur as two prominent longitudinal debris bands, up to 20 cm in thickness and disposed vertically (Figure 6.13). As on the eastern moraine the constituent particles are dominantly angular and sub-angular, but the sediment is more densely packed and there is at times a clear stratification, consisting of layers of clear and sediment-rich ice. Appreciable quantities of finer material (-1ϕ to $+4\phi$) are

FIGURE 6.12 Coarse angular debris emerging from longitudinal debris band, Haut Glacier d'Arolla

also present, although the grain size distribution remains skewed towards the coarser size ranges.

The formation of the debris bands nourishing the western moraine is clearly related to the confluence of contributory ice streams from either side of La Vierge. At the latter point flow-lines adjacent to the two valley side-walls, associated with debris-enriched marginal ice, converge beneath the 'source' of the medial moraine. The contained debris is interpreted as (a) the product of

FIGURE 6.13   Longitudinal englacial debris bands, western medial moraine, Haut
Glacier d'Arolla

entrainment of fragments falling from the supraglacial valley walls and (b)
sediment which is derived either from glacial erosion or the comminution of
supraglacial clasts along the side-wall–ice interface.

(iii) At the junction of the Glacier de la Mitre and the Haut Glacier, a medial
moraine is developed, although within a short distance down-glacier this is
overwhelmed by falls of supraglacial debris from the western valley wall and
becomes a component of the western supraglacial moraine complex of the

Haut Glacier. Although the bulk of the overlying debris of this moraine complex displays supraglacial characteristics (coarse and angular fragments in abundance), rounded, sub-rounded and abraded clasts are also present. Moreover, clearance of superficial debris has revealed the presence of a 1 m thick longitudinal debris band which extends to the glacier snout (where in 1983 it was fortuitously revealed in cross-section, and could be observed dipping westwards beneath overlying ice in the manner of a large shear plane). The sediment within the band was very densely packed, strikingly stratified and consisted of rounded and sub-rounded clasts and an abundance of smaller particles in the silt to coarse sand range (-1ϕ to $+4\phi$). Down-glacier of the confluence with the Glacier de la Mitre, the Haut Glacier progressively narrows from a width of 750 m to less than 400 m, and as a result of the lateral compressive stresses engendered by this the 'inset' ice stream (Sharp, 1949) derived from the Glacier de la Mitre appears to be overriding the ice of the Haut Glacier. Basal ice and associated sediments are thus exposed at the glacier surface as a longitudinal debris band, thus accounting for (a) the thickness and detailed nature of the band (whose structure is indicative of successive freezings-on of ice and sediment by basal regelation) and (b) the manner in which debris is being squeezed out from the band (at some points compacted ridges of silt-sized material stand proud of the ice surface).

6.5 THE ROLE OF ENGLACIAL SEDIMENTS IN THE FORMATION OF SUPRAGLACIAL MORAINES

As has been shown in the preceding discussion, the contribution of englacial sediments to the formation and development of medial and supraglacial lateral moraines can be considerable. Sharp (1949) has noted that medial moraines frequently mark the outcrops of 'steep, debris-rich zones in glaciers,' adding that 'the debris comprising such moraines is largely a residue accumulation left on the surface by melting.' Thus only a small proportion of the debris cover of the medial moraine 'has always been superglacial.' In his study of the constitution of valley glaciers, Sharp (1948) proposed that these debris-rich zones (*longitudinal septa*) develop not only at the junctions of 'juxtaposed' ice streams but also where small tributary glaciers join major glaciers to form 'superimposed' and 'inset' ice streams. The field observations of the Haut Glacier d'Arolla (pp. 129–135) are fully consistent with Sharp's interpretation. Nevertheless, the role of englacial sediment supply is not invariably a major one. Thus, Loomis (1970), in his description of the major medial moraine of the Kaskawulsh Glacier states that 'the debris load is largely supraglacial with relatively clear ice lying below the sharp ice–debris interface. That is, by volume it is estimated less than ten percent of the rock debris of the medial moraine is englacial in nature.' The determining factor is clearly the point at which the supraglacial debris falls on to the glacier surface. If this is below the firn line, there will be 'direct' feed to the moraine, with limited 'recycling' where debris falls into crevasses and is re-exposed by ablational lowering of the ice. If debris fall is above the firn line, the debris will be incorporated, become englacial and will emerge on the ablation zone of the glacier. (See the discussion of englacial transport paths on p. 113.)

Eyles and Rogerson (1978) proposed a fundamental division of medial moraines into the *ablation dominant* type (AD) and *ice stream interaction* type (ISI). The former develop at the junction of two tributary glaciers, the supraglacial lateral moraines of which amalgamate to form the medial moraine. This will consist of a debris-covered ice ridge extending down the trunk glacier (sometimes for a distance of several kilometres), and terminating near the snout, where it may merge with the general spread of ablation moraine covering the lowermost parts of a debris-rich glacier. Ablation-dominant moraines, by contrast, form where debris is 'held englacially and revealed down-glacier by ablation.' Such moraines appear to grow out of the glacier surface on the ablation zone, and are sometimes characterized by a 'beaded' form in their uppermost sections (for example, the medial moraines of Austerdalsbreen, Norway, and the Glacier de Tsidjiore Nouve). Eyles and Rogerson recognize two sub-types of AD moraines: the *above firn line* sub-type, in which debris is incorporated on the accumulation zone, and the *below firn line* sub-type, in which surface debris enters the glacier by open crevasses on the ablation zone but is rapidly re-exposed at the glacier surface by summer melting.

It is apparent that the distinction between AD and ISI moraines is not an absolute one. For example, where debris falls on to the margins of ice streams within the accumulation zone, it becomes englacial and contributes to the bed-parallel debris septum (p. 116). Amalgamation of these ice streams may then produce longitudinal englacial septa containing debris which is revealed by ablation beyond the firn line. Such a process plays a significant role in the development of the western medial moraine of Haut Glacier d'Arolla, although at first sight this appears to be a classic ISI moraine.

6.6   CASE STUDIES OF MEDIAL MORAINE FORM AND GENESIS FROM THE VAL D'HÉRENS, SWITZERLAND

6.6.1   Medial moraines of Glacier de Tsidjiore Nouve

This glacier displays two AD-type medial moraines which emerge from the ice surface about 400 m from the base of the Pigne d'Arolla ice fall, linking the firn basin and the lower glacier tongue (Small and Clark, 1974). The larger southern moraine attained a maximum height of 22.25 m in 1971 (by 1976 this had risen to 30.8 m) and a maximum measured width of 101 m. The smaller northern moraine reached a maximum height of 6 m and a maximum width of 59 m in the 1971 survey. The moraines are each approximately 1 km in length, but merge with each other and supraglacial lateral moraines to form a continuous debris cover near the glacier snout.

The main morphological features of the moraines are as follows:

(i) When fully developed down-glacier, both constitute single ice-cored ridges, with a debris cover 5–10 cm in thickness. This debris, by retarding ablation (Østrem, 1959; Loomis, 1970), is the major factor in moraine formation, as is shown by the ablation measurements taken in the late summer of 1976 (Table 6.1).

(ii) Both moraines grow in height down-glacier over a distance of about 400 m, but then undergo some decline as the debris cover becomes attenuated as a result of lateral sliding on the flanks of the moraines. Nevertheless, the ridges do continue, albeit in subdued form, as far as the glacier snout. There is a clear resemblance to the much larger medial moraine of Kaskawulsh Glacier, Alaska (Loomis, 1970). This is an ISI-type moraine which grows in height from 2 m at the glacier confluence to 20 m at a distance of 1.4 km down-glacier; the moraine then declines to 7 m after a further 5 km, and thereafter remains approximately constant in height for several more kilometres to the snout. Loomis (1970) postulated that the main factors controlling the form of the Kaskawulsh moraine are (a) differential ablation (as between debris-covered and clean ice), (b) lateral ice compression below the confluence of contributing ice streams, (c) extending flow (which causes the debris cover to become thinner and less effective as an insulator downglacier, so that ablation rates increase) and (d) lateral sliding of debris (in this instance over maximum slopes of 39–40°, by comparison with the slopes of 27–29° on the Tsidjiore Nouve moraines).

(iii) Both moraines commence as a series of discontinuous transverse mounds and ridges, related to the release of debris from well defined englacial debris bands (p. 119).

The 'waxing' and 'waning' stages of development displayed by the Tsidjiore Nouve moraines appear to result from a number of closely interrelated

TABLE 6.1   Tsidjiore Nouve Glacier: ablation transect (24.7.1976 to 2.9.1976). Transect across peak of main medial moraine and declining section of subsidiary moraine

Site	Debris thickness (cm)	Total ablation (cm)	Days of observation	Mean diurnal ablation (cm)
S. face main moraine	6	64.3	39	1.6
Crest of main moraine	5	52.8	39	1.4
N. face main moraine	6	67.4	32	2.1
Clean ice	—	142.4	36	3.9
S. face subsidiary moraine	2	169.9	39	4.3
N. face subsidiary moraine	1	161.1	35	4.6
Clean ice	—	141.0	36	3.9

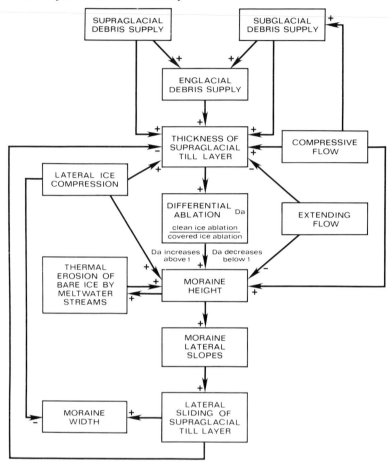

FIGURE 6.14   Factors controlling the development of medial moraines

factors (Figure 6.14). The principal determinant of the waxing stage here is differential ablation; a series of ablation measurement in 1971 showed that a 6 cm debris layer reduced ablation to only 78.3% of that of nearby clean ice [comparable figures are the 66% determined for Isfallsglaciären by Østrem (1959) and 74% for Kaskawulsh Glacier recorded by Loomis (1970)]. On the upper part of each Tsidjiore Nouve moraine there is a strong positive relationship between moraine height (increased down-glacier by approximately 1 m for each year of ablation), moraine slope, debris sliding and moraine width. Surveyed cross-profiles of the southern moraine reveal that within a few hundred metres of the 'source' in the emergent debris bands, lateral slopes gradually steepen to a maximum of 29° and moraine width (as a result of active sliding) increases from 38 to 101 m.

The waning stage of development appears to commence where lateral sliding has become so active that the debris cover to the moraine ridge becomes significantly thinned. Additional factors may be (a) a decrease in englacial debris supply as the debris bands of the 'source' zone, reaching a finite depth within the glacier, are consumed and (b) a tendency towards extending flow as the glacier surface steepens down-glacier. Whereas on the waxing stage differential ablation (defined as Da = clean ice ablation/covered ice ablation) has a value > 1.0, on the waning stage (for moraine decline to be explained) the Da value must become < 1.0 (or in other words the debris cover must be such as to increase ablation rates significantly, in relation to those of clean ice). Sharp (1949) has suggested that such indirect ablation (or accelerated ablation) is associated with a very thin debris layer, and results from the effective conduction downwards to the ice surface of solar heat absorbed at the surface of the debris. Ablation measurements by Østrem (1959) and Loomis (1970) indicate that the critical debris thickness, producing accelerated ablation, is 0.5–1 cm. Measurements made in September 1972 along the crest of the southern Tsidjiore Nouve moraine revealed that the debris cover changed from a continuous 4 cm layer to an increasingly patchy cover consisting in part of a 'one-particle' layer, and that as a result of this debris thinning ablation increased from 45.3% of that of nearby clean ice to 125.7% (Table 6.2).

It is suggested therefore that the waxing and waning stages of medial moraine development on the Glacier de Tsidjiore Nouve can be explained satisfactorily in terms of differential ablation and accelerated ablation, related to systematic variations in supraglacial debris thickness down-glacier produced by (a) melting out of englacial debris and (b) attenuation of the debris layer by sliding and other mechanisms (such as extending flow). The 'constant' stage of development, noted by Loomis on the lowermost section of the Kaskawulsh medial moraine, is represented on the Glacier de Tsidjiore Nouve by the final sections of the two moraines, where decline in height becomes much less pronounced. Here the mechanisms responsible for moraine 'waning' are clearly much reduced in effectiveness. Firstly, the onset of compressive flow as the glacier snout is approached causes some thickening of the supraglacial debris layer, and a return to differential rather than accelerated ablation. Secondly, as moraine slopes are

TABLE 6.2   Tsidjiore Nouve Glacier: ablation along moraine crest (September 1972)

Site	Description	Mean diurnal ablation (cm)	% of mean diurnal ablation of bare ice
1	4 cm debris	0.97	45.3
2	2–3 cm debris	1.72	80.3
3	Patches of debris (2–3 cm)	2.51	117.3
4	Bare ice between dispersed debris patches	2.69	125.7

reduced lateral sliding becomes less active. Thirdly, and perhaps most importantly, there is considerable 'recycling' of debris which is continually falling into open crevasses and then re-emerging a short distance down-glacier as a result of ablational lowering of the glacier surface.

An alternative explanation of the origin and form of the Tsidjiore Nouve medial moraines is that of Shaw (1980), who proposed the hypothesis that they are the product of longitudinal folding within the glacier, as a result of lateral ice compression [a subsidiary process on the Kaskawulsh moraine, and not regarded by Small and Clark (1974) as a significant factor on the Glacier de Tsidjiore Nouve]. Shaw refers in particular to (a) the overriding by the glacier margins of large lateral moraine embankments (termed Röthlisberger–Schneebeli moraines, after the work of these authors (see Chapter 8)) and (b) the strong 'convergence' at the base of each anticlinal structure, giving rise to concentrations of debris-rich basal ice. Deposition of the latter material, by lodgement or melt-out, will in time form a glacial fluting; however, at an earlier stage (represented by the present-day glacier) the folding may result in the upward transfer of basal sediments to the ice surface (rather as in the Posamentier model of transverse folding discussed on p. 118). Shaw argues that the hypothesis of moraine development based primarily on the differential ablation of debris-covered and bare ice (Small and Clark, 1974) can be dismissed. This carries the clear implication that the surface relief of the glacier is the direct product of longitudinal folding. However, in the light of the ablation studies referred to earlier in this chapter (pp. 138–139) it is impossible to believe that, whatever the factors responsible for the initiation of the moraine ridges, differential ablation (and associated processes such as lateral debris sliding) has not played an important role in the development of the moraines as relief features.

In fact, there is much evidence that can be invoked to counter Shaw's 'longitudinal folding' hypothesis, at least as applied to the Glacier de Tsidjiore Nouve. Firstly, the character of the covering debris is indicative of supraglacial derivation, from exposed rock faces mid-way down the firn zone of the glacier (p. 126). Shaw is not specific about the origin of the debris, although one can assume that he infers a subglacial source from his statement that 'flow vectors have a strong upward component beneath anticlines.' Secondly, Shaw's diagrams show that the longitudinal folding mechanism should produce debris-rich folia (possibly in the form of concentrated debris bands) running parallel to the axes of the anticlines and dipping steeply away on either side of each medial moraine ridge. However, detailed field examination has revealed only transverse debris bands, cutting across the supposed anticlinal axes, in the zone of englacial debris at the head of each moraine. Additionally, a detailed field map of the foliation pattern of the Glacier de Tsidjiore Nouve fails to support the folding hypothesis (Figure 6.15). Although much of the glacier tongue is covered by supraglacial debris to a depth of ca. 0.5 m, especially along the southern margins and towards the snout, the ice 'vales' adjacent to the medial moraines and marginal crevasses

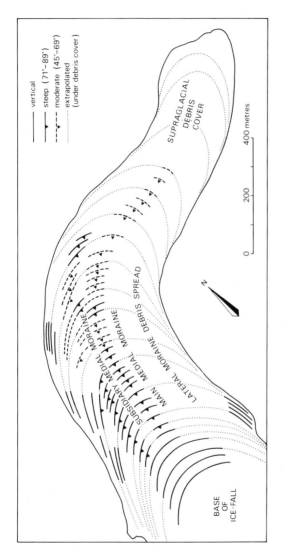

FIGURE 6.15   Foliation pattern of Glacier de Tsidjiore Nouve

afford adequate exposures of folia. The foliation pattern is in essence simple, and comparable to that of other 'single ice stream' (as opposed to 'multiple ice stream') glaciers (Taylor, 1963). On the upper glacier tongue, at the base of the Pigne d'Arolla ice fall, the foliation is intense and near vertical; down-glacier it becomes less intense but still strongly developed, with up-glacier dips maintained in excess of 45°. Along the centreline of the glacier (that is, across the medial moraines) the folia are transverse, and it is only close to the glacier margins that they turn to become longitudinal. The pattern as a whole conforms closely to the 'spoon-shaped' structure described by Rutter (1965). What is important in the present context is that there is no indication of the longitudinal foliations parallel to the medial moraine axes required either by Shaw's hypothesis of lateral compression, or by Boulton's hypothesis of subglacial rock obstacles producing upward division of flow-lines and associated transfer to the glacier surface of subglacial debris (p. 117).

6.6.2   Medial moraines of Bas Glacier d'Arolla (see p. 127)

These are very short AD-type medial moraines (approximately 200 m in length), developed from the emergence of englacial debris in low-level transport. The western moraine develops rapidly as a sharp crested ridge, reaching a maximum height of 6–8 m; however, very close to the snout a rapid decline to 3.5 m occurs, probably as a result both of lateral and longitudinal (towards the snout) sliding of supraglacial debris. The eastern moraine, nourished by well developed englacial debris bands (p. 127), is more prominent, growing to a maximum height of 14–16 m before again declining just above the glacier snout.

6.6.3   Medial moraines of Haut Glacier d'Arolla (see p. 129)

This glacier also supports two medial moraines, together with a medial–lateral moraine complex, which contrast strongly with each other. The western moraine appears to be an ISI type, but is also nourished to some extent by the emergence of sediment from longitudinal debris bands (p. 132). The moraine extends for about 2.6 km from source to snout. Both the height and width increase progressively over this distance, reaching 11 and 60.5 m, respectively, on a cross-profile surveyed immediately above the glacier terminus. Despite its considerable length, the moraine undergoes restricted development (by comparison with those of the Glacier de Tsidjiore Nouve), probably as a result of the limited initial input of supraglacial debris from La Vierge and a relatively small input from the englacial septum. As a result it displays only the waxing stage of development.

The eastern moraine is a complex and ultimately more prominent feature of the AD above firn-line sub-type. The initial individual moraine ridges, related to prominent longitudinal debris bands (p. 132), merge down-glacier to give one major moraine ridge, reaching a height of 15–18 m and a width of nearly 100 m. The waxing stage of development is again very evident and even close to the glacier snout there is no tendency towards reduction in relief.

6.6.4 Types of medial moraine in the Val d'Hérens

The classification of medial moraines by Eyles and Rogerson (1979) into ice stream interaction (ISI) and ablation dominant (AD) types is based on the origin and emergence of the supraglacial debris cover rather than on moraine morphology as such. An alternative approach might be to consider the extent to which individual medial moraines display waxing, waning and constant stages of development.

The six medial moraines of the Glacier de Tsidjiore Nouve, Bas Glacier d'Arolla and Haut Glacier d'Arolla (the medial–lateral moraine complex of the western margin of the latter is omitted) exhibit the considerable diversity of form that can occur even within a small area; this diversity reflects, of course, the complexity of the interactions shown in Figure 6.14. The failure of four of the moraines described even to approach 'full' morphological development appears to reflect the following three main factors, all related to debris supply (the prime cause of all moraines), and overall glacier morphology (Small, 1982).

(i) *Volume of debris.* If the debris inputs are considerable, differential ablation will cause a sequence of rapid moraine growth in height, effective lateral sliding of debris, attenuated debris cover, accelerated ablation and (at a point down-glacier) the onset of moraine decline. However, if debris inputs are less substantial, differential ablation will be less pronounced and the stage of moraine growth much more protracted — even to the point that moraine decline will not be initiated before the glacier snout is reached.

(ii) *The balance between direct supraglacial debris supply and indirect englacial supply.* If the moraine is nourished by debris falls within the ablation zone, it will grow in height rapidly and at a rate commensurate with the volume of debris increments. If, however, it is supplied by falls in the firn zone, the debris will become englacial and eventually be exposed only on the ablation zone. If derived from the headwall, and therefore following a low-level transport path through the glacier, the debris will emerge close to the glacier snout (as on the Bas Glacier d'Arolla), forming short moraines with only the waxing stage developed. If derived from rock outcrops some distance down the firn zone (as on the Glacier de Tsidjiore Nouve), the debris will emerge further up-glacier, and the resultant moraines are likely to experience more advanced development. Possible relationships between debris supply and moraine morphology are shown in Figure 6.16.

FIGURE 6.16 *(overleaf)* Debris supply and moraine development. *Reproduced with permission from Small, Clark and Cawse (1979)*

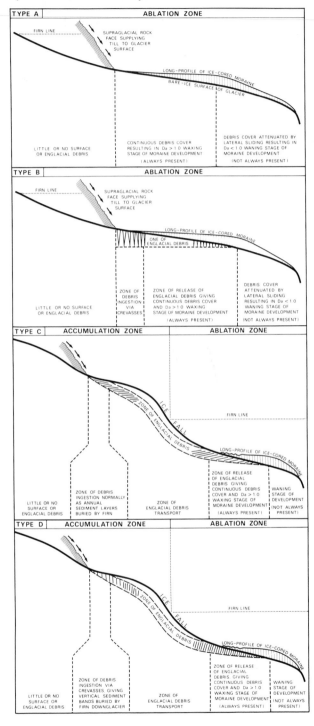

(iii) *The length of the glacier.* This is important in the sense that *ceteris paribus* the longer the glacier the greater is the opportunity for full morphological development of moraines. However, even on long glaciers the precise point at which debris accumulates, and also the actual amount of debris, is critical. This is illustrated by the two moraines of the Haut Glacier d'Arolla, neither of which approaches full morphological development.

It remains to be added that the observations made on the form and development of medial moraines can be applied equally well to supraglacial lateral moraines. These develop as a result of differential ablation, debris sliding, accelerated ablation, etc., and are produced as a result both of direct debris feed from marginal outcrops below the firn line and of indirect feed, by the release of englacial debris (from the bed-parallel debris septum) incorporated above the firn line. Indeed, it is appropriate to recognize lateral moraines of the ablation dominant above firn line sub-type and ablation dominant below firn line sub-type (where debris falls directly into marginal crevasses on the ablation zone, and is subsequently re-exposed at the glacier surface, to produce an ice-cored ridge by way of differential ablation), as well as those that might be termed direct feed lateral moraines (the nearest equivalent to the ice stream interaction type of medial moraine).

REFERENCES

BOULTON, G. S. (1967), The development of a complex supraglacial moraine at the margin of Sørbreen, Ny Friesland, Vestspitzbergen, *Journal of Glaciology*, **6**, 717–735.

BOULTON, G. S. (1978), Boulder shapes and grain-size distribution of debris as indicators of transport paths through a glacier and till genesis, *Sedimentology*, **25**, 773–799.

BOULTON, G. S., and EYLES, N. (1979), Sedimentation by valley glaciers: a model and genetic classification, in SCHLUCHTER, C. (Ed.), *Moraines and Varves: Origin, Genesis, Classification*, A. A. Balkema, Rotterdam, pp. 11–23.

EYLES, N., and ROGERSON, R. J. (1978), A framework for the investigation of medial moraine formation: Austerdalsbreen, Norway, and Berendon Glacier, British Columbia, Canada, *Journal of Glaciology*, **20**, 99–113.

FISHER, J. E. (1962), Ogives of the Forbes type on Alpine glaciers and a study of their origins, *Journal of Glaciology*, **4**, 53–61.

GOMEZ, B., and SMALL, R. J. (1985), Medial moraines of the Haut Glacier d'Arolla, Valais, Switzerland: debris supply and implications for moraine formation, *Journal of Glaciology*, **31**, 303–307.

GROVE, J. M. (1960), The bands and layers of Vesl Skautbreen, in LEWIS, W. V. (Ed.), *Norwegian Cirque Glaciers*, Royal Geographical Society Research Series No. 4, pp. 11–23.

KING, C. A. M., and LEWIS, W. V. (1961), A tentative theory of ogive formation, *Journal of Glaciology*, **3**, 913–939.

LOOMIS, S. R. (1970), Morphology and ablation processes on glacier ice, in BUSHNELL, V. C., and RAGEL, R. H. (Eds.), *Icefield Ranges Research Project, Scientific Results*, Vol. 2, American Geographical Society, New York; Arctic Institute of North America, Montreal, pp. 27–31.

SHAW, J. (1980), Drumlins and large-scale flutings related to glacier folds, *Arctic and Alpine Research*, **12**, 287–298.

Glacio-fluvial Sediment Transfer
Edited by A. M. Gurnell and M. J. Clark
©1987 John Wiley & Sons Ltd.

Chapter 7

The Subglacial Sediment System

R. A. SOUCHEZ and R. D. LORRAIN

ABSTRACT

This chapter deals with processes by which sediments are produced, incorporated and transported at the base of alpine glaciers. The main theme is the detailed study of the basal ice layer (BIL), which plays a prominent role not only in transferring debris at the glacier base down-glacier but also as an effective tool for glacier abrasion. The mechanisms responsible for the formation of this BIL are discussed in the light of chemical and isotopic analyses of the constituting ice. The characteristics of grain size distributions resulting from abrasion are interpreted. Finally, glacial plucking and the role of meltwater at the ice-rock interface are considered.

7.1 THE SUBGLACIAL SEDIMENT SYSTEM AND THERMAL CONDITIONS AT THE BED

Debris production at the glacier bed and its transfer by ice constitute an important question in analysing the subglacial sediment system of alpine glaciers. Glacier erosion needs substantial sliding at the bed; this condition implies a bed at the pressure-melting point. This last expression is in fact misleading because impurities as well as pressure affect the melting temperature of ice. However, its use is so general that we will continue to define a glacier sole as being at the pressure-melting point when the melting temperature of ice is reached. Possible coexistence of the liquid and solid phases of water at this temperature is responsible for a reduction of friction against the bed. Therefore, glacier sliding may occur. Abrasion, plucking and particle displacement may be considered to be a consequence of glacier sliding.

The ice–bedrock interface of Alpine glaciers is in general at the pressure-melting temperature but sometimes ice is frozen to the bed. This is the case at Grubengletscher (Saastal, Swiss Alps), where the basal ice temperature in a tunnel dug for controlling the level of an ice-dammed lake is a few degrees below the melting point as a result of the occurrence of permafrost at this site. In this situation, heat from the refreezing of meltwater at the glacier sole can be removed and a basal ice layer is likely to form. The open thermal system required for its formation also exists if cold air reaches the bed through crevasses or subglacial cavities. Cold ice patches that may act as heat sinks for the refreezing process at the glacial sole can also occur in the basal part of a glacier as a consequence of the heat pump mechanism described by Robin (1976). Independently, another heat sink for the refreezing can be the glacier itself. Indeed, Haeberli (1975) reviewed negative temperature data from glaciers in the Alps. On Mont Blanc, for example, the 15 m temperature is $-16.7\,°C$ at 4785 m a.s.l. and $-7.3\,°C$ at 3960 m and only accumulation areas below 3800 m are temperate (Lliboutry *et al.*, 1976). Grenzgletscher, the mainstream of Gornergletscher near Zermatt in the Swiss Alps, has cold firn in the higher part of the accumulation area. It reappears as bubble-rich ice in the glacier tongue where the temperature of the ice is still below 0 °C (Blatter and Haeberli, 1984).

Thus, for different reasons, thermal conditions at the bed of alpine glaciers are such that ice accretion may be a frequent phenomenon at the glacier sole.

7.2   THE FORMATION OF THE BASAL ICE LAYER

Alpine glaciers do in general exhibit at their base an ice layer up to 1 m thick which is different from glacier ice above. This basal ice layer is characterized by a much larger debris content than glacier ice which is nearly particle-free. Its debris content may reach 10% by volume and consists mainly of dirt or fine sand with pebbles or boulders from place to place. The basal ice layer is generally stratified, and it is composed of a sequence of different types of layers with variable thickness. The following elementary layers constitute the basis for such sequences but are not always present together:

(a) a bubble-poor ice layer about 0.2–3 cm thick which is nearly particle-free;
(b) a debris-rich layer of about the same thickness which is composed of fine mineral particles in an ice cement;
(c) a very bubbly but particle-free ice layer, the thickness of which is very variable from one sequence to another. This third type of layer is generally absent from the central bottom part of alpine glaciers and its formation seems to be related to its position along the marginal part of the glacier.

Bubble trains have been observed in basal ice layers and have led to the recognition of regelation phenomena (Kamb and LaChapelle, 1964). When boulders are present, the different layers forming the basal sequence are often interrupted at their contact and are not deformed around them. In the lee of some boulders, a pocket of crushy ice crystals is sometimes present, indicating the filling by ice accretions of a small cavity. The debris in the layers consists of rock fragments, whose shape indicates the effects of abrasion. Sometimes this basal ice layer in the marginal part of the glacier is contorted and even shows recumbent folding although this is not frequent in the Alpine region. Ice-fabric analyses of basal ice layers are often deceptive since the pattern of the poles of *c*-axis on a Schmidt net is very often not significantly different from the three or four maxima pattern obtained for glacier ice above. This is due to the fact that crystal orientation seems to adjust relatively quickly to the stress field involved and no legacy of previous conditions is retained. On the basis of field observations, several authors have confirmed that the basal ice layer is due to phase changes at the ice–bedrock interface and that bottom freezing is the main process by which rock fragments are incorporated at the sole of the glacier (Boulton, 1970; Clapperton, 1975; Vivian, 1975). Isotopic analyses of the basal ice layer clearly demonstrate that it is due to refreezing of meltwater circulating at the glacier base.

Freezing of water is accompanied by an isotopic fractionation. Therefore, basal ice has been studied for its isotopic composition in order to gain information on processes occurring at the sole of the glacier. Different authors describe oxygen isotope profiles from the bottom parts of glaciers (Gow *et al.*, 1979; Lawson and Kulla, 1978; Hooke and Clausen, 1982). More information can however be obtained if, instead of focusing on a single isotopic ratio, both D/H and $^{18}O/^{16}O$ are considered. Recently, basal ice has been studied in this way by Jouzel and Souchez (1982). The stable isotope fractionation process by freezing has been developed theoretically for closed and open systems and has been experimentally tested (Souchez and Jouzel, 1984). On a graph where $\delta^{18}O$ is the abscissa and δD is the ordinate, two linear trends can be observed. Points representing glacier ice samples lie on a straight line with a slope of approximately 8. This trend is recognized in the literature as an important feature of the isotopic fractionation of water occurring during condensation or sublimation in simple equilibrium processes (Craig *et al.*, 1963; Dansgaard, 1964; Merlivat and Jouzel, 1979). This relationship is generally true for fresh precipitation and is thus valid for glacier ice which has not been submitted to major isotopic changes during and since its formation. $\delta^{18}O$ and δD values for basal ice samples plotted on the same graph do not fall near the line with a slope of 8 but are located on another straight line with a slope of 6.4 for the samples of the basal ice layer from Tsanfleuron glacier in the Swiss Alps and of 4.9 for the samples of the basal ice from Aktineq glacier on Bylot Island (Arctic Canada). This situation appears to be the result of melting without isotopic change and refreezing with isotopic fractionation. The differentiation occurs because the isotopic fractionation coefficients are such that the $\alpha - 1/\beta - 1$ ratio is different from 8, α and β being the equilibrium fractionation coefficients for deuterium and oxygen-18, respectively. Water before refreezing has an isotopic composition given on the δD–$\delta^{18}O$ diagram by the point where the two lines cross; it is called the initial liquid. The slope on which ice samples resulting from refreezing of the initial liquid are aligned depends on the isotopic composition of this initial

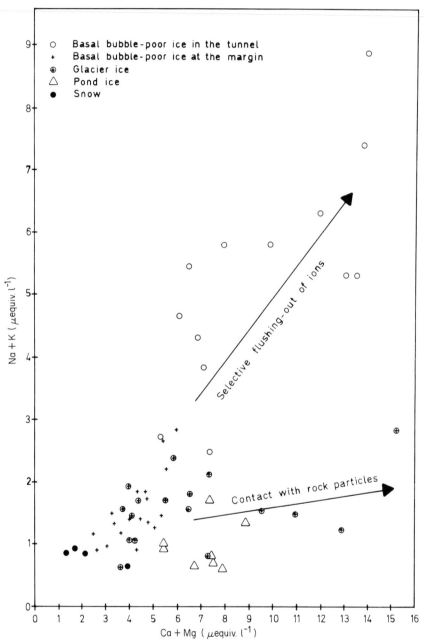

FIGURE 7.1   Relationship between alkali and alkaline-earth metal content of the types of ice studied at the Glacier de Tsidjiore Nouve. *Reproduced by permission of the International Glaciological Society from* Annals of Glaciology, *1981,* **2,** 65

liquid and decreases if its isotopic composition in δD and δ¹⁸O becomes more negative. This is the reason for the slope difference between alpine and polar glaciers. Since such a situation is displayed by other alpine glaciers investigated by members of our laboratory, there seems to be little doubt that the basal ice layer of alpine glaciers is due to refreezing of meltwater circulating at the base. This process incorporates debris within the basal ice layer and this debris is transferred down-glacier by the displacement of the ice.

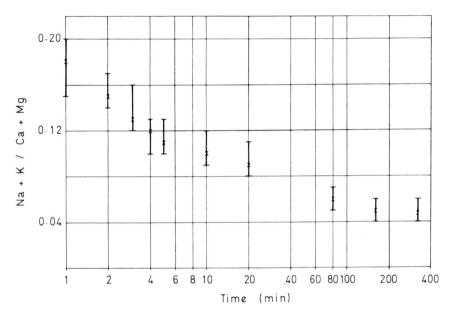

FIGURE 7.2   Ratio of alkali to alkaline-earth metals as a function of time in the experiment (see text). Time in this figure is the time elapsed between the contact of the first drop with the particles to the completed filtering. *Reproduced by permission of Macmillan Journals Limited from* Nature (London), *1978, 273, 456. Copyright © 1978*

This transfer of particles by the basal ice layer is not the only role played by it in the subglacial sediment system. Glacier abrasion is dependent on the presence and concentration of debris at the glacier sole. This abrasion is likely to occur down-glacier from a site of formation of a basal ice layer. It is therefore critical to study the mechanism for the formation of such a basal ice layer to increase our knowledge of abrasional processes by alpine glaciers. The chemical composition of the ice constituting the basal ice layer sheds some light on the origin of water that refreezes at the ice–rock interface. In Figure 7.1 the alkali versus the alkaline earth metal content of different types of ice from the Tsidjiore Nouve glacier is indicated. Each point on the diagram represents one sample. Samples of snow, some samples of glacier ice and most samples of basal ice at the margin of the glacier are located in the lower left part of the diagram. They represent, from a chemical point of view, a single population. Two trends can be observed. If, during ice formation, prolonged contact with rock particles occurs, the $Na + K/Ca + Mg$ ratio is greatly reduced because of a preferential contribution of alkaline earths by the particles. This can be shown by the very low $Na + K/Ca + Mg$ ratio in subglacial flow (0.05–0.09), in supraglacial streams (0.11–0.14), and by the following experiment. If we leave the fine particles ($< 50\mu m$) of morainic deposits from the Tsidjiore Nouve glacier in contact with melted ice from the same glacier at 0 °C for various periods (1–320 min), the $Na + K/Ca + Mg$ ratio decreases with increasing contact time (Figure 7.2), a value of 0.18 being recorded after 1 min of contact and a value of 0.05 after 320 min (Souchez *et al.*, 1978). Some glacier ice and pond ice samples fall in this category, as is indicated by the lower arrow in Figure 7.1. If, on the other hand, selective flushing-out of ions occurs as a consequence of water squeezing, ice formed from refreezing of this water will have a much higher $Na + K/Ca + Mg$ ratio (Souchez *et al.*, 1978). This trend is represented by the upper arrow in Figure 7.1. Basal bubble-poor ice investigated in a subglacial tunnel falls into this category, showing $Na + K/Ca + Mg$ ratios varying from 0.34 to 0.85 with a mean value of 0.58. A fractional melting experiment showed a similar situation (see Section 11.3). Robin (1976) suggested that pressure-melting within the basal ice mass, as distinct from processes at the ice–bedrock interface, may be responsible for the formation of excess water in zones of high-pressure ice upstream of glacier-bed obstacles and that this water is squeezed out of the ice by the pressure. A simple heat pump exists that will tend to cool the basal ice, and this mechanism can produce cold patches at the ice–rock interface.

   The formation of the basal ice layer up-glacier from the tunnel investigated at the Tsidjiore Nouve glacier is related to refreezing of water squeezed out from the ice. Indeed, meltwater circulating at the base has a much lower $Na + K/Ca + Mg$ ratio and is even lower for the ice produced from its refreezing since the divalent ions are preferentially incorporated into the growing ice (Malo and Baker, 1968). The chemical composition of basal bubble-poor ice in the tunnel implies squeezing of water from the ice and limited contact with rock particles. This situation is discussed by Souchez and Tison (1981). Bubble-poor ice layers in the tunnel are interbedded with debris layers. Near the end of the tunnel, this basal sequence makes contact with a rock protuberance on the side wall. The polished bedrock is partly covered with fine silt resulting from crushing. Striae are impressed in the silt and in the rock. One of them can be traced towards the tool which caused the striation, a rock fragment still present in the basal ice layer at the glacier sole. This provides evidence of sliding and present-day grinding. Thus a situation in which bed protuberances exist immediately down-glacier of a site where a debris-rich basal ice layer is formed seems favourable for abrasion. The example described above does not imply that basal ice layers are only formed from water squeezed out from the basal ice mass. Investigations in natural subglacial cavities of the Tsanfleuron glacier show that water flowing slowly on polished bedrock floors freezes in the up-glacier part of the cavities, forming ice layers with variable dirt content. Once incorporated at the base of the glacier by subsequent change in the form of the cavity, this ice gives rise to a basal ice layer (Lemmens *et al.*, 1982).

7.3 ALPINE GLACIER ABRASION

It has long been recognized that glacial abrasion is due to rock fragments forced against and dragged along the glacier bed by sliding ice. Hallet (1979) developed a theoretical model of glacial abrasion that has been widely accepted. The rate of abrasion depends primarily on the effective force with which individual rock fragments are pressed against the bed, the flux of fragments that make contact with the bed and the relative hardness of rocks incorporated in the ice and of the bed. Striations are formed by the motion of rock asperities that locally fracture the rock of the glacier bed. This requires large stress differences between the contact point and areas immediately adjacent to it. Where abrasion occurs, rock fragments are enveloped by ice or water except where they are in contact with the bed. This implies that the basal pressure does not affect the stress differences so that the effective force of contact is independent of basal pressure and, hence, of the glacier thickness. It is only dependent on the buoyant weight of rock particles and on the viscous drag induced by ice flow towards the glacier bed due to basal melting and longitudinal straining of basal ice. The primary control on the contact force and, hence, of the abrasion rate is ice flow towards the bed due to basal melting. In a later paper, Hallet (1981) took account of the significant impeding effect of rock fragment concentration on glacier sliding. An increase in the debris concentration above a certain value can reduce the abrasion rate by decreasing the sliding rate. Therefore, abrasion is probably most rapid for glaciers containing only scattered debris at their base, much as was anticipated by Röthlisberger (1968).

If we exclude debris bands where ice is only a cement, the basal ice layer from alpine glaciers has in general a mean debris concentration of a few percent by volume, reaching 10% in a few cases. It is to be expected, therefore, that large abrasional features will result when the basal ice layer is progressively melted against the bed. The occurrence of such a phenomenon has been demonstrated by an isotopic study of the sole of Tsanfleuron glacier. A laminated carbonate deposit is locally present on polished bedrock floors in natural subglacial cavities which temporarily exist at the margin of this glacier. The deposit is fluted and furrowed parallel to the ice-flow direction. In some places the laminations, 10–100 μm thick, are visible in cross-section. Glacial striae sometimes scour the smoothed bedrock and the deposit and, in places, glacial abrasion is responsible for a smoothing of the carbonate deposit itself (Figure 7.3). To account for these characteristics, it is apparent that during its formation this deposit must

FIGURE 7.3   A subglacial calcite precipitation with flutes and furrows parallel to the former ice-flow direction. The finger points to a striae scouring the bedrock and the deposit itself. *Reproduced by permission of Universitätsverlag Wagner, Innsbrück, from* Zeitschrift für Gletscherkunde und Glazialgeologie, *1982*, **18**, *153*

have been surrounded by ice with only a microscopic sheath of water separating it from the ice. This type of carbonate deposit is explained by Hallet (1976) as being due to calcium carbonate precipitation at the eutectic point in freezing subglacial water on the lee-side of a bedrock protuberance. Samples of the precipitate have been analysed for their isotopic composition: their $\delta^{18}O_{smow}$ values range from 22.42 to 24.62°/oo with a mean of 23.40°/oo. The basal ice layer from the glacier in the subglacial cavities, the ice layers coating the floor of the cavities, glacier ice and different types of meltwater were also sampled and analysed for their isotopic composition in stable isotopes. Calcite precipitation must have occurred in isotopic equilibrium with water since it is due to a concentration increase by freezing. It is thus accompanied by a preferential incorporation of oxygen-18 in the calcite that corresponds to a fractionation factor of 1.0347 for calcite and water at 0 °C. Using this factor and the $\delta^{18}O_{smow}$ for the subglacial calcite, one obtains $\delta^{18}O_{smow} = -10.92$°/oo on average for subglacial water in equilibrium with it. Water in equilibrium with calcite is the residual water that has been enriched by $CaCO_3$ by partial freezing until precipitation occurs. Souchez and Jouzel (1984) have shown that, in the course of freezing, water in equilibrium with the growing ice is progressively impoverished of heavy isotopes. Therefore, the residual water in equilibrium with calcite considered here must have a lower $\delta^{18}O_{smow}$ value than initial subglacial water at the beginning of freezing. Thus initial water or melted ice giving rise to the precipitate by freezing must have a $\delta^{18}O_{smow}$ value higher than $-10.92°/oo$. Such a value can only be produced in this environment by melting of the basal ice layer, all other samples having too negative a value (Lemmens *et al.*, 1982). The calcite precipitates are thus formed by refreezing of meltwater from the basal ice layer. Therefore, evidence exists that this layer is partially melted and that it is responsible for abrasion immediately down-glacier. The role of the basal ice layer as an abrasive ice layer is thus demonstrated by this isotopic study.

Present-day grinding, because rock fragments incorporated into the basal ice layer are brought into contact with the bed by basal melting, is probably of widespread occurrence. However, because of the existence of a subglacial water film, the finest grain sizes are probably ineffective in abrading bedrock. Examination of thin sections shows that subglacial precipitates contain varying amounts of fine rock fragments incorporated during precipitation. It is likely that these fragments have been transported by water flowing to lee surfaces. It is assumed that particles with a diameter smaller than the film thickness can be transported through the film; hence the size distribution of the particles in precipitates offers a mean of estimating the thickness of the subglacial water film. Hallet (1979) showed that the majority of the fragments are smaller than 30 μm and very few exceed 50 μm. Vivian (1975) determined the size distribution of fragments at the ice–rock interface of the Glacier d'Argentière and observed a deficiency of particles smaller than 200 μm. He too indicated that the finest particles had been removed by water flowing through a subglacial film roughly 200 μm thick. Temporal variability of film flow was suggested by Hallet (1979). A multiplying effect has to be expected in abrasion, as indicated by Röthlisberger (1979). Indeed, if one visualizes a stone of the same material as the underlying rock embedded in the ice, then one would expect that the stone would be able to break off one or more pieces of bedrock before becoming completely fragmented. Small rock particles will become detached from the tool and the bedrock alike and finally the grinding effect would be equal on the grains which move and on the grains of the bedrock below. Finally, the importance of abrasion can be judged from the fact that a much larger amount of small particles than coarse ones is carried in glacial streams.

7.4 PLUCKING CONDITIONS AND THE DISPLACEMENT OF BOULDERS

Plucking is usually the result of more than one mechanism. A rock fragment has first to become loose, then it has to be extracted from the bed and finally it has to be entrained by the ice. Recent studies suggest that, in this series of events, water pressure plays an important role at the base of alpine glaciers. Lliboutry (1962) indicated that variable pressure can cause rock fatigue and Robin (1976) suggested that plucking may occur at cold patches when ice freezes to the rock as a result of changes of stress distribution induced by water pressure. Short-term fluctuations of velocity and water pressure were reported by Iken (1978). The same author observed an uplift of the Unteraargletscher surface in spring and early summer beside the enhanced horizontal displacement. The formation of large water-filled cavities at the bed is considered to be the explanation for this phenomenon. Röthlisberger and Iken (1981) developed the concept of the hydraulic-jack mechanism. Rapid transient glacier sliding occurs at the time when large water-filled cavities open up under increased water pressure. A large pressure drop takes place at the apex of bed undulations if lee-side cavities are expanding by a small increase of water pressure. Rock fatigue will thus be more pronounced at the apex. The plucking mechanism is efficient if the rock fragments to be extracted adhere to the ice. This happens just downstream of high pressure areas as a consequence of the heat-pump effect proposed by Robin (1976). Rock fragments which have already become loose may be extracted from the bedrock and lifted out into a cavity. Laboratory experiments have shown that small pressure variations of a few bars are sufficient for rock fragments to be extracted.

FIGURE 7.4   One of the protogine boulders protruding at the ceiling of a subglacial cavity of the Glacier d'Argentière. The boulder is 60 cm long

The nature of the bedrock, for example its jointing, is of great importance in relation to plucking under alpine glaciers. Vivian (1975) cited experiments carried out by Lütschg in Switzerland between 1920 and 1925. Holes were drilled on glacier floors and resurveyed after glacier advance. At the Allalingletscher the rock, a gneiss containing amphiboles, was shattered and debris was exported. A layer of at least 3 cm thickness was removed over a distance of 17 m in 5 years while, at upper Grindelwaldgletscher, the limestone rock was abraded up to 3.9 cm in 4 years with a mean value of 4.5 mm for the fourteen holes drilled. The effectiveness of plucking versus abrasion is lithologically controlled.

Investigations of enclosed subglacial cavities reached by tunnels carved into the rock by hydroelectric companies at the Glacier d'Argentière and at the Mer de Glace have led to the recognition of various debris entrainment mechanisms. In the subglacial cavities of the Glacier d'Argentière, blocks of granite (protogine) up to 1 m long are progressively expelled from the basal ice in which they are incorporated (Figure 7.4) and are displaced toward the lower end of the cavities where they are again entrained down-glacier. This process is connected with stress distribution effects, and ice accretions resulting from freezing of water diffusing around the block are visible between it and the frozen mud layer plastered at the sole of the glacier through which the boulder protrudes. By this process, rock fragments are concentrated at the ice–rock interface. Once surrounded by ice, each boulder moves if the viscous drag due to ice flowing past it, together with the downslope component of the buoyant weight, exceeds

FIGURE 7.5   A typical basal ice layer (BIL) sequence from which a block was removed just before the picture was taken. The scale is 10 cm. Glacier ice is clearly visible above the BIL

the frictional resistance offered by the glacier bed. At the Mer de Glace, measurements made in the excavated tunnels indicated that rounded granite boulders lying on the bed were moving at a mean velocity of 10 mm per day, while the mean horizontal component of ice velocity was 65 mm per day (Charpentier *et al.*, 1972). This process of entrainment is thus so effective that whole blocks may be dragged along. Basal ice layers visible in subglacial cavities sometimes contain incorporated blocks up to 40 cm long (Figure 7.5). The basal sequence of bubble-clear ice layers and dirt layers is interrupted around these blocks and there seems little doubt that they were progressively frozen on to the glacier bottom.

Present-day plucking is occurring under alpine glaciers, particularly in regions of favourable lithology. However, it remains difficult to estimate its importance, although most authors consider that abrasion is much more effective in removing rocks beneath temperate valley glaciers (see general discussion of the symposium on glacier beds: the ice–rock interface held in Ottawa, 15–19 August 1978, *Journal of Glaciology*, 1979, **23**, 381–400).

7.5   ALPINE SUBGLACIAL SEDIMENTS

Surfaces recently exposed by the retreat of alpine glaciers do not show a continuous sediment layer of appreciable thickness lying on the floor. Observations in natural subglacial cavities in the Alps also reveal the same trend: most of the surface is bare bedrock or bedrock that is veneered by a thin layer of basal till. The thickness of this basal till does not generally exceed a few centimetres and only occasionally does it reach a few decimetres. Sedimentological analyses of the basal till gives some insight on its origin. Every lithologic component of basal till has a bimodal particle size distribution if the rock is monomineralic or consists of minerals of similar physical properties. One of the modes is in the clast-size group, the other reflects the mineral fragments in the till matrix. The clast-size mode is relatively larger than that of the matrix mode near the source, where the glacier has picked up the rock fragments. With increasing distance of glacial transport from the source, the matrix modes become larger and the clast-size modes are reduced. The matrix modes are restricted to certain particle size grades which are typical for each mineral. These mineral grades are called terminal grades by Dreimanis and Vagners (1971). The particle size grade produced depends on the original size of the mineral grains in the rock and on the resistance of each mineral to comminution during glacial transport. Terminal grades for quartz range from 0.004 to 0.25 mm, for feldspars from 0.034 to 0.25 mm and for calcite from 0.002 to 0.062 mm. In this last case, solution during and after glacial transport must also be considered as well as the original particle size of the mineral and the effect of crushing or abrasion. Some grain size curves of basal till from alpine glaciers flowing on limestone outcrops or on gneissic rocks exhibit a bimodal distribution. However, in the same environment, other distribution types are also present. On phi-probability paper, a grain-size distribution conforming to a straight line will have a log-normal distribution. This type of distribution for clastic sediments like basal till results from a specific kind of fracturing process that is composed, as suggested by Epstein (1947), of different breakage events. The model developed by Epstein leading to a log-normal distribution has two prerequisites: (a) the probability of fracture for each particle is a constant, independent of the size of the particle and of the number of breakage events that have occurred previously; and (b) the size distribution obtained by a breakage event to a single particle is independent of the initial size of the particle.

Very few grain-size distributions of present-day subglacial tills in the Alps (either in the limestone area or in the gneissic area of western Switzerland) conform to a log-normal distribution. However, if we consider restricted particle size ranges, log-normal distributions are displayed between 4 and 250 μm in the Tsanfleuron glacier (on limestone) and between 4 μm and 1 mm in the Tsidjiore Nouve and Ferpècle glaciers (on gneisses). Kittleman (1964) indicated that clastic materials derived through mechanical disintegration or crushing might be described better in relation to Rosin's law of crushing. Rosin's law was originally derived by Rosin and Rammler (1934) as a mean of analysing artificially crushed material. This function is similar to the log-normal distribution in that it is strongly skewed but becomes more nearly symmetrical upon logarithmic transformation of the size variates. Rosin's law may be expressed as a straight line by twice taking the logarithm of the inverse of the cumulative weight percent and plotting the result as the ordinate and the logarithm of the size as the abscissa. The grain-size distribution of present-day subglacial tills in the area investigated in the Swiss Alps that have a multi-lithologic composition obeys Rosin's law of crushing. Dreimanis and Vagners (1971) suggest that, if several bimodal distribution curves are superimposed one upon the other, the resulting curve becomes more or less straight on Rosin and Rammler paper. If, on the other hand, a monomineralic rock predominates in the till, the cumulative curve does not approach a straight line. If the probability of fracture depends on the size of the particle, then crushing and abrasion will lead to a power function between cumulative weight percent and size. This power function can be transformed into a straight line on a bilogarithmic graph so that the prerequisite can be easily tested. The majority of the grain-size distributions from the present-day subglacial tills examined at our laboratory exhibit a straight line on a bilogarithmic graph up to 35 μm (on gneisses) or 62 μm (on limestones). Thus, the role of size in the crushing and abrasion process seems important for the fine particles.

A comparative study of the grain-size distributions of tills collected in subglacial cavities and of the particles embedded in the basal ice layer above the sole of the glacier fails to distinguish between these two types of sediment.

This implies that particles in the basal ice layer are picked up from the ice–rock interface by the freeze-on process, that further downstream they make contact again with the bedrock by basal melting, and they can be recycled several times.

When the basal till has a thickness of a few decimetres, it can be fluted. These flutings were exposed recently in the proglacial zone of the Tsanfleuron glacier by recent retreat of this glacier. Boulton (1976) considers that the flute is formed by lateral displacement of material behind a boulder sitting on the floor toward a lee-side cavity. The same is true for a bedrock protuberance if a cavity exists down-glacier. This theory is particularly suitable for explaining the small flutings occurring in some basal tills of alpine glaciers.

7.6 THE ROLE OF SUBGLACIAL WATER CHANNELS

Meltwater originating at the base of an alpine glacier flows at the glacier bed in the form of a film which primarily accommodates the local transport of water associated with regelation sliding, but may also carry water melted from the base of the ice by geothermal and frictional heat (Weertman, 1972). However, Walder and Hallet (1979) propose that water produced by geothermal and frictional heat reaches the channel system at the glacier bed. Surface meltwater and rain that find their way within the glacier to the bed are also confined in channels. Two types of channels exist: Röthlisberger channels (R-channels) similar to conduits within the ice except that they are located at the ice–rock interface, and Nye channels (N-channels) carved in the bedrock at the bed (Paterson, 1981). Sometimes large subglacial water-filled cavities open up under increased water pressure. Water in these channels is usually rich in particles in suspension. These particles are picked up by the flowing water from the bed and often represent the product of glacier erosion against the bed. For this reason measurements of sediment concentration in glacial meltwaters in the frontal zone of the glacier are presumed to give a representative idea of the importance of glacier erosion. However, as indicated by Collins (1979), the situation is more complicated. The interaction of basal meltstreams with the products of glacier erosion must be more carefully investigated. Temporal variations of suspended sediment concentration during a period of summer sustained ablation were studied by Collins (1979) on the Gornera, the only meltwater stream draining from Gornergletscher. Every day, the peak discharges were accompanied by minimum concentrations of suspended sediment. Irregularity of sediment concentration suggests rapid injection of sediments into subglacial streams followed by exhaustion, recurring independently of rhythmic variations of diurnal discharge.

Clockwise hysteresis loops occur in suspended sediment concentration–discharge relationships. They result from greater concentration of suspended sediments on the rising limb of the diurnal hydrograph than at equivalent discharges during the falling stage, owing to the flushing out of sediment collected at margins and beds of subglacial streams during low flows. The migration of basal streams during reorganization of drainage channels allows the flowing water to impinge on unworked pockets of fine sediment stored at the ice–rock interface, providing a short-lived supply of suspended load. The development of the channel cross-sections by melting of their ice-walls as a consequence of the higher water pressure during increased discharge has the same result. Sediment concentration in meltwaters not only depends on hydrological conditions but also on sediment supply. Depending on erosion conditions at the ice–rock interface, sediments can be stored subglacially. Their evacuation in meltwater streams is highly variable with time; a large proportion of total annual transport may occur in a period of a few days. This great variation in transport suggests that rate of supply and rate of removal of sediment from storage at the bed are not equal over a period of a few years. This makes the estimation of glacial erosion rates from stream discharge problematic if measurements are limited, as they usually are, over such a time span.

N-channels carved into the subglacial bedrock represent erosion by subglacial streams. R-channels are in constant danger of being closed, either by the overlying weight of ice or by the forward movement of the ice which might pinch off part of them near an obstacle (Nye, 1973). On the other hand, a channel in bedrock, once formed, is much more permanent. Its frequent presence at the bottom of the glacier valley favours its continued use by meltwater. It also represents an area of locally low basal pressure which tends to attract meltwater flow. For these reasons, the R-channel can develop into a subglacial gorge into the upper part of which ice penetrates, but at the base of which a channel is maintained by water circulation. The erosional activity of subglacial waters is mainly related to the coarse fraction of sediment carried in suspension in meltwater streams. Bezinge and Schafer (1968) showed, for example, that, if particles greater than 200 μm in diameter are removed by settling in a lake, the erosion of surfaces in contact with the meltwater is greatly reduced. The load of finer rock flour which does not settle has little erosional effect. The erosional activity of the particles can be visualized as a form of wet sand-blasting. Examination of the grains shows fractures reflecting the violence of transport. Vivian (1975) cited corrasion on a quartzite pavement by the meltwater from the Mer de Glace of 10 cm in 5 years. In more soluble substrata, solution may also play a role but this is beyond the scope of this chapter. Thus meltwaters participate in the subglacial sediment system of alpine glaciers both in evacuating the debris produced by subglacial erosion and thus permitting a renewed action on the bed and in promoting erosional activity mainly by the presence of particles of sand sizes in suspension.

7.7 CONCLUSIONS

Present-day sediment transfer and glacier erosion at the base of alpine glaciers are dependent on the two main processes:

(i) The formation of a basal ice layer by phase changes at the ice–rock interface is likely to incorporate debris into the glacier. Abrasion is dependent on the presence and concentration of debris at the glacier sole. If, because of changing conditions at the interface, the basal ice layer is partially melted, then it may act as an abrasive ice layer. The result is debris comminution and bedrock erosion. The debris formed may itself be later incorporated in the ice, once released by melting, and so on. This is an efficient means of subglacial sediment production and transfer.

(ii) The other main process is the role of meltwater that reaches the bed and is confined to conduits and channels. This water exports debris lying on the bed and, because of the transport of such debris mainly in suspension, is able to 'carve' the glacial bed.

The importance of these two processes is demonstrated here in relation to temperate valley glaciers in a mountainous environment such as present-day Alpine glaciers.

REFERENCES

BEZINGE, A., and SCHAFER, F. (1968), Pompes d'accumulation et eaux glaciaires, *Bulletin technique de Suisse romande*, **20**, 282–290.

BLATTER, H., and HAEBERLI, W. (1984), Modelling temperature distribution in Alpine Glaciers, *Annals of Glaciology*, **5**, 18–22.

BOULTON, G. S. (1970), On the origin and transport of englacial debris in Svalbard glaciers, *Journal of Glaciology*, **9**, 213–229.

BOULTON, G. S. (1976), The origin of glacially fluted surfaces — observations and theory, *Journal of Glaciology*, **17**, 287–309.

CHARPENTIER, G., COLLIOUD, M., and VIVIAN, R. (1972), *Observations glaciologiques sous les glaciers d'Argentière et de la Mer de Glace (Mont Blanc)*, Réunion des 2 et 3 mars 1972 de la Société Hydrotechnique de France, Section de Glaciologie, unpublished, 17pp.

CLAPPERTON, C. M. (1975), The debris content of surging glaciers in Svalbard and Iceland, *Journal of Glaciology*, **14**, 395–406.

COLLINS, D. N. (1979), Sediment concentration in melt waters as an indicator of erosion processes beneath an alpine glacier, *Journal of Glaciology*, **23**, 247–257.

CRAIG, H., GORDON, L. I., and HORIBE, Y. (1963), Isotopic exchange effects in the evaporation of water, *Journal of Geophysical Research*, **68**, 5079–5087.

DANSGAARD, W. (1964), Stable isotopes in precipitation, *Tellus*, **16**, 436–446.

DREIMANIS, A., and VAGNERS, U. J. (1971), Bimodal distribution of rock and mineral fragments in basal till, in Goldthwait, R. E. (Ed.), *Till, a Symposium*, Ohio State University Press, Columbus, OH, pp. 237–250.

EPSTEIN, B. (1947), The mathematical description of certain breakage mechanisms leading to the logarithmico-normal distribution, *Journal of the Franklin Institute*, **244**, 471–477.

GOW, A. J., EPSTEIN, S., and SHEEHY, W. (1979), On the origin of stratified debris in ice cores from the bottom of the Antarctic ice sheet, *Journal of Glaciology*, **23**, 185–192.

HAEBERLI, W. (1975), Eistemperaturen in den Alpen, *Zeitschrift für Gletscherkunde und Glazialgeologie*, **11**, 203–220.

HALLET, B. (1976), Deposits formed by subglacial precipitation of CaCo₃, *Geological Society of America Bulletin*, **87**, 1003–1015.

HALLET, G. (1979), Subglacial regelation water film, *Journal of Glaciology*, **23**, 321–334.

HALLET, B. (1979), A theoretical model of glacier abrasion, *Journal of Glaciology*, **23**, 39–50.

HALLET, B. (1981), Glacial abrasion and sliding: their dependence on the debris concentration in basal ice, *Annals of Glaciology*, **2**, 23–28.

HOOKE, LEB. R., and CLAUSEN, H. B. (1982), Wisconsin and Holocene $\delta^{18}O$ variations, Barnes Ice Cap, Canada, *Geological Society of America Bulletin*, **93**, 784–789.

IKEN, A. (1978), Variations of surface velocities of some alpine glaciers measured at intervals of a few hours. Comparison with arctic glaciers, *Zeitschrift für Gletscherkunde und Glazialgeologie*, **13**, 23–35.

JOUZEL, J., and SOUCHEZ, R. A. (1982), Melting–refreezing at the glacier sole and the isotopic composition of the ice, *Journal of Glaciology*, **28**, 35–42.

KAMB, B. and LACHAPELLE, E. (1964), Direct observation of the mechanism of glacier sliding over bedrock, *Journal of Glaciology*, **5**, 159–172.

KITTLEMAN, L. R., JR. (1964), Application of Rosin's distribution in size–frequency analysis of clastic rocks, *Journal of Sedimentary Petrology*, **34**, 483–502.

LAWSON, D. E., and KULLA, J. B. (1978), An oxygen isotope investigation of the origin of the basal zone of the Matanuska Glacier, Alaska, *Journal of Geology*, **86**, 673–685.

LEMMENS, M., LORRAIN, R., and HAREN, J. (1982), Isotopic composition of ice and subglacially precipitated calcite in an alpine area, *Zeitschrift für Gletscherkunde und Glazialgeologie*, **18**, 151–159.

LLIBOUTRY, L. (1962), L'érosion glaciaire, *International Association of Scientific Hydrology Publication*, 59 (Symposium on Land Erosion, Bari), pp. 219–225.

LLIBOUTRY, L., BRIAT, M., CRESEVEUR, M., and POURCHET, M. (1976), 15 m deep temperatures in the glaciers of Mont Blanc (French Alps), *Journal of Glaciology*, **16**, 197–203.

MALO, B. A., and BAKER, R. A. (1968), Cationic concentration by freezing, in GOULD, R. F. (Ed.), *Trace Inorganics in Water*, Advances in Chemistry Series No. 73, American Chemical Society, Washington, DC, pp. 149–171.

MERLIVAT, L., and JOUZEL, J. (1979), Global climatic interpretation of the deuterium–oxygen-18 relationship for precipitation, *Journal of Geophysical Research*, **84**, 5029–5033.

NYE, J. F. (1973), Water at the bed of a glacier, in *Symposium on the Hydrology of Glaciers (Proceedings of the Cambridge Symposium, 7–13 September 1969)*, International Association of Scientific Hydrology Publication 95, pp. 189–194.

PATERSON, W. S. B. (1981), *The Physics of Glaciers*, 2nd ed., Pergamon Press, Oxford.

ROBIN, G. DE Q. (1976), Is the basal ice of a temperate glacier at the pressure melting point?, *Journal of Glaciology*, **16**, 183–196.

ROSIN, P., and RAMLER, E. (1934), Der Kornzusammensetzung des Mahlgutes im Lichte der Wahrescheinlichkeitslehre, *Kolloid Zeitschrift*, **67**, 16–26.

RÖTHLISBERGER, H. (1968), Erosive processes which are likely to accentuate or reduce the bottom relief of valley glaciers, *International Association of Scientific Hydrology Publication*, 79 Commission of Snow and Ice, General Assembly of Bern, pp. 87–97.

RÖTHLISBERGER, H. (1979), General Discussion of the Symposium on glacier beds: the ice–rock interface, Ottawa, 15–19 August 1978, *Journal of Glaciology*, **23**, 381–400.

RÖTHLISBERGER, H., and IKEN, A. (1981), Plucking as an effect of water-pressure variations at the glacier bed, *Annals of Glaciology*, **2**, 57–62.

SOUCHEZ, R. A., and TISON, J. L. (1981), Basal freezing of squeezed water: its influence on glacier erosion, *Annals of Glaciology*, **2**, 63–66.

SOUCHEZ, R. A., and JOUZEL, J. (1984), On the isotopic composition in δD and $\delta^{18}O$ of water and ice during freezing, *Journal of Glaciology*, **30**, 369–372.

SOUCHEZ, R., LEMMENS, M., LORRAIN, R., and TISON, J. L. (1978), Pressure melting within a glacier indicated by the chemistry of re-gelation ice, *Nature (London)*, **273**, 454–456.

VIVIAN, R. (1975), *Les Glaciers des Alpes Occidentales*, Imprimerie Allier, Grenoble.

WALDER, J., and HALLET, B. (1979), Geometry of former subglacial water channels and cavities, *Journal of Glaciology*, **23**, 335–346.

WEERTMAN, J. (1972), General theory of water flow at the base of a glacier or ice sheet, *Reviews of Geophysics and Space Physics*, **10**, 287–333.

Glacio-fluvial Sediment Transfer
Edited by A. M. Gurnell and M. J. Clark
©1987 John Wiley & Sons Ltd.

Chapter 8

Moraine Sediment Budgets

R. J. SMALL

ABSTRACT

The implications of large Neoglacial dump moraines for rates of sediment transport
and deposition by Alpine glaciers are briefly stated. Studies of sediment output,
based mainly on the dimensions and age of moraines and measurements of
suspended sediment load in proglacial streams, are presented for selected glaciers
in Baffin Island, Norway and Colorado, and the problems of inferring rates of
glacial erosion from such evidence are examined critically. A detailed study of
the sediment budget of the Glacier de Tsidjiore Nouve, Switzerland, is preceded
by a general review of the form, origin and sedimentological characteristics of
lateral moraines. The Tsidjiore Nouve study is based primarily on the inference
of past and present rates of sediment accumulation on massive Neoglacial moraine
embankments, related to a series of minor glacial advances from ca.5000 BP.
Techniques for determining the origin of this sediment (from glacial erosional
processes or supraglacial rock-wall weathering) are presented and employed in the
analysis. In addition, detailed field monitoring of suspended sediment and bed
load transport by the proglacial stream from the Glacier de Tsidjiore Nouve has
been made by I. R. Beecroft for 1981 and 1982. A provisional sediment budget
for the Glacier de Tsidjiore Nouve is calculated; this indicates (i) the considerable
relative importance of supraglacial sediment inputs in the recent past and at present
and (ii) a high overall rate of geomorphological activity, evidenced by an 'erosion
rate' for the glacier catchment as a whole of 1.55–2.29 mm yr^{-1}.

8.1 INTRODUCTION

Moraine ridges and accumulations marginal to glaciers represent the output of
a proportion of the sediment contained within the glacial system. Of particular
importance in this context are the prominent Neoglacial lateral and latero-frontal
dump moraines (Boulton and Eyles, 1979), which were developed during several
periods of limited glacial advance, when the ice surface was slowly rising but
the ice margins did not advance sufficiently to override and destroy the moraines,
following the post-glacial 'climatic optimum.' The volume of sediment preserved

in these moraines provides some measure of the effectiveness of the glacier in terms of glacial erosion (*sensu lato*) and transportation over the past 4000–5000 years. If the age and volume of these moraines can be determined, an annual rate of sediment increment can be easily calculated. It may then be possible to identify the sources of the debris (whether supraglacial or subglacial), and to assess the relative importance of these sources as contributions of sediment to the glacial system as a whole. More specifically, the rates of weathering of exposed rock faces within the glacier catchment and the rates of erosion by plucking and abrasion beneath the glacier can be established and compared.

However, several complications inevitably arise in the study of glacier sediment budgets; these stem principally from the varied nature of the sediment inputs to, and outputs from, the glacial system (Figure 8.1). The former consist of (a) the products of supraglacial weathering, rock collapse and avalanches, (b) dust blown or washed by rain on to the ice surface, (c) fluviatile sediments

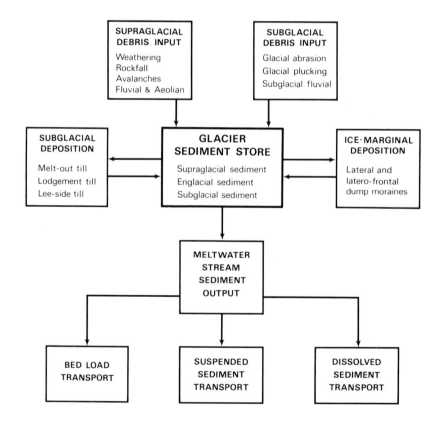

FIGURE 8.1   Components of the glacial sediment budget

washed in from adjacent slopes, (d) the products of glacial erosion of subglacial rock surfaces and (e) 'cannibalized' sediments derived from previously deposited glacial sediments, such as lateral moraines and basal deposits which are overridden in times of glacial advance. The outputs consist of (a) sediments which have accumulated (sometimes only temporarily) on marginal moraine ridges and (b) sediments which are transported from beneath, within and on the glacier by meltwater streams, in the form of suspended sediment load, bed load and solution load. In attempts to quantify the sediment budget of individual glaciers, it is the outputs rather than the inputs which are, for purely practical reasons, determined, either indirectly from previously formed sediments (for example, moraines of known age) or directly as they occur (for example, the day-to-day monitoring of the sediment loads of proglacial streams).

8.2 STUDIES OF THE OUTPUTS OF SEDIMENT FROM GLACIERS

A relatively simple study of the inferences to be drawn from moraines is that of the Akudlermuit glacier in East Baffin Island (Andrews, 1972). In front of this small cirque glacier is an area of ice-cored moraines approximately $400 \times 300 \, \text{m}^3$ in extent. The moraine cover of the ice-core is 0.5–1 m in thickness; with an assumed void space of 20%, the total volume of the debris is approximately 100 000 m, which, since the area of the glacier is $1 \times 10^6 \, \text{m}^2$, represents an average lowering of the cirque floor by 0.1 m over the period of moraine accumulation. Since the moraines are of Neoglacial age, representing about 4000 years of glacial activity, a mean annual erosion rate of $0.25 \, \text{mm yr}^{-1}$ is implied. To gain an idea of the true erosion rate, this needs to be increased (in the view of Andrews to $0.5 \, \text{mm yr}^{-1}$) to take account of unrecorded losses due to solution and suspended sediment loads removed by meltwater.

A more comprehensive study is that of a small cirque glacier at Kråkenes, western Norway, by Larsen and Mangerud (1981). Three types of associated sediment accumulation may be identified here: an *end-moraine* consisting mainly of glacially eroded material, plus a relatively small amount of debris weathered from the supraglacial headwall, and with a corresponding bed-rock volume (allowing for voids) of 94 000 m³; *fluvially transported sediments* on the floor of a lake beyond the glacier snout (corresponding bed-rock volume 12 600 m³); and a *deltaic formation* formed at the entrance of the proglacial stream into this lake (corresponding bed-rock volume 11 300 m³). Sediments derived from the glacier passing this lake, or escaping as dissolved load, were assumed to be negligible. Larsen and Mangerud propose, on the basis of radiocarbon dating of silts above and below the glaciolacustrine deposits, that the sediments as a whole represent the total output from the cirque glacier over a period of 700 years in the Younger Dryas period (10 900–10 200 BP). From planimetric measurements of the 'real' areas of the base of the former glacier and the supraglacial head-wall, a subglacial erosion rate of 0.6 mm yr^{-1} and a head-wall retreat rate of 0.1 mm yr^{-1} can be calculated. On the basis of these figures, Larsen and Mangerud suggest that a total period of erosion of from 83 000–125 000 years would be required for the development of the cirque to its present-day dimensions.

The problem of quantifying glacial sediment budgets is arguably greater when current rates of sediment output are being determined. Leaving aside the purely technical difficulties of measuring active sediment transport and deposition, there is the virtual impossibility of demonstrating conclusively that the recorded rates for individual years are representative of a longer period of time, or indeed reflect present-day rates of glacial erosion. The question of sediment storage is particularly difficult. Is the suspended sediment load within proglacial streams, and being washed out from beneath the glacier, the product of current abrasion and attrition, or is it at least in part released from a subglacial store which has accumulated over a period of years? The problem is exacerbated at the present time, when an increasing number of Alpine glaciers are advancing over previously laid-down proglacial sediments, which are now being 'eroded' by subglacial streams. Many would argue that it is clearly simplistic (a) to infer rates of erosion from relatively short-term measurements of sediment discharge or (b) to assume that the products of subglacial erosion will be removed in their entirety in the year of their formation—or even that an equivalent volume of sediment from the glacier store will be washed out within the year. As Collins (1979) pointed out, 'while the addition of sediment to subglacial storage depends on erosion conditions at the ice–sediment–rock interface, and occurs over large areas of the glacier bed, evacuation of sediment in meltwater streams depends on the interaction of temporally and spatially restricted hydrological conditions. Subglacial channels may migrate across the bed but the length of time necessary for access to all areas of the bed will be great. Observations over short periods (even if close-interval sampling is undertaken) will be unrepresentative of average conditions, since a large proportion of total annual sediment transport can occur in a period of 1 or 2 [days], which may or may not be in the observation period.' It may not be enough that the period of measurement of suspended sediment transport be extended to cover the whole ablation season, since 'great variations in transport between years suggests also that rate of supply and rate of removal of sediment from storage are not equal over two- or three-year periods.'

Nevertheless, despite these cautionary remarks it is possible that current glacial erosion rates may be estimated with a reasonable degree of accuracy, provided that observations are carried out over a sufficiently long period. Of particular value in this respect are the studies of selected Norwegian glaciers between 1967 and 1980 (Kjeldsen, 1981). For each glacier the sediment yield, via the proglacial stream, was measured over summer periods of up to 4 months. In the case of Nigardsbreen both suspended sediment load and the increment of coarse material (the result of bed-load transport) on developing deltas in front of the glacier

TABLE 8.1 Sediment yield at Norwegian glaciers (after O. Kjeldsen)

	1968	1969	1970	1971	1972	1973	1974	1975	1976	1977	1978	1979	1980
A. NIGARDSBREEN:													
(i) Observation period (days)	76	95	110	132	101	115	104	111	114	106	107	106	104
(ii) Estimated annual transport (tonnes)	10800	21800	31800	14300	31800	31100	13600	28000	21800	10700	13100	32400	17100
(iii) Sediment yield (tonnes km^{-2})	225	455	665	300	665	650	280	585	455	225	275	675	355
(iv) Erosion rate (mm yr^{-1})	0.08	0.17	0.25	0.11	0.25	0.24	0.10	0.22	0.17	0.08	0.10	0.25	0.13
B. ENGABREEN:													
(i) Observation period (days)	—	57	85	95	113	112	98	103	110	100	111	106	106
(ii) Estimated annual transport (tonnes)	—	15800	22500	28000	35000	19500	22500	22500	19700	20800	18300	19500	24100
(iii) Sediment yield (tonnes km^{-2})	—	415	590	735	920	515	590	660	520	550	480	515	635
(iv) Erosion rate (mm yr^{-1})	—	0.15	0.22	0.27	0.34	0.19	0.22	0.24	0.19	0.20	0.18	0.19	0.23

were recorded. From the total annual sediment transport both the yield (in tonnes km^{-2}) and the corresponding glacial erosion rate (in $mm\,yr^{-1}$) were then calculated. The results of a 13-year study of Nigardsbreen are shown in Table 8.1. Over this period the mean erosion rate was $0.16\,mm\,yr^{-1}$ with rates for individual years ranging between 0.08 and $0.25\,mm\,yr^{-1}$. Comparable figures for a 12-year study of Engabreen were $0.22\,mm\,yr^{-1}$ with a range from 0.15 to $0.34\,mm\,yr^{-1}$. Clearly, erosion rates calculated in this way need to be adjusted upwards where dumping of glacial sediments on moraine ridges occurs. However, if realistic gross erosion rates, in the strict sense of the term, are to be established, the problem of the provenance of moraine sediments needs to be solved.

A simple interpretation of the sediments produced by active glaciers might rest on the assumptions that (a) the finer products (mainly in the sand–silt fraction) are the product of glacial abrasion of subglacial rock surfaces, plus the related process of comminution of the clasts used in abrasion, and (b) the coarser fragments (washed out as bed-load by meltwater streams or deposited on moraine ridges) result from the group of processes (joint-block fracturing, crushing, discontinuous rock-mass failure) conveniently referred to as plucking. However, this ignores not only processes such as subglacial fluvial erosion but also, more importantly, the proportion of the glacier sediment load derived from supraglacial rock projections, valley head-walls and marginal rock outcrops. This debris of supraglacial origin is either transported on the ice surface or becomes englacial (p. 113); some fragments may actually penetrate to the glacier bed, where they will assist the process of glacial abrasion. Much supraglacial debris, particularly that weathered from valley sides and occupying supraglacial lateral moraines, is deposited ultimately on latero-frontal dump moraines, together with (possibly) material produced by glacial erosion *sensu stricto*. The question of the relative contributions from supraglacial and subglacial sources was briefly discussed by Boulton (1978), who in general considered the latter to be overwhelmingly dominant. However, as stated on p. 200, there are reasons to believe that, under present-day conditions in the Alps, the balance may have been shifted significantly towards the former.

Techniques for differentiating debris of subglacial and supraglacial origin have undergone some development during the past decade. In their analysis of the end moraine at Kråkenes (p. 167), Larsen and Mangerud (1981) demonstrated from a roundness analysis of included fragments that only 9% of the fraction between 20 and 80 mm in size consists of angular material. They suggest that whilst some of this debris may have been derived from rock falls on the cirque headwall, with the resulting fragments being transported supraglacially and englacially to the end moraine, 'angular fragments are also produced subglacially;' thus 'we estimate that 5% of the volume of the end moraine is derived from rock-fall.' However, this interpretation does not take into account (a) that some supraglacial debris is liable to have fallen into the head-wall gap of the glacier (McCall, 1960) to become part of the traction load, in which situation it would be subject to rounding processes, and (b) that angular debris released subglacially would immediately begin to experience rounding in basal transport. For example, Goldthwait (1971) states that rounding of stones will increase from 0.1 to 0.5 (on the Powers scale) in only 1.6 km of subglacial transport. It is probable that the 5% figure adopted by Larsen and Mangerud is an underestimate rather than an overestimate of the supraglacial contribution.

A more detailed analysis of moraine sediments is that undertaken by Reheis (1975) for the Arapaho Glacier, a small cirque glacier in the Colorado Front Range. The debris transported by this glacier is derived from three sources: rock fall from the cirque walls, rock debris contained within avalanches and glacial quarrying and abrasion of bedrock (in effect, the first two are indistinguishable in sedimentological analysis, and are therefore treated together). Roundness, presence of striations and degree of polishing are the three associated characteristics likely to prove useful in the differentiation of subglacially derived and transported stones from rock fall and avalanche debris transported supraglacially and/or englacially. Of these, roundness was taken by Reheis to be particularly significant. 'A stone is considered round if its roundness is greater than 0.1. It is assumed that debris which shows even a slight degree of rounding has been subglacially transported at some time during its history, while completely angular debris was derived from rock fall and has undergone only englacial, supraglacial, rock fall or avalanche transport.' Reheis does not therefore consider that 'slight' rounding may result from weathering of individual fragments subsequent to their deposition on moraines.

In the analysis of the Arapaho moraines, Reheis distinguishes between 'present' transport and deposition (represented by a recent ablation moraine formed over the last 30–50 years) and 'past' transport and deposition (represented by moraines formed over a 300-year period). Present deposition amounts *in toto* to 470 m^3 (or 9.4–15.7 m^3 yr^{-1}); of this, rounded debris constitutes 35% and unrounded debris 65%. However, since the ablation moraine lies only half-way between the head-wall and the main terminal moraine, and rounding increases with distance of subglacial transport, Reheis infers a present production of '70% subglacially transported debris and 30% rock-fall debris carried englacially or supraglacially.' Past deposition amounts *in toto* to 728 000–1 214 000 m^3 (or 2430–4050 m^3 yr^{-1}); of this, rounded debris constitutes 88%, with the implication that the production rate for 'subglacial' debris has been 2140–3560 m^3 yr^{-1} and for 'supraglacial' debris 290–485 m^3 yr^{-1}. The large discrepancy between present and past rates of production can be explained in a number of ways (for example, a real acceleration of rock-fall rates and glacial erosion during the period of maximum extension of the Arapaho Glacier, or a serious underestimation of the age of the terminal moraine, which may be 1200 rather than 300 years old). Reheis is of the opinion that, taking all the possibilities into account, the Arapaho Glacier was in geomorphological terms five times more active at times of maximum extent than at the present day (past and present denudation rates are estimated to be 1.26–2.04 and 0.14–0.24 mm yr^{-1}, respectively).

In a study of the erosion rates of ten sub-polar cirque glaciers in the Pangnirtung Fiord area of Baffin Island, Anderson (1978), employing a methodology basically similar to that of Reheis, calculated the volume of moraines of Neoglacial age (mainly formed since 4000 BP, but possibly including sediments incorporated from earlier Neoglacial moraines). Angularity of contained clasts was the sole criterion used to identify source. Using a sampling procedure, clasts were classified as *angular* (all edges sharp), *subangular* (one third of edges smooth and rounded), *subrounded* (two thirds of edges smooth and rounded) or *rounded* (all edges smooth and rounded). Anderson followed Reheis in accepting that subangular, subrounded and rounded clasts are of subglacial origin, 'as any amount of clast rounding should indicate subglacial transport (and hence erosion),' and that angular clasts are derived from rock falls. At Pangnirtung the proportion of subglacially derived cobble- and larger-sized debris ranges from 84 to 96%, and supraglacial debris from 4 to 16% — a result in line with that from Arapaho Glacier.

Nevertheless, rates of debris production are very much lower at Pangnirtung. The most productive glacier (designated D127) deposited subglacial material at a *maximum* rate of 96–133 m^3 yr^{-1} (assuming the shorter 4000 year period of moraine formation), in comparison with 2140–3560 m^3 yr^{-1} at Arapaho. Moreover, rates of rock-fall debris release in the Pangnirtung cirques as a whole ranged from 1 to 22 m^3 yr^{-1}, in contrast with 290–485 m^3 yr^{-1} at Arapaho. These results appear to contradict strongly the suggestion by Boulton (1970) that sub-polar and polar glaciers produce subglacial debris at a greater rate than temperate glaciers, and also the view (Boulton, 1972) that most of the debris cover on ice-cored moraines is of supraglacial rather than subglacial origin.

In addition, Anderson undertook grain-size analyses of sediments from the crests of five moraines at Pangnirtung. This revealed a very low content of silt (average 4.6%) and an even smaller amount of clay (average 1.2%). These figures again contrast strikingly with those derived from the Arapaho moraines (19–32% and 4–10% respectively) and other Colorado Front Range sediments. Although lithology may be a contributory factor, glacier thermal regime may be primarily responsible. Anderson argued that temperate glaciers (as in Colorado) flow partly by basal sliding, have relatively high velocities and are more effective agents of abrasion, whereas sub-polar glaciers (as in Baffin) are 'wholly or partly frozen to their beds and, hence, should have a very low potential for abrasion.' The latter may, however, be more powerful agents of plucking — hence the large proportion of coarse, at least partially rounded subglacial debris and the low silt and clay content of the Pangnirtung moraines.

In terms of overall erosion rates there also appears to be a contrast between sub-polar and temperate cirque glaciers. Erosion rates at Pangnirtung were calculated by Anderson (1978) to range between 0.008 and 0.076 mm yr^{-1} [in comparison with calculations by Andrews (1971) for the same area of 0.025–0.09 mm yr^{-1}, and by Reheis (1975) for the recent past at Arapaho as 1.26–2.04 mm yr^{-1}]. In other words, polar or sub-polar glaciers such as those on Baffin Island have produced only 10–50% of the debris, at a rate one or two orders of magnitude lower than temperate glaciers such as those of the Colorado Front Range.

In recent years a field study of the sediment output from the Glacier de Tsidjiore Nouve, Switzerland, has been initiated by the author (Small, 1983; Small *et al.*, 1984). This is based mainly on an analysis of the massive lateral moraine embankments of the glacier, together with monitoring by I. R. Beecroft of sediment transport by the proglacial meltwater stream. Some consideration of the characteristics and formation of lateral and latero-frontal moraines is a necessary prelude to a more detailed discussion of the Tsidjiore Nouve study.

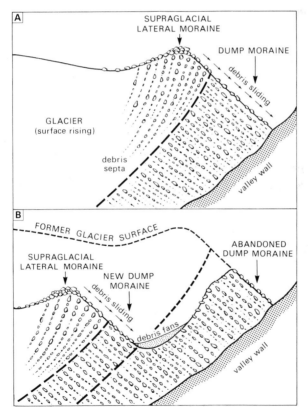

FIGURE 8.2   Formation of lateral dump moraines. *Reproduced with permission from Small (1983)*

8.3   THE CHARACTERISTICS AND FORMATION
OF LATERAL MORAINES

Lateral moraines were defined by Flint (1971) as end-moraines built along the lateral margins of glaciers occupying valleys. The implication is that the moraines consist of debris which has slid off the ice surface into the ablation valley between the glacier and valley wall. This is supported by Boulton and Eyles (1979) who referred to 'lateral and latero-frontal dump moraines from supraglacially derived debris.' A clear distinction needs to be drawn between such dump moraines and supraglacial lateral moraines, which consist of debris in transit on the ice surface. Such supraglacial moraines consist of ice-cored ridges, often transected by marginal crevasses and reaching heights of 10 m or more as a result of the differential ablation of debris-covered ice and bare ice. The debris cover of the moraines results from (a) direct rock fall on to the glacier margins and (b) melting out of sediment contained within the bed-parallel debris septum (p. 116) Figure 8.2 shows the probable relationships between supraglacial and dump moraines. The latter are progressively built up, largely as a result of successive increments of debris from a rising glacier surface during a period of glacial advance. Subsequently, the moraines may be abandoned during a phase of glacial recession. Although the moraine distal slope remains intact, the exposed proximal face may be severely eroded, with the result that the upper moraine face is commonly deeply gullied, and the lower face masked by coalescent debris fans.

Lateral dump moraines commonly display pronounced till fabrics, and in particular may be marked by a crude stratification of slabby boulders, in a matrix of finer sediments, dipping towards the valley wall at 10–40°. Lateral moraine fabric was measured by Galloway (1956) in Lyngsdalen, Norway, where the disposition of stones with long axes transverse to the directions of lateral moraine crests was attributed to the action of laterally spreading ice. Mills (1977) also found, on a lateral moraine of Athabasca Glacier, Canada, a strong fabric mode on the distal flank, transverse to the moraine crest and dipping away from the glacier. This was assumed to result from the lateral sliding of debris from a supraglacial situation, and the associated build up of a bank of debris along the glacier margin, as depicted in Figure 8.2. Mills also demonstrated the presence, on the proximal flank of the moraine ridge, of two weak nodes transverse to the crest, and one weak node parallel to the crest and orientated in the direction of glacier flow.

In a study of large lateral moraines marginal to Bethartoli Glacier, Garhwal Himalaya, Osborn (1978) drew attention to a prominent till fabric, consisting of disc- and blade-shaped stones whose *a*- and *b*-axes define a plane dipping away from the glacier at an angle of 30–35°; this plane also parallels the distal flank of the ridge. Osborn inferred that 'each of the oriented stones was accreted on the distal flank of the moraine, or, in other words, the moraine grew mainly by accretion of debris on the distal flank. The accreted material was probably mostly superglacial debris originally; the material may have slid down the sloping margin of the glacier and on to the moraine flank.' However, Osborn added that 'some of the material may have been subglacial debris, lodged between ice and already existing moraine at the extreme margin of the glacier, which, when exposed by ablation, slid down the distal flank.' Osborn found no fabric on the moraine proximal face represented by *a/b* planes parallel to the ice–moraine interface and reflecting coarse debris being sheared between ice and moraine. 'If till with that fabric ever existed, it has been eroded off.' In fact, field evidence suggests that, since ice retreat, proximal face erosion has been minimal, so that the moraine, even before erosion, consisted almost entirely of distal flank debris.

'Layered lateral moraines,' similar to those of the Bethartoli Glacier, were reported from the Alps by Humlum (1978), for example, adjacent to the Jamtal Ferner, Guslarferner and Waxeggkees glaciers in Austria. These possess a conspicuous layered structure, consisting of 'zones of alternating high and low content of large blocks.' Where exposed on the proximal slope, the large blocks are orientated with their *a/b* plane parallel to the moraine distal slope. Humlum stated that 'moraine-building periods must have been periods of large glacier volume, as the glacier surface must have been at least as elevated as the moraine ridges to make the dumping mechanism possible.'

Another important feature of the Bethartoli moraines is the asymmetry of their development (Osborn, 1978). On the eastern side of the valley bottom there is a single large moraine ridge, whilst on the western side there are thirteen smaller ridges, each of which stands as high as or higher than its adjacent outer neighbour. There is thus evidence of moraine development during several successive readvances of the glacier, each of which led to the formation of a new debris ridge on the western side (thus giving rise to a series of nested moraines) and to the addition of a new layer of debris on the superimposed moraine on the eastern side. Osborn proposed that this pattern of moraines reflects a shift in the glacier mid-line to the east each time an advance occurred, although the reasons for this are not wholly clear. Similar 'within-valley asymmetry' of lateral moraines has been noted elsewhere, for example at Findelengletscher in the Swiss Alps (Schneebeli, 1976) and in the Jotunheim of Norway (Matthews and Petch, 1982), where the volume of material comprising the north-facing lateral moraine at Hurrbreen is at least three times that in the south-facing lateral. In the latter instance (where the asymmetry is of volume rather than morphology), the larger moraine is the product either of greater supraglacial and subglacial debris supply from north-facing head-walls in the glacier catchment, or of the pushing forward of available valley-side regoliths (till and scree) by the advancing glacier margin.

8.3.1 The sources of lateral moraine sediments

The form and structure of lateral dump moraines both point to the important contribution to development made by the sliding of supraglacial debris from the glacier flanks. Humlum (1978) argued that layered moraines reflect the overwhelming dominance of supraglacial transport (as opposed to both englacial and subglacial transport), and suggested that the moraines are composed mainly of material derived from rock walls above the glacier. Each block-enriched layer is seen as developing 'during a period with frequent rock falls on the glacier surface ... above the equilibrium line.' During these periods, the glacier probably transported exceptionally large amounts of supraglacial material which was subsequently dumped along the glacier margins. The more fine-grained intervening layers may comprise a mixture of material derived partly from rock-walls above the glacier, and 'partly from the glacier itself.'

From a study of clast shapes (involving a six-fold classification into 'very angular,' 'angular,' 'subangular,' 'subrounded,' 'rounded,' and 'well rounded'), Matthews and Petch (1982) demonstrated a significant 'decrease in roundness' (or, more realistically, an increase in angularity) with increasing distance from the former glacier snout along the crests of lateral moraines (where the fragments are predominantly angular or very angular). However, it needs to be emphasized that even at 'low-altitude sites' (on the latero-terminal or terminal moraines) the clasts tend to remain subangular, rather than subrounded, rounded or very rounded.

Matthews and Petch discussed the possibility that the bulk of the lateral moraine sediments are of supraglacial origin (rock fall in the accumulation zone, followed by supraglacial and englacial transport). The large volumes of many moraines could perhaps have resulted from intensified supraglacial rock-wall recession during the Little Ice Age climatic deterioration. However, the tendency towards down-moraine increase in roundness remains a difficulty. Progressive modification of clast shape by microgelivation can probably be dismissed on the grounds that such weathering should *increase*, not decrease, with altitude. A more realistic hypothesis is that subglacial clasts (partially rounded by basal transport) are deposited in increasing numbers towards the glacier snout, as flow-lines within the ice 'turn upwards' and carry basal sediment towards the surface. However, the present-day basal layer in many Jotunheim glaciers is weakly developed, and it is therefore difficult to envisage that the contribution of subglacial debris to the formation of these large lateral moraines (which are in any case dominated by debris with 'supraglacial' characteristics) can have been large. A final possibility is the incorporation of clasts from older till deposits by glacier advance in the latero-frontal zone.

The hypothesis of the construction of lateral moraine embankments by successive increments of supraglacial debris, under conditions of a rising glacier surface, was accepted by Boulton and Eyles (1979). However, they emphasized the role of englacial transport via the bed-parallel debris septum, consisting of debris which is 'derived supraglacially from flanking valley walls, and is entrained along foliation planes parallel to the glacier bed and for some distance above it.' As grain contacts within the septum are rare, little comminution can occur, so that the debris retains its original 'supraglacial' characteristics. Release of this marginal englacial sediment in increasing quantities in the glacier ablation zone will contribute to the formation of supraglacial ice-cored moraines and, ultimately, the growth of dump moraines. However, Boulton and Eyles also postulated (a) the addition of lodgement till, from the basal debris layer, on the lower proximal faces of lateral dump moraines, and (b) the incorporation within lateral moraines of finer sediments ('the voids within these open-textured bouldery deposits are slowly filled by finer material which is washed in or flows in').

A much more significant contribution of subglacial sediment to lateral moraines was envisaged by Shaw (1980). This results from the development of a transverse flow component at the bed of a glacier, which causes basal ice and associated debris (often containing faceted and striated clasts) to rise upwards and even over the summits of previously formed lateral moraine ridges. 'Lateral moraines have been attributed generally to the accumulation of supraglacial debris by gravitational sliding, and accumulation by basal transport of debris to the moraine crests which had not been previously described represents a distinctive genetic moraine type.' Such Röthlisberger–Schneebeli moraines (so-called by Shaw from the research carried out by these authors on lateral moraines in Valais, Switzerland, including those at the Glacier de Tsidjiore Nouve) are, in effect, 'half drumlins or flutings' (see also the discussion of medial moraine formation on p. 140).

8.4   THE LATERAL MORAINE EMBANKMENTS OF THE GLACIER DE TSIDJIORE NOUVE

To the north and south of the Glacier de Tsidjiore Nouve the lateral dump moraines constitute massive debris embankments, up to 60 m in height on their

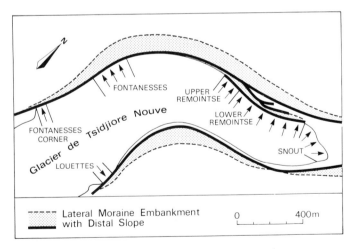

FIGURE 8.3   Sites of current dumping of sediment, lateral moraines of the Glacier de Tsidjiore Nouve. *Reproduced with permission from Small, Beecroft and Stirling (1984)*

distal flanks and confining the glacier tongue like giant levées. In large measure they are fossil features, dating from the Little Ice Age (ca.1500–1850) and earlier glacial advances of the Post-Glacial period. Röthlisberger (1976) and Röthlisberger and Schneebeli (1979) have demonstrated from the evidence of fossil soils and buried larch trunks that moraine walls in this area of Valais were 'built up at least by the advances of the last 3500 years' and probably

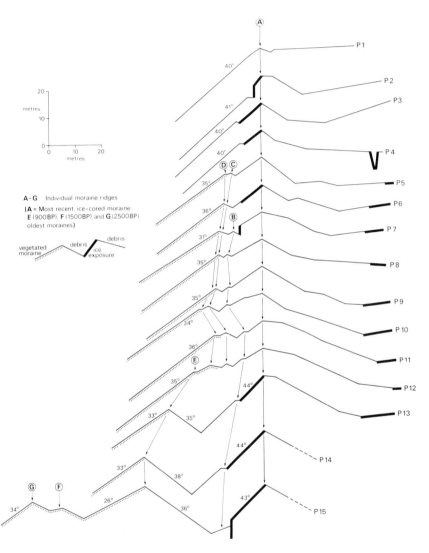

FIGURE 8.4   Cross profiles, northern lateral moraine embankments, Glacier de Tsidjiore Nouve. *Reproduced with permission from Small (1983)*

'during the last 9000 years in some instances.' However, in places (notably along a 400 m stretch of the northern moraine) the moraine summits are being actively surmounted by glacier ice, and supraglacial debris is being shed on to the distal moraine faces (Figure 8.3). This provides a rare example of dump moraine formation in action, and an opportunity to study the processes that, in the Neoglacial period, contributed to the development of the Tsidjiore Nouve moraines.

8.4.1 The southern moraine

This remains largely a fossil feature, although it is overtopped at its head over a distance of approximately 30 m. In the field the moraine is revealed as a composite feature. In its middle section the main ridge is surmounted by a smaller, partly vegetated ridge, forming a superimposed moraine. On the proximal side there is a small unvegetated ridge (coincident with the line of shear along the glacier margin) and a much larger ice-cored supraglacial moraine ridge (which has a counterpart along the northern glacier edge). These moraines combine with the main superimposed ridge to give a weakly developed sequence of nested moraines, successively younger in the direction of the glacier.

8.4.2 The northern moraine

This is longer than the southern moraine, and morphologically more complex, consisting of at least seven superimposed and nested moraine ridges. Figure 8.4 shows a series of surveyed cross-profiles from the head of the embankment (P1) to its terminus near the glacier snout (P15). In its upper part the moraine includes two main elements: the main debris ridge, with distal slopes of 31–41°, and the overtopping glacier margin (part of supraglacial lateral moraine A, with numerous ice exposures on the distal face, as on profiles P2, P3, P4, P5 and P7). In its middle section the moraine consists of several small superimposed and nested ridges (note that moraines C and D are separate on profiles P5 and P10, but that C obscures the underlying D on profiles P6–P9). Moraine B, as a small ridge at the base of the distal face of the supraglacial lateral moraine A, coincides with the shear line along the glacier margin but is at least partly composed of debris which has slid from the summit of A. In the lower section of the lateral moraine a series of large and well defined nested ridges is formed (see E, F and G on profile P15). Moraines C and D are replaced here by a major gully between ridge E and the glacier edge. Profiles P13–P15 show the continued existence of moraine B, which is now a more substantial debris accumulation being overridden by the glacier but receiving large increments of debris which slides over the bare ice margin from the glacier surface.

On *a priori* grounds it seems safe to assume that the moraines G–A are progressively younger. Röthlisberger and Schneebeli (1979) give radiocarbon dates as follows: G, 2500 BP; F, 1500 BP; and E, 900 BP. Moraines D and C are possibly of Little Ice Age date. Moraines B and A are undergoing development at the present time.

8.4.3   Processes of lateral moraine formation

The current phase of moraine development appears to have been initiated in the late 1960s, as the Glacier de Tsidjiore Nouve began to experience what has proved to be a substantial advance lasting through the 1970s into the 1980s. The overtopping of the northern lateral moraine was first noted in 1970 (as an area of bare ice 10 m long and a maximum of 5 m down-slope) and reported in 1973 (Whalley, 1973) as a singular phenomenon ('I have been unable to find a record of a similar occurrence, either on this glacier or on others'). Since 1973

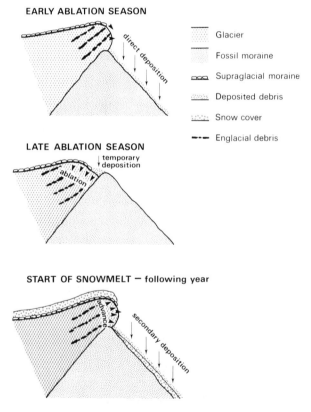

FIGURE 8.5   Annual rhythm of deposition, northern lateral moraine embankments, Glacier de Tsidjiore Nouve. *Reproduced with permission from Small, Beecroft and Stirling (1984)*

exposures of ice above the moraine have become more numerous and extensive, so that by 1984 approximately two thirds of the northern moraine was experiencing active development. The debris being dumped on to the moraine appears to be derived largely from a cover of supraglacial material, about 0.2–0.8 m thick, on the sharp-crested supraglacial moraine A. Additionally there is some release of debris, from marginal englacial debris bands, either to the supraglacial cover or directly, as the exposed ice margins melt back, on to the lateral dump moraine. No clear evidence has been found of the deposition of subglacial debris on or close to the moraine crests.

Observation has shown that there is a simple annual rhythm to the depositional process (Figure 8.5). Each summer the ice marginal faces, despite their forward motion over the moraine crest, melt back by up to 5 m. As a result undermined supraglacial debris plus melted-out englacial sediments slide on to the crest or proximal face of the embankment, building up a temporary debris ridge or a series of debris fans. In winter, with the cessation of ablation, the ice advances on to or even beyond the embankment summits, bulldozing the accumulated sediments on to the distal face. Thereafter, until ablation has again caused ice-face recession, debris is shed directly from the glacier on to the moraine outer slopes. Under present-day conditions of dumping the processes leading to the formation of a moraine fabric cannot be observed. Owing to the height (60 m) and steepness (ca. 40°) of the distal face, many large boulders released by the glacier gain considerable momentum, and roll and bounce to the base of the embankment. These boulders can actually cause 'erosion' of the moraine slope at some points, although overall there is a net gain of smaller clasts. Meltwater rivulets, fed by the melting ice exposures, have developed at several points; these both transport finer sediments and wash them into the coarser debris and — where the release of water is rapid — cut gullies to a depth of 1–2 m into the moraine.

8.4.4   Characteristics of the moraine sediments

In an attempt to determine the origin of the debris constituting the Tsidjiore Nouve lateral moraines, both clast shape analysis and particle-size analysis have been undertaken.

(1) At the present time, the Glacier de Tsidjiore Nouve receives large increments of fresh angular debris from the rock walls north of the Pigne d'Arolla ice fall (on to the northern supraglacial lateral moraine) and the ridge of Louette Econdoi to the south of the glacier (on to the southern supraglacial moraine). Following transport down-glacier, a proportion of this material is transferred rapidly by lateral sliding on to the moraine embankment crests. In an attempt to assess whether other important debris sources contribute to moraine growth, surface clasts with maximum dimensions in excess of 15 cm were selected along five transects of the crests and upper slopes of the northern

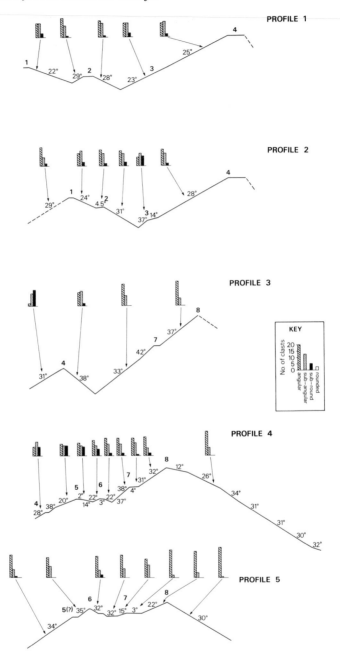

FIGURE 8.6   Variations in clast shape, northern lateral moraine embankments, Glacier de Tsidjiore Nouve

moraine complex (Figure 8.6), and their 'roundness' was determined by the technique proposed by Anderson (p. 172). The study was confined to the lowermost 0.8 km of the moraine, where on *a priori* grounds there is a greater likelihood of debris contributions from englacial and/or subglacial sources. A number of individual sites were identified along each transect, depending on the number of nested ridges present, and at each site 25 clasts were measured. The resultant data were analysed to reveal the following:

(i) There is an overwhelming dominance of angular and subangular clasts, and a complete absence of rounded clasts. Thus of the 800 clasts measured at all the sites 456 (55.7%) were angular, 256 (32.0%) were subangular and 98 (12.3%) were sub-rounded. The initial conclusion must be that supraglacial sources of debris are dominant, and that subglacial sources contribute a relatively limited amount of partially rounded clasts (it may be significant that in several ice caves in the area basal debris was seen to contain large numbers of well rounded clasts).

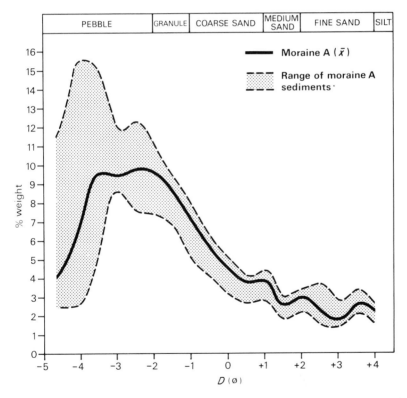

FIGURE 8.7   Particle size graph of sediment, supraglacial lateral moraine, Glacier de Tsidjiore Nouve. *Reproduced with permission from Small (1983)*

(ii) There is a tendency for angular clasts to increase in importance as the nested moraine ridges become successively younger towards the present glacier margin (in fact, the youngest moraine consists of over 80% angular clasts and no sub-rounded clasts).

(iii) There is also some trend towards reduced angularity in a down-moraine direction. Thus, over 75% of the clasts on profile 5 (Figure 8.6) are angular, whereas on profiles 3, 2 and 1 less than 50% are angular; conversely, there is an increase in subangular clasts from 25% on profile 5 to 42% on profile 1.

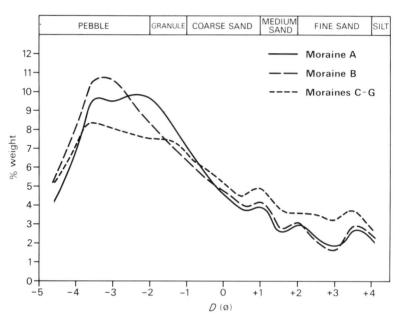

FIGURE 8.8   Particle size graphs of sediment from lateral moraines A, B and C–G, Glacier de Tsidjiore Nouve. *Reproduced with permission from Small (1983)*

In the interpretation of these results, the question inevitably arises as to whether all 'rounding' of clasts (even the slight changes from angular to subangular) should be regarded as diagnostic of subglacial transport [as inferred by Reheis (1975), Anderson (1978), Larsen and Mangerud (1981) and Matthews and Petch (1982)]. In the author's view this is not necessarily the case. Moraines 1, 2 and 4 at Tsidjiore Nouve have been dated by Röthlisberger and Schneebeli (1979) as 2500 BP, 1500 BP and 900 BP, respectively. There is a distinct probability that over such a long period of time surface clasts have experienced some *in situ* modification, involving the rounding and blunting of sharp edges by microgelivation or chemical weathering. This would help to account for the increased 'roundness' both away from the glacier (as the sediments become older) and also in a down-glacier direction (it should be noted that the up-glacier profile 5 consists mainly of the younger moraine ridges 5–8, whereas on the down-glacier profiles 1 and 2 the older nested ridges 1–4 are present). It is also possible that some modification of angular clasts may have occurred during the process of supraglacial transport over 1–2 km. This would have arisen both from weathering and mechanical abrasion, caused by continual shifting and sliding of the surface debris cover due to irregular ablation of the subjacent ice.

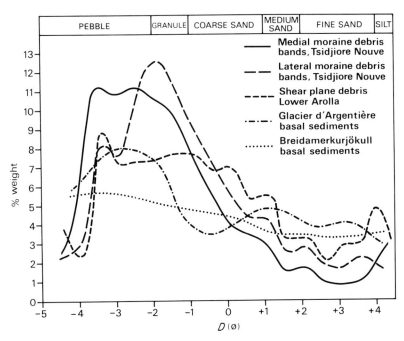

FIGURE 8.9   Particle size graphs of sediment from englacial debris bands, Glacier de Tsidjiore Nouve, with selected comparisons. *Reproduced with permission from Small (1983)*

(2) The results of particle size analysis of sediments taken from the supraglacial moraine ridge A (Figure 8.7) and the nested moraine ridges B–G are shown in Figure 8.8. As expected, the particle size graph for A supports the hypothesis of supraglacial derivation of the moraine debris. The predominantly coarse nature of the sediments (which would have been further emphasized had the abundant clasts in excess of 32 mm been included) and the relative paucity of fine sand and silt are clearly evident. In fact, the graph is closely similar to those produced by Boulton (1978) for sediment in high-level transport on the Glacier d'Argentière and Breidamerkurjökull, and regarded by him as typically supraglacial. The supraglacial lateral moraine A is also associated with the release

of englacial debris from steeply dipping or vertical bands running parallel to the glacier margin. These are best exposed within transverse crevasses or on the ice exposures where the glacier is overriding the main moraine embankment. Particle size graphs of sediment samples from six of these debris bands are shown in Figure 8.9. When compared with the graph of 24 medial moraine debris-band samples (p. 125), the lateral moraine samples again display a predominantly 'supraglacial' character, although there is some lack of material in the -3ϕ to -1ϕ range. There is certainly a strong contrast with graphs of shear plane (= subglacial) debris from the Bas Glacier d'Arolla and zone of traction sediment from the Glacier d'Argentière and Breidamerkurjökull. Particle size graphs for the older dump moraine ridges C–G (Figure 8.8) show, in comparison with the more youthful moraines A and B, a relative lack of sediment in the -2ϕ to -4ϕ range and a greater content of fine sand and silt [which may result from the washing in of fines at a late stage of moraine formation, as suggested by Boulton and Eyles (1979)]. When the individual particle size graphs for moraines C–G (excluding E) are plotted (Figure 8.10), it becomes evident that the debris of moraines C, D and G are more characteristically 'supraglacial,' and that moraine F (with its high content in the $+1\phi$ to -2ϕ range) has a distorting

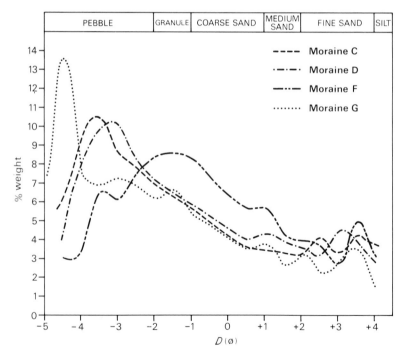

FIGURE 8.10   Particle size graphs of sediment from lateral moraines C, D, F and G, Glacier de Tsidjiore Nouve. *Reproduced with permission from Small (1983)*

effect. In this instance sediment samples were taken from both the proximal and distal moraine slopes, and the possibility of the inclusion of lodgement till from the former must be considered.

The conclusion from this analysis of clast shape and particle size on the northern lateral moraine embankment at the Tsidjiore Nouve is that, in contrast to the findings of other studies of moraine debris provenance cited in this discussion, the bulk of the sediment is supraglacially derived, supraglacially and englacially transported, and dumped primarily from above on to the crest and distal slope of the moraine. There is little evidence of a major contribution from subglacial sources. It is proposed, as a first approximation, that between 60 and 80% of the moraine debris is of supraglacial origin, with the former figure being almost certainly too conservative.

8.5 THE LATERAL MORAINE SEDIMENT BUDGET, GLACIER DE TSIDJIORE NOUVE

8.5.1 Rates of sediment input to the lateral moraines during the Neoglacial advances

Radiocarbon dating indicates that the Tsidjiore Nouve lateral moraine embankments have been formed within the past 4000–5000 years, during a series of glacial advances (reflected particularly by the pattern of nested moraines on the northern embankment) culminating in those of the Little Ice Age (ca.1550–1850). The volumes of the embankments have been calculated from surveyed cross-profiles located at intervals of 100 m. One problem arises from the fact that the proximal faces are concealed by the glacier, except for a 1 km section at the lowermost end of the northern embankment, where melt-back of the glacier margins has exposed part of the inner slope (at a 30–40° angle). It seems reasonable to infer, on a number of grounds, that this exposed face has not been seriously eroded and approximates to the former subglacial surface. Therefore, in the calculations of volumes of debris contained within the lateral moraine embankments (Table 8.2), limits to the steepness of the proximal slopes are assumed to lie within the range of 30° (minimum) to 45° (maximum), with the former being more likely than the latter. It is also assumed that the moraines are not ice-cored, although in fact dead-ice masses have recently been found to occupy the lower parts of lateral dump moraines marginal to the Mont Miné and Haut-Arolla glaciers in this area. However, these moraines are in many respects morphologically contrasted to those at Tsidjiore Nouve, and consist of sediments 'plastered on to' constricting valley walls and standing up to 180 m above the present glacier surfaces; the dead ice is comparatively youthful, and seems to have been abandoned during the retreat stages of the Little Ice Age since 1850 (Small, 1983). Certainly at Tsidjiore Nouve, where the moraine has at some points been subjected to localized erosion (for example, where sudden

TABLE 8.2   The neoglacial moraine embankments, Glacier de Tsidjiore Nouve

A. Estimated volumes of moraine embankments

	Gross volumes (m³)		Corrected volumes (m³)	
	30° proximal slope	45° proximal slope	30° proximal slope	45° proximal slope
Northern moraine	8037600	6028200	6430200	4822560
Southern moraine	4228240	3171180	3382592	2536844
Total	12265864	9199380	9812792	7359504

B. Estimated annual rates of accumulation during the Neoglacial period (m yr^{-1})

	Assuming 5000 years accumulation		Assuming 2500 years accumulation	
	30° proximal slope	45° proximal slope	30° proximal slope	45° proximal slope
Northern moraine	1286	964	2572	1928
Southern moraine	668	507	1336	1014
Total	1954	1471	3808	2942

flows of water have incised gullies into the distal slope) no ice has been exposed. However, it is admitted that the question must remain an open one to some extent.

As Table 8.2 shows, the gross volume of the Tsidjiore Nouve embankments lies within the range $9.2-12.3 \times 10^6 \, \text{m}^3$ (with the volume of the northern moraine being approximately twice that of the southern). The moraine debris is now well packed, owing to the settling of dumped sediments and the washing of fines into voids. In calculating the corrected volumes a relatively low void space of 20% has therefore been assumed. This gives a range of moraine volume from $7.4 \times 10^6 \, \text{m}^3$ (with proximal slope 45°) to $9.8 \times 10^6 \, \text{m}^3$ (with proximal slope 30°). Annual rates of debris accumulation by volume can be calculated as follows. If the accumulation had been constant over the 5000-year period of formation, the average annual rate would have been between 1500 and 2000 m^3. However, with a known history embracing several Neoglacial advances and retreats, it is likely that permanent dumping of sediment on the distal moraine slopes would have occupied approximately half this period (2500 years). This would allow a more 'realistic' mean annual accumulation rate of 2900–3800 m^3 to be calculated.

8.5.2   Present rates of sediment input to the lateral moraines

At locations where the Glacier de Tsidjiore Nouve is overtopping the lateral moraine embankments, and active deposition is occurring in the manner described on p. 181, there is the possibility of determining annual increments on debris from precise surveys of moraine profiles. The main sites of such dumping at present are as follows:

(i) Towards the head of the northern moraine (at a site designated for convenience *Fontanesses*), the ice margins are overriding the embankment summits over a distance of at least 200 m (Small, 1983). This is a very important site for the deposition of both supraglacial and englacial debris (the latter derived from a discontinuous ice face several metres in height).

(ii) At the very head of the northern moraine (designated *Fontanesses Corner*), there is a much shorter section (about 40 m) of glacial overriding and deposition of supraglacial debris.

(iii) Towards the lower (eastern) end of the northern embankment, recession of the glacier margin has left a valley between the ice face and the proximal slope of the embankment. Deposition is very active along the uppermost 130 m section (designated *Upper Remointse*) of the glacier margin, and is leading to the formation of a new 'inner' moraine ridge (Figure 8.11). The crests of this ridge are being overridden by the ice in a manner broadly similar to that at Fontanesses.

(iv) Down-glacier from the Upper Remointse site, a further section of the glacier margin (designated *Lower Remointse*), about 525 m in length, consists

of a partially overhanging ice face up to 20 m or more in height. From the crest of this ice slope supraglacial debris falls and slides almost continuously, although at the base there is no longer a moraine ridge comparable to that at Upper Remointse. Rather, the deposits accumulate on the proximal slope of an 'inner' branch of the lateral moraine embankment. Some debris at the base of this slope is removed by an ice marginal stream which seasonally develops along the glacier edge.

(v) Towards the head of the southern moraine embankment (at a site designated *Louettes*), the ice margin again overrides the crests over a distance of 40 m, and dumping of supraglacial debris is moderately active.

(vi) At the glacier snout (designated *Snout*), deposition is very active as the 250 m long ice face ablates back, undermining the cover of supraglacial debris and releasing englacial sediment. This is a much more complex depositional environment than at sites 1–5. Recent accumulations of debris in the proglacial zone are being overridden and concealed by the advancing glacier. However, towards the southern end of the snout supraglacial debris in particular is accumulating in large quantities on the lower ice slopes. This debris cover restricts ablation of the underlying ice; at the same time, the largely bare upper ice face is melting back relative to the lower slope, with the result that a debris-covered 'ice pediment' has formed.

In the design of a programme for monitoring active sediment accumulation at these sites, certain difficulties need to be surmounted. The distal moraine slopes are very steep and by their nature highly unstable, and there are numerous falls of large boulders and stones from the overriding ice. Survey of successive moraine profiles with the required degree of precision by conventional levelling techniques is therefore inpracticable. I. R. Beecroft has adapted a technique, devised by Stirling (1982), for the construction of very accurate cross-profiles from ground-based stereo photographs. The method was applied at the Glacier de Tsidjiore Nouve during the summer of 1982, and for comparative purposes earlier profiles were derived from aerial photographs taken in September 1977. The technique, and its requirements and limitations, were described in detail in Small *et al.*, (1984).

The findings at each site can be summarized as follows:

(i) At Fontanesses five profiles were constructed (at 50 m intervals), revealing maximum accumulation on the upper distal face. The mean cross-sectional area of accretion for the profiles was calculated as $40.7 \, \text{m}^3$ ($8.14 \, \text{m}^3 \, \text{yr}^{-1}$); a volumetric increase at the site as a whole of $8138 \, \text{m}^3$ ($1627 \, \text{m}^3 \, \text{yr}^{-1}$) has been computed.

(ii) At Louettes six profiles were drawn along the 40 m depositional margin, allowing a volumetric increase of $707 \, \text{m}^3$ ($141 \, \text{m}^3 \, \text{yr}^{-1}$) to be calculated.

FIGURE 8.11   *(opposite)* Active deposition of sediment at Upper Remointse site, where the margin of the Glacier de Tsidjiore Nouve is overriding the lateral moraine ridge

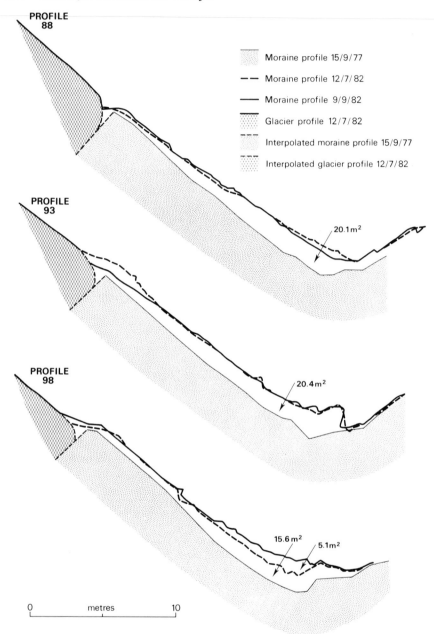

FIGURE 8.12   Photogrammetric profiles of lateral moraine embankment at the Upper Remointse site, Glacier de Tsidjiore Nouve. *Reproduced with permission from Small, Beecroft and Stirling (1984)*

TABLE 8.3 Estimated rates of annual marginal accumulation, Glacier de Tsidjiore Nouve

Site	Margin length (m)	Annual accumulation (m³)	Annual accumulation (m/m of glacier margin)
1. Fontanesses	200	1627	8.14
2. Fontanesses Corner	40	141–325	3.54–8.14
3. Upper Remointse	130	530	4.08
4. Lower Remointse	525	1286	2.45
5. Louettes	40	141	3.45
6. Snout	250	884–2034	3.54–8.14
Total	—	4611–5946	—

(iii) At Upper Remointse three profiles were constructed (at 5 m intervals on the central part of this section), allowing the calculation of a total debris increment of 204 m^3 over a 10 m length of the moraine; this implies a total accretion of 2653 m^3 (530 m^3 yr^{-1}) over the whole 130 m section (Figure 8.12).

At the more 'difficult' sites of Fontanesses Corner, Lower Remointse and Snout, a combination of profile construction using photogrammetry and measurements of ice-margin ablation (volume of sediment = seasonal recession × thickness of supraglacial debris undermined) was used to calculate deposition rates (for full details, see Small *et al.*, 1984). At Fontanesses Corner the annual deposition was estimated to be between 141 and 325 m^3, with the former figure more 'probable' than the latter; at Lower Remointse, where dumping is occurring actively over a 525 m ice margin, the annual deposition was calculated as 1286 m^3; and at Snout (a particular problematic site, since rapid forward movement of the debris-covered ice renders the determination of accretion from successive profiles impossible) the annual deposition was estimated, largely on the basis of ablation measurements, as lying between 884 and 2034 m^3 yr^{-1}.

Table 8.3 depicts rates of annual marginal accumulation at each site, and the total volume of sediment now being dumped annually by the Glacier de Tsidjiore Nouve (between 4611 and 5946 m^3 yr^{-1}). For purposes of comparison between the individual sites the annual accumulation in cubic metres per metre length of the glacier margin is also shown; this lies between 2.45 and 8.14, with a mean rate of 4.8 m^{3}m^{-1} yr^{-1}. The figures in Table 8.3 are not adjusted to take account of void space.

8.6 CONCLUSIONS

At the present time between 4600 and 5900 m^3 yr^{-1} of sediment are being dumped either on the lateral moraine embankments or on to the proglacial zone beneath the glacier snout. Most of this debris is derived from the supraglacial cover of the glacier, although some is released from debris bands exposed along the ice margins. During the 5000 years since the Post-Glacial Optimum, sediment has evidently been deposited on the lateral moraines at a rate of 2900–3800 m^3 yr^{-1}. If the estimates of current rates of deposition in Table 8.3 are adjusted to take account of void space (30%, since these sediments are as yet unpacked), figures of 3220–4130 m^3 yr^{-1} are arrived at. There is thus a good measure of agreement between the past and present estimated rates of deposition.

Sedimentological analysis (clast shape and particle size analysis) suggests strongly that the bulk of the debris on the lateral moraine embankments is supraglacially derived. It is assumed that, of the debris currently being dumped, 80% is of supraglacial origin, then the annual contribution from rock falls to the glacier surface can be calculated as 2576–3328 m^3 yr^{-1} (if a balance between 'inputs' and 'outputs' is being maintained). A planimetric study of the

Glacier de Tsidjiore Nouve has shown the area of rock faces within the catchment to be approximately 1.55 km²; to provide the input of supraglacial debris these would have to experience a recession rate of 1.7–2.1 mm yr⁻¹ — seemingly a high figure, but one not difficult to reconcile with (a) the numerous falls occurring from highly shattered and oversteepened rock exposures above the glacier at the present time, and (b) the vast amount of angular surface debris (estimated at 115 000 m³) resting on the glacier tongue.

In recent years a programme of monitoring suspended sediment concentrations in the proglacial meltwater stream from the Glacier de Tsidjiore Nouve has been carried out by I. R. Beecroft, who, for the ablation seasons of 1981 and 1982, calculated total sediment outputs of 3515 and 3223 m³, respectively. Beecroft has also estimated (from sediment traps in a water intake constructed by Grande Dixence S. A.) that bed-load discharge for these two summers amounted to 1835 and 1113 m³, respectively. However, it must be emphasized that a substantial proportion of this bed load consists of supraglacial and englacial debris from the glacier snout, so that there is some overlap between marginal dumping of debris and proglacial stream transport. If it is assumed that only the suspended sediment load is derived from glacial erosion *sensu stricto*, affecting a subglacial surface with a measured area of approximately 3.6 km², then a glacial erosion rate of 0.89–0.99 mm yr⁻¹ is implied. If it is assumed that *all* the stream load is derived from glacial erosion, and that 20% of the dumped morainic sediments should also be included as being the products of glacial erosion, the estimated glacial erosion rate rises to between 1.38 and 1.71 mm yr⁻¹. These figures nevertheless suggest that the Glacier de Tsidjiore Nouve has recently been more active in transporting and depositing supraglacial debris than in glacial erosion *sensu stricto*. Such a change of role since the Pleistocene could be due to glacial thinning and increased exposure of valley walls and rock faces during the Post-Glacial period.

TABLE 8.4   Glacier de Tsidjiore Nouve: Estimated annual deposition (m³ yr⁻¹)

	Maximum	(%)	Minimum	(%)
Deposition on lateral moraines	2738	28.8	2609	40.4
Deposition at glacier snout	1424	14.9	619	9.6
Proglacial stream transport:				
(a) suspended sediment	3515	37.0	3223	50.00
(b) Bed load	1835	19.3	—	—
Total	9512	100	6451	100

If no attempt is made to differentiate between supraglacial and subglacial sources, and all the outputs of sediment are summed, a figure lying between 6451 and 9512 m³ yr⁻¹ is arrived at (Table 8.4). Since the total catchment area of the glacier is approximately 4.15 km², an 'erosion rate' of 1.55–2.29 mm yr^{-1} is implied. Such figures, together with those quoted above, add support to the overwhelming impression gained in the field (given from the extensive supraglacial debris cover, the numerous englacial debris bands and the massive lateral moraine embankments) that the Glacier de Tsidjiore Nouve is unusually active in geomorphological terms.

REFERENCES

ANDERSON, L. W. (1978), Cirque glacier erosion rates and characteristics of Neoglacial tills, Pangnirtung Fiord area, Baffin Island, N.W.T., Canada, *Arctic and Alpine Research*, **10**, 749–760.

ANDREWS, J. T. (1971), Estimates of variations in glacial erosion from the volume of corries and moraines, *Geological Society of America Abstracts with Programs*, **3**, 493.

ANDREWS, J. T. (1972), Glacier power, mass balances, velocities and erosion potential, *Zeitschrift für Geomorphologie*, **13**, 1–17.

BOULTON, G. S. (1970), On the origin and transport of englacial debris in Svalbard glaciers, *Journal of Glaciology*, **9**, 213–229.

BOULTON, G. S. (1972), The role of thermal regime in glacial sedimentation, in PRICE R. J., and SUGDEN, D. E. (Eds.), *Polar Geomorphology*, Institute of British Geographers Special Publication, No. 4, pp. 1–19.

BOULTON, G. S. (1978), Boulder shapes and grain-size distribution of debris as indicators of transport paths through a glacier and till genesis, *Sedimentology*, **25**, 773–799.

BOULTON, G. S., and EYLES, N. (1979), Sedimentation by valley glaciers: a model and genetic classification, in SCHLUCHTER, C. (Ed.), *Moraines and Varves: Origin, Genesis, Classification*, A.A. Balkema, Rotterdam, pp. 11–23.

COLLINS, D. N. (1979), Sediment concentrations in melt waters as an indicator of erosion processes beneath an Alpine glacier, *Journal of Glaciology*, **23**, 247–258.

FLINT, R. F. (1971), *Glacial and Quaternary Geology*, Wiley, New York.

GALLOWAY, R. W. (1956), The structure of moraines in Lyngsdalen, north Norway, *Journal of Glaciology*, **2**, 730–733.

GOLDTHWAIT, R. P. (1971), Introduction to till, today, in GOLDTHWAIT, R. P. (Ed.), *Till: a Symposium*, Ohio State University Press, pp. 3–26.

HUMLUM, O. (1978), Genesis of layered lateral moraines: implications for palaeoclimatology and lichenometry, *Geografisk Tidsskrift*, **77**, 65–72.

KJELDSEN, O. (1981), *Materialtransportundersøkelser Norske i Bre-elver 1980*, Vassdragsdirecktorat Hydrologisk Avdeling, Oslo, Rapport Nr. 4–81.

LARSEN, E., and MANGERUD, J. (1981), Erosion rate of a Younger Dryas cirque glacier at Kråkenes, western Norway, *Annals of Glaciology*, **2**, 153–158.

MATTHEWS, J. A., and PETCH, J. R. (1982), Within-valley asymmetry and related problems of Neoglacial lateral moraine development at certain Jotunheim glaciers, southern Norway, *Boreas*, **11**, 225–247.

MCCALL, J. G. (1960), The flow characteristics of a cirque glacier, in LEWIS W. V. (Ed.), *Norwegian Cirque Glaciers*, Royal Geographical Society Research Series, No. 4, pp. 35–62.

MILLS, H. H. (1977), Differentiation of glacier environments by sediment characteristics, *Journal of Sedimentary Petrology*, **47**, 728–737.

OSBORN, G. D. (1978), Fabric and origin of lateral moraines, Bethartoli Glacier, Garhwal Himalaya, India, *Journal of Glaciology*, **20**, 547–553.

REHEIS, M. J. (1975), Source, transportation and deposition of debris on Arapaho Glacier, Front Range, Colorado, U.S.A., *Journal of Glaciology*, **14**, 407–420.

RÖTHLISBERGER, F. (1976), Gletscher- und Klimaschwankungen im Raum Zermatt, Ferpecle und Arolla, *Die Alpen, Zeitschrift des Schweizer Alpen Club*, 59-152.

RÖTHLISBERGER, F., and SCHNEEBELI, W. (1979), Genesis of lateral moraine complexes, demonstrated by fossil soils and trunks; indicators of proglacial climatic fluctuations, in SCHLUCHTER C. (Ed.), *Moraines and Varves: Origin, Genesis, Classification*, A.A. Balkema, Rotterdam, pp. 387-419.

SCHNEEBELI, W. (1976), Untersuchungen von Gletscherschwankungen im Val de Bagnes, *Die Alpen, Zeitschrift des Schweizer Alpen Club*, pp. 5-57.

SHAW, J. (1980), Drumlins and large-scale flutings related to glacier folds, *Arctic and Alpine Research*, 12, 287-298.

SMALL, R. J. (1983), Lateral moraines of Glacier de Tsidjiore Nouve: form, development and implications, *Journal of Glaciology*, 29, 250-259.

SMALL, R. J., BEECROFT, I. R., and STIRLING, D. M. (1984), Rates of deposition on lateral moraine embankments, Glacier de Tsidjiore Nouve, Valais, Switzerland, *Journal of Glaciology*, 30, 275-281.

STIRLING, D. M. (1982), Measuring short term glacial fluctuations by aerial and terrestrial photogrammetry—a comparative study, *International Archives of Photogrammetry*, 24, 484-496.

WHALLEY, W. B. (1973), An exposure of ice on the distal side of a lateral moraine, *Journal of Glaciology*, 12, 327-329.

Glacio-fluvial Sediment Transfer
Edited by A. M. Gurnell and M. J. Clark
©1987 John Wiley & Sons Ltd.

Chapter 9

The Glacial Sediment System:
An Alpine Perspective

R. J. SMALL

ABSTRACT

It is proposed, on the basis of the large size of Neoglacial latero-frontal dump
moraines and measurements of supraglacial and englacial sediment, that at present
supraglacial sediment sources are of particular importance in Alpine environments.
This reflects changes in glacier basin configuration in the post-Würm period, leading
to a greater incidence of rock falls and more effective gelifraction. Some conflicting
evidence (for example, from the Front Range, Colorado) is also presented and
discussed.

Under present-day conditions, many Alpine glaciers are typified by considerable
loads of supraglacial and englacial sediment. This high debris content is reflected
by the formation both of numerous and extensive supraglacial moraines and
of large latero-frontal dump moraines (p. 174). Although the latter are not
experiencing active development, except in isolated instances such as that of
the Glacier de Tsidjiore Nouve, there is clear evidence of several recent episodes
of dumping. These were associated with the series of minor glacial advances
of the Neoglacial (approximately 5000 BP to the present time), of which the
most recent was the Little Ice Age (ca. 1500–1850 AD). Where the Alpine glaciers
are largely confined within deep, comparatively narrow valleys the moraines
take the form of prominent valley-side deposits with clearly defined upper limits.
They stand up to 150 m or more above the ice surface (lowered by the important
glacial recession since 1850) and are often spectacularly gullied on their proximal
faces (as to the west of the Bas Glacier d'Arolla and Glacier du Mont Miné,
Valais, Switzerland). In situations where the glacier tongues extend into broad
depressions (for example, the Glacier de Tsidjiore Nouve, Feegletscher and
Findelengletscher, also in Valais), the Neoglacial moraines consist of massive
moraine embankments, often of complex form and with well preserved, partially
vegetated distal faces.

199

The large load of supraglacial and englacial sediment to which these recent moraines testify is not always typical of glaciers in other climatic environments, and may not have been such a pronounced feature of Alpine glaciers during the major Pleistocene advances. Thus Boulton (1978) has written of Breidamerkurjökull: 'on the basis of supraglacial and subglacial measurements at many localities in the terminal zone I would estimate that 30 to 100 times more debris is discharged in the zone of traction than in high-level transport.' Boulton added that the subglacial component will be 'proportionally smaller in valley glaciers dominated by high mountain peaks, but is still likely to be the major component, and in ice caps and ice sheets the tractional-type distribution is likely to account for almost 100% of the debris discharged.'

However, field observation in the Alps today indicates that, whilst abrasion and plucking remain active (as shown in Chapter 7), the contribution of sediment from supraglacial sources is very considerable. For example, in the case of the Glacier de Tsidjiore Nouve, estimates of the current rate of sediment output (Table 8.4) imply an overall erosion rate within the glacier catchment as a whole of 1.55–2.29 mm yr^{-1}. However, when this sediment is broken down, by the methods outlined in Chapter 8, into supraglacial and subglacial components, it emerges that the former (at up to 3328 m^3 yr^{-1} or possibly more) amounts to at least 54% of the latter (a maximum of 6184 m^3 yr^{-1}, although this is almost certainly an overestimate, since it includes proglacial stream bedload which in large measure is derived from supraglacial and englacial sediment of supraglacial origin). Moreover, when the areal extent of supraglacial and subglacial surfaces (1.55 and 3.6 km^2, respectively) within the Tsidjiore Nouve catchment are taken into account, glacial erosion *sensu stricto* can be calculated as up to 1.71 mm yr^{-1}, whilst supraglacial rock-wall retreat is actually higher, at up to 2.16 mm yr^{-1}.

Whilst it is accepted that, in the context of the Swiss Alps as a whole, the Glacier de Tsidjiore Nouve may with its considerable supraglacial debris mantle be the exception rather than the rule, there are many Alpine glaciers in which the balance between supraglacial and subglacial sediment production and transport appears to be substantially different from that inferred by Boulton from Breidamerkurjökull. For example, Humlum (1978) noted that Guslarferner (in the Ötztaler Alps, Austria) 'is currently transporting material supraglacially in medial moraines, while inglacial transport and transport at the glacier sole quantitatively seems to be of less importance. Especially the left part of the glacier is supplied with talus from the surrounding high rock walls.' Nor does this situation seem to be confined to the Alps. In his pioneer study of Vesl-Skautbreen, Norway, McCall (1960), considering the basal, englacial and supraglacial fragments transported by this small cirque glacier, stated that 'the impression was gained that (subaerial erosion) is now the chief contributor, and because of this an appreciable portion of the moraine is in reality an accumulation of talus or protalus.' Again, in an examination of lateral moraine formation in the Jotunheim, Matthews and Petch (1982) argued against a primarily subglacial origin for the sediments contained within these moraines on the grounds that the large volumes of sediment are incompatible with the small quantity of debris found in the basal zone of temperate glaciers. 'Our observations in the ice caves at Hurbreen revealed debris-rich basal ice, but this was not more than 0.5 m thick.' The situation is neatly summarized in the correspondence between Boulton and Andrews (Boulton, 1972). In response to suggestions by Andrews to the effect that (i) cold glaciers (in Baffin Island) contain very small quantities of englacial debris (0.5% by volume in the ablation zone) and (ii) late glacial end and lateral moraines of temperate glaciers in Colorado, Norway and Italy are considerably larger than those of Arctic glaciers, Boulton stated that 'it is true that many temperate glaciers have large terminal and lateral ice-covered moraines. I believe this stems from the fact that many such glaciers are valley glaciers in which englacial debris is introduced not from the bed but from valley-sides, nunataks and cirque headwalls. Where these latter features do not occur, there is almost no englacial debris above the basal layer.'

Any change in the balance betwen supraglacial and subglacial sediment sources must reflect (i) the occurrence and extent of supraglacial outcrops of rock susceptible to gelifraction and/or discontinuous rock mass failure (Addison, 1981) and (ii) the efficiency of glacial erosional processes (which are controlled by factors such as ice thickness, glacier mass balance, thermal regime related to rates of basal sliding and regelation mechanisms and rock fracture density). In areas such as the Alps which have experienced large-scale deglaciation in the post-Würm period, a major relative increase in supraglacial sediment input is to be expected. A comparison of the *Atlas der Schweiz* (1970) reconstruction of Switzerland during the last glacial period with the present-day maps of the *Landeskarte der Schweiz* serve to illustrate this point. As the glaciers have shrunk, oversteepened and potentially unstable valley walls have been increasingly exposed, leading to the proliferation of rock slides. It is also possible that, with the climatic amelioration, an increase in the number of diurnal frost cycles—and thus more effective gelifraction—has occurred. Whether the decrease in glacier thickness has led to a reduction in glacial erosion rates remains a matter for surmise, although on general grounds this seems likely. Boulton's suggestion that, beneath very thick glaciers (such as would have been characteristic of the Würm period in the Alps), bed contact forces would be so high that basal fragments would be transported at reduced velocity, thereby reducing abrasion rates and eventually inducing lodgement (Boulton, 1974) is, in Hallet's view (Hallet, 1981), a 'questionable prediction' and hardly compatible with existing geomorphological evidence (deep glacial valleys and fiords carved by glaciers of great size).

However, not all the evidence, for all glacial environments, supports the notion of a significant change in the operation of the glacial sediment transport system. Thus, in their study of Kråkenes, Larsen and Mangerud (1981) proposed, on the basis of a 'roundness analysis' of moraine deposits (p. 170), that this small Younger Dryas (10 900–10 200 BP) cirque glacier deposited sediment at a mean annual rate of 94 000 m³ (corresponding bedrock volume). Of this only 5000 m³ appears to have been derived from rock fall on the headwall. Indeed, the latter experienced a rate of headwall retreat of only 0.1 mm yr⁻¹, in comparison with a subglacial erosion rate of 0.6 mm yr⁻¹. From a study of the temperate Arapaho Glacier, Front Range, Colorado, Reheis (1975) concluded that during the Gannet Peak stade (the last 300 years) 88% of the deposited moraine was derived and/or carried subglacially. Estimates of the sources of debris in the 'present-day' ablation moraine of Arapaho Glacier indicate (i) a slight change in the relative contribution of supraglacial sediment (from 12 to 30%) but an actual fall in the rate of production of supraglacial sediment (from 290–485 to 35–50 m³ yr⁻¹). Although the roundness analysis employed by Reheis can be questioned (p. 171), it seems unlikely that the conclusions relating to the Gannet Peak stade are grossly inaccurate. However, the inferences drawn from the ablation moraine (which actually yield 65% unrounded, or 'supraglacial,' clasts from the sample sites) are more questionable, and it is not impossible that the reduced Arapaho Glacier is now transporting more supraglacial than subglacial sediment.

Of particular interest is the analysis of sediments deposited by ten sub-polar cirque glaciers in the Pangnirtung Fiord area, Baffin Island (Anderson, 1978). This has revealed a generally similar pattern of sediment production to that at Arapaho, although the actual rates of production are considerably reduced. The latter vary considerably from one glacier to another (subglacial debris production ranges from a maximum of 96–133 m³ yr⁻¹ to a minimum of 24–41 m³ yr⁻¹). However, supraglacial debris production (maximum 16–22 m³ yr⁻¹, minimum 1–2 m³ yr⁻¹) is always substantially less. Anderson noted a positive correlation between the percentage of subglacially derived debris and the percentage of total cirque area that is glacierized. 'Those cirques with the smallest relative area of headwall (source area for rock fall material) show the highest percentage of subglacially derived morainal debris,' and are generally 'the cleanest and most free of supraglacial debris.'

Although further quantitative studies of sediment transport by glaciers are needed before anything like a comprehensive picture can be obtained, the following interim conclusions can be drawn.

1. When viewed purely as sediment transport systems polar, sub-polar and temperate glaciers may not behave in inherently different ways—although there may be significant variations in rates of sediment production and even mechanisms of erosion (with abrasion possibly being more effective than plucking in temperate glaciers).
2. The balance between supraglacial, englacial and subglacial sediment transport is largely a function of catchment morphological variables (presence or

absence of valley walls, headwalls and nunataks, which are in turn related to the extent of glacierization). Where significant changes have occurred over a period of time (as in the post-Würm deglaciation of the Alps) it is inevitable—applying the relationships determined by Anderson—that supraglacial sediment inputs, and thus supraglacial and englacial transport, are substantially altered.

REFERENCES

ADDISON, K. (1981), The contribution of discontinuous rock-mass failure to glacier erosion, *Annals of Glaciology*, **2**, 3–10.

ANDERSON, L. W. (1978), Cirque glacier erosion rates and characteristics of Neoglacial tills, Pangnirtung Fiord area, Baffin Island, N.W.T., Canada, *Arctic and Alpine Research*, **10**, 749–760.

BOULTON, G. S. (1972), Englacial debris in glaciers: reply to the comments of Dr. J. T. Andrews, *Journal of Glaciology*, **11**, 155–156.

BOULTON, G. S. (1974), Processes and patterns of glacial erosion, in COATES, D. (Ed.), *Glacial Geomorphology*, Binghamton, New York, pp. 41–87.

BOULTON, G. S. (1978), Boulder shapes and grain-size distribution of debris as indicators of transport paths through a glacier and till genesis, *Sedimentology*, **25**, 773–799.

HALLET, B. (1981), Glacial abrasion and sliding: their dependence on the debris concentration in basal ice, *Annals of Glaciology*, **2**, 23–28.

HUMLUM, O. (1978), Genesis of layered lateral moraines: implications for palaeoclimatology and lichenometry, *Geografisk Tidsskrift*, **77**, 65–72.

LARSEN, E., and MANGERUD, J. (1981), Erosion rate of a Younger Dryas cirque glacier at Kråkenes, western Norway, *Annals of Glaciology*, **2**, 153–158.

MATTHEWS, J. A., and PETCH, J. R. (1982), Within-valley asymmetry and related problems of Neoglacial lateral moraine development at certain Jotunheim glaciers, southern Norway, *Boreas*, **11**, 225–247.

MCCALL, J. G. (1960), The flow characteristics of a cirque glacier, in LEWIS, W. V. (Ed.), *Norwegian Cirque Glaciers*, Royal Geographical Society Research Series, No. 4, pp. 35–62.

REHEIS, M. J. (1975), Source, transportation and deposition of debris on Arapaho Glacier, Front Range, Colorado, U.S.A., *Journal of Glaciology*, **14**, 407–420.

SECTION III

Fluvial Sediment Transfer

Glacio-fluvial Sediment Transfer
Edited by A. M. Gurnell and M. J. Clark
©1987 John Wiley & Sons Ltd.

Chapter 10
Glacial Hydrology

H. RÖTHLISBERGER and H. LANG

ABSTRACT

An overview of the hydro-climatic conditions of glaciers and their effects on the runoff is given in the first part of this chapter. The meaning of mass balance and water balance in glacierized areas is explained and methods for their determination are described. The main factors influencing accumulation and ablation are considered and their relative importance is assessed. The characteristics of glacier runoff (diurnal and annual cycles, aperiodic fluctuations related to weather or hydro-glaciological processes) are outlined and long-term effects of climatic variations are discussed.

The second part considers the englacial and subglacial drainage of meltwater. The physical background of various aspects of fast water flow in ice-walled pipes and slow seepage through permeable ice is described, and the consequences for discharge patterns and the location of englacial and subglacial conduits are discussed. Reciprocal effects of glacier sliding and water drainage are shown to explain various observations including water storage during glacier uplift. The problems of glacier-dammed lakes and the methods of investigating intra- and subglacial drainage are also briefly considered.

10.1 THE HYDRO-CLIMATIC CONDITIONS OF GLACIERS

The existence of glaciers is one of the fascinations of mountain areas such as the Alps. Their occurrence is related to climatic conditions and they have their origin in those parts of a mountain region situated above the climatic snow-line. During an average year more snow is accumulated than can be melted under existing climatic conditions and these upper parts of a glacier are called the accumulation areas. In the context of hydrology, the accumulation areas are places where over an average hydrological year a net gain in the annual mass balance (b positive) is observed at the glacier surface. At the climatic snow line there is on average just enough heat available to bring ablation and accumulation into balance ($b = 0$). Under the force of gravity and the control of the laws of ice physics, glacier masses move from the accumulation area down to the ablation area below the climatic snow line. In these low elevation zones of glaciers, the mass balance is negative. Here, on average, more ice and snow is melted and evaporated than is replaced by solid precipitation.

A glacier's motion, which reflects these differences in mass balance, approaches equilibrium conditions under stable climatic conditions. However, the timescale of the response of glaciers to climatic variations and the mode of adaptation of the glacier as a response to climatic change varies between glaciers and is mainly a function of the size of glacier and its dynamic behaviour. Under stationary conditions of glacier mass balance and geometry, the vertical component of the glacier flow at any point on the surface would compensate for the positive mass balance in the accumulation area and the negative mass balance in the ablation area. Thus the seasonal changes in storage (ΔS) in the water balance of the glacier ($P = R + E + \Delta S$, where P is precipitation, E is evaporation and R is runoff) would cancel out on an annual timescale, so that the annual runoff from the glacier basin would simply reflect the difference between annual precipitation input and evaporation loss with no storage change effect (i.e. $\Delta S = 0$). However, the 'stationary' glacier (i.e. a glacier whose geometry and mass balance are unchanging on an annual basis) is a very hypothetical concept because of the inherent variability of climate. Climatic

TABLE 10.1   Time scales of glacier hydrologic processes

Type of climatic processes	Time scale	Glacier–hydrologic processes
Micro-meteorological scales	0.01–1.0 h	Short time melt intensity, snowfall event, precipitation
Daily cycle (earth's own rotation), atmospheric synoptic scale weather processes	1 h–10 d	Daily cycle of melt water flow, temporary snow cover
Seasonal cycle (eclipse of earth's orbit)	10–365 d	Accumulation and ablation periods, seasonal and annual mass and water balance, seasonal cycle of glacier run-off (storage and yield)
Stochastic processes, processes not yet well understood, including ocean-atmosphere interactions, sun spot cycle, volcanism, anthropogenic influences (CO_2 and other 'greenhouse' gases), feedback mechanisms	10^0–10^3 yr	Glacier variations, Little Ice Ages
Cyclic changes in Earth–Sun orbital system, stochastic and unknown processes, including feedback mechanisms	10^3–10^5 yr	Big Ice Ages

variations occur over all timescales from the stochastic variability of day-to-day fluctuations in weather to long-term variations in incoming solar radiation resulting from cycles in the sun–earth orbital system (Table 10.1). Thus, an understanding of glacier mass balance is essential to the study of glacial hydrology.

10.2 WATER BALANCE AND MASS BALANCE OF GLACIERS AND THEIR DETERMINATION

10.2.1 The hydrological and the glaciological methods

The storage change component ΔS in the water balance of a glacierized area:

$$\Delta S = P - R - E \qquad (10.1)$$

where P = precipitation, R = runoff and E = evaporation, is equivalent to the mass balance b for the whole ice mass:

$$b = c - a \qquad (10.2)$$

where c = accumulation and a = ablation, if the change in the volume of liquid water storage over the computation period can be assumed to be negligible. The use of the water balance equation, with the assumption that $\Delta S = b$, is called the hydrological method and is only justified over periods of hydrological years, beginning and ending during the low flow season in winter.

The hydrological method has been employed in assessing the mass balance of the Gr. Aletschgletscher. The discharge records are reliable but there may be some bias in the estimation of areal values for precipitation and evaporation. The method allowed a mass balance record to be established back to 1922 (Kasser, 1959).

The glaciological method determines the mass balance b directly from measurements of accumulation and ablation of mass at the glacier surface. This method is usually based on a number of point measurements taken over the whole glacier surface. In principle, this technique permits the determination of mass balance over any time period, but it requires considerable manpower.

There are a number of glaciers in the Alps with long-term mass balance studies (for example, in Austria, Sonnblickkees since 1963, Hintereisferner since 1952 and Vernagtferner since 1965; in Switzerland, Silvretta since 1960, Limmern Gletscher since 1959, Griesgletscher since 1961 and Aletschgletscher since 1922; in France, Glacier de Sarennes since 1948) published in the reports of the ICSU/FAGS Permanent Service on the Fluctuations of Glaciers (Kasser, 1967, 1973a; Müller, 1977; Haeberli, 1985). Unfortunately, there is no known glacier study that shows the total accumulation and ablation over a whole year. The net annual balance values that are given for each hydrological year or the net values given for each balance year provide less informative climatological interpretations than would be possible with information on the components of the total mass balance.

10.2.2   The geodetic method

The geodetic method of mass balance determination is based on the repeated photogrammetric survey of a glacier over a period of at least 5–10 years. The surveys should be conducted under as near to identical minimum snow conditions as is possible in order to ensure comparable vertical density conditions in the accumulation area. This is necessary to allow accurate conversion of differences in glacier volume into differences in glacier mass.

The application of the geodetic method on the Gr. Aletschgletscher for the period 1927–57 (Kasser, 1967, 1973a) allows calibration of the mass balance determination by the hydrological method, implying an acceptable level of accuracy for the term $P - E$. However, there remains some uncertainty about the accuracy of the individual terms P and E. The estimation of P and E is one of the great problems of mountain hydrology.

A further comparative study of methods of mass balance determination was conducted for the Hintereisferner (Austrian Alps). The hydrological and glaciological methods were applied for the two years 1957–58 and 1958–59 (Hoinkes and Lang, 1962a,b). Precipitation was estimated from storage gauge measurements with correction for systematic error based on direct winter snow cover water equivalent measurements. The estimates of the hydrological balance for the Hintereisferner drainage basin (Steg Hospiz gauging station, catchment area 26.62 km², 58% glacier cover) are presented in Table 10.2. The Hintereisferner glaciological mass balance records were found to be in close agreement with the results from the hydrological method and also with the geodetic method which was employed for the period 1953–62 as an overall control (Lang and Patzelt, 1971).

TABLE 10.2   Hydrological water (= mass) balance of the Hintereisferner drainage basin (Rofenache/Steg Hospiz: 26.6 km², mean altitude 2891 m a.s.l., 58% glacier area) for the years 1957–58 and 1958–59. The storage change ΔS, precipitation P and runoff R are based on direct field observations; the areal evaporation E is calculated from $E = P - R - \Delta S)$

Parameter	1957–58 (mm)	1958–59 (mm)
P	1606	1533
R	1837	1744
E	214	145
$\Delta S = P - R - E$	− 445	− 356

10.2.3   Some results of water and mass balance studies in the Austrian and Swiss Alps

The mean specific mass (water) balances for the Aletschgletscher basin, derived from the hydrological balance equation, are presented in Figure 10.1 for periods of 10 (and more recently 5) years together with runoff data. The values are given in mm water depth related to the whole catchment area of the river Massa (catchment area 197.4 km², 67% covered by the Gr. Aletschgletscher). The extreme annual runoff values observed for individual hydrological years and corresponding storage change for the period 1923 to 1985 are: 1946/7, $R = 3177$ mm, $\Delta S = -1574$ mm; 1977/8, $R = 1590$ mm, $\Delta S = +1191$ mm.

The specific mass balances of the Hintereisferner, about 210 km to the east of the Aletschgletscher, show a similar year-to-year behaviour. The comparison gives some indication of the spatial scale of annual climate–glacier relationships. Both glaciers are typical central alpine valley glaciers. They are characterized by uninterrupted retreat of their glacier termini since 1920, even in years of positive mass balance. Their tongues can be regarded as relics of the Little Ice Age which have not yet adapted to current average climatic conditions. Other glaciers, particularly those which are located more closely to the level of the climatic snow line and generally smaller glaciers, seem to be well adapted to current climatic conditions. An example is the Kesselwandferner (catchment area 3.97 km²), which is in the immediate vicinity of Hintereisferner. Figure 10.2 shows how this glacier gained mass by + 40 cm water equivalent in the period 1953–73, whilst Hintereisferner lost 700 cm water equivalent. In the same

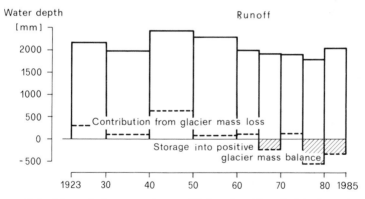

FIGURE 10.1   Mass balance and runoff in the Aletschgletscher river basin (Massa/Blatten, 195 km², 67% glacier area) for the period 1922–23 to 1984–85. The additional contribution of meltwater to the runoff in the period of strong glacier retreat is clearly reflected in the high average runoff, particularly between 1940 and 1950. The minimum runoff in the period 1975–80 is connected with the significant storage of water into positive glacier mass balance. [Data sources: Aellen (1985), and personal communication; Haeberli (1985); Kasser (1967, 1973a); Müller (1977).]

period Aletschgletscher lost 95 cm. The 'well adapted' Kesselwandferner showed a fairly immediate response of its terminus to the positive mass balances (Kuhn *et al.*, 1979, 1985; Hoinkes *et al.*, 1974).

The period of approximately 40 years of general glacier retreat which began around 1927 came to an end in 1965, the first year of a series of positive mass balances for most alpine glaciers (Kasser, 1973a, 1981).

It is clear that the areal distribution of a glacier above and below the climatic snow line is an essential parameter of its climate–mass balance relationship. Therefore, the accumulation area ratio (AAR) has come into use to describe the mass balance conditions of glaciers (AAR = accumulation area/total glacier area). At the same time, the local topographic characteristics are extremely

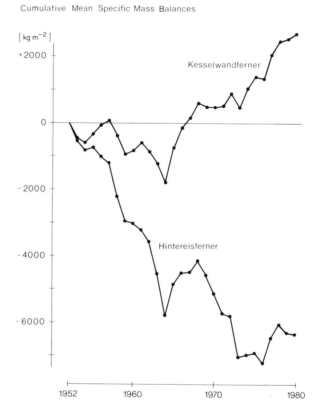

FIGURE 10.2   Cumulative mean specific mass balances of the two adjacent glaciers Hintereisferner (9.08 km², mean altitude 3050 m) and Kesselwandferner (4.44 km², mean altitude 3180 m) in the Oetztal Alps for the period 1952–80. The smaller and on average higher elevated Kesselwandferner is obviously more adapted to the present climate of the last 30 years. [Data sources: Haeberli (1985); Kasser (1967, 1973a); Müller (1977).]

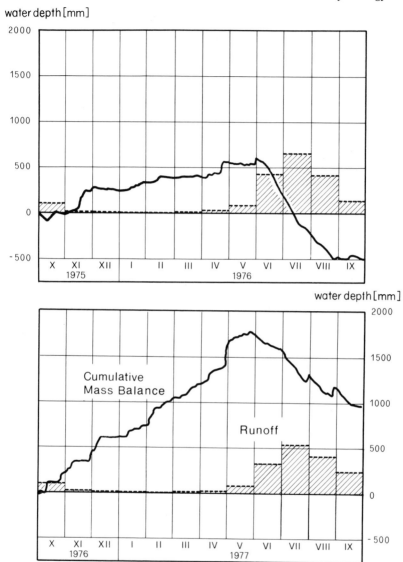

FIGURE 10.3 Seasonal processes of accumulation (storage) and ablation (yield) for the Aletschgletscher river basin, as shown from the cumulative hydrological mass balance $P - R - E$ in mm water depth. The vertical bars represent the monthly runoff from this basin (gauging station Massa/Blatten) at the same scale. There was very low precipitation and therefore little accumulation during 1975–76. This caused the mass balance to become negative as early as July. 1976–77 was characterized by ongoing and partly heavy mass accumulation from October to the end of May; the ablation period was, moreover, interrupted by two accumulation events. This resulted in one of the highest positive annual mass balances since the record in 1922. [Data sources: Hydrologisches Jahrbuch der Schweiz (1975–7); Kasser *et al.* (1983)]

important because they are decisive factors in the development of the glacier geometry.

Figure 10.3 gives an indication of the nature of seasonal variations in the water balance. Cumulative curves for the water balance of the Aletschgletscher basin are given for a maximum positive and maximum negative balance year and show the typical situation for the mid-latitude humid climate in the Alps (data source, Kasser *et al.*, 1983).

10.3   THE MELT PROCESSES AT THE GLACIER SURFACE

This section will consider the melting process at the glacier surface. First the energy or heat balance at the glacier surface will be considered and methods of estimating the components of this balance will be presented (Section 10.3.1), followed by some results of field studies on particular glaciers (Section 10.3.2).

10.3.1   The heat balance and its components

An understanding of the heat balance processes at the glacier surface is essential to an understanding of the runoff characteristics in glacier areas.

The energy (Q_M) available for melt on a horizontal snow or ice surface of unit area and over a unit of time consists of the following components:

$$Q_M = Q_{NR} + Q_S + Q_L + Q_P + Q_G \qquad (10.3)$$

where

Q_{NR} is net radiation;
Q_S is sensible heat;
Q_L is latent heat of condensation or evaporation;
Q_P is heat provided from precipitation;
Q_G is heat from heat conduction in the snow pack;
Q_M is heat used for melt or gained from refreezing of meltwater.

If all the components of the energy budget are known, the melt rate (melted water equivalent of snow or ice per unit area and per unit time) can be determined:

$$M = Q_M/S \qquad (10.4)$$

where S is the latent heat of melting ($333.7 \, J \, g^{-1}$ at $0\,°C$).

(*i*)   *Net radiation* (Q_{NR})

In a basin with heterogeneous surface conditions, it is preferable to determine each component of net radiation separately (i.e. incoming and outgoing, shortwave and longwave radiation fluxes), to facilitate accurate interpolation and extrapolation. In a flat, homogeneous basin, direct observations of total net radiation with a net radiometer are acceptable.

Net radiation can be estimated from the following equation:

$$Q_{NR} = G(1 - \alpha) + \epsilon_a \sigma T_a^4 - \epsilon_e \sigma T_e^4 \qquad (10.5)$$

where

G is global radiation;

α is albedo;

$\epsilon_{a,e}$ are emissivity of the atmosphere (a) and of the surface (e); for snow a value of 0.98–0.99 is used;

σ is the Stefan–Boltzmann constant $(5.67 \times 10^{-8} \, \text{J m}^{-2} \text{K}^{-4} \text{s}^{-1})$;

$T_{a,e}$ are the radiation temperatures of the lower atmosphere (a) and of the snow or ice surface (e).

If no direct measurements of the radiation fluxes are available, the shortwave net radiation, $G(1 + \alpha)$ can be estimated from sunshine duration records or from cloud observations, by taking into consideration latitude, time of year, exposure and slope. The albedo can be estimated from empirical relationships between α and the age of the snow surface layer [reviews on methods and procedures were given, for example, by Kuzmin (1961) and Male and Granger (1979); see also United States Army Corps of Engineers (1956)].

Longwave net radiation can also be estimated from empirical relationships based upon cloudiness. A compilation of results was given in Lang *et al.* (1977) and the problem was discussed in detail by Kondratyev (1969). A detailed treatment of the radiation conditions in the Alps was given by Müller (1984). In most cases of practical snow hydrology the snow and ice surfaces are at melting point (i.e. $T_e = 273$ K) and so the corresponding longwave emission $(\epsilon_e \sigma T_e^4)$ is well defined. Hence the problem of estimating longwave net radiation is essentially reduced to the estimation of longwave incoming radiation $(\epsilon_a \sigma T_a^4)$. Since the representative emission temperature of the atmosphere, T_a, is not known, empirical relationships for its estimation have been established using air temperature and vapour pressure observed at standard levels (1.5–2 m above the surface). An additional difficulty is the estimation of the atmospheric counter-radiation component of the longwave incoming radiation under cloudy conditions. There have been many investigations concerning appropriate empirical constants; a good discussion was provided by Kondratyev (1969).

(ii) Sensible heat (Q_S) and latent heat (Q_L)

Direct measurement of eddy heat fluxes is based on eddy correlation techniques. However, these techniques require sophisticated instrumentation which does not yet seem to be suitable for operational purposes. Therefore, the classical gradient method (or eddy diffusion method) is still widely employed because it performs well with accurately calibrated standard temperature and humidity sensors. The method is based to a considerable extent on the theoretical work of Prandtl (1904, 1934), Schmid (1925) and Lettau (1939). The application of the technique to ice and snow was primarily introduced by Sverdrup (1936).

According to theory, the sensible and latent heat fluxes are proportional to the gradients of potential air temperature (θ) and specific humidity in the atmospheric boundary layer:

$$Q = c_p \varrho K_S \cdot \frac{d\theta}{dz} \tag{10.6}$$

$$Q = L \varrho K_L \cdot \frac{dq}{dz} \tag{10.7}$$

where

c_p is the specific heat of dry air;
ϱ is air density;
K_S is the eddy diffusivity for sensible heat;
K_L is the eddy diffusivity for latent heat;
L is the latent specific heat of evaporation;
θ is the potential air temperature; for gradients over a few metres the actual air temperature can be used;
q is specific humidity (grams of water vapour per gram of moist air); $q = 0.622e/p$, where e is vapour pressure and p is atmospheric pressure;
z is the height of the sensor above the surface.

The equations are based on the assumption of constant flux with height. K_S and K_L are determined from wind profiles, which depend on the roughness of the surface (roughness parameter) and on the temperature stratification. Under neutral conditions in the constant flux layer the profiles may be expected to be logarithmic. From the analogy between eddy diffusion of momentum, heat and water vapour, it can be assumed that $K_S = K_L$ for neutral stability. Working equations for the estimation of sensible heat and latent heat have been derived as a result of these concepts. The equations that are most widely used are the following:

(a) According to the mixing length theory used by Sverdrup (1946):

$$Q = L \varrho k^2 \cdot \frac{u_2(e_2 - e_0)(0.622/P)}{[\ln(Z_2/Z_0)]^2} \tag{10.8}$$

where

k is the von Karman constant (usually taken to be 0.4);
u_2, e_2 are wind velocity (cm s^{-1}) and vapour pressure at level 2;

e_0 is the saturation vapour pressure at surface temperature at height Z;
Z_0 is the roughness parameter (the height above the surface at which $u = 0$);
Z_2 is the height of level 2.

Similarly, the sensible heat flux can be estimated from the following equation:

$$Q_S = C_p \varrho k^2 \cdot \frac{u_2(\theta_2 - \theta_0)}{[\ln(Z_2/Z_0)]^2} \tag{10.9}$$

(b) According to Thornthwaite and Holzman (1939), who used the gradient method for the determination of evaporation from land and water surfaces:

$$Q_L = L\varrho k^2 \cdot \frac{(u_2 - u_1)(e_2 - e_1)(0.622/P)}{[\ln(Z_2/Z_1)]^2} \tag{10.10a}$$

$$Q_S = c_p \varrho k^2 \cdot \frac{(u_2 - u_1)(\theta_2 - \theta_1)}{[\ln(Z_2/Z_1)]^2} \tag{10.10b}$$

In this case the wind velocities (u_2, u_1), vapour pressures (e_2, e_1) and temperatures (θ_2, θ_1) have to be observed at two levels (Z_2, Z_1) above the surface.

In both of the above cases, which assumed neutral stability, the eddy diffusivities for sensible and latent heat (K_S, K_L) are equal. Deviations from neutral conditions can be taken into account through the use of the Monin–Obukhov similarity theory (Monin and Obukhov, 1954). This theory was considered in detail by Male and Granger (1979). The classical method to correct for stability is to introduce the Richardson number. The magnitude of the correction increases substantially with decreasing wind speed.

(iii) Precipitation as a heat source (Q_p) (the release of heat by cooling and freezing)

Rainwater of mass m_p (g cm^{-2}) falling on to a melting glacier surface ($T_0 = 0\,°C$) provides heat proportional to the temperature of the rain drops arriving at the surface (T_p):

$$Q_p = C\dot{m}_p(T_p - T_0) \tag{10.11}$$

where C is the specific heat of water (4.1868 J g^{-1} °C^{-1}).

A simple example illustrates that even very heavy and warm rain does not constitute a very important heat source. For $m_p = 4$ g cm^{-2} with a temperature of $T_p = 10\,°C$, the maximum supply of heat can reach $Q = Cm_pT_p = 167.4$ J cm^{-2}. This heat is sufficient to melt only 5 mm of snow water equivalent!

If rain falls on a cold snow pack (i.e. temperature below $0\,°C$), thermal equilibrium is reached after freezing of the rain water and the release of its latent heat and the heat released by cooling to the snow pack. The heat released by freezing ($Q_M = Sm_p$, where S is the specific latent heat of freezing) can be an important factor in the ripening of a cold snow pack. The 'cold content' (C) of a snow pack (i.e. the heat necessary to warm the snow pack to $0\,°C$) is given by

$$C = c_S \int \varrho\, T_S(z) dz \qquad (10.12)$$

where
    c_S is the specific heat of snow or ice ($2.10\,\mathrm{J\,g^{-1}\,°C^{-1}}$ at ($0\,°C$);
    ϱ is the density of the snow pack at depth z;
    T_s is the temperature at depth z.
For example, considering a snow pack where $z = 100\,\mathrm{cm}$, $\varrho = 0.25\,\mathrm{g\,cm^{-3}}$ and $T = -10\,°C$, then under the assumption of homogeneous density, the 'cold content' is

$$C = (2.10\,\mathrm{J\,g^{-1}\,°C^{-1}})(0.25\,\mathrm{g\,cm^{-3}})(10\,°C)(100\,\mathrm{cm})$$
$$= 527.5\,\mathrm{J\,cm^{-2}}$$

This cold content could be removed by the release of latent heat from freezing of the following quantity of rain or meltwater;

$$m_p = Q_M/S = 527.5\,\mathrm{J\,cm^2}/333.7\,\mathrm{J\,g^{-1}} = 1.58\,\mathrm{g\,cm^{-2}}$$

This implies that 15.8 mm of rain at $0\,°C$ has sufficient thermal potential to bring the cold snow cover of the example to the melting point temperature.

In computations in glacial hydrology, the cold content of both the winter snow cover and the surface ice layers is an important retention component. For example, Ambach (1961) observed a cold content of the order of $4100\,\mathrm{J\,cm^{-2}}$ in the Hintereisferner towards the end of the winter. This is equivalent to the latent heat of fusion of 126 mm of meltwater. In the case of the Hintereisferner, about half of the cold content was compensated by refreezing of meltwater, which could be observed as ice layers in the snow pack or as superimposed ice on the glacier surface, and the remaining half was compensated by heat conduction.

10.3.2   Some results from field studies

Table 10.3 presents heat balance estimates for the Aletschgletscher and the associated mean and maximum melt rates. These rates are typical of alpine climatic conditions. Maximum melt rates of the order of 90 mm per day occurred in the lower parts of the ablation area on days with high total net radiation

TABLE 10.3 Results from heat balance measurements at the Aletschgletscher at two different altitudes: accumulation area (3366 m a.s.l.) and ablation area (2220 m a.s.l.). Heat flux values in J cm^{-2} d^{-1}; melt rates in mm d^{-1} water depth (Lang and Schönbächler, 1967; Lang et al., 1977)

| | | Shortwave incoming radiation, R_S (J cm^{-2} d^{-1}) | Net radiation (J cm^{-2} d^{-1}) | | | Heat flux (J cm^{-2} d^{-1}) | | Available Energy Q_M (J cm^{-2} d^{-1}) | Melt rates, M (mm d^{-1}) |
	Albedo, α		Shortwave, NR_S	Longwave, NR_L	Total, NR_T	Sensible, Q_S	Latent, Q_L		
Accumulation area (3366 m a.s.l.) **(3–19 Aug. 1973):**									
Daily values									
Mean:	0.74	2419	636	−251	385 92%	34 8.0%	−27	392 100%	11.8
Max.	0.88	2981	887	+54	569	10	18	572	17.1
Min.	0.66	1431	259	−620	155	−1.7	−323	116	3.5
Ablation area (2220 m a.s.l.) **(2–27 Aug. 1965):**									
Daily values									
Mean:	0.27	1863	1348	−230	1117 71%	326 21.0%	120 8.0%	1563 100%	46.9
Max.	0.42	2939	2168	+54	1704	656	741	3102	93.0
Min.	0.21	205	133	−578	188	42	−766	334	10.0

(NR_T) when fluxes from sensible heat (Q_S) and from latent heat (Q_L) were high. On average, the percentage contributions of these three components to the total heat available for melting were 71%, 21% and 8% for NR_T, Q_S and Q_L, respectively, whereas maximum melt rates were caused by contributions of 55%, 21% and 24%, respectively. In addition to the fact that net radiation is the most important of the three components, it is clear that the latent heat flux plays a very important role in governing day-to-day variations in the melt rate. This was also obvious from more detailed analyses of the melt processes (Lang, 1980). The relative contributions of the three components to the heat balance varies with altitude. Whereas net radiation increases systematically with altitude, Q_S and Q_L tend to decrease as a result of the vertical lapse rates of air temperature and vapour pressure. A review of current knowledge of the radiation conditions in the Alps was given by Müller (1985).

The great importance of net radiation in the ablation process at the glacier surface, and thus in the mass balance and hydrology of glaciers, means that the snow coverage of the glacier becomes very influential in the glacier's heat balance, because of the relatively high albedo of snow in comparison with glacier ice. The albedo values observed on the Aletschgletscher during the ablation season are presented in Table 10.3. For the snow-covered site in the accumulation area a mean daily value of $\alpha_a = 0.74$ was observed, whereas in the ablation area the mean daily value for the ice surface was $\alpha_b = 0.27$. This means that under the same conditions of incoming solar radiation G, the available energy from shortwave net radiation will be larger at the ice surface by a factor $G(1 - \alpha_b)/G(1 - \alpha_a) = 2.8$ in comparison with the snow surface of the accumulation area. The impact of snowfall during the ablation period in reducing melting and runoff, because of the snow's impact on glacier surface albedo, has long been recognized (Richter, 1888; Hoinkes and Rudolph, 1962; Tronov, 1962). For example, Lang (1966) estimated that one strong snowfall in July 1958 at the Hintereisferner, reduced the July runoff by 27% and the annual runoff by 7%.

The vertical lapse rates of air temperature and vapour pressure control the ranges at the glacier surface within which the sensible and latent heat flux are directed towards the glacier surface. Above this level the melting processes are strongly reduced or cease altogether. Mass loss by evaporation is relatively small because of the high specific latent heat, which is 8.6 times larger than the specific melt heat. At the same time, the large energy demand of evaporation strongly reduces total ablation, giving evaporation processes an indirect significance in mass balance and meltwater runoff formation.

In glacier catchment hydrology, the atmospheric elevation levels of 0 °C air temperature and 6.11 mb atmospheric vapour pressure (temperature and saturation vapour pressure for the melting glacier surface) become important, particularly in cases where these levels are observed in the elevation ranges of glaciers of great areal extent.

10.4 CHARACTERISTICS OF GLACIER RUNOFF

10.4.1 The compensating effect

Under the climatic conditions of the Alps, glacier discharge is dominated by the runoff of meltwater. The controlling factors are the heat fluxes and surface energy balance. Precipitation generally has a negative influence on glacier runoff because incoming solar radiation is reduced during precipitation events and because the new snow has a high albedo. Variations in glacier runoff follow approximately the reverse pattern of a rain-dominated runoff regime. As a result, it is possible to observe a compensation effect in mountain river basins where the upper parts of the basins experience a melt water runoff regime (nival and glacial regime) and at the same time in the lower parts of the basins runoff is dominated by rainfall. In alpine glaciated areas, much of the annual precipitation falls in solid form throughout the year, so that it contributes to mass storage rather than directly to runoff. The percentage of the annual precipitation falling as snow increases by 2.5–3.5% for every 100 m increase in elevation; at an altitude of 3500–4000 m a.s.l. in the Alps, approximately 100% of the precipitation is in solid form.

Very few storms provide liquid precipitation over the whole elevation range of an alpine glacier basin. Major rainfall-induced floods can result from such storms and they tend to occur during the second half of the ablation period when the winter snow cover and its retention capacity are at a minimum and

TABLE 10.4 Coefficient of variation for precipitation and runoff for July 1927 to 1947 in relation to percentage glacier area of different river basins (after Kasser, 1959)

River basin	Area (km²)	Glacier area (%)	Coefficient of Variation (%) Runoff	Precipitation
Massa/Massaboden	205	67.6	20	38
Drance de Bagnes/Châble	254	31.6	13	41
Rhone/Porte du Scex	5 220	16.2	5	33
Rhein/Rheinfelden	34 550	1.6	22	36
Emme/Emmenmatt	443	0.0	54	37

TABLE 10.5 Monthly coefficients of variation, CV, for the runoff from the Aletschgletscher basin (Massa/Blatten, drainage area 195 km², 67% glacier area) over the period 1964/65 to 1984/85

Month	10	11	12	1	2	3	4	5	6	7	8	9	10–9
CV (%)	37	28	18	30	25	31	58	35	24	20	18	31	11

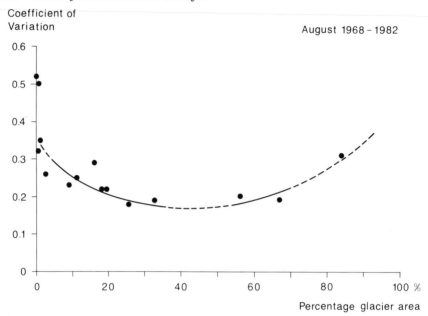

FIGURE 10.4   Coefficients of variation (CV) of runoff in August in relation to the percentage of glacier area for alpine river basins, fourteen of which are in the Swiss Alps (0–67% glacierized, period 1968–82) and one in the Austrian Alps (84% glacierized, period 1974–83). The compensation effect seems to be most effective at a percentage glacier area of between 30% and 60% where CV < 0.2. The river basins without glaciers have a CV which is similar to that of precipitation in the area (i.e. 0.3–0.5). The basins with a high percentage glacier area also show considerable variation in runoff

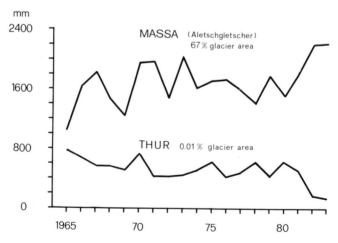

FIGURE 10.5   Mean summer runoff (May to September) during the period 1965–83 for two alpine river basins (Aletschgletscher, 67% glacierized; Thur, 0.1% glacierized). The opposing responses to hydro-climatic conditions are obvious

when, at the same time, the internal drainage system of the glacier is well developed. An example of such an extreme flood event has been described for the Hintereisferner region by Rudolph (1962).

The combined positive and negative effects of precipitation and melt processes on glacier river discharge reduce the variability of stream flow from alpine river basins that include some glacier cover within their catchment areas. It is of great benefit to water resources management of mountain river systems to have meltwater as a significant component of total discharge, particularly during dry, warm summer periods when the meltwater can augment 'dry weather flow' and so compensate for the low flow from the rain-dominated parts of the catchment area.

In the Alps, a minimum variation in annual runoff is observed from river basins with 30–40% glacier cover (Kasser, 1959), whereas during the main ablation season, the maximum compensation effect is obtained with a lower percentage glacier cover (Table 10.4). Whilst the coefficient of variation of July precipitation lies between 33 and 41%, the river Rhone/Port du Scex (16.2% glacier cover) has a coefficient of variation of July runoff of only 5% (during the period 1927–47, the influence of hydropower storage reservoirs was negligible). The maximum coefficient of variation of July runoff in Table 10.4 is associated with the lower alpine basin of the river Emme, which contains no glaciers.

The analysis of more recent runoff data (1968–82) from a sample of 15 basins in the Alps indicates that the minimum coefficient of variation of runoff for the main ablation season is associated with a glacier cover of 30–60% (Figure 10.4).

Table 10.5 presents the annual pattern of coefficients of variation of monthly runoff from the Aletschgletscher. The large variation in runoff in the spring months results from the great differences in the intensity of runoff at the beginning of the melt season from year to year. Figure 10.5 provides a further illustration of the compensation effect on the runoff regime which results from different degrees of glacier cover. It contrasts runoff under similar climatic conditions from two drainage basins located 120 km apart; one basin contains extensive glacier cover whereas the other is glacier-free.

10.4.2 Diurnal and annual cycle, storage processes

The diurnal and annual cycles of glacier runoff are primarily controlled by incoming solar radiation and the associated pattern of air temperature variations. Figure 10.6 illustrates the diurnal cycle of meltwater runoff for a 4480 m^2 area of the lower part of the Aletschgletscher. The hydrograph corresponds closely to the pattern of global radiation (incoming direct and diffuse solar radiation), which is the main source of melt energy, and the runoff maximum occurs immediately after the radiation maximum. Runoff from the whole basin (195 km^2, 67% glacier cover) is characterized by a considerable 'base-flow' upon which is superimposed a distinct diurnal cycle of runoff with a maximum lag of a few hours after the daily melt maximum.

Hydrograph analysis, tracer experiments and isotope studies permit a better understanding of the glacial drainage system (e.g. Lang, 1966, 1968, 1973; Elliston, 1973; Ambach *et al.*, 1974; Stenborg, 1970; Behrens *et al.*, 1971, 1982; Lang *et al.*, 1979; Tangborn *et al.*, 1975; Collins, 1977, 1982). From such

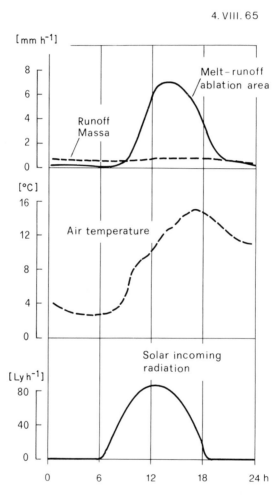

FIGURE 10.6   Typical diurnal courses of meltwater runoff from a 4500 m² area in the lower ablation zone of Aletschgletscher and of runoff from the whole drainage area of the Aletschgletscher basin (Massa/Blatten, catchment area 195 km², 67% glacierized), at the same scale, mm h⁻¹. Air temperature at a nearby climatological station and incoming solar radiation

studies it appears that the 'base-flow' consists of several different components:
— groundwater runoff;
— runoff from storage zones within the ice (e.g. water-filled cavities);
— runoff from the firn water aquifer which is fed by summer melt processes in the accumulation area of the glacier;
— regular drainage from lakes.

The superimposed diurnal cycles of runoff consist of rapidly draining components of that day's meltwater:
— meltwater from the lowest parts of the glacier basin, draining supra- and subglacially to the proglacial stream;
— meltwater from the snowfree part of the glacier which drains along short connections to the main subglacial conduits. The theory of the development

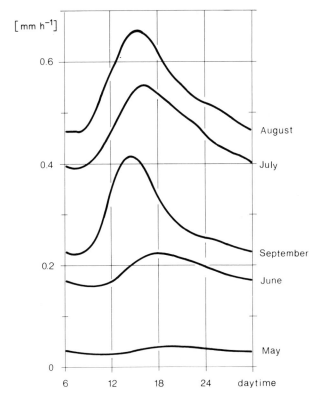

FIGURE 10.7 Mean diurnal variation of runoff, 1974–80, of Vernagtbach (Austrian Alps), drainage area 11.44 km², 84% glacier area. The progressive development of the glacier drainage system during the ablation period is clearly reflected in the changing characteristics of the diurnal variation in runoff from month to month (illustrated by changes in the time of daily maximum flow, the gradients of the rising and falling limbs of the hydrograph). [Data source: Oerter (1981).]

of the main subglacial drainage conduits as the result of water pressure and the release of potential energy was developed by Röthlisberger (1972) and is explored in detail later in this chapter. Analysis of the recession characteristics of mean daily hydrographs has provided evidence of the progressive development of the drainage system of the Aletschgletscher in the early part of the ablation season and of its significant increase in extent from June to July (Lang, 1968). An excellent example of the seasonal development of the glacial drainage system as reflected in the diurnal variations in runoff is given in Figure 10.7.

Seasonal water storage. The perennial storage of meltwater in firn aquifers in the accumulation areas of alpine glaciers was revealed by field studies on the Vernagtferner and the Ewigschneefeld (Aletschgletscher) (Oerter and Moser, 1982; Schommer, 1976, 1978; Lang *et al.*, 1977, 1979). Water level measurements and tracer studies revealed distinct firn aquifer recharge during the ablation season. The rate of drainage from the aquifer into the main glacial drainage conduits appears to vary considerably, ranging from a few days to over a year, and the total volume of water in storage is as yet unknown. Differences in the volume of water in storage may occur from year to year and these possible differences should be taken into account when the hydrological method is applied to determine the mass balance of glaciers.

Glacier runoff during the winter. Glacier runoff in winter is characterized by a 'dry weather' exponential decline in flow which begins immediately after the cessation of the ablation season in autumn. The beginning of the runoff season occurs well after the commencement of snowmelt as a result of the thermal and hydraulic retention capacity of the snow cover and of the intra- and subglacial drainage system.

10.4.3   Aperiodic variations and extreme floods

Aperiodic runoff variations are related to (a) the stochastic components of alpine weather conditions which determine rainfall, snowfall and the intensity of the ablation process (Figure 10.8), and (b) sudden changes in the unstable glacial drainage system, including the retention and release of water resulting from the collapse of tunnels and associated channels, and outbursts from ice-dammed lakes and from intraglacial and subglacial water pockets and cavities (Figure 10.9).

In addition to the daily periodic flow variations, glacier runoff exhibits extremely variable flow characteristics within the ablation season. The ratio of daily minimum to maximum flow is large for small glacier basins, but with increasing basin and glacier size, this ratio decreases as a result of the increasing 'base-flow' which results from the superposition of different lags in flow time.

There are three processes which can cause extreme floods in glacier basins: extreme melt rates, storm precipitation and sudden outbursts of stored water from ice-dammed lakes or from water pockets within the glacier.

Extreme melt rates are caused by maximum heat fluxes to the glacier surface. Maximum heat fluxes occur on the snow-free (low albedo) part of the glacier surface on radiation days combined with a maximum of sensible and latent heat flux. Such conditions result in the occurrence of maximum melt rates at all elevation zones of the glacier simultaneously. Maximum observed, daily, point

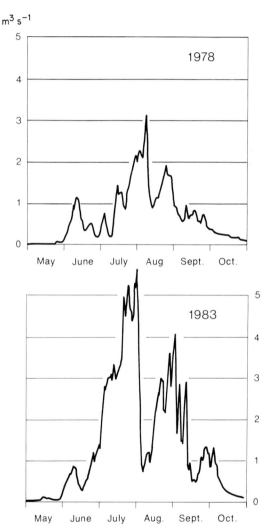

FIGURE 10.8   The 'irregular' character of glacier runoff exemplified by mean daily flows of the Vernagtbach (drainage area 11.44 km², 84% glacierized) during the ablation season of a 'minimum' (1978, positive mass balance) and a 'maximum' year (1983, negative mass balance). In both years the strikingly strong summer snowfalls interrupted the melt periods (an albedo effect). [Data source: Oerter (1984).]

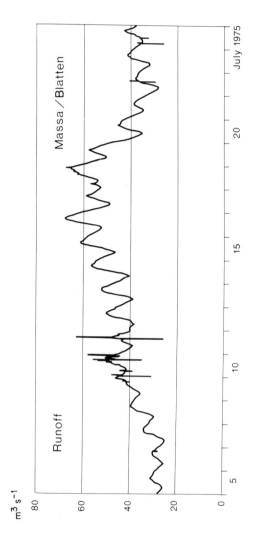

FIGURE 10.9   Diurnal variations of runoff from the Aletschgletscher river basin (Massa/Blatten, catchment area 195 km², 67% glacierized) during a typical melt period in July 1975 with high air temperatures and high net radiation. The base flow increases from day to day, mainly fed by meltwaters with long flow times from the accumulation area of the glacier. The sudden flows on the 10th/11th July are caused by calving glacier ice suddenly damming up the river for a short time and equally rapidly releasing the flow (after Emmeneger and Spreafico, 1979)

melt rates are in the range 90–100 mm d^{-1}, which corresponds to runoff of approximately 1100 l s^{-1} km^{-2}. The probable maximum heat flux on a horizontal ice surface can result in maximum hourly melt rates of up to 10 mm water equivalent on mid-summer days. This can yield a maximum specific runoff of 2800 l s^{-1} km^{-2} from the lower parts of valley glaciers (e.g. runoff coefficients of nearly 1.0 were recorded on an experimental plot of 4500 m^2 on Aletschgletscher at 2200 m a.s.l.). Maximum meltwater runoff from a large glacier basin occurs only after a 'high-pressure weather' period of a week or more with the 0 °C isotherm remaining at a high elevation. In such circumstances the 'base-flow,' which consists mainly of meltwater from the accumulation area which has a long time of concentration, can steadily increase over a considerable number of days (Figure 10.9). Table 10.6 illustrates some specific runoff values related to the whole glacier area from a sample of glaciers.

In high alpine regions there are few data available to reveal the characteristics of *storm precipitation induced floods*. Undoubtedly there have been rare, high-intensity rainfall events extending to elevations of over 2500 m a.s.l. in the Alps. In general, high-intensity rainfall occurs in periods of reduced receipt of solar radiation and thus reduced melt rates. However, it should be recognized that an extreme flood may occur when a period of maximum meltwater runoff from the lower part of a glacier combines with an intense rain storm in the late afternoon or early evening. For example, such a flood occurred on 17th September 1960 in the Oetztal Alps, causing great damage downstream. Unfortunately, there was no gauging station operative near a glacier to record the details of the proglacial discharge hydrograph. Qualitative observations of a similar event at the Hintereisferner were reported by Rudolph (1962).

Systematic observations of maximum floods caused by sudden *outbursts* of subglacial water pockets and glacier-dammed lakes have been reported by Röthlisberger (1981) and Haeberli (1983). The nature of such floods are discussed in detail later (Section 10.8), but it is worth noting here that these floods are

TABLE 10.6   Maximum specific meltwater runoff observed at various glacier river basins (related to the total glacier area)

Hintereisferner (Austrian Alps, 15.45 km^2) (Lang, 1966)	10th July, 1959 (period 1957–59) Max. daily mean: 640 l s^{-1} km^{-2} (55 mm d^{-1}) Max. 2-hourly mean: 950 l s^{-1} km^{-2}
Vernagtgletscher (Austrian Alps, 9.3 km^2) (calculated from data from Moser *et al.*, 1983)	July 1976 (period 1974–82) Max. hourly mean: 780 l s^{-1} km^{-2}
Aletschgletscher (Swiss Alps, 123 km^2) (calculated from data from Emmenegger and Spreafico, 1979)	August (period 1965–77) Max. daily mean: 688 l s^{-1} km^{-2} Max. instantaneous value: 854 l s^{-1} km^{-2}

characteristic of some glaciers and can reach catastrophic proportions far in excess of the estimated magnitudes of meltwater or rainfall induced floods. Some examples of this type of catastrophic flood include:

1. In the Oetztal valley, Austria, several sudden advances of the Vernagtferner dammed the main valley of the Rofenache glacier river during the Little Ice Age, resulting in catastrophic flooding (Hoinkes, 1969).
2. Outbursts from a lake in the Val de Bagnes, Switzerland, in 1599 and 1818 resulted in the destruction of parts of the community of Martigny.
3. There is a long list of flood events which have occurred in Valais, Switzerland, and which can be identified from chronicles dating back to 563 AD

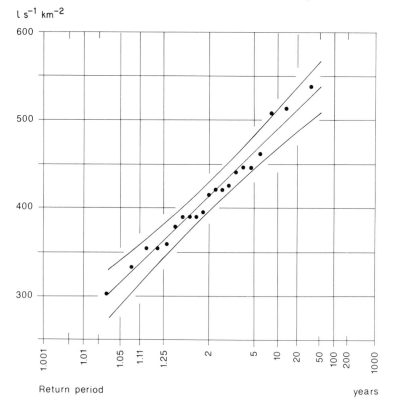

FIGURE 10.10   River basin of the Aletschgletscher, Massa/Blatten, 195 km² drainage area, 67% glacier cover: maximum annual instantaneous runoff values ($l\,s^{-1}\,km^{-2}$) for the period 1965–85 plotted on normal probability paper together with the fitted straight line and 80% confidence limits. There is clear evidence that all of the annual maximum values during the period were caused solely by meltwater flow with the exception of the three highest floods which appear to have been amplified by simultaneous rain storm activity and which do not fit the same statistical distribution well. (Data source: *Hydrologisches Jahrbuch der Schweiz.*)

[e.g. for the Aletsch area (Märjelensee), see Lütschg (1926, pp. 382–454); for the Mattmark area, see Lütschg (1915)].

An estimate of the return period of large flood events is frequently required for engineering purposes. This discussion has illustrated that great care should be taken when determining the flood frequency characteristics of glacierized basins. Floods from glacierized basins are usually induced by high-intensity rainfall or by high melt rates. These events should be treated separately in the statistical analysis of flood frequency from floods induced by outbursts from glacier-dammed lakes or subglacial water pockets, particularly if such outburst events form the largest recorded floods. If there are no flow records for a basin, a field survey could be invaluable in detecting the existence and nature of any glacial lakes and in identifying the storage and discharge mechanisms of such lakes. In Figure 10.10 the probability distribution of the maximum annual runoff for the Aletschgletscher basin is given, covering the period 1965–85.

10.4.4 Long-term variations

The long-term variations in glacier runoff consist of two components:

1. There exists a direct relationship between climatic variations and accumulation/ablation processes, water storage and runoff. Figure 10.1 presents runoff and specific mass balance records for the Aletschgletscher for the period 1923–85. The maximum loss of glacier mass occurred in the decade 1940–49, the warmest decade since the commencement of records. The extreme ablation rates of that period caused the maximum runoff values. In contrast, a series of 'bad weather' summers (cool with several periods of snowfall during the ablation season) favoured positive mass balances over the periods 1965–70 and 1975–80 and, as a consequence, very low annual runoff levels were recorded.

2. The predominance of negative glacier mass balances in this century in the Alps has resulted in the strong shrinkage of the lower sections of glacier tongues. Since these are the zones of highest melt rate, the progressive disappearance of these zones has resulted in a corresponding loss of potential for meltwater yield. An analysis of long-term runoff records for the Rhone river basin (Porte du Scex, 5220 km^2) revealed a clear declining trend in summer runoff as a result of the reduction in glacier area (Kasser, 1973b. Some of the results of this analysis are presented in Table 10.7. This example illustrates the order of magnitude and the great hydrological significance of recent secular glacier variations. Table 10.8 provides an indication of the magnitude of such changes over the entire region of the Swiss Alps.

The general retreat of Alpine glaciers from the last maximum of the Little Ice Age came to an end in 1965 (Kasser, 1981; Patzelt, 1973). At present it is

TABLE 10.7   Glacier variations and resulting hydrological effects in the river basin of the Rhone/Porte du Scex (drainage area 5220 km²), mean altitude 2130 m a.s.l., altitude range 373–4634 m a.s.l.)

Glacier area percentage in the years 1916 and 1968:
  1916: 16.8%
  1968: 13.6%

Change of glacier area in the period 1916–68: – 19% (of the area in 1916)

Influence of the glacier area loss in the period 1916–68 on the runoff conditions, estimated from trend analyses and from direct melt water estimations; with the assumption of equivalent meteorological conditions, the runoff shows at the end of this period a systematic reduction of 141 mm. This corresponds to a reduction of 16% of the mean summer runoff of 887 mm (April–September) (Kasser, 1973b)

TABLE 10.8   Hydrological significance of the glacier retreat in the Swiss Alps

  1876:   1817.6 km² glacier area
  1973:   1342.2 km² glacier area

1876–73:   475.4 km² loss of glacier area
           26% of the 1876 area (0.27% per year)

Estimated potential melt water yield from these lower areas of the glaciers in an average year: 3 m (water equivalent) or 1426×10^6 m³.
  Taken over the whole area of Swiss river basins, this glacier area loss causes a reduction of the annual specific runoff of 35 mm or 3.4% in an average year at the end of the stated period of glacier recession (Kasser, 1981)

generally agreed that mean air temperatures will increase in future decades more than ever before in historic times as a result of the increase in CO_2 and other 'greenhouse' gases in the atmosphere (UNEP WMO ICSU, 1985). As a response, the glaciers will probably again lose ice mass and area, inducing another declining trend in the meltwater runoff potential, although the problem is complex. The climate–glacier runoff system is nonlinear because of feedback mechanisms and it is also necessary to take account of the fact that the reaction of different glaciers to climatic change may vary considerably (Meier, 1983). Further, we cannot be sure of the impact of increasing air temperature on precipitation and other climatic parameters. Current climatic models indicate a tendency towards lower precipitation in the middle latitudes which, coupled with increasing temperatures, would result in a period of pronounced glacier retreat and thus a marked loss of meltwater runoff potential. Initial attempts to estimate the reactions of mid-latitude glacier mass balance to predicted climatic changes have been reported by Kuhn (1985).

10.5   INTRA- AND SUBGLACIAL DRAINAGE

The remainder of this chapter will consider in detail a number of aspects of the drainage of water through glaciers. However, before considering any of these aspects in detail, it is necessary to provide a brief overview of the sources of water which contribute to runoff from glacierized drainage basins and of the fundamental components of the drainage of the water through glacier systems.

As was indicated in Section 10.3, in the alpine glacial environment runoff originates mainly at the glacier surface from melting or as rain or dew. For simplicity, all the water from the surface will be referred to as meltwater in the present discussion, except when rainfall events are specifically mentioned. Much less water forms internally in a glacier as a result of the dissipation of energy in relation to the flow of ice or water. For example, a simple calculation shows that a block of ice at 0 °C has to fall 34 km vertically to melt as a result of loss of height, which means that a very small percentage of ice melts by loss of potential energy even in glaciers with a very large altitudinal range. The major part of frictional melting occurs at or very near to the glacier sole rather than internally. Additional melting at the contact between the ice and the substratum is caused by geothermal heat flow. For a typical Alpine glacier, surface melt rates vary between 0.1 and 10 m yr^{-1} depending on altitude (see, for example, Table 10.3), whilst the frictional and geothermal melt rates are of the order of only 10^{-2} m yr^{-1}. A consideration of the relative magnitudes of these various sources of meltwater indicates that, within the context of glacial hydrology, emphasis must be placed on the surface-derived meltwater and its drainage. In addition to meltwater, some water may also reach the glacier from the surrounding unglacierized area, either by surface flow or as groundwater, and so may subsequently contribute to runoff.

The mode of drainage varies substantially according to glacier surface conditions. If the surface layer consists of snow or firn, the meltwater will percolate through this layer as through an unsaturated porous medium (Meier, 1973; Ambach *et al.*, 1981) until it reaches a layer of considerably lower permeability, usually glacier ice. The snow or firn layer forms an aquifer in which water is stored and moves in the same way as ordinary groundwater, with a water table and flow under gravity and according to Darcy's law. The main difference between this snow or firn aquifer and an aquifer in regular unconsolidated sediments is likely to be its more rapid change in permeability with time in connection with the metamorphism of the snow. Water can drain from the firn aquifer in three main ways: outflow at the firn line into channels on the glacier surface, drainage into crevasses and seepage through the glacier ice.

In the ablation area of a glacier the path of the meltwater can be followed at the glacier surface, but after some distance that water also usually disappears into the glacier. Large surface streams which flow as far as the glacier margins or terminus are an exception. Instead, most of the water disappears down

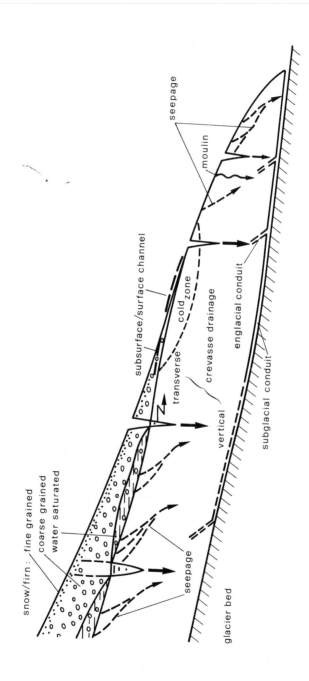

FIGURE 10.11   Schematic diagram of supraglacial, englacial and subglacial drainage passageways

crevasses or moulins, but seepage of water through the glacier ice should also be considered. The various drainage paths are sketched in Figure 10.11. For the englacial and subglacial drainage an arborescent system of passageways is a commonly accepted concept. Numerous individual conduits join together, eventually forming one or a few final subglacial tunnels that leave the glacier at a portal.

Although there are important differences between the drainage from the accumulation and the ablation areas of glaciers, the internal passages which convey the water are probably similar and obey the same laws. It is, therefore, appropriate to deal with the englacial drainage as a whole. When considering the question of the permeability of a glacier, it is important to be aware of the existence of a thin surface layer of the glacier which extends some distance above and below the equilibrium line and which may be permanently frozen (Figure 10.11). The extent and thickness of this layer depends on climatic conditions so that the frozen layer is thickest when dry and cold weather prevails in the autumn. For a limited period of the winter and spring into early summer, the whole of the ablation surface may be cold (if it is not covered by wet snow throughout the winter, as may occur in a humid maritime environment).

Since the interior of the glacier is virtually inaccessible, the character of the englacial and subglacial drainage system and its physical parameters have to be deduced from field measurements of water input and output and from theoretical deductions. It is scarcely surprizing, therefore, that many questions have remained unanswered but it is the aim of the following sections to summarize what is known and what remains to be explained.

10.6 WATER FLOW IN ICE CONDUITS AND VEINS

10.6.1 Basic principles of water flow in ice conduits

Many processes influence the size of englacial passageways. Some of these processes are important regardless of the size of the conduit whereas other processes may be relatively unimportant except when the diameters of the englacial passageways are either very large or very small. It is useful, therefore, to distinguish between large passages, usually referred to as tunnels, conduits or channels, and small tubes, tube-like features or veins. The term 'vein', in particular, is used by many authors to describe capillaries situated along three-grain intersections (Nye and Frank, 1973). These veins are triangular in cross-section with curved walls convex towards their centres (cf. Figure 10.18). Processes that become important in small passageways are related to heat flow in the ice and to surface energy problems at the ice–water interface (i.e. to wall curvature and capillary forces). The transition between 'large' and 'small' occurs at around 10^{-3}–10^{-4} m^3 s^{-1} for the discharge (cf. Figure 10.24).

FIGURE 10.12   Englacial drainage conduit (*photograph: H. Röthlisberger*)

Neglecting effects that are specific to small passageway, processes influencing the size of conduits are related to heat transfer between the transmitted water and the surrounding ice and to deformation of the ice. The relative importance of these two groups of processes varies considerably both with conduit size and with local conditions. The controlling factors on conduit size are usually the melting of the ice as a result of the frictional heat produced in the flowing water [also referred to as viscous dissipation of potential energy (Meier, 1973)], and the closure of the conduit by plastic deformation of the ice because of an overburden ice pressure that is larger than the water pressure. However, the gain or loss of energy when the water adjusts to the pressure-dependent melting point can be equally, or even more, important.

To facilitate computations, englacial passageways are assumed to be circular in cross-section when operating as pressure conduits. It appears that real conduits are usually very close to this circular cross-sectional shape, because there is a tendency for the ice flow to approach such a form (Nye, 1965). Deviations from the circular form may result from anisotropic ice, anisotropic stress in the surrounding ice, characteristics of the water flow (e.g. meandering or vortex flow) and sediment load. The strong effects which result from alternations between pressure flow and free-surface flow (i.e. under atmospheric pressure) and the special conditions at the glacier bed will be discussed separately later. Figure 10.12 provides an illustration of an englacial conduit.

Many authors have contributed to the development of the theory of englacial drainage. The first stage concerns the solution of some problems relating to stationary conditions (Röthlisberger, 1972; Shreve, 1972; Weertman, 1972). In this case the primary unknown variable to be computed is the water pressure (p_w) in the conduit, while the geometry of the glacier surface, the position of the conduit within the glacier and the discharge are assumed to be known. Figure 10.13 illustrates an arbitrary conduit element of length ds. The frictional

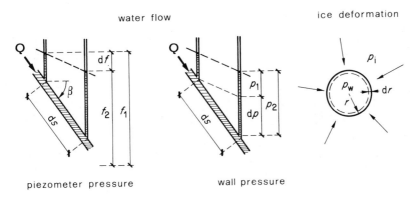

FIGURE 10.13 Differential elements for the computation of water pressure in ice conduits with a circular cross-section

pressure loss ($df = f_2 - f_1$) along the conduit element is indicated by the change of piezometer level (measured in relation to a horizontal datum), while $dp = p_2 - p_1$ refers to the change of local water pressure within the conduit (i.e. the water pressure exerted on the wall). For convenience a coordinate axis, s, is used which coincides with the axis of the conduit and points in the direction of water flow [i.e. opposite to the x-axis in Röthlisberger (1972)]. The relation to a cartesian coordinate system (x, y, z) is given by $ds = \sqrt{dx^2 + dy^2 + dz^2}$. The relation between df, dp and the slope (β) of the conduit follows from geometric considerations, given that $dp = df$ for a horizontal conduit, as

$$-dp = -df + ds\ \sin\beta \tag{10.13}$$

when the conduit is inclined, with β positive when the conduit slopes upwards in the positive direction of s (converse dip).

The frictional melt rate ($\dot{m}_f$) by volume in the section ds is given by

$$\dot{m}_f = c_m^{-1}\varrho_i^{-1}\varrho_w g Q\left(-\frac{df}{ds}\right)ds = C_1 Q\left(-\frac{df}{ds}\right)ds \tag{10.14}$$

where

c_m is the energy of fusion;
ϱ_i is the density of ice;
ϱ_w is the density of water;
g is the acceleration due to gravity;
Q is the discharge.

Using the numerical values given in Table 10.9, a value of $3.21 \times 10^{-5}\,\text{m}^{-1}$ is obtained for C_1.

The rate of melting (or freezing) ($\dot{m}_p$) that occurs when the flowing water adjusts to the new pressure melting point as the water flows through the element ds can be obtained from

$$\dot{m}_p = c_t c_w c_m^{-1}\varrho_w^2\varrho_i^{-1}g Q\left(-\frac{dp}{ds}\right)ds = C_2 Q\left(-\frac{dp}{ds}\right)ds \tag{10.15}$$

where

c_t is the change of pressure melting point with pressure;
c_w is the specific heat capacity of water.

Using the numerical values given in Table 10.9 for c_w and $c_t = c_{t,0}$ for pure water, one obtains $C_2 = -10^{-5}\,\text{m}^{-1}$, therefore $C_2/C_1 = -0.312$ (the same factors are valid when the concentration of impurities, i.e. air, does not change with pressure). Most authors (Röthlisberger, 1972; Shreve, 1972; Spring, 1980; Spring and Hutter, 1981, 1982; Hooke, 1984) have used these values. Lliboutry (1983) suggested taking $c_t = c_{t,s}$ for air-saturated water instead (cf. Harrison,

TABLE 10.9 Physical constants of ice and water at 0 °C and related properties

Symbol	Property	Quantity	Unit
g	Acceleration due to gravity	9.807	$m \, s^{-2}$
ϱ_w	Density of water	999.8	$kg \, m^{-3}$
ϱ_i	Density of ice	916	$kg \, m^{-3}$
c_w	Specific heat of water	4.217×10^3	$J \, kg^{-1}K^{-1}$
c_m	Energy of fusion	3.34×10^5	$J \, kg^{-1}$
$c_{t,0}$	Change of pressure meting point with pressure (pure water)	-7.4×10^{-8}	$K \, Pa_{-1}$
$c_{t,s}$	Change of pressure melting point with pressure (air-saturated water)	-9.8×10^{-8}	$K \, Pa^{-1}$
η	Viscosity of water	1.787×10^{-3}	$kg \, s^{-1}m^{-1}$
A	Ice deformation coefficient used by various authors:		
	Paterson (1981)	0.167	$bar^{-3}a^{-1}$
	Hooke (1984)	0.244	$bar^{-3}a^{-1}$
	Lliboutry (1983)	0.232	$bar^{-3}a^{-1}$
	Röthlisberger (1972)	1.00	$bar^{-3}a^{-1}$
n	Exponent in the power law of ice deformation	3.00	—
k	Hydraulic roughness parameter for various conduits:		
	Smooth	100	$m^{1/3}s^{-1}$
	Medium	50	$m^{1/3}s^{-1}$
	Very rough (torrent)	10	$m^{1/3}s^{-1}$

1972), obtaining a value of $C_2 = -1.33 \times 10^{-5} \, m^{-1}$, and therefore $C_2/C_1 = -0.414$. The use of $c_{t,s}$ implies that the gas concentration adjusts to the varying local pressure throughout the drainage system, which is hardly reasonable in an ice conduit, except, perhaps, under special conditions when sufficient air is dragged into a moulin. Since Metcalf (1984) has reported that summer meltwater from ice is strongly undersaturated in CO_2 (by about ten times), it can be implied that the same water is also undersaturated in air. No concentration change of impurities is therefore to be expected with a change of pressure, i.e. it is generally justified to use the first set of factors cited above, related to $c_t = c_{t,0}$.

The rate of conduit closure by radial ice flow can be estimated from the equation given by Nye (1953) (the factor A of Paterson, 1981, p. 39 is used in place of Nye's A^{-n}):

$$\frac{\dot{r}}{r} = A n^{-n} (p_i - p_w)^n \tag{10.16}$$

where r is the conduit radius and n and A are ice deformation parameters.

The rate of conduit closure by volume as a result of plastic ice flow into the conduit element of length ds is obtained from

$$\dot{m}_c = 2r\pi\dot{r}ds = 2\pi r^3 A n^{-n}(p_i - p_w)^n ds \qquad (10.17)$$

In order to link $\dot{m}_c$ with $\dot{m}_f$ and $\dot{m}_p$, a further relationship is needed linking water flow to conduit size. For laminar flow the following equation is used:

$$Q = \frac{\pi \varrho_w g}{8\eta} \cdot r^4 \left(-\frac{df}{ds} \right) \qquad (10.18)$$

where η is the viscosity of water at the pressure melting point.

In the case of turbulent flow, various equations can be used, for example the Gauckler–Manning–Strickler equation (Röthlisberger, 1972), which, with k as a roughness parameter, is expressed as

$$Q = 2^{-2/3} \pi k r^{3/8} \left(-\frac{df}{ds} \right)^{1/2} \qquad (10.19)$$

Alternatively, there is the Darcy–Weisbach law which was employed by Spring (1980) and Spring and Hutter (1982). In this case a relation of $Q \infty\ r^{2/5}$ instead of $r^{3/8}$ is introduced (Lliboutry, 1983). Since Q and not r is the known variable, r can be eliminated, leading to the following equation for $\dot{m}_c$:

$$\dot{m}_c = C_3 Q^q (p_i - p_w)^n \left(-\frac{df}{ds} \right)^m ds \qquad (10.20)$$

The factor C_3 depends on the mode of water flow, the channel roughness and the ice flow parameters. The exponent n is the exponent of the flow law of ice [equation (10.16)]. The other exponents (q and m) depend on the law of water flow and will have a magnitude of less than unity. Unlike C_2 and C_1, no universally applicable value of C_3 can be given because C_3 will vary with local conditions. These conditions may vary greatly both with time and also spatially within a drainage system. Particularly significant unknown factors include the channel roughness (which will vary greatly according to whether the water is flowing in an extremely smooth tube in the ice or a boulder-strewn tunnel at the glacier bed) and the ice flow parameters (which are badly defined and depend on the location of the conduit, for example in the interior of the ice mass or in a layer of basal ice undergoing intensive deformation). This question of unknown or poorly defined parameters, which is of particular concern when considering drainage at the glacier bed, has been discussed widely (Röthlisberger, 1972; Lliboutry, 1983; Hooke, 1984).

The differential equations defined above serve primarily to facilitate the estimation of hydraulic pressure (i.e. the hydraulic grade lines for the steady state by integration). Assuming stationary flow with constant discharge (Q), the steady state is defined by

$$\dot{m}_f + \dot{m}_p = \dot{m}_c \tag{10.21}$$

which expresses the equilibrium condition when the volume change resulting from melting (or freezing) is balanced by ice flow. One important general finding from the combination of the equations for $\dot{m}_f$, $\dot{m}_p$ and $\dot{m}_c$ [equations (10.13), (10.14), (10.15) and (10.20)] is that the hydraulic gradient ($-\mathrm{d}f/\mathrm{d}s$) is proportional to Q^{-b}. The negative exponent indicates that large discharges are associated with smaller hydraulic gradients than small discharges. This implies that after integration the conduits with the highest discharge show the lowest water head. This means that large conduits may draw water from small ones and also that water is more likely to drain in channels than in sheets in the case of drainage at the bed. Shreve (1972) has discussed this question in more detail. Based on Röthlisberger's (1972) equations (20a) and (42), it can be inferred that the magnitude of the exponent b varies between a small fraction of one and one, depending on the mode of flow and the magnitude of $\mathrm{d}f/\mathrm{d}s$. When b is small, which is usually the case, the indicated effect will not be very pronounced; it is nevertheless not negligible because Q may vary by orders of magnitude!

10.6.2 Results of water pressure computations for some typical configurations of ice surface and englacial conduits

The integration of the expression for $-\mathrm{d}f/\mathrm{d}s$ arising from equation (10.21) can be carried out explicitly for the simplest case when the ice surface and the conduit

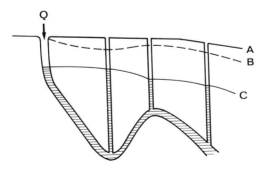

FIGURE 10.14 Hydraulic terms used to describe englacial and subglacial conduit flow: A = glacier surface; B = water equivalent line (ice overburden pressure shown as water column); C = hydraulic grade line (piezometer pressure, hydraulic head)

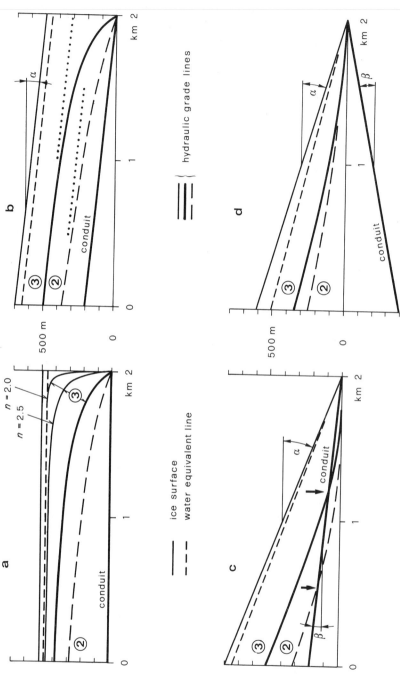

FIGURE 10.15   Hydraulic grade lines of some model cases with constant discharge $Q = 10 \, m^3 \, s^{-1}$. Ice flow parameters for curves 2 and 3 according to Table 10.9, Paterson (1981) and Röthlisberger (1972), respectively, and roughness $k = 10 \, m^{1/3} \, s^{-1}$ for a torrential-type conduit; exponent $n = 3$ for all curves except where indicated otherwise. *Reproduced by permission of the International Glaciological Society from Journal of Glaciology, 1972, 11, 186*

at a given depth are both horizontal. Solutions in other cases are obtained by numerical integration. Results from some typical situations that have been analysed by one of the authors (Röthlisberger, 1972) are presented in Figure 10.15 and some useful terms that are employed in these calculations are illustrated in Figure 10.14. All the computations for Figure 10.15 assume that where the ice body ends, whether at a vertical cliff or at the tip of a wedge-shaped terminus, the conduit also ends. The water pressure at the end of the conduit is atmospheric ($p_w = 0$; the atmospheric pressure is not taken into account in the computations). The hydraulic grade line ends at the outlet of the conduit, or else at a point some distance upstream where $-df/ds \geqslant -\beta$, below which point flow is atmospheric. Numerical computations are carried out backwards from that point using the physical constants of Table 10.9. Other combinations of the three parameters A, k and Q than those indicated in the figure caption could be adjusted to produce identical curves.

Case (a). Ice surface and conduit are both horizontal (e.g. a channel at the base of a horizontal ice sheet of constant thickness).

At some distance from the end of the conduit the hydraulic grade line is at or very close to the water equivalent line. Towards the outlet the hydraulic grade line becomes gradually lower and finally falls away steeply as the outlet is approached (Figure 10.15a). Depending on the value of n, the water equivalent line is either approached asymptotically at increasing distance from the outlet or it is reached tangentially at a finite distance from the outlet (the transition occurs at $n = 1.375$ when using the Gauckler–Manning–Strickler equation— [equation (10.19)] for the computation of water flow (cf. Röthlisberger, 1972, Figure 3). It is worth noting that at a sufficient distance from the outlet in the case of a small n, no conduit may stay open. This implies that the water will remain in a sheet and will not drain because of the lack of a pressure gradient. Although such a situation may appear to be academic, it can be taken to indicate slow drainage under particular conditions similar to those at the centre of an ice sheet where the ice surface is horizontal; small values of n have been suggested to be realistic at very low stress (at high water pressure in a conduit the stress causing plastic closure *is* very low).

Case (b). An inclined ice body with the conduit parallel to the surface (e.g. a channel at the inclined base of an ice body of constant thickness).

Figure 10.15b shows that the hydraulic head is located at some depth below the ice surface and it falls at an increasing rate towards the outlet of the conduit. The hydraulic grade line at a distance from the outlet can be approximated by a line parallel to the ice surface (i.e. parallel to the water equivalent line). This is indicated by dotted lines in the figure. The depth of these lines below the water equivalent line can be readily obtained by taking account of the equilibrium conditions $\dot{m}_p = 0$, $\dot{m}_f = \dot{m}_c$. From equations (10.13), (10.14) and (10.20), this depth (expressed as the difference between the water equivalent of the ice overburden, $p_i = (\varrho_i/\varrho_w)h\cos\beta$, and the water head, p_w, in the conduit) amounts to

$$p_i - p_w = \frac{C_1}{C_2} \cdot Q^{(1-q)/n} (-\sin\beta)^{(1-m)/n} \tag{10.22}$$

Since the exponents are all positive, q and m are less than 1 and n is larger, the depth of the hydraulic grade line below the ice surface increases with the discharge (Q) and the slope ($-\beta$). The numerical computations have shown that the limit of the water head as expressed by equation (10.22) seems to be reached at a relatively short distance from the outlet, although whether the equilibrium is reached at a finite distance or asymptotically is not known. The physical meaning of these results is that when both the conduit and surface are inclined (a very common situation), pressure flow can only occur when the conduit is located at sufficient depth to allow compensation of the melting resulting from frictional heat by the ice flow, otherwise free-surface flow would occur at atmospheric pressure (or under a vacuum). Therefore, steady-state pressure flow may never occur in thin glaciers (Figure 10.24; cf. Hooke, 1984).

Case (c). The ice surface and conduit are both inclined but at a different angle (e.g. a conduit at the base of a wedge-shaped terminus).

Figure 10.15c shows the case when the ice surface and the conduit slope in the same direction (i.e. when β is negative). In this case the hydraulic grade line can also be approached by a straight line for limiting water head, which is, as before, parallel to the water equivalent line. However, the hydraulic grade line is now located above the straight line and it levels out towards the conduit outlet as a result of the diminishing ice overburden. At some distance before reaching the outlet the hydraulic grade line drops below the conduit indicating negative water pressure. Since negative pressure cannot exist with free access of air, free flow must occur in the final section of the conduit between the intersection of the hydraulic grade line with the conduit and the conduit outlet. This intersection is indicated by arrows in Figure 10.15c. Even if no air could gain access to the conduit, the negative pressure would be limited to less than 10 m of water head, and consequently the section previously indicated to have free flow but now considered to be under a vacuum would not be greatly reduced. It is also possible to express by an explicit equation the limiting depth of water head below the water equivalent line but this becomes considerably more complicated than equation (10.22) because the angle (α) of the ice surface slope must be included as an additional variable. The limiting value of $p_i - p_w$ increases with discharge (Q) and the conduit slope ($-\beta$) as in the previous example, but it also increases with the surface slope ($-\alpha$) [note that case (c) is a special case of (b) when $\beta = \alpha$].

When the conduit is ascending (i.e. when β becomes positive), the hydraulic head rises closer to the water equivalent line (Figure 10.15d) or, when β is sufficiently large, the two lines coincide. This is the case when

$$\tan\beta = \frac{\varrho_i}{\varrho_w} \cdot \frac{C_1 + C_2}{C_2} \cdot \tan\alpha \tag{10.23}$$

This shows that the ascending flow has become independent of the discharge and so it is independent of conduit size. Flow in any size of conduit or in a sheet is equally favoured, because the water pressure is equal to the ice overburden pressure throughout. Nevertheless, a hydraulic gradient exists and is given by the slope of the water equivalent line. Frictional heat is balanced by the heat necessary to warm up the ascending water to adjust to the higher pressure melting point. There is no closure of conduits by plastic ice flow (if conduits pre-exist) and no new conduits can form because the ice and water pressure are equal and so no energy is available for melting. The numerical value of the factor $(\varrho_i/\varrho_w)[(C_1+C_2)/C_2]$ is -2.02 for pure water and -1.30 for air-saturated water. The above conditions are fulfilled for pure water when the slope of the ascending conduit amounts to twice the magnitude of the ice surface gradient, and for air-saturated water when the ascending conduit is 30% steeper than the slope of the ice surface. At larger ascending conduit gradients, supercooling of the water would occur, probably followed by freezing and the closure of the conduit. One of the authors has observed ice mounds around artesian wells situated a few hundred metres from the terminus of the Griesgletcher in spring. The inability of the water to maintain ascending conduits under certain conditions is of great importance in relation to subglacial drainage, particularly in overdeepened sections of glaciers.

10.6.3   Time-dependent discharge

Of the many simplifications introduced to establish the above model, one of the most important is to ignore heat transfer. For example, there is a discrepancy between the pressure melting point within the ice and in the water at the channel wall. This leads to a temperature difference between ice and water, although this is only significant for small conduits (Figure 10.24). In large conduits it is the delay in the transfer of the heat produced in the water to the ice wall which is important (i.e. the distance that the water travels while giving off the heat). For steady-state conditions this may not be very serious because the heat given off for melting becomes continuously resupplied over a reasonably short distance. Therefore, the neglect of heat transfer should result in acceptable approximations in steady conditions. In non-steady situations, however, it can be very important to take account of water temperature.

The constant discharge assumed in the basic concept of steady conditions in Sections 10.6.1 and 10.6.2 hardly ever exists in reality. For a full understanding of intraglacial drainage it is, therefore, important to consider the fluctuations of discharge with time. In the case of steady flow conditions when neglecting water temperature it was possible to carry out the integration along s starting at the lower end of the conduit where pressure is atmospheric. In the general case, where variations of discharge with time are included, a repeated integration over the length of the channel is necessary for each time interval to satisfy the

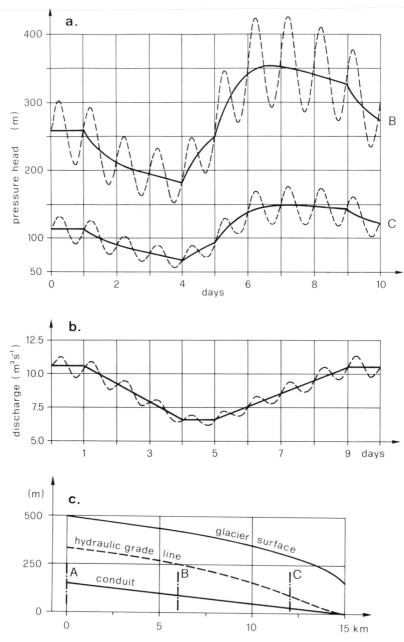

FIGURE 10.16   Computer simulation of the piezometric water pressure (a) for a given discharge (b); the longitudinal section of the glacier is shown at the bottom (c). The graph at the top (a) shows the piezometric pressure head at locations B and C, respectively.

boundary conditions at both ends of the channel. Three conditions have to be satisfied: water pressure and water temperature at the upper end of the conduit and water pressure (usually atmospheric) at the lower end of the conduit. Simplified equations for non-steady water and ice flow have been presented by Nye (1976), and a more general theory has been elaborated by Spring and Hutter (1981, 1982).

Typical examples of temporal variations in discharge are harmonic diurnal discharge fluctuations and outbursts from ice-dammed lakes. If the discharge exhibits high-frequency fluctuations, with periods ranging from a few hours to a day, the geometry of the conduit cannot respond to the fluctuations and so the pressure head varies in phase with the discharge. For very long periods (of the order of a year), the conduit adjusts to the near steady state of current conditions and as a result the pressure is out of phase with the discharge in accordance with simple steady-state theory (smaller pressure gradient and, therefore, lower integrated pressure accompanies a larger discharge and vice versa; see p. 241). Figure 10.16 illustrates that conduit adjustments occur within a few days.

So far the computations have assumed a given discharge, first constant, then varying with time. A fundamentally different situation is encountered when the water originates from a large reservoir with a fixed water level (Nye, 1976). If the water from the reservoir drains through an existing intraglacial conduit system, the initial discharge is determined by the pressure head between the reservoir and the portal. If the discharge is smaller than is needed to keep the conduit in equilibrium with closure by overburden pressure, then the conduit will contract and may eventually close, so reducing the discharge to zero. If, in contrast, the discharge is larger than is necessary, then the ice at the conduit walls will melt at a faster rate than it is moving in towards the conduit and as a result the conduit will enlarge and the discharge will increase with time. Between these two possibilities, an unstable equilibrium condition will exist at a specific discharge and a specific conduit size.

However, under natural conditions the reservoir water level would not remain fixed and so the implications of a changing water level will now be considered. As a starting point it is useful to conceive of a reservoir with a constant inflow

---

(continued) Solid lines in (a) refer to the pressure changes caused by the slow change of discharge defined by the solid line in (b); dashed lines show the effect of a superimposed diurnal oscillation. The dashed hydraulic grade line in (c) refers to a constant discharge of $10 \, m^3 \, s^{-1}$. Note that the pressure changes increase with increasing distance from the terminus and that the diurnal pressure fluctuations are considerably smaller when the mean discharge decreases with time than when it increases. Adjustments to new conditions occur within several days. *Reproduced by permission of Dr D. Vischer, VAW, from U. Spring, Intraglazialer Wassabfluss: Theorie und Modellrechnungen, pp. 156 and 166 (1980)*

rate and with a drainage conduit in equilibrium with the inflow. If the inflow now increases, the water level will rise and this will result in an increase of the pressure gradient, i.e. in outflow. The water level that would be in equilibrium with the increased outflow would, however, be lower because of the $-df/ds \propto Q^b$ relationship (see p.241)! The enhanced discharge will cause additional melting and enlargement in the conduit system. This enlargement will be likely to continue beyond the equilibrium conduit size for the new inflow, resulting, at a later stage, in the reservoir level falling too low so that an oscillation in reservoir level is expected to ensue. Nye (1976) has theoretically analysed the conditions for oscillation using the simplified theory where the heat transfer delay is neglected, while Spring (1980) has shown that oscillations also occur when heat transfer is included.

Spring (1983) has shown that heat transfer can be important in two ways: (1) at the outlet from a reservoir, the water temperature is a decisive factor such that if water is at the pressure melting point, there is no heat available to keep the beginning of the channel open!; and (2) at the end of a conduit, the water transports some heat away, since the water leaves the glacier at a temperature higher than 0 °C. The loss of frictional energy under (2), which is no longer available for melting, increases with increasing discharge to the point where

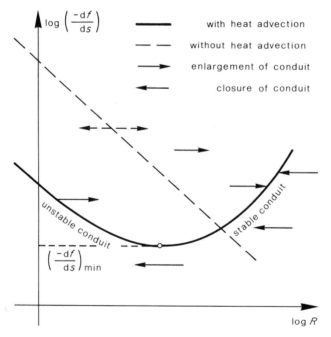

FIGURE 10.17   Instability of ice conduits in the drainage of large reservoirs (Spring, unpublished work)

the enlargement of the conduit ceases. This is illustrated schematically in Figure 10.17. Each curve in this graph separates two regions in the pressure gradient—discharge diagram. Below the curve the frictional heat is insufficient and so the conduit closes with time, above the curve is the region where surplus heat leads to an enlargement of the conduit (conduit closure or enlargement are indicated by arrows). The dashed line shows the result of employing the theory which neglects heat transfer. A uniform decrease exists in the critical pressure gradient separating the two regions (of closure and enlargement) so that an unlimited increase of discharge is indicated on the right side of the line (provided that the flow can be maintained by a sufficient supply of water!) The solid line, which represents the result of employing the theory which includes heat transfer, is curved so that the slope reverses. Above the curve the conduit expands with time, whereas below it contracts. Thus the expansion is now limited. The descending branch of the curve represents an unstable equilibrium, whereas the ascending branch represents a stable equilibrium. Below the minimum point on the curve there is no stable conduit. However, when considering heat transfer, the initial water temperature should be taken into account and this generates a series of similar curves shifted higher for lower initial temperatures and vice versa. These findings are fundamental to the understanding of the outbreak of glacier lakes.

10.6.4 Seepage

The question of the permeability of temperate glaciers to water was debated by researchers during the last century (Lliboutry, 1971). More recently, Nye

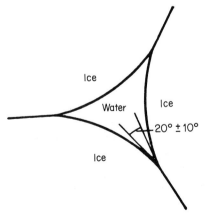

FIGURE 10.18 Cross-section of a vein of liquid situated at a grain edge, where three grain boundaries meet. *Reproduced by permission of the International Association of Hydrological Sciences from* Hydrology of Glaciers, IASH Publ. 95, p. 158 (1973)

and Frank (1973) have taken up the question and have presented arguments in favour of permeable ice (in a paper which was presented in 1969 at a Symposium on the Hydrology of Glaciers held in Cambridge, England). Their concept of free drainage through a system of veins at three-grain boundaries, where the boundary lines represent the contact between three adjacent ice crystals (Figure 10.18), has been questioned by Lliboutry (1971). Lliboutry's reasoning was that a permeable glacier would disappear by frictional melting, but Nye (1976) has pointed out that the necessary water supply for this to happen is lacking. Nevertheless, mechanisms at the ice grain scale discussed by Lliboutry, such as air bubbles blocking the veins or the stress effects related to ice deformation and recrystallization, may profoundly affect seepage.

Experimental evidence for intraglacial seepage has been presented by Raymond and Harrison (1975), who investigated ice cores taken from a site slightly below the usual late-summer snow line on Blue glacier, Mount Olympus, Washington, USA. No frozen surface layer was encountered at the sample site on this maritime-type glacier, in spite of the proximity to the equilibrium line. Based on measurement of the size of the veins (about 25 μm), the grain size of the ice crystals, the abundance of air bubbles blocking the passageways and *in situ* temperatures (Harrison, 1972), it was concluded that seepage through the vein system was negligible. Raymond and Harrison (1975) gave an upper limit for annual water flux in veins of less than 10^{-1} m y^{-1} (with a corresponding velocity of 60 m yr^{-1}). Much more effective drainage is likely to take place through larger tubular conduits and such tube-like features were observed by the same authors in the ice cores, one with a diameter of 1.9 mm at 9.6 m depth and one a few millimetres thick at 20 m. The total summer ablation of 25 mm d^{-1} (or 9 m yr^{-1}) could drain through a surface layer a few metres thick with enlarged veins and through a network of tubular conduits below. One to a few conduits per square metre would be adequate to achieve this and would be compatible with observed ice temperature. Although Raymond and Harrison's observations may not be representative for other locations, they serve as a guide for further considerations and observations.

An arborescent network of veins combining into larger tubes and then joining larger conduits originating from moulins has also been envisaged by Shreve (1972). He formulated some general rules and equated general principles of englacial and subglacial drainage in terms of the water pressure potential (ϕ) within the network. On the assumption that the flow pattern is primarily influenced by the ice pressure, Shreve has been able to apply, with suitable caution, the concept of the potential throughout the ice body. From this he concludes that the general direction of drainage will be perpendicular to the equipotentials of ϕ, which in turn will dip up-glacier at an average angle

$$\delta = \arctan \frac{\varrho_i \cos\alpha}{\varrho_w - \varrho_i \cos\alpha} |\tan\alpha| \qquad (10.24)$$

where α is the slope of the ice surface. Shreve (1972) omitted the $\cos\alpha$ factor, so that the equation

$$\delta = \arctan\left[\varrho_i|\tan\alpha|/(\varrho_w - \varrho_i)\right] \qquad (10.24a)$$

is an approximation for the case of gentle surface slopes. The same is true for his statement (Shreve, 1972, p. 207) that the water moves steeply downwards in the direction of the surface slope, and also (p. 211) that the equipotentials of ϕ will slope downwards in the up-glacier direction with a gradient approximately 11 times that of the surface; this factor becomes smaller with increasing surface slope. Figure 10.19 illustrates the slopes of the equipotentials (solid lines) and direction of drainage (dashed arrows) for various surface slopes from 1° to 30°.

Nye (1976) analysed the flow in individual small tubes. He considered the steady-state condition where the rate of melting at the channel wall is balanced by the rate of closing by plastic deformation, and he further assumed a constant cross-section of the circular tube independent of depth. From these assumptions it follows that $p_i - p_w$ does not change with depth. A definite relationship between $p_i - p_w$ and tube size is obtained, which for laminar flow is

$$(p_i - p_w)^n = \frac{\varrho_w - \varrho_i - (C_2/C_1)\varrho_i}{16g\eta c_m A n^{-n}\varrho_w^3} \cdot \frac{\varrho_w - \varrho_i}{\varrho_i} \cdot r^2 \qquad (10.25)$$

The gradient of the water pressure (dp/ds) is equal to the gradient of the ice pressure (i.e. $dp/ds = \varrho_i/\varrho_w$). For a tube of circular, 20 μm diameter,

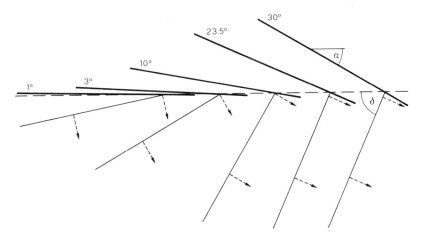

FIGURE 10.19   Inclination δ of equipotentials (solid lines) and seepage gradients (dashed lines) according to equation (10.24) for various surface slopes α

cross-section, Nye obtained a pressure difference between ice and water of 6 m (corresponding to a piezometric height of 6.5 m below the surface). For any other pressure difference there would be another definite value of discharge and related tube size, but as discharge increases, a change from laminar to turbulent flow will occur, necessitating an appropriate change of the factors in the equation. Other equilibrium conditions apart from the ones just described will exist, because other combinations of ice over-pressure and discharge can occur. Where the tube reaches the bed the pressure difference $p_i - p_w$ is determined by the water pressure in the subglacial drainage system. For all discharge values other than the one determined by equation (10.25), $p_i - p_w$ will not be constant along the vertical tube, neither will the tube diameter be constant nor the gradients $-dp/ds$ and $-df/ds$. Generally the equilibrium solutions have to be obtained by numerical integration (Röthlisberger, 1972).

10.6.5   The formation of englacial passageways

So far in this discussion, the existence of englacial conduits has been taken for granted. Little has been said about their location within the ice, except that it has been implicitly assumed that water would drain from the upper part of a glacier to the terminus and that there would be a tendency for water to move from the surface of the glacier to the bed. Since the glacier bed represents a major discontinuity, it is reasonable to assume that some of the water-bearing passageways, and probably the most important ones, are located at the glacier bed. Since the meltwater is largely generated at the glacier surface, there has to be a means by which it may reach the glacier bed. In this context there are two groups of processes which require consideration: drainage through large conduits which originate where surface meltstreams disappear into moulins and seepage through veins and tube-like passages in the permeable ice. Seepage through the permeable ice was considered in Section 10.6.4 and so the role of moulins will be discussed in this section.

Moulins are sink holes in the glacier surface. Meltwater draining into moulins plunges vertically down into the body of the ice. Its subsequent path has always aroused the curiosity of glaciologists and it has occasionally been considered in connection with subglacial water capture for hydroelectricity schemes.

As early as 1943, Carol (1945) attempted to reach the glacier bed via moulins, but he had to give up at a depth of 72.5 m below the glacier surface because the passageway was too narrow. Down to that depth horizontal or gently inclined sections separated vertical shafts. As a result of many observations since that time, it is known that moulins regularly develop in conjunction with crevasses. Active moulins are usually located at the upstream end of crevassed zones, where the crevasses are merely cracks in the ice. Old inactive moulins are often located downstream from the active moulins and abandoned sections of surface meltwater streams are frequently still visible nearby. Even where there is no obvious crevassed area, it is usually possible to link moulins to structural lines in the ice which can be interpreted as former cracks. It is conceivable, therefore, that moulins in these areas have developed when a crack has temporarily opened. Under these conditions moulins typically occur in small numbers, even singly, and they may be extremely large, indicating that they have existed for years.

From the observation that moulins are linked to crevasses, it is a small step to infer that water-filled crevasses may propagate all the way to the glacier bed, in view of the higher density of water compared with ice. Robin (1974) and Weertman (1974) have provided a theoretical basis for this concept, which has been applied in Switzerland since 1973 when applying hydro-fracturing. Artificially drilled piezometer holes have been blasted at depth so that, after

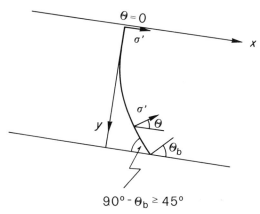

FIGURE 10.20   Orientation of the principal stress deviator $\acute{\sigma}$ as a function of depth y in the glacier

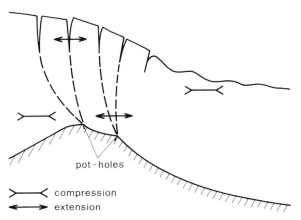

FIGURE 10.21   Hypothetical relationship between the position of potholes and the stress field in the glacier

some delay, they have become linked to the subglacial drainage system (Röthlisberger *et al.*, 1979). The propagation of a water-filled crack can be expected to be governed in two ways, namely by the excess pressure in the water compared with ice, and by the stress field within the ice. The increasing excess pressure with depth causes the crack to propagate preferentially downwards, while the stress field influences the direction that the crack is going to take. In the same way as crevasses point downstream as they approach the edge of the glacier, they should also dip in the downstream direction at depth and become flatter as they approach the bed (Figure 10.20; see also Paterson, 1981, Figure 6.3). Although a water-filled crack propagates downwards, it also spreads sideways, so that the lateral drainage towards the valley side is a likely possibility (Stenborg, 1973; Holmlund, in preparation). Sideways spreading of the water-filled crack, however, will also occur where the crack reaches the bed at depth. Sooner or later it will thereby intersect a major subglacial channel, and the water will start to drain. With the now falling water pressure the crack will tend to close again, but with continuing water supply from the surface one or more passages will remain open by the action of the flowing water, eventually developing into englacial circular conduits. Water falls freely in cascades at atmospheric pressure for the first few tens of metres within a moulin and often series of cascades exist within a moulin. At the bottom of each cascade pools of water rapidly form as a result of the excess of heat gained from the free fall. A speculative side remark can be made in relation to the formation of potholes in the rock. It is conceivable that, because of the stress fields generated at prominent bed irregularities, water-filled cracks may invariably terminate at the same location at the glacier bed regardless of the exact origin of the crevasse at the glacier surface or changes of ice thickness. Such a situation could well lead to pothole development (Figure 10.21). This speculative concept needs further investigation. Observations indicate that at least some potholes are aligned along former subglacial streams!

This basic concept of the crevasse-related origin of moulins is corroborated by the available scant field observations and experience. When surface meltstreams follow medial moraines they frequently transport sediment. Sand- and gravel-sized particles are washed into moulins where they become deposited in the basins at the foot of cascades within the upper sections of the moulins. These sediments later become exposed at the glacier surface in the form of debris cones as a result of ablation. These cones are characteristically found very close to the medial moraines, rather than being lined up on paths leading to the edge of the glacier. The scarcity or even lack of sediment traces at some distance from the medial moraines supports the concept of steeply sloping moulins. More important evidence is provided by Iken's (1974) observations in a moulin on White glacier, Axel Heiberg Island, N.W.T., Canada. She obtained the slope of moulin conduits by direct observation from lowering pressure transducers. She experienced difficulties in lowering the pressure transducers in the upper vertical sections of the moulin where pools of water form during low water level. However, once the transducer had passed into the more regular conduit below, it continued readily down the inclined conduit. The inclination of the conduit was determined from the difference between the cable length supporting the transducer and the observed pressure increase. As it was not possible to separate the pressure change resulting from friction and depth, the depth may have been slightly underestimated by this method (i.e. the conduit may have been slightly steeper than estimated), but repeating the readings at high and low discharge showed that the error was negligible. An example of Iken's moulin inclination curves is presented in Figure 10.22. Figure 10.22 shows the gradual change from vertical to inclined orientation as one expects from the general characteristics

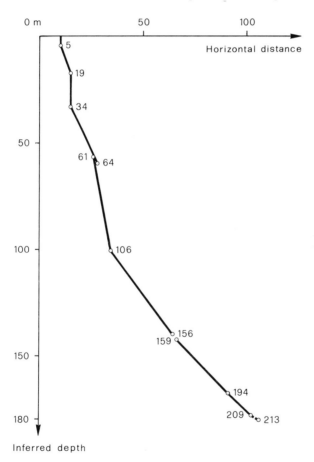

FIGURE 10.22 Approximate slope of a moulin conduit measured by Iken on White glacier, Axel Heiberg Island, Canadian Arctic Archipelago. *Reproduced by permission of Axel Heiberg Island Expeditions from* Glaciology, *1974, 5, 85*

of the stress field (Figure 10.20). If this interpretation is correct, then the lower part of the water passage is sloping downstream in relation to the glacier; alternatively it might be tilted upstream if the shape of the trajectory reflects a long period of ice deformation related to the flow of the glacier.

While moulins seem generally to be crack-induced at the glacier surface, the continuing englacial conduits at depth may form in a different way. Holmlund (in preparation) suggests that when water-filled crevasses cut across veins and tube-like passageways referred to in Section 10.6.4, the water starts to feed into them. Those carrying sufficient water will enlarge and develop into major englacial conduits. The moulins investigated by Holmlund on Storglaciären, Sweden, consisted of 30–40 m deep initial cascades. The conduits leading down from the bottoms of these had inclinations of 0–45°, showing a trend in the direction of the original crevasse rather than becoming normal to the equipotential planes defined by equation (10.24). It is conceivable that the Holmlund-type origin of englacial moulin conduits is typical of slowly moving glaciers, while crack-induced moulin conduits are more likely to develop on larger, more active glaciers. The latter mechanism only allows explanation of the existence of typical moulins of the type described by Iken (1974; cf. Figure 10.22) in cold glaciers. (Further typical features for the polar environment are sloping near-surface englacial streams, which are merely extensions of surface melt streams covered by an accumulation of iced snow drifts and superimposed ice.)

10.7   DRAINAGE AT THE GLACIER BED

10.7.1   Specific conditions at the bed

Once the water reaches the glacier bed, it is most likely to follow the discontinuity between the ice and the rock on its way towards the glacier terminus. According to Weertman (1972), two different types of flow may occur side by side in the form of sheet flow and channel flow, with the surface-derived water showing a preference for channel flow. For flow in channels, Weertman has made a sharp distinction between R and N channels, depending upon whether they are incised upwards into the ice or they consist of grooves incised in the rock [R stands for Röthlisberger, N for Nye, with reference to papers presented by these two authors at the Symposium on the Hydrology of Glaciers in Cambridge, England, 7–13th September 1969; Röthlisberger (1972); Nye (1973)]. However, when applying the theory for intraglacial channels to channels at the glacier bed, the distinction between the two types of channel is of secondary concern, since the principle of balance between melting and plastic ice deformation holds unconditionally. The movements of both water and ice are likely to be very strongly disturbed at the glacier bed, so that some of the numerical factors used in intraglacial conduit computations are no longer valid. Further, it is necessary to take additional effects into account.

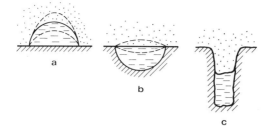

FIGURE 10.23 Various forms of channels at the glacier bed

The first necessity in considering conduits at the glacier bed is that various shapes of channel cross-section must be considered and that these will depend strongly on the existence of grooves in the bedrock. Figure 10.23 illustrates several possible symmetrical profiles. Symmetrical profiles would be expected where the ice moves in the same direction as the water or where there is no ice movement, whereas asymmetric profiles would be likely to exist for channels running at an angle to the ice motion. In his computation of water pressure at the glacier bed, Lliboutry (1983) suggested a cross-section in the shape of a sector of a circle (when the groove in the bedrock is V-shaped), whereas Hooke (1984) based his computations on a semi-circular cross-section. In either case the difference from a circular cross-section is small in comparison with the many other uncertainties.

Another important difference between an intraglacial and a subglacial conduit is their hydraulic roughness, which can increase by a factor of 10 between a smooth ice pipe and a boulder-strewn bed. The roughness depends on the type of substratum and the channel size.

Even more important are the factors governing ice flow. Lliboutry (1983) discussed at some length the discrepancy between generally applicable ice flow coefficients and those used by Röthlisberger (1972) when explaining observed piezometric water levels at Gornergletscher (cf. Table 10.9). Lliboutry gave two main explanations for these differences, namely the low viscosity of the ice because of its high water content and the enhanced ice deformation where the water passage crosses high ice pressure zones on the upstream side of bed undulations. (The effect in the high pressure zones is not compensated by that in the low pressure zones where no deformation takes place, because of the strong increase in deformation at high pressure according to the power law.) It is worth adding that also as a result of the power law of ice deformation, the mean water pressure during times of discharge oscillations will be higher than the corresponding water pressure for the average discharge, because of the very strong ice deformation during the time of the lowest water pressure.

Heat transfer is a further factor that should be considered. Even more heat will be transported downstream and away from the glacier than in the case of

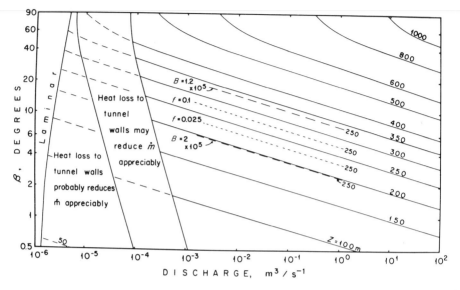

FIGURE 10.24   Critical values of discharge, bed slope and ice thickness. If the discharge in a conduit is greater than the value read on the abscissa for a given bed slope and ice thickness, the conduit is likely to be open. *Reproduced by permission of the International Glaciological Society from* Journal of Glaciology, *1984, 30, 182*

intraglacial conduits, because of the reduced contact with the ice. However, as Lliboutry (1983) pointed out, heat may also be added to the basal drainage system from external sources. The inflow of groundwater that is warmer than the temperate ice will cause additional melting, hence lowering the piezometric pressure. Further complications should be expected as a result of sediment transport.

Lliboutry (1983) and Hooke (1984) have both emphasized that channel flow occurs under many circumstances at atmospheric pressure rather than as pressure-conduit flow. This is the case wherever the ice is thinner than the depth of the piezometric level given by pressure theory (i.e. when the hydraulic grade line drops below the conduit). Hooke (1984) computed the critical values of discharge, slope and ice thickness which separate free flow from pressure flow for semi-circular channels parallel to the glacier surface. His diagram is reproduced in Figure 10.24. Although he used an ice deformation coefficient of 0.244 bar^{-3} yr^{-1}, representing relatively 'soft' ice, the depth to pressure flow appears to be larger than has been experienced with most of the drillings carried out on Swiss glaciers, where piezometric levels deeper than about 150 m have occurred rarely and only for short periods of time (Röthlisberger *et al.*, 1979). This implies that in reasonably large glaciers with appreciable movement at the bed, the closure of the channels is enhanced in comparison with Hooke's computations as pointed out above. A closer fit between computed and measured piezometric pressures can be achieved by introducing an apparent ice deformation coefficient that seems unreasonably high. In thin glaciers with little mechanical activity at the bed, the lower coefficient of Hooke's diagram may be adequate.

10.7.2 Location of channels

One of the important questions concerning water drainage at the glacier bed is related to where, on the scale of the whole glacier, such water actually flows and, in particular, whether it is at the centreline, near the glacier margins, or somewhere in between. Theoretical considerations do not give a straightforward answer because there are different mechanisms which may determine the location of the flow. A consideration of the gravity potential alone, as defined by Shreve (1972) (see p. 250–251) would suggest that water should flow downslope to the lowest point as a result of its higher density in comparison with ice. However, in most situations water will flow preferentially in existing channels rather than melting out new ones. Hence the formation of the first conduits is a central issue when considering the location of drainage passages at the glacier bed. Before considering the formation of these first conduits, it is helpful to consider some basic principles of channel preference.

Röthlisberger (1972) introduced the concept of channels at the hydraulic grade line (gradient channels). The objective of introducing these channels was not to imply that such channels exist, nor to consider how they could form. The computations served to investigate the relative elevations of hypothetical hydraulic grade lines in order to learn something about the possible tendencies for migration which might exist in subglacial drainage systems. Hooke (1984), in discussing the possibility of channel migration in the zone above the line defining the position of the 'gradient channel,' expected that channels incised upwards into the ice would run diagonally from the glacier margin towards the centreline as a result of the interaction between the tendency of water to flow straight down a slope and the tendency of the moving ice to bring the channel into alignment with the direction of glacier flow. As the excess of frictional heat decreases with the increase in plastic closure of the conduit with depth, the channel would be expected to describe a curve turning towards the direction of the centreline. Hooke further noted the possibility of an instability related to a change in discharge. Hooke's model is not much more realistic than that of 'gradient channels' because of the assumption of a completely smooth bed. On a bed with bumps, hollows, gullies and ribs, water would be expected easily to find its way to the deep parts of the glacier bed along passages determined as much by the bed topography as by the processes of channel migration. For example, the tendency of water to flow straight down the subglacial slopes is probably pronounced in the Alps, where, as a result of the presence of at least two systems of steeply inclined joints in the rock, one of these systems usually cuts across the glacial valleys. An additional important aspect of channel position is related to the very strong diurnal discharge oscillations, as well as the longer term non-cyclic variations related to the weather. Wherever two conduits should exist, only the lower one will carry water at times of low discharge! In consequence, the lowest passage would expand at the expense of the higher ones.

However, drainage at the glacier bed cannot be discussed satisfactorily without consideration of the influence of the glacier's surface topography. This influence can be illustrated with reference to the three simple cross-sections I–III presented in Figure 10.25 and representing concave, flat and convex surfaces, respectively. In Figure 10.25, the glacier surface and bed are represented by solid lines, whereas the water equivalent line is dashed (the dotted lines pass through the lowest points of the dashed lines and are horizontal). The water equivalent line, which is at roughly 10/11ths of the ice thickness, represents the piezometric height of the ice overburden pressure at the glacier bed. Water in a thin film moves from high to low piezometric pressure. This is indicated by arrows (Weertman, 1972). Similar tendencies can be inferred, although with some reservation, for conduits. The most favourable position of a conduit is then indicated by the lowest point of the water equivalent line, which is located at the centre for profiles I and II, but at the glacier margin for profile III. Thus, given the opportunity, channels may migrate according to the directions of the arrows in Figure 10.25. Björnsson (1982) has mapped the water pressure vectors and the water divides at the glacier bed for a large area of Vatnajökull, Iceland, using surface and bed topography, the latter based on radio echo soundings. He found that some of the subglacial drainage basins are distinctly different from those inferred from surface topography alone.

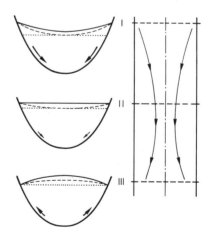

FIGURE 10.25   Effect of the surface relief of the glacier on the tendency of drainage at the bed

One question has puzzled glaciologists for a long time, namely, how does the meltwater traverse the zones of overdeepening of a valley floor? Lliboutry (1983) suggested that water will bypass the overdeepened sections at the sides of the basin. In a comment on one of his equations [equation (22), Llibourty, 1983], he states that in these sections the highest waterway carries almost all of the water. However, since the equation is meant to hold for the equilibrium condition, its true meaning must be the opposite, namely that more water is needed to maintain an equilibrium for the highest waterway than for the lower ones, not that more water is actually flowing. The lowest passage would, therefore, be the most favoured one. A similar argument holds for Lliboutry's equation (23). Lliboutry's reasoning that most water should flow along the banks as gradient conduits is therefore not proved, although it is not *a priori* wrong. The answer can only be obtained by numerical integration. In view of the flood event model proposed below, it is obvious that at maximum bed separation the water may well follow the glacier axis right across the overdeepening and there is no reason why the flow should not remain there (except on a steeply ascending bed on the upstream side of a riegel, where Weertman-type sheet flow is to be expected according to equation (10.23)). Whereas in the case of an overdeepened section of a simple trunk glacier the location of the main subglacial drainage conduit may be ambiguous, there is little alternative to a path in the form of an inverted siphon when the overdeepening is located at the junction of two or more tributaries, unless the water follows an englacial conduit, which is extremely unlikely. If a path from the inner side of a two glacier junction was to avoid the area of overdeepening, it would have to cross one of the glacier branches at some distance upstream of the junction where the pressure potential would be very unfavourable to such a crossing because of the higher ice surface elevation! Note that a dissymmetry exists for descending and rising water in such a way that the frictional pressure loss is less for descending than for rising water, because of heat advection related to the change in pressure melting point (valve effect, Röthlisberger, 1972).

10.7.3 Impact of spring melts and floods on the subglacial drainage system

During winter, when discharge drops to a small fraction of summer runoff, most of the water passages decrease considerably in size. Therefore, the capacity of the drainage system is insufficient to transmit the supplied water when fresh meltwater suddenly starts to flow in the following spring and in consequence high water pressure results. This high water pressure in turn affects the movement of the glacier at the bed, so that new passageways can open up during enhanced bed slip, and the drainage system of the new summer season develops (Elliston, 1966; Röthlisberger and Aellen, 1967; Haefeli, 1970; Müller and Iken, 1973; Röthlisberger, 1980; Iken *et al.*, 1983; Hooke *et al.*, 1985).

Various observations can be attributed to what may be termed the 'spring event.' These observations (cf. Röthlisberger *et al.*, 1979; Deichmann *et al.*, 1979; Iken *et al.*, 1983) include in particular high water levels close to the glacier surface in moulins, sudden drainage of surface water in connection with an ice velocity peak, a simultaneous uplift of the glacier, enhanced seismicity, large sediment loads with a high percentage of fines in the proglacial streams and eventual sudden changes in the sub- and proglacial drainage pattern. Since exactly the same phenomena occur in relation to outbreaks from ice-dammed water bodies, the more general term 'flood event' might be used as a synonym for 'spring event.' Some details of the phenomena associated with these events will be discussed in the following paragraphs.

On the Gornergletscher and Unteraargletscher, Switzerland (cf. Röthlisberger *et al.*, 1979), early ablation season drainage through some of the previous year's moulins was inferred from the presence of slush bands, which showed at the snow surface during periods of intense melting and which ended abruptly at the location of old moulins. An attempt was made to measure the water level in one giant moulin on the Unteraargletscher by inserting a piezometric pipe into the void below the snow bridge at the mouth of the moulin. Although this attempt failed, the reasons for the failure are instructive. The pipe had presumably become obstructed by slush floating on top of the water column in the moulin. It was implied that the slush rose temporarily to within a few metres of the glacier surface, that is, higher than the piezometric overburden pressure, which at the ice thickness of 350–400 m at the observation site would have been about 35 m lower than the ice surface (even if we allow for the effect of the moulin's location within a trough at approximately 15 m below the average level of the surrounding ice, a depth of about 20 m remains, which is considerably deeper than the level of the floating slush). A pressure drop may have occurred in the steep moulin conduit between the glacier surface and the bed, but it is still highly probable that a water pressure close to the overburden pressure was reached temporarily at the bed. Since an increase in water pressure at the bed, even to a much lower level than the overburden pressure, can have a pronounced effect on the sliding motion of the glacier (Iken, 1974, 1981; Iken and Bindschadler, 1986), the observation of the water and slush levels on Unteraargletscher, together with similar observations on other glaciers (e.g. Gorner), leaves hardly any doubt that significant hydraulic action may take place at the glacier bed.

As a plausible explanation of the 'spring event,' it is proposed that water accumulates on the glacier surface in a saturated layer at the base of the snow, from where it drains into surface depressions to form slush bands and pools. Because of the sub-zero temperatures of the glacier's surface layers at the end of the winter, water can only drain into the ice in a few places where the routes have not become blocked during the winter. When water reaches the glacier bed as a result of drainage along these isolated routes, the hydraulic pressure will rise to a level which is dictated by drainage conditions at the bed. Drainage is likely to be poor because of the degeneration of the drainage system through the winter. Thus, water pressure may rise to the ice overburden pressure, or even to about 8–10% higher than the ice overburden pressure where a moulin fills to the top. An overpressure of that magnitude does not guarantee free drainage at the ice–rock interface because of local stress variations at the glacier sole, but, nevertheless the opening of a crack or the expansion of a cavity in the ice may eventually provide access for the water to a low-pressure area at the bed. From this point onwards the expansion of bottom cavities will progress more rapidly as the water spreads out at the glacier bed, provided that sufficient inflow is maintained at sufficient pressure from above. As the water pushes into the stress depressions at the glacier sole, the net effect on the glacier can be compared to the effect of a hydraulic jack which drives the ice forward and upward along those segments of the bed irregularities which face upstream. Figure 10.26 illustrates the situation schematically. Assuming a rigid ice body on a frictionless bed, the critical water pressure p_c necessary for the hydraulic jack mechanism to become operative is given by

$$p_c = p_0 \frac{\sin\psi}{\cos\alpha\sin(\alpha+\psi)} = p_0 - \tau/\tan(\alpha+\psi) \qquad (10.26)$$

where $p_0 z \varrho g h \cos^2\alpha$ is the mean overburden pressure at the bed, ϱ is the density of the ice, g is the acceleration due to gravity and τ is the basal shear stress. Figure 10.26 also illustrates the opening up of cavities related to the displacement c by sliding and the respective uplift d (cf. Röthlisberger, 1980; Röthlisberger

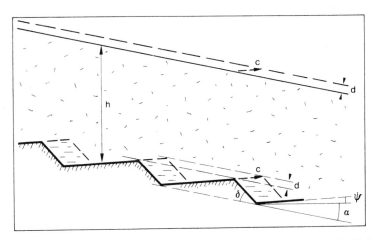

FIGURE 10.26 Basic rigid-body model of translatory glacier displacement by high water pressure in cavities (hydraulic jack effect). *Reproduced by permission of the International Glaciological Society from* Annals of Glaciology, *1981, 2, 57*

and Iken, 1981; Iken *et al.*, 1983). The process of cavity formation commences slowly and locally, but once the affected area has become sufficiently large to influence the stress field in the ice so as to cause crevassing, numerous water links rapidly develop between the glacier surface and the bed. A cascading process may evolve and with an unlimited supply of water the glacier would soon be flushed down the valley. However, because of the limited water supply this flushing effect soon terminates, although in the meantime some regional flooding of the bed may occur. This explains the abundance of fine-grained sediments which are flushed out of the glacier in the meltwater. These sediments are rock flour which has been milled at the glacier sole during the preceding months. Very high sediment concentrations are also observed during the drainage of ice-dammed lakes like Gornersee (Collins, 1979a).

The potential effectiveness of the hydraulic jack mechanism depends on three main factors: how fast the water can spread out at the glacier bed (from a few scattered moulins), how much (and at what rate) water is available at the glacier surface and how rapidly the water drains away at the glacier bed. On one occasion at Unteraargletscher, an unusually spectacular 'spring event' was observed following heavy rainfall and intense melting. Within the space of a few hours a number of meltwater pools at scattered locations on the glacier surface had drained and numerous fresh crevasses had developed. Particularly remarkable was the fact that these cracks were visible at the surface of about 2 m thickness of wet, partly saturated snow! Loud bangs were audible as the ice cracked and the seismicity was correspondingly high (Deichmann *et al.*, 1979). The opening-up of one of the major cracks was of particular interest. It extended from a large moulin near the centre of the glacier over a distance of several hundred metres to the edge of the glacier. In the opinion of the authors, it originated at the moulin, possibly at depth. Sources of seismic events have indeed been located at depth. Unfortunately, no velocity measurements could be taken at the peak of the event because of avalanche danger, but from the observations of the subsequent velocity decrease it is clear that the velocity was high. A remarkable increase in both the discharge and the turbidity of drainage was observed at the glacier snout.

At Unteraargletscher, it has been possible to estimate the proportion of glacier movement that was attributable to the high water pressure effect during one year (1974–75). Iken *et al.* (1983) showed that 8 m out of the annual movement of 27 m could be accounted for by enhanced sliding during the melt season. More than half of the sliding occurred during four different flood events (see Figure 10.27). From Flotron's (1973) measurements, these flood events were estimated to have lasted a few days (the time base on the automatic camera was either one or four days), while Iken (1977) observed shorter high velocity peaks lasting from a few hours to a day on various glaciers.

Based on the proposed model of the 'spring event,' it is possible to suggest the way in which the subglacial drainage system might become established at

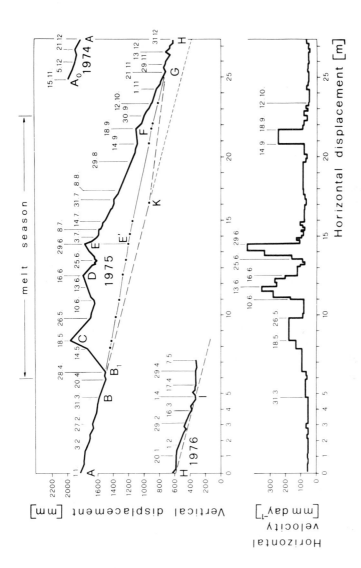

FIGURE 10.27 Motion of a surface point on Unteraargletscher from 15th November 1974 to 7th May 1976 showing characteristic uplift during velocity peaks. Top: trace of flow line in a vertical plane (vertical scale five-fold exaggeration). Bottom: plot of horizontal component of velocity versus horizontal displacement. *Reproduced by permission of the International Glaciological Society from Journal of Glaciology, 1983, 29, 32*

the beginning of the ablation season. As soon as an area with high water pressure connects to one of the main subglacial drainage conduits that has remained operative through the winter, water can escape causing the drainage conduit to enlarge. The drainage pattern then follows essentially the pattern of the preceding year. An alternative mechanism would occur if areas of high water pressure enlarged to the point where they interconnected. Under these circumstances, the pre-existing pressure differences between the individual areas would cause the water to move from one area to another following the large-scale pressure gradients. At the commencement of these flows there would probably be a highly complex network of interconnecting passageways with many different possibilities for water routing. However, because the greatest amount of melt will occur where the most water flows, flow will eventually become concentrated into one conduit whose location will depend, to a certain extent, on chance. With water drainage, the water pressure decreases and the ice settles back on to the substratum. Thus, the pressure distribution depends not only on the local bed topography but also on the topography of the ice sole, which has been shaped by the preceding movements of ice and water (i.e. the whole history of the event). Drainage patterns as well as flood events could be expected to depend on weather conditions. A complete lack of channels incised into the bedrock across some riegels supports the concept of fairly frequent changes in drainage conduit location, otherwise the water would have cut into the rock. In other cases, however, deep gorges exist across the riegels.

Although the detailed routing of drainage may be poorly defined and partly determined by chance factors, this is not the case over larger areas, where the major pressure gradient is determined by the slope of the glacier surface, the ice thickness and the slope of the bed at a scale which is closer to the ice thickness than to conduit diameter. Evidently, at maximum bed separation the water follows the main pressure gradient, and thus the overall drainage system develops according to the Shreve potential [equation (10.24)], as it is defined at the bed (see also Björnsson, 1974, 1975, 1982; Nye, 1976). When the glacier condition changes, the drainage pattern will accordingly change. This is evident when the glacier portal changes its position, for example from left to right and back again as has happened on the Glacier de Tsidjiore Nouve (Figure 10.28). It was also evident at the subglacial intake of the Argentière glacier (Hantz and Lliboutry, 1983), where the water flow was lost in the spring after it had been flowing regularly for years and still throughout the preceding winter (personal communication from Electricité d'Emosson S.A.). It can be concluded that major adjustments occur during flood events. In the case of the Glacier d'Argentière, the new condition was a larger increase in ice thickness on the left than on the right side of the glacier, which caused the conduit to move from the left to the right. A change of drainage from one year (1960) to another (1968) has been reported by Stenborg (1973). A sudden change of the sub- and proglacial drainage system is, therefore, possible under the proposed model.

FIGURE 10.28   The arrows indicate different positions of the exit of the principal melt stream at the terminus of the Glacier de Tsidjiore Nouve on two different dates (4th July 1972 and 19th August 1974). The melt stream has since returned to the former position on the left-hand margin (*photographs: H. Röthlisberger*)

A further hydraulic, or rather hydrological, implication related to the flood events concerns water storage. Iken *et al.* (1983) have shown that bed separation and the storage of a considerable amount of water is the most plausible explanation for an uplift of several tenths of a metre which was observed on Unteraargletscher by Flotron (1973). The phenomenon is illustrated in Figure 10.27. One of the essential features of the graph is that during the melt season the glacier moves up and down as it moves forward. This is shown by the upper trace, which represents the spatial trajectory of the movement (i.e. the movement, with five-fold vertical exaggeration, in a vertical plane) of a surface marker fixed in the ice. Several peaks occur in the trace and these represent flood events. The lower trace shows corresponding horizontal velocities. From the combination of the two traces it can be seen that, with the exception of the first event, the glacier moves much faster while it is rising than when it has settled down again. In addition to the sharp peaks, a general rise of the glacier surface can be observed at the beginning of the melt season which is followed by a slow decline lasting from July until November. In comparison with the summer, the winter motion is slower and more uniform. In the opinion of the authors, the sharp rise in the glacier surface at the commencement of each of the movement events can be explained by the action of water at relatively high pressure pushing the glacier forward and upward (the hydraulic jack effect previously described), accompanied by the development of water-filled cavities behind bed irregularities. During the sequence of falling ice surface that follows, water drains slowly out of the cavities. Because the formation and closure of the cavities and the accompanying injection and drainage of water are not entirely reversible, part of the water remains entrapped in the cavities which have become closed off by the ice. Drainage continues but at a slower rate. The water storage at the bed which is related to these drainage and glacier movement processes probably accounts for at least part of the water storage inferred by Tangborn *et al.* (1975) from hydrological water balance measurements. Hooke *et al.* (1983) have drawn similar conclusions.

10.8 GLACIER-DAMMED LAKES AND OUTBURST FLOODS

A characteristic feature of some glacierized basins is floods caused by the outbreak of water reservoirs. On a world-wide scale, the size of such reservoirs and the frequency and magnitude of the floods, as a rule, increase with the intensity of glacierization. The most spectacular outbursts, however, seem to be related to volcanic activity (Björnsson, 1974, 1975). From the extensive literature existing on this subject only a few references will be cited, but the reader is referred to the more extensive reference lists given by Björnsson (1976), Liestøl (1955), Meier (1973), Post and Mayo (1971), Whalley (1971) and further papers cited below. A few catastrophic floods have occurred in the Alps (Haeberli, 1983; Hoinkes, 1969; Lütschg, 1926; Mougin annd Benard, 1922; Röthlisberger, 1981).

Numerous sources of outbursts exist: the reservoirs of water can be located at the glacier margin, at the ice surface, in the interior or at the bed of the glacier. The mechanism of the outbursts may also vary, depending mainly on the location of the drainage passageways and the topographical situation of reservoir and ice dam. Ice dams may break, or the water may flow over the dam and melt out a cut or gorge. Much more commonly, however, the drainage occurs below the surface across the dam, usually underneath it at the glacier bed. Most outbursts can therefore be understood, at least partially, on the basis of the instability of the drainage of a reservoir through a conduit in ice (cf. Figure 10.17 with the associated explanatory text). The principle is that flowing water can enlarge a pre-existing conduit by melting, the necessary heat being provided by advection from the reservoir and dissipation of potential energy. Liestøl (1955) suggested that both heat sources are important and that a progressive opening of conduits is to be expected because of the proportionality of the widening of the cross-section y to the discharge ky: $dy/dt = ky$. Meier (1973) referred to an exponential growth of a typical hydrograph, while Clague and Mathews (1973) gave an empirical relationship between the discharge, Q_t, at time t and the volume, V_t, drained up to the time t of the form

$$Q_t = K(V_t)^b \qquad (10.27)$$

where K and b are constants that vary significantly from one ice-dammed lake to another. Equation (10.27) implies a power law for $Q = f(t)$ except when $b = 1$, i.e. the increase of Q with time is exponential. Most exponents given by Clague and Mathews are less than unity, except for one case where $b = 1.5$. Differential equations for non-steady water flow have been derived and applied to the 1972 Grimsvötn outburst by Nye (1976), where a deep lake of uniform area can be assumed. Nye's theoretical solution gives a value of $b = 4/3$. With this value he obtained a very close agreement with the observed values of $Q = f(t)$, that is, in his terms, 'not far from exponential, but nevertheless very distinguishably different.' Clarke (1982) has extended the theoretical study of lake outbursts by systematically analysing the relative importance of reservoir temperature, viscous dissipation in the moving water and creep closure of the conduit, and by taking the lake geometry into account.

The theoretical solutions for the development with time of the outburst conduit do not give an answer to the question of how and when outbursts start. It has been observed in many cases that drainage can set in before the water in the lake reaches the overburden pressure of the ice dam. For Grimsvötn, Nye (1976) suggested a cantilever effect, while Björnsson (1974) considered that passageways may exist at a low elevation in the rugged volcanic rock forming the rock ridge underneath the ice dam. A mechanism can also be contemplated (Fisher, 1973) in which drainage conduits may link a glacial lake to the regular drainage system beyond the ice dam in such a way that part of the water drains into the lake at high flow rates, but some water may be fed back out of the lake into the outlet system at low flow rates; only when the lake level rises to a sufficient height, the water temperature becomes sufficiently high, or when the piezometric level in the glacier drainage system falls sufficiently low for a sufficient time span, does the outburst become self-sustaining. In the case of Summit Lake (Fisher, 1973) the outbursts occur in autumn, while in other cases (e.g. Gornersee) it may follow temporary high discharge or possibly be connected to an active phase of glacier motion in conjunction with a 'flood event.' A further possibility for outburst initiation is seepage through permeable material below the glacier sole; when the temperature of the water rises in summer, ice melts at the sole and an initial channel develops at the contact.

In the case of the outbursts (in Iceland referred to as 'jökulhlaups') of Grimsvötn, the investigators were first puzzled by the fact that the outburst became shut off at a time when the lake level was still well above the surface elevation of the rock ridge underneath the ice dam. Björnsson (1974) favoured the idea that, apart from the fast creep closure—possible collapse—of the conduit, the lowering of a flat floating ice lever on to a relatively flat rock ledge may speed up the termination of the water flow considerably, whereas Nye (1976) based his theory on creep closure alone. Clarke (1982) investigated the combination of a variety of conditions for which the outburst stops before the lake becomes empty. Spring and Hutter (1981) have shown, also for Grimsvötn, that by taking a suitable temperature profile in the lake into account, the entire hydrograph can be modelled on both the ascending and descending limbs.

Although it is better to know the $Q = f(t)$ function, it is often sufficient for practical purposes to predict the peak discharge Q_{max}. Clague and Mathews (1973) gave the empirical function for Q_{max} in $m^3 s^{-1}$

$$Q_{max} = 75(V_{max})^{0.67} \tag{10.28}$$

where V_{max} is the total volume drained during the outburst, given in $10^6 m^3$. Clarke (1982) derived a number of equations that allow the estimation of Q_{max} by taking into consideration the geometry of the lake, the geometry of the glacier and the water temperature. It is disturbing that in the cases investigated so far, equation (10.28) has given results as good as the more sophisticated theory, although there is no obvious relationship to deduce the empirical equation from theory (cf. Clarke, 1982).

So far a stable rock bed has been assumed to exist. However, in a number of cases the ice barrier has been resting on loose material such as ground moraine. Subglacial erosion has then played a major role in the outburst event. Typical examples are Mattmarksee (Lütschg, 1926) and Lac de Mauvoisin, which was dammed up by an ice-debris cone originating at the bottom of a deep valley from ice avalanches of the advancing glacier de Giétro (Röthlisberger, 1981). In the latter case, drainage was slowly progressing through an artificial ice tunnel in which water was working its way continuously downwards, when basal water flow at the ice–talus boundary started to take over. The sudden rupture of the remaining intermediate ice lamella separating the two channels caused a catastrophic flash flood that was then responsible for severe loss of life and property all along the valley of Bagnes down to the town of Martigny in the Rhone valley. Haeberli (1983) made a clear distinction between the glacier floods caused by the progressive enlargement of conduits (primarily in ice) and the sudden break of a barrier (ice or moraine). For the latter case he recommended the estimation of Q_{max} (in $m^3 s^{-1}$) by dividing the outbreak volume (in m^3) by 1000–2000s.

Some of the most catastrophic floods that have occurred all around the world are related to the strong glacier retreat of recent decades. They have been caused by the rupture of terminal moraines and have thereby only indirectly been caused by the glacier, except in those cases where the rupture was induced by an ice avalanche (Lliboutry *et al.*, 1977). Ice-cored moraines, nevertheless, can be involved and they represent a type of ice dam. It seems that the ice core is occasionally more resistant to erosion than the moraine itself (Haefeli, 1963).

10.9 METHODS OF INVESTIGATION OF INTRA- AND SUBGLACIAL DRAINAGE

The investigation of the internal drainage of glaciers is less straightforward than the study of meltwater input at the surface or the discharge at the terminus, considered earlier in this chapter. Meier (1973) summarized various procedures for gaining information on the passage of water through the glacier, listing selected references at an early stage. An overview of methods will be given below with more recent references.

Direct investigation of natural ice caves is easiest shortly after outbursts of glacier lakes, but it can be extremely dangerous. Less risks are involved when rock tunnels reach the glacier bed in association with mining operations (Mathews, 1964) or hydro projects (Vivian and Zumstein, 1973; Hooke *et al.*, 1985). Direct observations of conduit geometry have been achieved, in addition to registration of piezometric water pressure over extended periods of time, the water pressure having been measured in drill holes through the rock from tunnels to the glacier bed (see also Bezinge *et al.*, 1973). Information on the subglacial sediment load has been reported from Norway (Hooke *et al.*, 1985), together with observations on discharge variations in different conduits and peculiar discharge oscillations.

In the névé field of Ewigschneefeld (Aletschgletscher, see p. 226), where groundwater flow exists in the porous surface firn layer, standard methods for groundwater studies have been employed to obtain information on the depth of the groundwater table (up to about 30 m), its variations with time and the permeability of the water-soaked layer where Darcy-type flow exists (Schommer,

1976, 1978). A numerical simulation of the water-level variations gave a satisfactory agreement between observations and computation. Drainage into crevasses has been implied in an area where no crevasses are visible at the surface. In the Austrian Alps a comprehensive study of the percolation of the meltwater through the unsaturated firn and the flow in the saturated layer below the water table was carried out in a 30 m deep shaft on the Kesselwandferner (Behrens *et al.*, 1979; Ambach and Eisner, 1979). Also on Kesselwandferner, very promising results on the depth and continuity of the water table and on other inhomogeneities in the glacier have been obtained by radio echo-sounding (Thyssen *et al.*, 1980).

   Because of the importance of the water pressure at the glacier sole in most of the theories on basal sliding, great efforts have been made to measure the piezometric water pressure at the glacier bed. Apart from the observations in rock tunnels mentioned above, this was first achieved by introducing pressure transducers into moulins (Iken, 1972; Müller and Iken, 1973). Because of the rare occurrence of suitable moulins for such measurements, piezometers have more recently been widely established by drilling, using either electrothermal 'hot points' (Hodge, 1976) or techniques employing a hot-water jet (Iken *et al.*, 1976; Hodge, 1979). Drilling with a hot point has the advantage that something can be learned about the nature of the ice (i.e. about the nature of air or water inclusions, or dirt bands), but this is at the expense of a low drilling speed of 5–6 m h^{-1}. In comparison, the hot-water jet advances at speeds of more than 100 m h^{-1} and it has the additional advantage that fine-grained dirt is washed away. Of course, none of these methods permits penetration of rocks. When a piezometer hole reaches the glacier bed, the water level may drop immediately, with a delay of several hours or days, or sometimes not at all within a field season in spite of the higher density of the water column compared with the surrounding ice. Whether or not a hole connects with the low-pressure basal drainage system undoubtedly depends on local conditions of ice pressure and the permeability of the ice and the basal sediments (Hodge, 1979). Connections with basal cavities or conduits have been forced to open up by blasting (Röthlisberger *et al.*, 1979). It is also necessary to realize that holes tend to lose connection to the drainage system at low piezometer levels, owing to the large pressure difference between ice and water which causes the ice to creep and the water to freeze. It is then necessary to refill the piezometer holes with water at least once a day until they drain again, or to apply a continuous flow of water, which should be temporarily halted to obtain true piezometer readings. Such 'active' piezometer holes give the water pressure in one of the nearby basal (or englacial) conduits or cavities whose exact location is unknown. The management of water flow into piezometer holes needs to be undertaken subtly and can be an extremely tedious process. However, if the connection of the piezometers is not assured, the measurements are of doubtful use. Parts of the conclusions of Hantz and Lliboutry (1983) may therefore be questioned. One of the best chances of operating a piezometer over an extended time period is found by drilling in the vicinity of the moulin of a major melt stream close to a small tributary that can later be fed into the piezometer hole. This results in the formation of an artificial moulin which serves to record the fluctuations of one of the main subglacial conduits with little pressure loss between the transducer and the main conduit. The correctness of the results can be checked by temporarily diverting the tributary melt stream away from the artificial moulin. Figure 10.29 gives an example of such pressure registrations obtained on Gornergletscher. Note the discharge, subglacial water pressure and ice velocity peaks during the outburst of Gornersee between 28th and 31st August. The rapid drop and rise of water

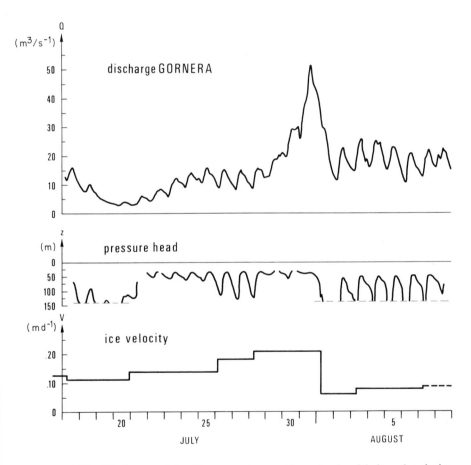

FIGURE 10.29 Discharge of the Gornera, piezometric water head below the glacier surface and horizontal velocity component from 17th July to 8th August 1974 on the tongue of Gornergletscher. *Reproduced by permission of Schweizerischer Wasserwirtschaftsverband from* Wasser, Energie, Luft, 1980, *72, 291*

pressure during a more gentle decrease and increase of discharge between 17th and 22nd July are in good qualitative agreement with the computer simulation of Figure 10.16.

Salt and dye tracing have also made a major contribution to the understanding of water flow through glaciers (Stenborg, 1969, 1973; Ambach *et al.*, 1972; Krimmel *et al.*, 1973; Lang *et al.*, 1979; Aschwanden and Leibundgut, 1982; Behrens *et al.*, 1982; Burkimsher, 1983). Very fast flow indicative of pressure and free flow in main conduits comparable to the river flows of mountain torrents have frequently been found, but also extended delays and complete loss of dye have occurred. Where two or more melt streams emerge from one terminus, overflow from one drainage area to another changing with the time of day and the seasons has also been observed and, in addition, double peaks from a single tracer input have indicated the existence of braided conduits. Various peculiarities of tracer results remain to be explained and one problem which has not yet been completely solved is the absorption of the dye by glacial flour. In addition to the work undertaken with artificial tracers, numerous investigations have been conducted using natural isotopes and tritium (Ambach *et al.*, 1973, 1976; Behrens *et al.*, 1971; Oerter *et al.*, 1985), and studying solids and solutes that were picked up by the water subglacially (Collins, 1979a,b,c). Gas content and gas composition of the ice in the lower part of the ablation area have been simulated with a model based on seepage through the ice and internal water production (Berner *et al.*, 1977).

10.10   CONCLUSIONS

Glacial runoff depends on the climatic processes acting at the glacier surface and on the physical mechanisms controlling the flow of water within the glacier and at the bed. Hydrological methods permit modelling of the discharge at the glacier terminus, using meteorological data and characteristic delay factors, which are specific to the individual glacier and which change with time of year, as inputs (Stenborg, 1970; Fountain and Tangborn, 1985). Temporary water storage, varying at time scales from an hour to the complete ablation season, is a typical feature of glacial hydrology and runoff. Where and how this water is stored are still not completely understood. Porous firn layers, veins, conduits, crevasses including possibly bottom crevasses, englacial and subglacial cavities of various sizes can all contain water but the relative amount of water stored in different places is difficult to estimate. Groundwater storage below and adjacent to glaciers must also be considered.

These aspects have been treated mainly with a view to the alpine characteristics of glacier hydrology which are the result of the present climatic conditions. The economic use of glacier water resources is based to a great extent on models and on forecasting and prediction techniques, which make use of the physical relationships between climatic variables and glacier runoff; however, the operational models are abstractions of the physical reality in order to be compatible with the availability of input data. It was beyond the scope of this chapter to discuss details of such techniques; a recent compilation of forecasting techniques was given by Young (1985). An important aspect of future research will be the comparative analysis of the processes of heat balance and of meltwater runoff under various climatic conditions such as moderate humid (Alps), oceanic humid (western Norway, northwestern North America), monsoon (Himalayas) and dry continental (Tianshan) climatic regimes; this will help to adjust the model structures to particular conditions and to improve understanding of climate–glacier relationships. Another aspect of research in glacier hydrology will be the long-term variations in glacier water resources under the different scenarios of probable future climatic variations.

Other runoff phenomena are related to instabilities in the internal drainage processes. These can only be understood by dealing with the complex physical mechanisms of intra- and subglacial drainage. A typical example is the outbursts of ice-dammed lakes. Closely related are the outbursts of so-called water pockets (Mougin and Bernard, 1922); so far it is virtually impossible to locate such englacial and subglacial water bodies in spite of the fact that their existence can mean impending disaster. Radio echo-sounding techniques provide some hope, at least where an englacial source is suspected.

There are other, both scientific and practical, reasons for studying the internal water flow in a glacier. Subglacial water intakes for hydro power or irrigation, for example, pose a major practical problem (Hantz and Lliboutry, 1983; Hooke *et al.*, 1983). Scientifically, the role of water in the sliding process is generally accepted. However, scientists have discussed the relative importance — or even existence — of a variety of plausible processes for some time, and they will probably continue to do so in the future (Lliboutry, 1968; Iken, 1981; Walder, 1982; Bindschadler, 1983; Hooke *et al.*, 1983; Iken *et al.*, 1983; Iken and Bindschadler, 1986; Weertman and Birchfield, 1983).

Particularly obvious is the importance of subglacial water flow in conjunction with sediment processes. A full range of situations exists from where the water is mainly an agent of sediment transport (Collins, 1979a) to those where it actively erodes, or where the water pushes the glacier hydraulically, whereupon the glacier in turn may remove rocks (Röthlisberger and Iken, 1981). Other geomorphological implications like those suggested by Shreve (1972) or Hooke (1984) in relation to eskers are of interest. Further conclusions can be drawn from conduit-flow theory, for example in connection with sediment transport across overdeepened basins. When we ask the question, can coarse pebbles be transported in a main basal conduit along the thalweg?, then the answer is 'yes and no.' It is 'yes' when the uphill slope from the bottom of the basin to the crest of the riegel is gentle (i.e.

$$|\tan\beta| < \frac{\varrho_i}{\varrho_w} \cdot \frac{C_1 + C_2}{C_2} |\tan\alpha|$$

in equation (10.23)). In that case conduit flow exists, and because uphill flow is faster than downhill flow (cf. Röthlisberger, 1972), pebbles may readily be carried upslope. When the slope is steeper than that shown in the above expression, however, conduits will freeze up and the water flows in sheets, in which case no pebbles at all can be transported by the water (the glacier, however, may incorporate the sediments and carry them up over the riegel, particularly in the case of a very steep slope, as part of the water will freeze). It is an intriguing question whether eskers reflect the larger channel sizes on downhill stretches than on uphill ones or, on existing glaciers, whether pronounced riegels might affect the composition of the sediment load of the glacial stream. Although such reflections may be discarded as mere speculations, one can nevertheless hope that progress will be possible by means of an appropriate mixture of theoretical insight and purposeful observations.

REFERENCES

AELLEN, M. (1985), Die Gletscher der Schweizer Alpen im Jahr 1983/84, *Die Alpen*, **61**, 188–213.

AMBACH, W. (1961), Die Bedeutung des aufgefrorenen Eises für den Massen- und Energiehaushalt eines Gletschers, *Zeitschrift für Gletscherkunde*, **4**, 169–189.

AMBACH, W., BEHRENS, H., BERGMANN, H., and MOSER, H. (1972), Markierungs-versuche am inneren Abflusssystem des Hintereisferners (Oetztaler Alpen), *Zeitschrift für Gletscherkunde und Glazialgeologie*, **8**, 137–145.

AMBACH, W., EISNER, H., and URL, M. (1973), Seasonal variations in the tritium activity of run-off from an alpine glacier (Kesselwandferner, Oetztal Alps, Austria), in *Symposium on the Hydrology of Glaciers (Proceedings of the Cambridge Symposium, 7–13 September 1969)*, International Association of Scientific Hydrology Publication 95, pp. 199–204.

AMBACH, W., ELSÄSSER, M., BEHRENS, H., and MOSER, H. (1974), Studie zum Schmelzwasserabfluss aus dem Akkumulationsgebeit eines Alpengletschers (Hintereisferner), *Zeitschrift für Gletscherkunde und Glazialgeologie*, **10**, 181–187.

AMBACH, W., EISNER, H., ELSÄSSER, M., LÖSCHHORN, U., MOSER, H., RAUERT, W., and STICHLER, W. (1976). Deuterium, tritium and gross-beta-activity investigations on alpine glaciers (Oetztal Alps), *Journal of Glaciology*, **17**, 383–400.

AMBACH, W., and EISNER, H. (1979), Ein Tracerexperiment zum Schmelzwasserfluss in der Wassertafel eines temperierten Gletschers, *Zeitschrift für Gletscherkunde und Glazialgeologie*, **15**, 229–234.

AMBACH, W., BLUMTHALER, M., and KIRCHLECHNER, P. (1981), Application of the gravity flow theory to the percolation of melt water through firn, *Journal of Glaciology*, **27**, 67–75.

ASCHWANDEN, H., and LEIBUNDGUT, C. (1982), Die Markierung der Wasser des Gornerseeausbruchs mit drei Fluoreszenztracern, in LEIBUNDGUT, C., and WEINGARTNER, R. (Eds.), *Tracermethoden in der Hydrologie*, Beiträge zur Geologie der Schweiz Hydrologie, 28, pp. 535–549.

BEHRENS, H., BERGMANN, H., MOSER, H., RAUERT, W., STICHLER, W., AMBACH, W., EISNER, H., and PESSL, K. (1971), Study of the discharge of alpine glaciers by means of environmental isotopes and dye tracers, *Zeitschrift für Gletscherkunde und Glazialgeologie*, 7, 79–102.

BEHRENS, H., MOSER, H., OERTER, H., BERGMANN, H., AMBACH, W., EISNER, H., KIRCHLECHNER, P., and SCHNEIDER, H. (1979), Neue Ergebnisse zur Bewegung des Schmelzwassers im Firnkörper des Akkumulationsgebietes eines Alpengletschers (Kesselwandferner—Oetztaler Alpen), *Zeitschrift für Gletscherkunde und Glazialgeologie*, 15, 219–228.

BEHRENS, H., OERTER, H., and REINWARTH, O. (1982), Results of tracer tests with fluorescent dyes on the Vernagtferner from 1974 to 1982, *Zeitschrift für Gletscherkunde und Glazialgeologie*, 18, 65–83.

BERNER, W., STAUFFER, B., and OESCHGER, H. (1977), Dynamic glacier flow model and the production of internal meltwater, *Zeitschrift für Gletscherkunde und Glazialgeologie*, 13, 209–217 (printed 1978).

BEZINGE, A., PERRETEN, J. P., and SCHAFER, F. (1973), Phénomènes du lac glaciaire du Gorner, in *Symposium on the Hydrology of Glaciers (Proceedings of the Cambridge Symposium, 7–13 September 1969)*, International Association of Scientific Hydrology Publication 95, pp. 65–78.

BINDSCHADLER, R. (1983), The importance of pressurized subglacial water in separation and sliding at the glacier bed, *Journal of Glaciology*, 29, 3–19.

BJÖRNSSON, H. (1974), Explanation of Jökulhlaups from Grimsvötn, Vatnajökull, Iceland, *Jökull*, 24, 1–26 (printed 1975).

BJÖRNSSON, H. (1975), Subglacial water reservoirs, jökulhlaups and volcanic eruptions, *Jökull*, 25, 1–14 (printed 1976).

BJÖRNSSON, H. (1976), Marginal and supraglacial lakes in Iceland, *Jökull*, 26, 40–51 (printed 1977).

BJÖRNSSON, H. (1982), Drainage basins on Vatnajökull mapped by radio echo soundings, *Nordic Hydrology*, 13, 213–232.

BRAUN, L. N. (1985), Simulation of snowmelt-runoff in lowland and lower alpine regions of Switzerland, *Zürcher Geographische Schriften*, Department of Geography, ETH, Zürich, No. 21, 166pp.

BURKIMSHER, M. (1983), Investigations of glacier hydrological systems using dye tracer techniques: observations at Pasterzengletscher, Austria, *Journal of Glaciology*, 29, 403–416.

CAROL, H. (1945), Über einen Versuch, den Gletschergrund mittels Einstiegs durch ein Strudelloch zu erreichen, *Die Alpen*, 21, 180–184.

CLAGUE, J. J., and MATHEWS, W. H. (1973), The magnitude of jökulhlaups, *Journal of Glaciology*, 12, 501–504.

CLARKE, G. K. C. (1982), Glacier outburst floods from 'Hazard Lake,' Yukon Territory, and the problem of flood magnitude prediction, *Journal of Glaciology*, 28, 3–21.

COLLINS, D. N. (1977), Hydrology of an alpine glacier as indicated by the chemical composition of meltwater, *Zeitschrift für Gletscherkunde und Glazialgeologie*, 13, 219–238.

COLLINS, D. N. (1979a), Sediment concentration in melt waters as an indicator of erosion processes beneath an alpine glacier, *Journal of Glaciology*, 23, 247–256.

COLLINS, D. N. (1979b), Quantitative determination of the subglacial hydrology of two alpine glaciers, *Journal of Glaciology*, 23, 347–361.

COLLINS, D. N. (1979c), Hydrochemistry of meltwaters draining from an alpine glacier, *Arctic and Alpine Research*, 11, 307–324.

COLLINS, D. N. (1982), Water storage in an alpine glacier, in GLEN, J. W. (Ed.), *Hydrological Aspects of Alpine and High Mountain Areas (Proceedings of the Exeter Symposium, July 1982)*, International Association of Hydrological Sciences Publication 138, pp. 113–122.

DEICHMANN, N., ANSORGE, J., and RÖTHLISBERGER, H. (1979), Observations of glacier seismicity on Untaargletscher (abstract), *Journal of Glaciology*, **23**, 409.

ELLISTON, G. R. (1966), Glaciological studies on the Gornergletscher: II, water movement through the glacier; III, changes in surface speed with time (unpublished).

ELLISTON, G. R. (1973), Water movement through the Gornergletscher, in *Symposium on the Hydrology of Glaciers (Proceedings of the Cambridge Symposium, 7–13 September 1969)*, International Association of Scientific Hydrology Publication 95, pp. 79–84.

EMMENEGGER, C., and SPREAFICO, M. (1979), La station hydrométrique fédérale de la Massa-Blatten au front du glacier d'Aletsch, *Mitteilungen der VAW/ETH Zürich*, **41**, 23–38.

FISHER, D. (1973), Subglacial leakage of Summit Lake, British Columbia, by dye determinations, in *Symposium on the Hydrology of Glaciers (Proceedings of the Cambridge Symposium, 7–13 September 1969)*, International Association of Scientific Hydrology Publication 95, pp. 111–116.

FLOTRON, A. (1973), Photogrammetrische Messung von Gletscherbewegungen mit automatischer Kamera, *Vermessung, Photogrammetrie und Kulturtechnik*, **71**, 15–17.

FOUNTAIN, A. G., and TANGBORN, W. (1985), Overview of contemporary techniques, in YOUNG, G. J. (Ed.), *Techniques for Prediction of Runoff from Glacierized Areas*, International Association of Hydrological Sciences Publication 149, pp. 27–41.

HAEBERLI, W. (1983), Frequency and characteristics of glacier floods in the Swiss Alps, *Annals of Glaciology*, **4**, 85–90.

HAEBERLI, W. (1985), *Fluctuations of Glaciers*, Vol. 4, Permanent Service on the Fluctuations of Glaciers of the IUGG–FAGS/ICSU, International Commission of Snow and Ice of the International Association of Scientific Hydrology (ICSI/IAHS)/UNESCO.

HAEFELI, R. (1963), Note on the history of the Steingletscher lake, *Bulletin of the International Association of Scientific Hydrology*, **8**, 123–125.

HAEFELI, R. (1970), Changes in the behaviour of the Unteraargletscher in the last 125 years, *Journal of Glaciology*, **9**, 195–212.

HANTZ, D., and LLIBOUTRY, L. (1983), Waterways, ice permeability at depth, and water pressures at glacier d'Argentiere, French Alps, *Journal of Glaciology*, **29**, 227–239.

HARRISON, W. D. (1972), Temperature of a temperate glacier, *Journal of Glaciology*, **11**, 15–29.

HODGE, S. M. (1976), Direct measurement of basal water pressures: a pilot study, *Journal of Glaciology*, **16**, 205–217.

HODGE, S. M. (1979), Direct measurement of basal water pressures: progress and problems, *Journal of Glaciology*, **23**, 309–319.

HOINKES, H. (1955), Measurements of ablation and heat balance on alpine glaciers, *Journal of Glaciology*, **2**, 497–501.

HOINKES, H. (1969), Surges of the Vernagtferner in the Oetztal Alps since 1599, *Canadian Journal of Earth Sciences*, **6**, 853–861.

HOINKES, H., and LANG, H. (1962a), Der Massenhaushalt von Hintereis- und Kesselwandferner 1957/58 and 1958/59, *Archiv für Meteorologie, Geophysik und Bioklimatologie, B*, **12**, 284–320.

HOINKES, H., and LANG, H. (1962b), Winterschneedecke und Gebietsniederschlag, *Archiv für Meteorologie, Geophysik und Bioklimatologie, B*, **11**, 424–446.

HOINKES, H., and RUDOLPH, R. (1962), Variations in the mass balance of Hintereisferner (Oetztal Alps), 1952–1961, and their relation to variations of climatic elements, *International Association of Scientific Hydrology Publications*, 58, pp. 16–28.

HOINKES, H., DREISEITL, E., and WAGNER, H. P. (1974), Mass balance of Hintereisferner and Kesselwandferner 1963/64 to 1972/73 in relation to the climatic environment, IHD activities in Austria 1965–1974, *Report to the International Conference on the Results of the IHD, 2–14 September 1974, Paris*, Institute of Meteorology, University of Innsbruck.

HOLMLUND, P. (in preparation), Geometry of the near-surface drainage system on Storglaciären, Sweden (International Workshop on Hydraulic Effects at the Glacier Bed, September 1985, abstract).

HOOKE, R. L. (1984), On the role of mechanical energy in maintaining subglacial water conduits at atmospheric pressure, *Journal of Glaciology*, 30, 180–187.

HOOKE, R. L., BRZOZOWSKI, J., and BRONGE, C. (1983), Seasonal variations in surface velocity, Storglaciären, Sweden, *Geografiska Annaler*, 65A, 263–277.

HOOKE, R. L., WOLD, B., and HAGEN, J. O. (1985), Subglacial hydrology and sediment transport at Bondhusbreen, southwest Norway, *Geological Society of America Bulletin*, 96, 388–397.

Hydrologisches Jahrbuch der Schweiz (1965–1982), Landeshydrologie, Bern.

IKEN, A. (1972), Measurement of water pressure in moulins as part of a movement study of the White glacier, Axel Heiberg Island, Northwest Territories, Canada, *Journal of Glaciology*, 11, 53–58.

IKEN, A. (1974), Velocity fluctuations of an arctic valley glacier, a study of the White glacier, Axel Heiberg Island, Canadian Arctic Archipelago, *Axel Heiberg Island Research Reports*, Glaciology Vol. 5, 116pp., McGill University, Montreal.

IKEN, A. (1977), Variations of surface velocities of some alpine glaciers measured at intervals of a few hours. Comparison with arctic glaciers, *Zeitschrift für Gletscherkunde und Glazialgeologie*, 13, 23–35 (printed 1978) (Munich Symposium on the Dynamics of Temperate Glaciers and Related Problems, European Geophysical Society, September 1977).

IKEN, A. (1981), The effect of the subglacial water pressure on the sliding velocity of a glacier in an idealized numerical model, *Journal of Glaciology*, 27, 407–421.

IKEN, A., RÖTHLISBERGER, H., and HUTTER, K. (1976), Deep drilling with a hot water jet. Zeitschrift für Gletscherkunde und Glazialgeologie, 12, 143–156 (printed 1977).

IKEN, A., RÖTHLISBERGER, H., FLOTRON, A., and HAEBERLI, W. (1983), The uplift of Unteraargletscher at the beginning of the melt season—a consequence of water storage at the bed?, *Journal of Glaciology*, 29, 28–47.

IKEN, A., and BINDSCHADLER, R. A. (1986), Combined measurements of subglacial water pressure and surface velocity of the Findelengletscher, Switzerland. Conclusions about drainage system and sliding mechanism, *Journal of Glaciology*, 32, 101–119.

KASSER, P. (1959), Der Einfluss von Gletscherrückgang und Gletschervorstoss auf den Wasserhaushalt, *Wasser- und Energiewirtschaft*, 6, 155–168.

KASSER, P. (1967, 1973a), *Fluctuations of Glaciers*, Vol. 1, 1959–65; Vol. 2, 1965–70, Permanent Service on the Fluctuations of Glaciers of the IUGG–FAGS/ICSU, International Commission of Snow and Ice of the International Association of Scientific Hydrology (ICSI/IAHS)/UNESCO.

KASSER, P. (1973b), Influence of changes in the glacierised area on summer runoff in the Porte du Scex drainage basin of the Rhone, in *Symposium on the Hydrology of Glaciers (Proceedings of the Cambridge Symposium, 7–13 September 1969)* International Association of Scientific Hydrology Publication 95, pp. 221–225.

KASSER, P. (1981), *Rezente Gletscherveränderungen in den Schweizer Alpen*, Gletscher und Klima, Jahrbuch der Schweizerischen Naturforschenden Gesellschaft, Wissenschaftlicher Teil—Annuaire de la Société Helvétique des Sciences Naturelles, Partie Scientifique, 1978, Birkäuser Verlag, Basle, Boston, Stuttgart, pp. 106–138.

KASSER, P., AELLEN, M., and SIEGENTHALER, H. (1983), Die Gletscher der Schweizer Alpen 1975/76 und 1976/77, *Glaziologisches Jahrbuch der Gletscherkommission der Schweizerischen Naturforschenden Gesellschaft (SNG)*, 97. and 98. Bericht, 208pp.

KONDRATYEV, K. (1969), *Radiation in the Atmosphere*, Academic Press.

KRIMMEL, R. M., TANGBORN, W. V., and MEIER, M. F. (1973), Water flow through a temperature glacier, in *The Role of Snow and Ice in Hydrology (Proceedings of the Banff Symposium, September 1972)*, International Association of Hydrological Sciences Publication 107, pp. 401–416.

KUHN, M. (1985), Reactions of mid-latitude glacier mass balance to predicted climatic changes, in *Glaciers, Ice Sheets and Sea Level: Effect of a CO_2 Induced Climatic Change. Report of a Workshop in Seattle, 13–15 September 1984*, National Technical Information Service, United States Department of Commerce, Springfield, VA, pp. 248–254.

KUHN, M., KASER, G., MARKL, G., WAGNER, H. P., and SCHNEIDER, H. (1979), *25 Jahre Massenhaushaltsuntersuchungen am Hintereisferner*, Institut für Meteorologie und Geophysik der Universitat Innsbruck, 80pp.

KUHN, M., MARKL, G., KASER, G., NICKUS, U., OBLEITNER, F., and SCHNEIDER, H. (1985), Fluctuations of climate and mass balance: Different responses of two adjacent glaciers, *Zeitschrift für Gletscherkunde und Glazialgeologie*, **21**, 409–416.

KUZMIN, P. P. (1961), *Melting of Snow Cover*, Gidrometeorologiskoe Izdatel'stvo Leningrad 1961, translated by Israel Programme for Scientific Translations, Jerusalem, 1972.

LANG, H. (1966), Hydrometeorologische Ergebnisse aus Abflussmessungen im Bereich des Hintereisferners in den Jahren 1957–59, *Archiv für Meteorologie, Geophysik und Bioklimatologie B*, **14**, 280–302.

LANG, H. (1968), Relations between glacier runoff and meteorological factors observed on and outside the glacier, in IUGG General Assembly, Berne, *International Association of Scientific Hydrology Publication*, 79, pp. 429–439.

LANG, H. (1973), Variations in the relation between glacier discharge and meteorological elements, in *Symposium on the Hydrology of Glaciers (Proceedings of the Cambridge Symposium, 7–13 September 1969)*, International Association of Scientific Hydrology Publication 95, pp. 85–94.

LANG, H. (1980), Theoretical and practical aspects in the computation of runoff from glacier areas, in *Proceedings of the Symposium in Tbilisi 1978*, Academy of Sciences of the USSR, Geophysical Committee, Data of Glaciological Studies, Publication 38, pp. 187–194.

LANG, H., and SCHÖNBÄCHLER, M. (1967), *Heat Balance Studies and Runoff at the Aletschgletscher, 1965*, VAW, ETH Zurich, Internal Report.

LANG, H., and PATZELT, G. (1971), Die Volumenanderung des Hintereisferner im Verleich zur Massenänderung im Zeitraum 1953–64, *Zeitschrift für Gletscherkunde und Glazialgeologie*, **7**, 39–55.

LANG, H., SCHÄDLER, B., and DAVIDSON, G. (1977), Hydroglaciological investigations on the Ewigschneefeld–Grosser Aletschgletscher, *Zeitschrift für Gletscherkunde und Glazialgeologie*, **12**, 109–124.

LANG, H., LEIBUNDGUT, C., and FESTEL, E. (1979), Results from tracer experiments on the water flow through the Aletschgletscher, *Zeitschrift für Gletscherkunde und Glazialgeologie*, **15**, 209–218.

LETTAU, H. (1939), *Atmosphärische Turbulenz*, Akademische Verlagsgesellschaft, Leipzig.

LIESTØL, O. (1955), Glacier dammed lakes in Norway, *Norsk Geografisk Tidsskrift*, **15**, 122–149 (published 1956).

LLIBOUTRY, L. (1968), General theory of subglacial cavitation and sliding of temperate glaciers, *Journal of Glaciology*, **7**, 21–58.

LLIBOUTRY, L. (1971), Permeability, brine content and temperature of temperate ice, *Journal of Glaciology*, **10**, 15–29.

LLIBOUTRY, L. (1983), Modifications to the theory of intra-glacial waterways for the case of subglacial ones, *Journal of Glaciology*, **29**, 216–226.

LLIBOUTRY, L., MORALES ARNAO, B., PAUTRE, A., and SCHNEIDER, B. (1977), Glaciological problems set by the control of dangerous lakes in Cordillera Blanca, Peru, *Journal of Glaciology*, **18**, 239–290.

LÜTSCHG, O. (1915), Der Märjelensee und seine Abflussverhältnisse, *Annalen der Schweizerischen Landeshydrologie*, **1**, 168–203.

LÜTSCHG, O. (1926), Über Niederschlag und Abfluss im Hochgebirge. Schweizerischer Wasserwirtschaftsverband—Verbandsschrift 14, *Veröffentlichungen der Hydrologischen Abteilung der Schweizerischen Meteorologischen Zentralanstalt*, Zürich, 479pp.

MALE, D. H., and GRANGER, R. J. (1979), Energy mass fluxes at the snow surface in a prairie environment in COLBECK, S. C. and RAY, M. (Eds.), *Proceedings Modelling Snow Cover Runoff*, CRREL, Hanover, NH, USA, pp. 101–124.

MATHEWS, W. H. (1964), Water pressure under a glacier, *Journal of Glaciology*, **5**, 235–240.

MEIER, M. F. (1973), Hydraulics and hydrology of glaciers, in *The Role of Snow and Ice in Hydrology (Proceedings of the Banff Symposium, September 1972)*, International Association of Hydrological Sciences Publication 107, pp. 353–369.

MEIER, M. F. (1983), Snow and ice in a changing hydrological world, *Journal of Hydrological Sciences*, **28**, 3–22.

METCALF, R. C. (1984), Field pH determinations in glacial melt waters, *Journal of Glaciology*, **30**, 106–111.

MONIN, A. S., and OBUKHOV, A. M. (1954), *Basic Laws of Turbulent Mixing in the Ground Layer of the Atmosphere*, Geophysics Institute, Academy of Sciences USSR, No. 24, 151, 163–187 (translated by the American Meteorological Society, T-R-174, 1959).

MOSER, H., BAKER, D., BEHRENS, H., ESCHER-VETTER, H., OERTER, H., RAUERT, W., REINWARTH, O., RENTSCH, H., and STICHLER, W. (1983), Abfluss in und von Gletschern, *Bericht 1980/1983*, *Sonderforschungsbereich 81*, Technische Universität München, Munich, 63pp.

MOUGIN, P., and BERNARD, C. (1922), Étude sur le glacier de Tête-Rousse, in *Études Glaciologiques*, Service des Grandes Forces Hydrauliques, Ministère de l'Agriculture, Paris, Vol. 4, pp. 3–90.

MÜLLER, F. (1977), *Fluctuations of Galciers, Volume 3: 1970–1975*, Permanent Service on the Fluctuations of Glaciers of the IUGG–FAGS/ICSU, International Commission of Snow and Ice of the International Association of Scientific Hydrology (ICSI/IAHS)/UNESCO.

MÜLLER, F., and IKEN, A. (1973), Velocity fluctuations and water regime of arctic valley glaciers, in *Symposium on the Hydrology of Glaciers (Proceedings of the Cambridge Symposium, 7–13 September 1969)*, International Association of Scientific Hydrology Publication 95, pp. 165–182.

MÜLLER, H. (1984), Zum Strahlungshaushalt im Alpenraum, *Mitteilungen der VAW/ETH Zürich*, **71**, 167pp.

MÜLLER, H. (1985), On the radiation budget of the Alps, *Journal of Climatology*, **5**, 445–462.

NYE, J. F. (1953), The flow law of ice from measurements in glacier tunnels, laboratory experiments and the Jungfraufirn borehole experiment, *Proceedings of the Royal Society, Series A*, **219**, 477–489.

NYE, J. F. (1965), Stability of a circular cylindrical hole in a glacier, *Journal of Glaciology*, **5**, 505–507.

NYE, J. F. (1973), Water at the bed of a glacier, in *Symposium on the Hydrology of Glaciers (Proceedings of the Cambridge Symposium, 7–13 September 1969)*, International Association of Scientific Hydrology Publication 95, pp. 189–194.

NYE, J. F. (1976), Water flow in glaciers: jökulhlaups, tunnels and veins, *Journal of Glaciology*, **17**, 181–207.

NYE, J. F., and FRANK, F. C. (1973), Hydrology of the intergranular veins in a temperate glacier, in *Symposium on the Hydrology of Glaciers (Proceedings of the Cambridge Symposium, 7–13 September 1969)*, International Association of Scientific Hydrology Publication 95, pp. 157–161.

OERTER, H. (1984), *Der Abfluss an der Pegelstation Vernagtbach (Oetztaler Alpen) 1974–1983*, Gesellschaft für Strahlen- und Umweltforschung mbH, Munich, GSF— Bericht R 363, pp. 1–27.

OERTER, H., and MOSER, H. (1982), Water storage and drainage within the firn of a temperate glacier (Vernagtferner, Oetztal Alps, Austria), in GLEN, J. W. (Ed.), *Hydrological Aspects of Alpine and High Mountain Areas (Proceedings of the Exeter Symposium, July 1982)*, International Association of Hydrological Sciences Publication 138, pp. 71–81.

OERTER, H., BAKER, D., STICHLER, W., and RAUERT, W. (1985), Isotope studies of ice cores from a temperate alpine glacier (Vernagtferner, Austria) with respect to the meltwater flow, *Annals of Glaciology*, **7**, 90–93.

PATERSON, W. S. B. (1981), *The Physics of Glaciers*, 2nd edn., Pergamon Press, Oxford, New York, Toronto, Sydney, Paris, Frankfurt.

PATZELT, G. (1973), Die neuzeitlichen Gletscherschwankungen in der Venedigergruppe (Hohe Tauern, Ostalpen), *Zeitschrift für Gletscherkunde und Glazialgeologie*, **9**, 5–57.

POST, A., and MAYO, L. R. (1971), *Glacier Dammed Lakes and Outburst Floods in Alaska*, United States Geological Survey Hydrologic Investigations Atlas, HA-455.

PRANDTL, L. (1904), On fluid motions with very small friction, in *Proceedings of the Third Mathematical Congress, Heidelberg 1904*, Teubner, Leipzig, pp. 484–491.

PRANDTL, L. (1934), *The Mechanics of Viscous Fluids (Aerodynamic Theory)*, Durand (Ed.), Vol. III, Division G, Berlin.

RAYMOND, C. F., and HARRISON, W. D. (1975), Some observations on the behaviour of the liquid and gas phases in temperate glacier ice, *Journal of Glaciology*, **14**, 213–233.

RICHTER, E. (1888), *Die Gletscher der Ostalpen*, Verlag Engelhorn, Stuttgart.

ROBIN, G. DE Q. (1974), Depth of water filled crevasses that are closely spaced (correspondence), *Journal of Glaciology*, **13**, 543.

RÖTHLISBERGER, H. (1972), Water pressure in intra- and subglacial channels, *Journal of Glaciology*, **11**, 177–203.

RÖTHLISBERGER, H. (1980), Gletscherbewegung und Wasserabfluss, *Wasser, Energie, Luft–Eau, Energie, Air* **72**, 290–294.

RÖTHLISBERGER, H. (1981), Eislawinen und Ausbrüche von Gletscherseen. In P. Kasser (Ed.), *Gletscher und Klima—glaciers et clima*, Jahrbuch der Schweizerischen Naturforschenden Gesellschaft, wissenschaftlicher Teil—Annuaire de la Société Helvétique des Sciences Naturelles, partie scientifique 1978, Birkhäuser Verlag Basel, Boston, Stuttgart, 170–212.

RÖTHLISBERGER, H., and AELLEN, M. (1967), Annual and monthly velocity variations on Aletschgletscher. Unpublished paper presented at the general assembly of the International Union of Geodesy and Geophysics, September 1967, Berne, Switzerland; conference abstract.

RÖTHLISBERGER, H., IKEN, A., and SPRING, U. (1979), Piezometric observations of water pressure at the bed of Swiss glaciers (abstract). *Journal of Glaciology* **23**, 429.

RÖTHLISBERGER, H. and IKEN, A. (1981), Plucking as an effect of water-pressure variations at the glacier bed. *Annals of Glaciology* **2**, 56–62.

RUDOLPH, R. (1962), *Abflussstudien an Gletscherbächen*, Veröffentlichung des Museum Ferdinandeum, Innsbruck, No. 41, pp. 118–266.

SCHMID, W. (1925), *Der Massenaustausch in Freier Luft und Verwandte Erscheinungen*, Henri Grand, Hamburg.

SCHOMMER, P. (1976), Wasserspiegelmessungen im Firn des Ewigschneefeldes (Schweizer Alpen) 1976, *Zeitschrift für Gletscherkunde und Glazialgeologie*, **12**, 125–141 (printed 1977).

SCHOMMER, P. (1978), Rechnerische Nachbildung von Wasserspiegelganglinien im Firn und Vergleich mit Feldmessungen im Ewigschneefeld (Schweizer Alpen), *Zeitschrift für Gletscherkunde und Glazialgeologie*, **14**, 173–190.

SHREVE, R. L. (1972), Movement of water in glaciers, *Journal of Glaciology*, **11**, 205–214.

SPRING, U. (1980), *Intraglazialer Wasserabfluss: Theorie und Modellrechnungen*, Mitteilungen der Versuchsanstalt für Wasserbau, Hydrologie und Glaziologie, No. 48, Zürich, 197pp.

SPRING, V. (1983), Turbulent flow in intraglacial conduits: Temperature induced instabilities (abstract), paper presented at Euromech 172, *Mechanics of Glaciers*, Interlachen, Switzerland, 19–23 September 1983, p. 25.

SPRING, U., and HUTTER, K. (1981), Numerical studies of jökulhlaups, *Cold Regions Science and Technology*, **4**, 221–244.

SPRING, U., and HUTTER, K. (1982), Conduit flow of a fluid through its solid phase and its application to intraglacial channel flow, International Journal of Engineering Sciences, **20**, 327–363.

STENBORG, T. (1969), Studies of the internal drainage of glaciers, *Geografiska Annaler*, **51A**, 13–41.

STENBORG, T. (1970), Delay of runoff from a glacier basin, *Geografiska Annaler*, **52A**, 1–30.

STENBORG, T. (1973), Some viewpoints on the internal drainage of glaciers, in *Symposium on the Hydrology of Glaciers (Proceedings of the Cambridge Symposium, 7–13 September 1969)*, International Association of Scientific Hydrology Publication 95, pp. 117–129.

SVERDRUP, H. U. (1936), The eddy conductivity of the air over a smooth snow field, *Geophysical Publication*, No. 11, 1–69.

SVERDRUP, H. U. (1946), The humidity gradient over the sea surface, *Journal of Meteorology*, **3**, 1–8.

TANGBORN, W. V., KRIMMEL, R. M., and MEIER, M. F. (1975), A comparison of glacier mass balance by glaciological, hydrological and mapping methods, South Cascade glacier, Washington, in JOHNSON, A., MEIER, M. F., and WARD, W. H. (Eds.), *Snow and Ice (Proceedings of the Moscow Symposium, August 1971)*, International Association of Scientific Hydrology Publication 104, pp. 185–196.

THORNTHWAITE, C. W., and HOLZMAN, B. (1939), The determination of evaporation from land and water surfaces, *Monthly Weather Review*, **67**, 4–11.

THYSSEN, F., EISNER, H., BLINDOW, N., and AMBACH, W. (1980), Kartierung von wassergesättigten Firnschichten auf dem Kesselwandferner mit dem EMR-Verfahren, *Polarforschung*, **50**, 9–16.

TRONOV, M. V. (1962), On the role of summer snowfall in glacier variations, *International Association of Scientific Hydrology Publication*, 58, pp. 262–269.

UNEP/WMO/ICSU (1985), An assessment of the role of carbon dioxide and of other greenhouse gases in climate variations and associated impacts, in *Joint UNEP/WMO/ICSU Conference, Villach, 9–15 October 1985*, Conference Statement, 4pp.

UNITED STATES ARMY CORPS OF ENGINEERS (1956), Snow hydrology, *US Army Corps of Engineers Summary Report of Snow Investigations*, North Pacific Corps of Engineers, Portland, Oregon.

VIVIAN, R. A., and ZUMSTEIN, J. (1973), Hydrologie sous-glaciaire au glacier d'Argentière (Mont-Blanc, France), in *Symposium on the Hydrology of Glaciers (Proceedings of the Cambridge Symposium, 7–13 September 1969)*, International Association of Scientific Hydrology Publication 95, pp. 53–64.

WALDER, J. S. (1982), Stability of sheet flow of water beneath temperate glaciers and implications for glacier surging, *Journal of Glaciology*, **28**, 273–293.

WEERTMAN, J. (1972), General theory of water flow at the base of a glacier or ice sheet, *Reviews of Geophysics and Space Physics*, **10**, 287–333.

WEERTMAN, J. (1974), Depth of water-filled crevasses that are closely spaced (correspondence), *Journal of Glaciology*, **13**, 544.

WEERTMAN, J., and BIRCHFIELD, G. E. (1983), Stability of sheet water flow under a glacier, *Journal of Glaciology*, **29**, 374–382.

WHALLEY, W. B. (1971), Observations of the drainage of an ice-dammed lake— Strupvatnet, Troms, Norway, *Norsk Geografisk Tidsskrift*, **25**, 165–174.

YOUNG, G. (Ed.) (1985), *Techniques for the Prediction of Runoff from Glacierized Areas*, International Association of Hydrological Sciences Publication 149.

Glacio-fluvial Sediment Transfer
Edited by A. M. Gurnell and M. J. Clark
© 1987 John Wiley & Sons Ltd.

Chapter 11

Solutes

R. A. Souchez and M. M. Lemmens

ABSTRACT

The aim of this chapter is to analyse the different mechanisms responsible for the solute content of alpine glacial meltwaters. The chapter begins by considering the characteristics of the alpine glacier environment which may influence the water chemistry. This is followed by an evaluation of atmospheric inputs to alpine drainage basins. The central theme of the chapter is the study of the mechanisms acting on the solute content of glacial meltwaters either englacially or subglacially. This discussion is concluded with an assessment of chemical activity in alpine glacierized drainage basins in comparison with drainage basins in other environments.

11.1 THE ALPINE GLACIER ENVIRONMENT AND WATER CHEMISTRY

The alpine glacier system can be distinguished in terms of chemical activity from any other system. Its peculiarities are the consequence of the mountainous environment and of the release of water by the glacier.

Alpine glaciers are usually developed between 2000 and 4000 m a.s.l. This altitudinal zone is situated above the tree limit and presents only a scanty vegetation cover; very few glaciers flow, like the Glacier des Bossons in the French Alps, into the forest zone. As a consequence, soils rich in organic matter are generally not present where glacier meltwaters reach the ground surface. Carbon dioxide dissolved in water is of direct atmospheric origin and is not the result of the local decomposition of organic matter by soil micro-organisms or of root respiration.

The amount of dissolved carbon dioxide in water depends on the carbon dioxide pressure in the atmosphere. At sea level, $p\mathrm{CO_2}$ is 0.00035 atm, whereas it only reaches 0.00025 atm at 2800 m a.s.l. if the partial pressure remains unchanged. This low value of $p\mathrm{CO_2}$ has a direct influence on chemical weathering but we must bear in mind the increased solubility of CO_2 in cold waters and the peculiar subglacial situation where water pressure may be well above atmospheric pressure.

Glacial meltwater is always near its freezing point and, at an altitude of more than 2000 m a.s.l., freezing is common. Phase changes bring additional characteristics to the alpine glacier system.

In this environment, we may expect that physical erosion by glacier grinding is rapid and that fresh mineral grains are constantly being exposed so that the initial alteration of fresh surfaces has a significant influence on water chemistry.

Finally, water may be released in great quantities either because of snow and ice melting during the summer or, sometimes, by the drainage of water stored in englacial or subglacial cavities.

11.2   THE ATMOSPHERIC CONTRIBUTION

In considering the solute load of glacial meltwater streams, it is useful to assess the proportion of this output of ions that is of atmospheric origin. The atmospheric contribution to the solute content of the melt stream is a result of the chemical composition of snow which falls on to the glacier and its immediate surroundings during most of the year and of the chemical composition of rainwater which may fall during part of the summer. Supraglacial streams are fed by the melting of snow and ice and by surface drainage of rain. Since the water in supraglacial streams is able to acquire solutes by contact with the dispersed particles present at the glacier surface, a chemical investigation of these streams provides the possibility of defining a maximum atmospheric contribution to the solute load of meltwater. Results of water sampling in supraglacial streams of the Glacier de Tsidjiore Nouve in the Swiss Alps give the following range of variation: 0.3–1.7 μequiv.l^{-1} Na, 0.4–1.3 μequiv.l^{-1} K, 4.0–13.0 μequiv.l^{-1} Ca and 0.8–5.8 μequiv.l^{-1} Mg. On average the sum of the concentrations of the four dissolved cations is 50 times lower in these supraglacial streams than in the proglacial stream at a low discharge of 0.2 m^3 s^{-1} and 18 times lower when the discharge rises to 2 m^3 s^{-1}. Similar results have also been obtained for the Tsanfleuron glacier about 60 km to the north of the Tsidjiore Nouve glacier (Lemmens and Roger, 1978). Thus, it is clear that the chemical composition of precipitation plays only a minor role in explaining the major cation content of the meltstream.

As the Alps lie within the populated region of north western and central Europe, dissolved sulphates in proglacial streams might originate from the atmosphere. Indeed, the anthropogenic sulphur dioxide from the combustion of fossil fuels is oxidized and gives rise to sulphate and hydrogen ions. This source of sulphates and acidity has considerably increased since the Second World War. In their work on solute acquisition in glacial meltwaters in the Argentière glacier area, Thomas and Raiswell (1984) deduced a weak atmospheric contribution to the total sulphate flux on the basis of a much lower Cl^-/SO_4^{2-} molar ratio in the subglacial and bulk meltwaters than in average precipitation. The mean value of 1.6 ppm sulphate in Swiss precipitation (Zobrist and Stumm, 1979) was obtained without taking the high alpine region into account and, therefore, is likely to be influenced by the proximity of industrial sites and by a reduced proportion of snow in total precipitation (the snow being less able to capture salts in the atmosphere than rainwater). Thus, 1.6 ppm of SO_4^{2-} is probably a maximum value for alpine precipitation and this is seven times less than the concentration in the proglacial stream of the Tsidjiore Nouve glacier.

11.3 SELF-PURIFICATION IN ALPINE GLACIERS, SELECTIVE FLUSHING OUT OF IONS AND LEACHING

Shreve (1972) showed that a glacier at the pressure melting point develops an internal arterial drainage system with cavities and channels throughout the glacier. This internal system is unlikely to be able to purify the ice. If water moves through part of the vein system, then impurities are removed from the ice but the meltwater is also removed. Glen *et al.* (1977) suggested that the major purification mechanism occurs during recrystallization when ice is deformed. The ice grain boundaries move through the material as the grain shape changes. The impurities reach the grain boundary by such a process and it seems likely that such impurities will be retained within the boundary as it migrates. Freezing and melting accompany the recrystallization. As ice rejects most of the impurities to the meltwater when it freezes, the new ice that is formed will be purer. Finally, the impurities reach larger cavities and channels by which they are flushed out of the glacier. The preferential loss of impurities from the ice crystals leads to a self-purification mechanism in alpine glaciers. This mechanism is likely to occur in the alpine glaciers that have been studied by the present authors but its importance is difficult to assess.

The flushing-out of cations from the ice is selective. Fractional melting experiments (Souchez and Tison, 1981) led to the recognition of this selective flushing-out process. In a suitable hot-water bath, sections from Tsidjiore Nouve glacier ice cores, devoid of visible mineral particles, were heated at the base. As soon as 10 ml of meltwater was produced, it was recovered with the aid of a polyethylene syringe and immediately press-filtered. Successive 10 ml samples of filtered water were taken and analysed. Figure 11.1 shows a sharp decrease in the $Na + K/Ca + Mg$ ratio during melting. This indicates a selective migration of ions, alkali metals being more easily flushed out than alkaline earth metals. Moreover, a systematic decrease in cation content takes place from the first to the last filtered sample. This is connected with dilution of the interstitial water which is mainly located at three- and four-grain intersections. These experiments probably reproduce the situation occurring in the glacier when water is squeezed out from the ice. The selective flushing-out of alkali metals in comparison with

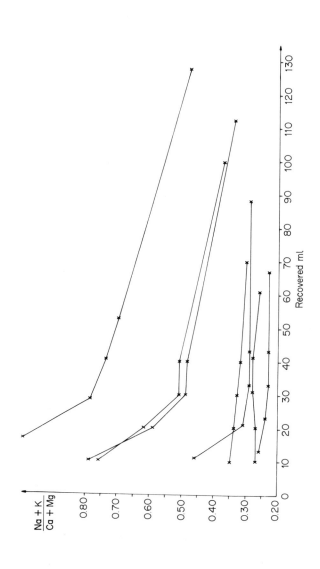

FIGURE 11.1   Alkali metal to alkaline earth metal ratios for seven glacier ice samples subjected to fractional melting. *Reproduced by permission of the International Glaciological Society from Annals of Glaciology, 1981,2, 65*

alkaline earth metals with the meltwater was proposed in the study of basal ice layers at the Tsidjiore Nouve glacier (Souchez and Lorrain, 1978). This process results from the higher diffusion coefficients of the alkali metals.

Free water, either rain or meltwater from the surface layers, percolates through snow and firn in the glacier's accumulation zone. As it moves downward into deeper layers, water-soluble substances originally present in the snow or firn as aerosols or dusts are entrained. This gives rise to a leaching process. In the Glacier du Mont de Lans in the French Alps, Ricq-de Bouard (1977) studied the distribution of major cations in a 6 m profile from a crevasse wall at 3200 m, representing at least 20 years accumulation. If the highly variable concentration of the first metre is not taken into account, the author found an approximately exponential decrease of Na, K, Ca and Mg with depth. The exponents indicate that alkaline earth metals are less rapidly leached than the alkali metals.

11.4 MEMBRANE FILTRATION

Dark horizons corresponding to summer layers with increased dust and aerosol content are often visible in the firn of alpine glaciers. Such horizons may act as filters retaining substances from overlying deposits which are transported downwards with meltwater or rain. For example, this was indicated by Prantl *et al.* (1972) for caesium-137.

In certain circumstances, particle layers may act as semi-permeable membranes, as demonstrated by Souchez and Lorrain (1975) for a frozen mud layer plastered at the base of the Argentière glacier at the ceiling of an enclosed natural subglacial cavity. A large pressure gradient existed across this mud layer: interstitial water in the glacier at its upper surface was under hydrostatic pressure of about 90 m of ice and, at its lower surface 2–3 cm below, under the atmospheric pressure of the cavity. This interstitial water was thus forced through the mud layer under a pressure gradient of about 3 bar cm^{-1} and then refroze in the form of ice accretion hanging at the ceiling. This refreezing results from the pressure dependence of freezing point. A comparison (Table 11.1) between the chemical composition of the samples of ice accretions and water held at their surface (a) and the samples of water in contact with the upper surface of the mud layer (b) provides the following significant characteristics:

1 . A lower content of the four cations in the water sampled at (a) than at (b), showing a desalinating effect as water is forced from (b) to (a).
2 . A higher Na + K/Ca + Mg ratio for (a) than for (b), indicating a monovalent –divalent cation species separation.
3 . A higher Na/K and a lower Ca/Mg ratio for (a) than for (b), indicating sorting effects between the alkali metals as well as between the alkaline earth metals.

These characteristics show that the frozen mud layer plastered at the base of the glacier may be considered as an ion-exchange membrane causing electrolyte filtration and relative retardation of cations (Helfferich, 1962). The mud layer is frozen at a temperature close to the melting point, so the large pores are filled with ice and unfrozen water exists partly at the interface between the mineral grains and ice and partly in crevices and capillaries when the radius of curvature is sufficiently small (Anderson, 1967). In such a situation, the ions must migrate through thin films of unfrozen water which are continuous through the frozen mud (Hoekstra and Chamberlain, 1964). The membrane properties

TABLE 11.1   Concentrations of Na, K, Ca and Mg in the samples collected at the base of the Glacier d'Argentière. *Reproduced by permission of the International Glaciological Society from* Journal of Glaciology, *1975, **14**, 263*

Ionic species or ratios	Water in contact with the mud layer (17 samples) (μequiv.l^{-1}) (b)*			Ice accretions and water held at their surface (9 samples) (μequiv.l^{-1}) (a)*		
	Mean	Min.	Max.	Mean	Min.	Max.
Na	71.6	28.3	139.2	55.8	34.8	80.0
K	176.4	61.4	296.6	58.7	35.3	81.8
Ca	460.0	249.5	774.4	38.9	6.5	64.4
Mg	50.0	17.3	95.4	17.6	9.1	32.9
Na + K + Ca + Mg	758.0	371.5	1263.7	171.0	89.7	242.2
(Na + K)/(Ca + Mg)	0.50	0.21	0.90	2.35	1.50	4.59
Na/K	0.41	0.23	0.61	0.96	0.81	1.12
Ca/Mg	10.02	6.49	15.83	2.37	0.49	5.51

*See text.

indicated above are explained by the peculiar electrochemical situation in these films in which the ion concentration distribution is determined by the overlap of adjoining double layers. This distribution would also be approximately true in thin films of water adsorbed on single solid surfaces with air on the other side of the film (Kemper, 1960). In these situations a Donnan exclusion exists and an electrolyte filtration effect occurs. This effect is selective; the relative passage rates will depend on the resultant of all the forces affecting their selective adsorption and their selective transport (Karaka and Berry, 1973). Kemper *et al.* (1970) and Berry (1969) indicate that with a low hydraulic pressure gradient, such as that existing across the frozen mud layer at the base of the Argentière glacier, the relative retardation sequence to be considered is Na < K < Mg < Ca. This retardation sequence leads to a higher Na + K/Ca + Mg ratio, a higher Na/K ratio and lower Ca/Mg ratio for the effluent solution, which is the case at the base of the Argentière glacier. A chemical sorting effect is thus taking place at the base of an alpine glacier that can be of importance for the understanding of the chemistry of the thin water film present at the base of a temperate glacier.

11.5 PHASE CHANGES

Water at the glacier bed can be in the form of a film which primarily accommodates the local transport of meltwater associated with regelation sliding. Through-flowing water appears to drain through a different hydraulic system, distinct channels at the ice–rock interface (Röthlisberger, 1972; Shreve, 1972). These waters produce a basal ice layer when they refreeze totally or partially. The selective rejection of solutes into the meltwater by the growing solid phase is a general characteristic of most freezing systems. It is well documented for many freezing bulk aqueous solutions, including those of NaCl (Terwilliger and Dizio, 1970) and $CaCO_3$ (Hallet, 1976). The effective distribution coefficient, defined as the ratio of the total concentration of solutes frozen into the ice to that remaining in the water, varies from 10^{-3} to unity. It is generally considerably less than unity in water freezing at rates comparable to those expected at the base of alpine glaciers. Subglacial chemical deposits, such as calcite precipitates, can be formed as a consequence of solute rejection by growing ice. These deposits appear to form in intimate association with the pressure melting and refreezing process, known as regelation, which is an essential component of glacier sliding. In this process, basal ice melts because of relatively high pressures along the up-glacier surfaces of bedrock protuberances, and the water so produced flows along thin subglacial films to neighbouring lee surfaces where it refreezes. During this freezing, solutes present in the water tend to be preferentially rejected by the growing ice. Solutes, therefore, tend to accumulate in the liquid phase at sites where refreezing is localized, such as at the lee side of bedrock obstacles overridden by a sliding alpine glacier. This mechanism of concentrating solutes is sufficiently effective to increase the concentration until it exceeds the saturation value and solutes precipitate. Once precipitation has occurred, the concentration is fixed by the solubility of the precipitate and the temperature at that site, and by the corresponding degree of freezing point lowering due to solutes in solution. These unique values for the composition and temperature, at which slow freezing occurs simultaneously with and in close proximity to solute precipitation, approximate the eutectic values for simple binary solutions. Removing heat from the system at the eutectic point causes ice growth on the glacier sole and calcite deposition on the bedrock. These subglacial deposits are commonly furrowed and shaped into small but distinct spicules aligned in the down-glacier flow direction. To account for this orientation, it is apparent that during the formation of these deposits, they must be completely enveloped in regelation ice with only a microscopic sheath of water separating them from the ice. These deposits have been reported from a number of past and present temperate glaciers (Hallet, 1976) and are particularly abundant on bedrock surfaces recently exposed by the retreat of the Tsanfleuron glacier in the Swiss Alps. It is very likely that similar deposits have been and are still continually forming under this glacier. From this, it can be inferred that a thin film of water at the eutectic composition for $CaCO_3$ exists at the base of this glacier. That this film is present can be demonstrated by the chemical composition of regelation ice. The calcium concentration of the regelation ice can be estimated by calculating the eutectic Ca concentration, ignoring effects of solutes other than dissolved $CaCO_3$ (they are unimportant in this case in controlling the calcite solubility). If the eutectic Ca concentration determined from the pCO_2 is multiplied by the effective distribution coefficient, an estimated Ca concentration of regelation ice can be obtained (Hallet *et al.*, 1978). This predicted Ca concentration is in close agreement with the measured Ca concentration in regelation ice.

Phase changes at the glacier–bedrock interface are likely to concentrate water flowing as a thin film to the saturation level for calcium carbonate. Through-flowing water draining in distinct channels is less affected by this influence of freezing on water chemistry. On the other hand, owing to the limited occurrence of calcite in the Tsidjiore Nouve glacier basin, which is located on gneissic rocks, the eutectic composition is not reached in the thin subglacial water film since the chemical analyses of regelation ice indicate a Ca concentration level well below the value that would be obtained if the above-described process was going on under that glacier. It seems probable that a saturated thin water film at an alpine glacier base with respect to $CaCO_3$ only exists if the bedrock is relatively rich in calcite minerals.

11.6 ION EXCHANGE

Clay minerals surfaces are usually negatively charged and are surrounded in water by a Gouy layer. This layer is the region in the liquid around the particle where cations are in excess of anions. The structure of this layer depends on the density of negative charges at the surface of the particle, on the nature of the ions, on the temperature and on the concentration of the solution. Its thickness ranges, for example, from 50 to 400 Å if the concentration of the solution varies from 0.1 to 0.01 N. An individual cation of the Gouy layer can be exchanged with one of the external solution. This process is called ion exchange. The substitution of H^+ ions from the water with cations sorbed in the Gouy layer surrounding a particle is a particular type of ion exchange and is often considered as the first step in weathering processes. Desorption is here considered as the process by which major cations (Na, K, Ca and Mg) leave Gouy layers and are distributed in the solution.

This process can, in certain cases, explain the origin of cations present in glacial meltwaters. Because of the shortage of organic matter in this system, this ion exchange is mainly due to clay minerals in the fine fractions of morainic deposits. In the Tsanfleuron proglacial zone, located in a limestone area at nearly 2400 m a.s.l., morainic material is very rich in clay minerals. They represent more than 50% of the non-carbonate $< 50 \, \mu m$ fraction and are mainly composed of illite and chlorite. Whereas Ca and Mg in the meltstreams are essentially the result of limestone dissolution, Na and K are connected with these clay materials. Alkali metals are very dilute in supraglacial streams: the values for Na are between 0.9 and 1.7 μequiv.l^{-1} and for K between 0.3 and 0.8 μequiv.l^{-1}. After flowing for a distance of about 10 m on morainic material, the concentration in water rises to levels of 3.6–12.2 μequiv.l^{-1} for Na and 8.00–12.4 μequiv.l^{-1} for K. After 30 m, the respective values are 5.0–22.0 μequiv.l^{-1} for Na and 6.7–12.9 μequiv.l^{-1} for K (Lemmens and Roger, 1978). These values are nearly the same as those obtained from samples taken 1 km downstream in the channel formed by the confluence of different meltstreams. The pH of the water also increases considerably during the first metres of contact with morainic material, rising from 6.2–6.7 in supraglacial streams to 8.4–9.2 after 10 m of water contact with morainic deposits. A rapid increase in the two alkali metals is thus a characteristic of the proglacial zone. This high rate of enrichment leads us to believe that here ion exchange, and desorption in particular, plays a major role. Indeed, Luce *et al.* (1972) showed, in experiments on weathering of silicate minerals, that a rapid increase in dissolved Mg occurs in the water before any rise in silica. They ascribed this initial increase to an ion exchange between surface Mg^{2+} ions and H^+ ions present in the water. This exchange was completed in about 15 min and subsequent weathering of the mineral itself led to a further increase in both Mg and silica in the solution. Wollast (1967) also showed, in the alteration of K-feldspar, the existence of a first phase of exchange reaction of K^+ ions with an accompanying high rise in the pH. These observations indicate that, in the time period involved in the Tsanfleuron proglacial stream, ion exchange is the most likely process operating at the surface of the clay minerals (Lemmens and Roger, 1978). Because of concentration increase downstream, the gradient between the Gouy layers and the solution is diminished so that diffusion is slowed. This can explain why, after a rapid enrichment in the frontal zone of the glacier, the proglacial stream does not significantly change its composition in Na and K over a 1 km distance. The role of ion exchange is also significant if we study the Na/K ratio. Whereas the ratio has a very low value in bottom sediments (0.07–0.25), its value always exceeds 0.45 in water samples and may even reach 1.87. This is due to the fact that Na desorbs relatively more than K because of its lower atomic number and the greater size of its hydrated ion (Caroll, 1959).

On the other hand, the Na/K ratio increase in the downstream direction can only be explained if the kinetics of ion exchange are considered. Samples taken during three days with very different weather conditions showed an increase in their Na/K ratio from 0.45 after 10 m of contact with morainic material to 0.85 after 30 m on the first day, from 0.64 to 1.67 on the second day and from 1.01 to 1.19 on the third day. The rate of exchange is therefore not the same for the two cations. Migration of cations from the particles into the solution depends on diffusion phenomena (Helfferich, 1962). The rate-determining

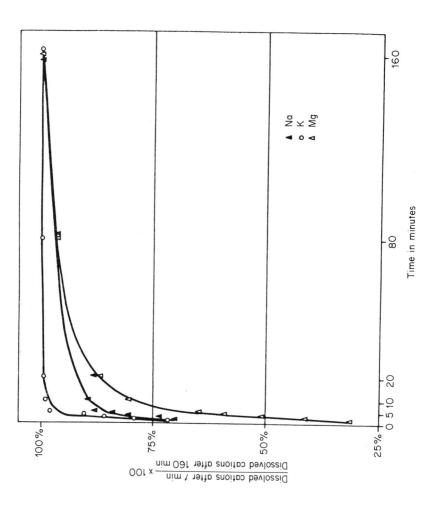

FIGURE 11.2   Percentage of dissolved Na, K and Mg as a function of time in the experiment. *From* Earth Surface Processes, *1978,* *3, 185*

control for this process, if no weathering of the mineral itself exists, is ion diffusion in the Gouy layer (film diffusion) which is slower than diffusion from the Gouy layer into the solution (Helfferlich, 1962). Clearly, diffusion coefficients in water cannot be used in this case and we do not know film diffusion coefficients. In order to solve this problem in a particular case, experiments were conducted at $0\,°C$ using fine particles from the morainic deposits of the Tsidjiore Nouve glacier and supraglacial water from the same glacier. Water used in these experiments has 0.13 μequiv.l^{-1} Na, 0.13 μequiv.l^{-1} K, 6.99 μequiv.l^{-1} Ca and 0.82 μequiv.l^{-1} Mg. Calcium was not taken into account. Its concentration reaches values 10–20 times those of the other major cations. Robinson and Stokes (1965) showed that diffusion coefficients are concentration dependent and such a concentration difference between calcium and the other major cations is likely to affect the relative order of migration rates into the solution. A series of mixtures of 20 ml of supraglacial water and 1 g of fine morainic particles were stirred at $0\,°C$ for periods lasting from 1 to 160 min. The waters were then filtered and analysed for the dissolved cations. The results indicated a very fast enrichment of the waters (Figure 11.2) such as is found at the Tsanfleuron site (Lemmens and Roger, 1978). After the first 10 min of contact, the concentrations reached values which were much higher than their initial level and which did not vary much more until the end of the experiment after 160 min. The last values of concentration indicated that the increase in dissolved cations was very slow; the initial and rapid cation exchange could thus be considered as having practically finished, with an equilibrium being approached. To compare the rates of diffusion of the different cations, Lemmens and Roger (1978) expressed the levels of dissolved cations reached at different times during the experiment as a percentage of the maximal value obtained for the whole experiment. The results, plotted in Figure 11.2, show that potassium diffuses more quickly than sodium and that magnesium is the slowest. As the concentrations of these three cations are of the same order of magnitude, their relative rates of diffusion give the sequence of the diffusion coefficients. It should be noted that this sequence is similar to that of the diffusion coefficients in water, viz. $Mg < Ca < Na < K$.

The rapidity of the enrichment of the supraglacial waters and its characteristics presumably indicate the essential role of ion-exchange processes as a source of ions in glacial meltwaters over a short time scale.

11.7 SUBGLACIAL WEATHERING

When water is flowing at the ice–rock interface, weathering of the bedrock or of the mineral particles constituting the basal till is likely to occur. The hydrogen ions responsible for the destruction of the minerals can come from various sources. Hydrogen carbonate is the dominant anion in the glacial meltwaters

(Rainwater and Guy, 1961; Reynolds and Johnson, 1972; Thomas and Raiswell, 1984). Carbonation is thus the most important chemical mechanism controlling the composition of these waters. The carbon dioxide which drives the carbonation reaction $(CO_2 + H_2O \rightarrow H^+ + HCO_3^-)$ is derived from prolonged contact of drainage waters with atmospheric CO_2. A second mechanism producing hydrogen ions responsible for the weathering is the oxidation of sulphides (commonly pyrites). The reaction can be written in the following way: $4FeS_2 + 15O_2 + 8H_2O \rightarrow 2Fe_2O_3 + 16H^+ 8SO_4^{2-}$. Substantial quantities of sulphates in glacial meltstreams may indicate that this reaction is going on in the drainage basin. Another source of sulphate ions can be the atmospheric SO_2 of anthropogenic origin but, as quoted in Section 11.2, this source can often be neglected in the alpine glacial environment. In the Tsidjiore Nouve glacier basin (Table 11.2), hydrogen carbonate is the dominant anion in the meltwaters but sulphate is abundant owing to the presence of pyrites; its oxidation is responsible for more than 50% of the H^+ produced if the atmospheric contribution is neglected.

TABLE 11.2   Composition of the Glacier de Tsidjiore Nouve meltstream. Concentrations in cations and anions are expressed in μequiv.l^{-1}. All samples were collected during the three summer months. Cation concentrations were measured on 25 samples, anion concentrations on 7 samples.

	Mean	Min.	Max.
Na	15.6	7.0	23.1
K	12.7	7.0	16.8
Ca	353.2	204.6	536.4
Mg	40.7	18.1	74.9
Σ^+	422.2	236.6	643.5
HCO_3	368.8	349.9	444.3
SO_4	233.5	195.7	291.5
Σ^-	602.3	545.6	735.8
Cation–anion balance (error in %)	2.3	0.7	4.9
SiO_2 (ppm)	1.87	1.52	2.25

Hence the anionic components of meltwaters demonstrate the acid source used in the dissolution and weathering of rock minerals. However, it must be borne in mind that the same is true if ion exchange is the dominant process, for instance at a short time scale. Fixation of hydrogen ions in the Gouy layers in place of other cations brings new carbon dioxide into solution if the atmospheric source is present.

The hydrogen ions produced react either with carbonate, silicate or aluminosilicate minerals. Following Raiswell (1984), the reactions can be written in the following forms:

1. For carbonates:
   $$CaCO_3 + H^+ \rightarrow Ca^{2+} + HCO_3^-$$
   (calcite)
2. For silicates:
   $$Mg_2SiO_4 + 4H^+ \rightarrow 2Mg^{2+} + H_4SiO_4$$
   (forsterite)
3. For aluminosilicates:
   $$2NaAlSi_3O_8 + 2H^+ + 9H_2O \rightarrow 2Na^+ + 4H_4SiO_4 + Al_2Si_2O_5(OH)_4$$
   (albite)

Dissolution of carbonates is more rapid than silicate or aluminosilicate breakdown. On the other hand, feldspars containing calcium are also among the most abundant aluminosilicates. This explains why calcium is in general the dominant cation in a glacial meltstream even if the drainage basin only contains very small amounts of carbonate rocks. This was indicated by Slatt (1972) for nine Alaskan glaciers. In the Tsidjiore Nouve glacier basin, calcium makes up more than 80% of the four major dissolved cations in a nearly carbonate-free environment (Table 11.2).

Raiswell (1984) made a distinction between open and closed system models in the evolution of meltwaters. A closed system is defined by the requirement that there is no dissolution of gaseous carbon dioxide to replace the dissolved carbon dioxide used in weathering. The consequence of rapid H^+ removal is to reduce the calculated equilibrium carbon dioxide pressure of the gas phase in contact with the water. This pCO_2 will be below atmospheric pressure. This does not imply a physical constraint on access to atmospheric CO_2 but can be simply a kinetic effect which is maintained as long as rates of CO_2 removal exceed rates of supply. By contrast, open-system meltwaters retain equilibrium with atmospheric CO_2. In this case, the rate of H^+ supply by CO_2 dissolution must approach the rate of H^+ consumption by weathering. This occurs with poorly reactive minerals where the rate of weathering is slow. In the supraglacial environment where water flows on the glacier surface, the rates of H^+ consumption by weathering are low in general and supraglacial melt should maintain open-system weathering characteristics. Although the englacial system may be isolated from physical contact with atmospheric carbon dioxide, the limited capacity for mineral–water reactions should ensure that these waters also display open-system behaviour. By contrast, the subglacial environment is in many respects the antithesis of the supraglacial and englacial systems. The residence time of subglacial water is likely to be at least several days, since subglacial recession flows can be maintained in the absence of ablation and after the drainage of englacial reservoirs (Collins, 1977, 1979a). The reactivity of the abundant mineral grains is also maintained at an optimal level by abrasion and crushing. Rapid weathering occurs and the subglacial waters exhibit closed-system characteristics. Observations of subglacial melt at the Argentière glacier by Thomas and Raiswell (1984) confirm this.

The englacial and subglacial components probably mix in the lower part of the glacier. This mixing cannot be considered as conservative. The initial factor which determines the characteristics of the bulk meltwater is the mixing ratio between the englacial and subglacial components. During recession flows the subglacial component predominates and the bulk meltwater may be expected to retain subglacial characteristics. During sustained ablation and typical diurnal discharge variations, the bulk meltwater may be 60–80% englacial (Collins, 1979a) and mixing will initially result in a bulk meltwater with open-system characteristics. However, because of the presence of abundant reactive minerals, further weathering is possible and the waters may exhibit more closed-system behaviour. Open-system bulk meltwater will usually result if post-mixing reactions cannot occur when, for instance, no reactive rock debris is available.

11.8   DISSOLVED LOAD–DISCHARGE RELATIONSHIPS AND THE ORIGIN OF SOLUTES

The relationships between discharge and dissolved load in glacial meltstreams are sometimes defined by a straight line on a bilogarithmic graph. An inverse relationship of the type $C = kQ^{-n}$ is exhibited, where C is the concentration of dissolved substances, Q the discharge, k a constant and n an exponent having, for example, a value of 0.6 in the Chamberlain glacier area in Alaska (Rainwater and Guy, 1961). In the case of the Tsidjiore Nouve glacier meltstream, similar equations have been obtained for the concentration in Na, K, Ca, Mg of samples collected in June, July and August. If discharge is expressed in $m^3 s^{-1}$ and concentration in μequiv.l^{-1}, then:

for Na, $C = 11.22 \, Q^{-0.59}$
for K, $C = 10.47 \, Q^{-0.37}$
for Ca, $C = 263.02 \, Q^{-0.54}$
for Mg, $C = 27.54 \, Q^{-0.72}$

Thus, higher discharges were characterized by a lower concentration in the four cations. The correlation coefficients were significant at a confidence level of 0.999 and were, respectively, -0.83 for Na, -0.70 for K and -0.95 for Ca and Mg (Lemmens and Roger, 1978; Lemmens, 1978).

The concentration range is not connected with the suspended load. Indeed, variations of the concentration per litre of cations sorbed on suspended particles, obtained by multiplying the concentration per gram by the weight of suspended sediment per litre of streamwater, were an order of magnitude less than those for the corresponding dissolved cations expressed in the same units (Lemmens and Roger, 1978). Fluctuations of suspended load cannot, therefore, explain variations in dissolved cations.

Such a relationship can be explained by mixing of waters from two different sources. In a hydrological model proposed by Collins (1977, 1979a), mainly for the Gornergletscher, two main components of glacial discharge were recognized: a dilute, surface meltwater which flows rapidly with little alteration through englacial channels and solute-rich waters flowing more slowly through the subglacial system. A simple two-component mixing model can be used to explain the temporal variations of the bulk meltwater emerging at the portal.

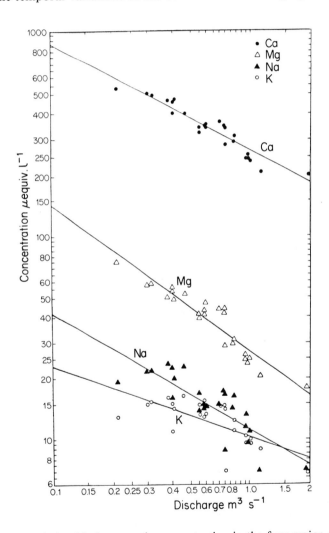

FIGURE 11.3 Relationship between the concentration in the four major cations and discharge for the Glacier de Tsidjiore Nouve meltstream. *Reproduced by permission of Catena Verlag from* Catena, *1978, 5, 232*

The different slopes of the concentration–discharge regression lines for the four major cations can now be examined (Figure 11.3), bearing in mind that water at low discharge, essentially subglacial water, is closest to equilibrium with mineral particles. If the concentration of dissolved cations is controlled by diffusion, the concentration ratio of a less mobile versus a more mobile cation (i.e. of a cation with a lower diffusion coefficient versus a cation with a higher diffusion coefficient) is lower when concentrations are distant from equilibrium values than when equilibrium values are approached. This concentration ratio must then increase when the discharge decreases. The slope of the regression line in the bilogarithmic coordinate system used in Figure 11.3 must, therefore, be steeper for a less mobile cation (i.e. a cation with a lower diffusion coefficient). Results of the experiments using fine particles from the Tsidjiore Nouve glacier basin, quoted in Section 11.6, indicate that K diffuses more quickly than Na and Mg is the slowest. The calculated values of the slopes of the regression lines are -0.72 for Mg, -0.59 for Na, -0.54 for Ca and -0.37 for K. The slope sequence for Na, K and Mg corresponds exactly to the sequence of the diffusion coefficients. The slope of Ca would be intermediate between those of Na and Mg if we consider the diffusion coefficients in water and this is almost the case with the Tsidjiore Nouve data. In fact, the slope for Na (-0.59) is slightly steeper than that for Ca (-0.54) but this difference is probably not significant.

The linear relationship, on a bilogarithmic graph, between discharge and dissolved load is not observed for all alpine meltstreams studied. Indeed, Collins (1979b) studying the chemical composition of water emerging from the snout of the Gornergletscher, pointed out a great range of ionic concentration associated with a given discharge and showed clockwise hysteretis loops of conductivity and discharge on a diurnal time basis.

11.9   COMPARISON OF CHEMICAL ACTIVITY BETWEEN AN ALPINE GLACIER BASIN AND OTHER BASINS

The Tsidjiore Nouve glacier basin provides an example of an alpine glacier basin (Lorrain, 1978). This small basin, about $4.82\,km^2$, has 67% of its area glacierized; its mean altitude is about 3000 m a.s.l. At its lower extremity, discharge is measured continuously from May to October by the Grande Dixence hydroelectric company. On the basis of the values plotted in Figure 11.3 and values obtained in September and October, a very good relationship exists between discharge and the total dissolved load for the four major cations. The simple regression relationship is $\log C = -0.4089 \log D - 0.5081$, where C is the concentration of the four major cations in mequiv.l^{-1} and D is the discharge in $m^3\,s^{-1}$. The correlation coefficient between the two variables is -0.81, which is significant at a confidence level of 0.999. Using the regression relationship and the calculated frequency of occurrence of measured hourly discharges, a flux of exported major cations can be obtained from May to October. For the winter period, discharge does not exceed $0.01 \, \text{m}^3 \, \text{s}^{-1}$ (Bezinge, personal communication) and the mean is probably around $0.005 \, \text{m}^3 \, \text{s}^{-1}$. A study of winter dissolved concentration was undertaken by Lutschg-Loetscher *et al.* (1950) on thirteen Swiss glaciers including the Arolla and Ferpècle glaciers, which are very close to the basin under consideration, and are on the same bedrock. The mean winter concentration was found to be about four times the mean concentration in August. On this basis, it is possible to estimate the winter contribution to the annual flux of cations.

The mean annual flow is $37.9 \, \text{l} \, \text{s}^{-1} \, \text{km}^{-2}$ and the total annual output of major cations is 2450 kequiv. This corresponds to a cationic denudation rate of $508 \, \text{mequiv.m}^{-2} \, \text{yr}^{-1}$. This value can be compared with that obtained by Reynolds and Johnson (1972) in the South Cascade glacier basin located in the Rocky Mountains. The estimated value in the latter case is $930 \, \text{mequiv.m}^{-2} \, \text{yr}^{-1}$, indicating a higher cationic denudation rate. This is partly due to the heavier annual rainfall in that region of the Rocky Mountains than in the Alps near Arolla ($4090 \, \text{mm} \, \text{yr}^{-1}$ instead of $1340 \, \text{mm} \, \text{yr}^{-1}$). In the Berendon glacier basin in British Columbia ($5200 \, \text{mm} \, \text{yr}^{-1}$), Eyles *et al.* (1982) obtained a value of $947 \, \text{mequiv.m}^{-2} \, \text{yr}^{-1}$. The cationic denudation rate of the Tsidjiore Nouve glacier basin is well above the values for the North American continent as a whole ($380 \, \text{mequiv.m}^{-2} \, \text{yr}^{-1}$) and for all the continents as a whole ($390 \, \text{mequiv.m}^{-2} \, \text{yr}^{-1}$). This gives weight to the assertion that alpine glacier basins are environments of high chemical activity. Chemical weathering in the broad sense of the term, including ion exchange, seems to be very effective in spite of the near absence of normal biological, and pedogenic processes.

REFERENCES

ANDERSON, D. M. (1967), The interface between ice and silicate surfaces, *Journal of Colloid and Interface Science*, **25**, 174–191.

BERRY, F. A. F. (1969), Relative factors influencing membrane filtration effects in geologic environment, *Chemical Geology*, **4**, 295–301.

CARROLL, D. (1959), Ion exchange in clays and other minerals, *Geological Society of America Bulletin*, **70**, 749–780.

COLLINS, D. N. (1977), Hydrology of an alpine glacier as indicated by the chemical composition of meltwater, *Zeitschrift für Gletscherkunde und Glazialgeologie*, **13**, 219–238.

COLLINS, D. N. (1979a), Quantitative determination of the subglacial hydrology of two alpine glaciers, *Journal of Glaciology*, **23**, 347–362.

COLLINS, D. N. (1979b), Hydrochemistry of meltwaters draining from an alpine glacier, *Arctic and Alpine Research*, **11**, 307–324.

EYLES, N., SASSEVILLE, D. R., SLATT, R. M., and ROGERSON, R. J. (1982), Geochemical denudation rates and solute transport mechanisms in a maritime temperate glacier basin, *Canadian Journal of Earth Sciences*, **18**, 1570–1581.

GLEN, J. W., HOMER, D. R., and PAREN, J. G. (1977), Water at grain boundaries: its role in the purification of temperate glacier ice, in *Isotopes and Impurities in Snow and Ice (Proceedings of the Grenoble Symposium, August–September 1975)*, International Association of Hydrological Sciences Publication 118, pp. 263–271.

HALLET, B. (1976), Deposits formed by subglacial precipitation of $CaCO_3$, *Geological Society of America Bulletin*, **85**, 1003–1015.

HALLET, B., LORRAIN, R., and SOUCHEZ, R. (1978), The composition of basal ice from a glacier sliding over limestones, *Geological Society of America Bulletin*, **89**, 314–320.

HELFFERICH, F. (1962), *Ion Exchange*, McGraw-Hill, New York, 624 pp.

HOEKSTRA, P., and CHAMBERLAIN, E. (1964), Electro-osmosis in frozen soil, *Nature (London)*, **203**, 1406–1407.

KARAKA, Y. K., and BERRY, F. A. F. (1973), Simultaneous flow of water and solutes through geological membranes. I. Experimental investigation, *Geochimica et Cosmochimica Acta*, **37**, 2577–2603.

KEMPER, W. D. (1960), Water and ion-movement in thin films as influenced by the electrostatic charge and diffuse layer of cations associated with clay mineral surfaces, *Soil Science Society of America Proceedings*, **24**, 10–16.

KEMPER, W. D., SILLS, I. D., and AYLMORE, L. A. (1970), Separation of adsorbed cation species as water flows through clays, *Soil Science Society of America Proceedings*, **34**, 946–948.

LEMMENS, M. (1978), Relation entre concentration en cations dissous et débit de l'èmissaire du glacier de Tsidjiore Nouve (Valais), *Catena*, **5**, 227–236.

LEMMENS, M., and ROGER, M. (1978), Influence of ion exchange on dissolved load of alpine meltwaters, *Earth Surface Processes*, **3**, 179–187.

LORRAIN, R. D. (1978), *Estimation Quantitative de la 'Dénudation Cationique' dans un Bassin Glaciaire Alpin*, Réunion du 9 Mars de la Societe Hydrotechnique de France, Section de Glaciologie, unpublished, 3pp.

LUCE, R. W., BARTLETT, R. W., and PARKS, G. A. (1972), Dissolution kinetics of magnesium silicates, *Geochemica et Cosmochimica Acta*, **36**, 35–50.

LUTSCHG-LOETSCHER, O., BOHNER, R., HUBER, R., HUBER, P., and DE QUERVAIN, F. (1950), Zur Hydrologie, Chemie und Geologie des winterlichen Gletscherabflusse der Schweizer Alpen, in *Beitrage zur Geologie des Schweiz*, Geotechnische Serie, Hydrologie, 4, Band 1, Teil 2, Kummerly and Frey, Berne, 121pp.

PRANTL, F. A., AMBACH, W., and EISNER, H. (1972), Alpine glacier studies with nuclear methods, in *The Role of Snow and Ice in Hydrology (Proceedings of the Banff Symposium, September 1972)*, International Association of Hydrological Sciences Publication 107, pp. 435–442.

RAINWATER, F. H., and GUY, H. P. (1961), Some observations on the hydrochemistry and sedimentation of the Chamberlain Glacier Area Alaska, *United States Geological Survey Professional Paper* 414-c, pp. c1–c14.

RAISWELL, R. (1984), Chemical models of solute acquisition in glacial melt waters, *Journal of Glaciology*, **30**, 49–57.

REYNOLDS, R. C., and JOHNSON, N. M. (1972), Chemical weathering in the temperate glacial environment of the Northern Cascade Mountains, *Geochimica et Cosmochimica Acta*, **36**, 537–554.

RICQ-DE BOUARD, M. (1977), Migration of insoluble and soluble impurities in temperate ice: study of a vertical ice profile through the Glacier du Mont de Lans (French Alps), *Journal of Glaciology*, **18**, 231–238.

ROBINSON, R. A., and STOKES, R. H. (1965), *Electrolyte Solutions*, Butterworth, London, 571pp.

RÖTHLISBERGER, H. (1972), Water pressure in intra- and subglacial channels, *Journal of Glaciologie*, **11**, 177–203.

SHREVE, R. L. (1972), Movement of water in glaciers, *Journal of Glaciology*, **11**, 205–214.

SLATT, R. M. (1972), Geochemistry of meltwater stream from nine Alaskan glaciers, *Geological Society of America Bulletin*, **83**, 1125–1132.

SOUCHEZ, R. A., and LORRAIN, R. D. (1975), Chemical sorting effect at the base of an alpine glacier, *Journal of Glaciology*, **14**, 261–265.

SOUCHEZ, R. A., and LORRAIN, R. D. (1978), Origin of the basal ice layer from alpine glaciers indicated by its chemistry, *Journal of Glaciology*, **20**, 319–328.

SOUCHEZ, R. A., and TISON, J. L. (1981), Basal freezing of squeezed water: its influence on glacier erosion, *Annals of Glaciology*, **2**, 63–66.

TERWILLIGER, J. P., and DIZIO, S. F. (1970), Salt rejection phenomena in the freezing of saline solutions, *Chemical Engineering Science* **25**, 1331–1349.

THOMAS, A. G., and RAISWELL, R. (1984), Solute acquisition in glacial melt waters. II. Argentière (French Alps), *Journal of Glaciology*, **30**, 44–48.

WOLLAST, R. (1967), Kinetics of the alteration of K-feldspar in buffered solutions at low temperature, *Geochemica et Cosmochimica Acta*, **31**, 635–648.

ZOBRIST, J., and STUMM, W. (1979), Wie sauber ist das Schweizer Regenwasser?, *Separatdruck aus der Neuen Zürcher Zeitung, Beilage Forschung und Technik*, 146, 4pp.

Glacio-fluvial Sediment Transfer
Edited by A. M. Gurnell and M. J. Clark
©1987 John Wiley & Sons Ltd

Chapter 12

Suspended Sediment

A. M. GURNELL

ABSTRACT

This chapter considers the characteristics of suspended sediment transport in proglacial streams. A description of the techniques and sampling designs employed to study the suspended sediment concentration of proglacial streams (Section 12.2) is followed by an analysis of data from 43 glacierized catchments to identify the main catchment characteristics which influence their suspended sediment yield (Section 12.3). The remainder of the chapter considers, in increasing detail, the nature and controls of suspended sediment transport in proglacial streams including the relationship between stream discharge and suspended sediment concentration (Section 12.4), the anomalies from such a relationship (Section 12.5), spatial aspects of suspended sediment delivery to the proglacial stream network (Section 12.6), the role of proglacial sediment sources and sinks in influencing suspended sediment yield (Section 12.7) and the particle size characteristics of proglacial suspended sediment.

12.1 INTRODUCTION

Suspended sediment load forms part of the undissolved sediment load of a stream. It is transported and maintained within the flow by turbulent mixing processes which counteract the tendency of the particles to settle out of the flow under the influence of gravity. Suspended load may be arbitrarily separated from bedload, which consists of larger particles which are moved by rolling, sliding or bouncing along the stream bed. Individual particles may be suspended under some flow conditions and form part of the bedload at other times. Glacial meltwater streams carry particularly high loads of sediment in the form of both suspended sediment load (including washload) and bed load in response to the plentiful supply of sediment to the streams by the glacial processes (see Chapters 6 to 9). The proportion of fine sediments in suspension as washload might be expected to be higher in the proglacial environment than in other fluvial environments because of the supply of fine material from glacial erosion

305

TABLE 12.1   Methods employed for monitoring suspended sediment concentration in some published studies of proglacial streams

Source	Turbidity meter	Filtration of water samples				Initial penetration pore size of filter paper	Sediment weight determination		
		Hand sampling		Automatic pumping sampler	Sample size (ml)		Pre and post-		
		Depth integrated	Point integrated				Ashing	weighing	Evaporation
Bezinge (1978)					50 000	Settlement of sediment and conversion to concentration by a density estimate of 1.6 kg l^{-1}			
Bryan (1972)		USDH-59	*						
Burrows *et al.* (1981)		P-61 or D-49							
Collins (1979a)		USDH-48		ALS	700–900	3 μm		*	
Ferguson (1984)				ALS Mk4B	300–800	3 μm		*	
Fahnestock (1963)		USDH-48							
Fenn (1983)		USDH-48	USDH-48	ALS Mk4B	200–400	8 μm		*	
Gilbert (1975)		*							
Gustavson (1975)			*						
Hammer and Smith (1983)		USDH-48				0.8 μm		*	* (after chemical flocculation)

			Pre-filter		
Hasholt (1976)	Swedish sampler		Munktell OOM then Munktell OOH	*	
Lanser (1958)	?	2000	?		*
Mathews (1964)	Bottles and bailing		Whatman 4		*
Østrem (1975a)	Plastic bottles	1000		*	
Østrem et al. (1967)	* Calibrated bottle	1000–2000	Munktell OA	*	
Smith et al. (1982)	* Depth int. sampler and bottles		0.8 μm		*
Tomasson et al. (1980)					
Weirich (1982)	USDH-48	500			

Wherever possible details of the technique employed by different researchers are specified. Where a technique has been used but the details have not been specified, this is represented on the table by an asterisk.

processes. The turbulent nature of the flow in proglacial streams might also be expected to bring coarser material into suspension and to cause the suspended sediment to be more evenly distributed in the stream cross-section than in other fluvial environments. The large load of suspended sediment in proglacial streams presents many problems when the water resources of glaciated catchments are developed and has been the impetus for many studies of suspended sediment concentration in alpine proglacial streams.

12.2   METHODS OF MONITORING SUSPENDED SEDIMENT

Information on suspended sediment concentration in a stream will reflect the method used for sampling the suspended sediment, the location chosen for sampling and the frequency with which the concentration is monitored both in time and in the stream cross-section. There are numerous reviews of all these aspects of suspended sediment monitoring in rivers (e.g. American Society of Civil Engineers, 1969a,b; Gregory and Walling, 1973; Hadley and Walling, 1984) and so the present discussion will be confined to the methods employed in proglacial stream studies.

Information on suspended sediment concentration in river water may be obtained either by the filtration of water samples to determine the weight of the contained suspended sediment or by monitoring the turbidity of the water. Attenuation of light passing from a light source through the water to a photoelectric cell in a turbidity meter is usually calibrated to provide quantitative estimates of sediment concentration by employing the water sampling and filtration technique. Table 12.1 gives an indication of the methods that have been employed in determining suspended sediment concentration in studies on proglacial streams.

There are four main ways in which water samples have been collected: hand sampling by depth integrating sampler, point integrating sampler or plastic bottle and repeated automatic sampling by pumping sampler which provides point integrated samples. The method of sampling, sample size, calibre of the filter paper and the method of determining the amount of suspended sediment employed in different studies are very variable and it is important to assess the impact that these differences will have on the suspended sediment concentration estimates obtained. Tómasson *et al.* (1980) stressed the great differences in estimates obtained by using plastic bottles and a properly designed depth integrating sampler, whereas Østrem (1975a) did not consider that the sampler was of particular importance and recommended that a relatively narrow-necked plastic bottle immersed at an angle of 45° upstream and held in the water for 10–30 s should provide a representative sample if the flow is turbulent. Figure 12.1 shows the results of calibration studies of different sampling techniques undertaken on the proglacial stream of the Glacier de Tsidjiore Nouve,

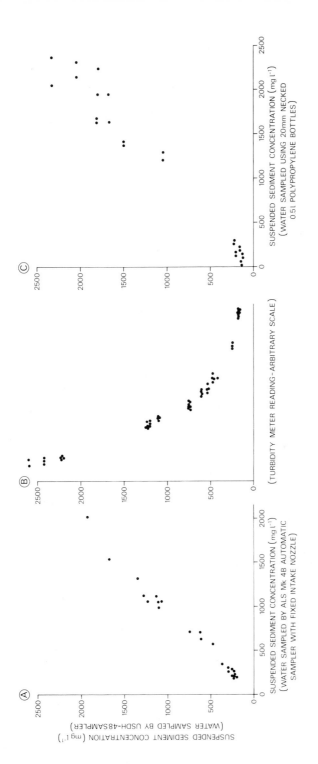

FIGURE 12.1 Calibration relationships for suspended sediment concentration determined by filtration of samples gathered using a USDH-48 sampler in comparison with (A) ALS Mk4B sampler; (B) turbidity meter; (C) 20 mm diameter neck polypropylene bottles

Switzerland. It appears that there is a very close relationship between suspended sediment concentration estimated from samples taken by a carefully designed, hand-held sampler (the USDH-48) and both the ALS Mk4B automatic pumping sampler, which samples the water by releasing a partial vacuum in a glass sample bottle (Figure 12.1A), and water turbidity (Figure 12.1B) but the estimates from samples taken by plastic bottles with a 20 mm diameter neck may overestimate the concentration (Figure 12.1C). The slight overestimation from using plastic bottles is probably a result of trapping coarse particles. Given the very turbulent nature of the flow in many proglacial streams, it could be argued that these larger particles form a true component of the suspended sediment load and the fact that other samplers are not trapping these particles in such large quantities could suggest that they are not sampling the full range of sediment in suspension representatively. Whatever the cause of the differences in the estimates of suspended sediment concentration and in the distribution of the particle sizes trapped, water samples can be pre-filtered through a mesh of appropriate pore size to remove coarse particles in order to yield comparable concentrations from different sampling devices.

The great variability in the initial penetration pore size of the filter paper used for filtering the water samples may also cause differences in the results obtained by different studies. The filter papers listed in Table 12.1 range in pore size from 0.8 to 8 μm and the choice of paper rests on achieving a balance between the rapid filtration of samples (often using crude techniques in the field) and trapping as much of the transported sediment as possible. It has been found that the use of relatively coarse filter papers on the Tsidijore Nouve proglacial stream is adequate for two reasons. Firstly, particles of less than 8 μm in size have been found to constitute less than 5% of the volume of suspended sediment at low suspended sediment concentration (e.g. suspended sediment concentration, 100 mg l^{-1}; median particle size of suspended sediment, 16 μm) and even less at higher suspended sediment concentration (e.g. suspended sediment concentration, 608 mg l^{-1}; median particle size, 32 μm) in a Coulter Counter analysis of hourly samples taken over a 24 h period from the proglacial stream of the Tsidijore Nouve glacier (abstracted from data presented by Beecroft, 1981). Similar results are presented by Bezinge (Chapter 17), where less than 10% of the suspended sediment deposited (Figure 17.18) and approximately 5% of the sediment in suspension (Figure 17.17) in the Z'Mutt lake were found to have a particle size of less than 8 μm. Secondly, the filter papers rapidly clog during filtering so that their effective pore size is very much smaller than the calibre of the filter paper would suggest. If similar circumstances are experienced on other proglacial streams, small differences in filter paper calibre should not cause significant differences in the results between studies.

Table 12.1 shows that turbidity meters have been used in addition to water sampling in some published studies of proglacial suspended sediment transport. Turbidity meters can be used to provide a continuous record of suspended sediment concentration, so that the representativeness of point samples can be assessed and the details of the sediment pattern can be identified. Turbidity readings should be carefully calibrated in relation to suspended sediment concentration, preferably by comparing readings with suspended sediment concentrations estimated from concurrent water sampling in the field rather than in the laboratory. Great care is needed in using a turbidity meter to quantify suspended sediment concentration because the turbidity reading may be affected by bubbles in the water, by ambient light conditions, by the colour and particle size distribution of the sediment and by accumulation of algae and sediment on the turbidity probe, particularly on the light source. Nevertheless, with careful selection of a monitoring site and meticulous maintenance of the instrument, a turbidity meter can provide an excellent source of information to back up a water sampling programme and, in particular, can provide information on the sudden and very marked variations in suspended sediment concentration that are so characteristic of proglacial streams (see Section 12.5) and that are virtually impossible to monitor by any method other than by continuous turbidity recording techniques.

Although the flow in proglacial streams is usually very turbulent and so should cause good mixing of the suspended sediment, the sediment concentration might be expected to vary across a stream cross-section, and in particular to increase with depth. The techniques of hand sampling of water employed in different studies reflects the degree to which different researchers considered that sediment was mixed in their streams. Østrem *et al.* (1971) undertook detailed investigations of sampling techniques and strategies. They noted that for practical reasons water samples are often taken near the stream banks and so they tested the validity of such samples for indicating the mean suspended sediment concentration in the cross-section. They found that there was no significant variation in concentration near the stream bank for distances 25 m upstream and 50 m downstream of their sampling site. They also found that although sediment concentration varied with depth in the cross-section and to some degree with distance from the stream bank, much of this variation resulted from variations in coarse material in suspension. They concluded that a careful choice of sampling site at a turbulent section and the exclusion of coarse particles when estimating suspended sediment concentration would minimize the differences in concentration estimates derived from different locations in the cross-section. Fenn (1983) undertook detailed surveys of suspended sediment concentration variations across the stream cross-section at his main suspended sediment monitoring site on the proglacial stream of the Tsidjiore Nouve glacier. He measured flow velocity and sampled the water using a USDH-48 sampler at regular depth increments on seven equally spaced vertical profiles across the stream at twelve different levels of discharge. Although there were no consistent patterns in suspended sediment concentration across the stream cross-section, there were great variations in the concentration which appeared to follow

TABLE 12.2   Sampling designs adopted in some studies of proglacial suspended sediment transport

| Source | Time scales encompassed by sampling design (sampling frequency indicated) | | | | Spatial scale |
	One hour to one day	Several days	One week to full ablation season	More than one ablation season	Single proglacial station
Borland (1961)				60 samples/6 yr‡	
Rainwater and Guy (1961)		97 samples/5d‡			
Mathews (1964)	5/d		2/wk		
Østrem et al. (1967)	8/d		3/d		
Church and Ryder (1972)			1/d		
Church (1974)	1/h				
Guymon (1974)			1/d		
Hasholt (1976)	1/h		1/d		
Lemmens (1978)			32 samples/ 4 months‡		
Smith (1978)	16–24/d‡		117 samples/2 seasons		
Collins (1979)			1/h		
Metcalf (1979)		>23 samples/6 months‡?			
Mills (1979)		20 samples/2.5 months‡?			
Bogen (1980)	17–19/d‡		66 samples/2 seasons?		
Kieldsen and Ølstrem (1980)*			5/d		
Tómasson et al. (1980)		1/d§			
Burrows et al. (1981)			35 samples/3 seasons‡		

Gurnell (1982) — 1/h — 6/d →

Fenn (1983) — 1/h →

Beecroft (1983) — 1/h§ →

Gurnell and Fenn (1984a,b) — 1/h →

More than one sampling station on proglacial stream

Mathews (1964) — 5/d →

Metcalf (1979) — >23 samples/6 months†? →

Mills (1979) — 11 samples/2.5 months →

Bogen (1980) — 17–19/d‡ →

Tómasson et al. (1980) — 1/d§ — 35 samples/3 seasons‡ →

Burrows et al. (1981) — 1/h — 1/2d →

Gurnell (1982) — 1/h →

Østrem (1975a) — 5/d →

Smith (1978)† — 117 samples/2 seasons‡ →

52 samples/2 seasons‡ →

Sample sites on more than one proglacial tributary

Gurnell (1982) — 1/h — 1/2d →

Fenn (1983) — 1/h — 1/d →

Kjeldsen and Østrem (1980)* — >1/d →

Fenn (1983) — 1/d →

Vivian and Zumstein (1973) — 51 samples/2 yr‡ →

Dowdeswell (1982) — 1/h — 1/d →

Sites on supra- or subglacial streams

* This single source represents the numerous reports from Norges Vassdrags-og Elektrisitetsvesen.
† Streams issue from different glaciers.
‡ Irregular sampling; total number of samples in sampling period.
§ Samples during a meltwater outburst.

'threads' in the flow. It appeared that a single point sample taken near to the centre of the cross profile of flow gave an adequate estimate of both the average concentration in the vertical section near the stream centre (equivalent to a depth integrated sample) and also of the average concentration across the entire section, but the confidence limits on both the derived regression relationships were wide. For example, a point sample providing an estimated mean cross-section concentration of $700 \, \mathrm{mg} \, l^{-1}$ had 99% confidence limits of approximately 600 and $800 \, \mathrm{mg} \, l^{-1}$.

This brief review of the sampling procedures and methods used for monitoring suspended sediment concentration in proglacial streams suggests a number of general conclusions. Firstly, the choice of monitoring site is very important if representative estimates of concentration at that site are to be achieved with minimum complexity of sampling. A well mixed turbulent site is required if only a single sampling location is to be used in the cross-section. The sampling site should be located as near to the centre of the flow as possible, although a really turbulent location and the exclusion of coarse particles from the filtered water sample would permit sampling nearer to the stream banks without a major deterioration in the representativeness of the sample. Secondly, the method of sampling does not seem to affect seriously the quality of the observations, although a properly designed automatic or hand held sampler will give more consistent results than the use of plastic bottles and should be used if possible. Østrem *et al.* (1971) noted that the size of the neck of the plastic bottle is important if this method of sampling is to be used, because the neck size affects the filling rate and thus the degree to which the bottle integrates very short time fluctuations in sediment concentration. Finally, the choice of filter paper is also probably unlikely to affect the representativeness of the results provided that its initial penetration pore size does not exceed $10 \, \mu\mathrm{m}$. However, it is best to test the efficiency of a range of papers before making a final choice of paper for a particular field location. Each site is likely to have its own particle size characteristics, as was clearly illustrated by the significant variations in particle size found in time and space by Tómasson *et al.* (1980) for different locations on the same proglacial stream network.

Frequency and location of water sampling on proglacial streams has varied enormously between different studies. Table 12.2 summarizes the sampling designs that have been employed in some studies of suspended sediment concentration in proglacial streams and it illustrates the way in which both temporal and spatial aspects of sampling have increased in their frequency and complexity as the perceived variability of the process being monitored has increased. The following sections progressively emphasize the variability and possible controls on suspended sediment transport in glacierized catchments that have been identified by changing intensities of sampling.

12.3   SUSPENDED SEDIMENT YIELD—A GLOBAL PERSPECTIVE

It is clear from the preceding section that a range of different techniques and sampling strategies have been employed to monitor the suspended sediment transport of proglacial streams. This variety of approaches to data gathering makes the comparison of the results of different studies difficult. Nevertheless, it is useful to attempt a comparison of results in order to see whether, in spite of the differences in sampling methodology, any trends can be identified in the data.

Walling and Kleo (1979) reviewed previous work on global sediment yield patterns and analysed their own sediment yield data base, which contains records from 1246 sediment monitoring stations throughout the world. Their data exhibit enormous variability but they provide the most reliable datum against which the sediment yield of glaciated catchments may be judged. The aim in the present study was to assemble information from the published literature for as many drainage basins containing glaciers as possible for which at least one ablation season's sediment yield and discharge data were available. The search was not confined to alpine drainage basins because this would have yielded an unacceptably small sample of basins. Many stations could not be included because their suspended sediment records were too short or because the associated discharge data had not been published. Forty-three gauging sites with suitable records were finally identified and their mean annual sediment and discharge yield and catchment characteristics [including catchment area, percentage glacier cover, relief and distance of the gauging site from the (nearest) glacier snout] were recorded. It was impossible to obtain values for all of these catchment characteristics for each of the 43 gauging sites, but inspection of topographic maps and of papers by other authors on the same field areas produced as complete a set of data as was possible. Appropriate information on the characteristics of catchment rock type, which would be indicative of its susceptibility to a whole range of processes which would culminate in fluvial suspended sediment yield from the catchment, was impossible to gather and was a major gap in the range of independent variables which could potentially be employed to explain differences in suspended sediment yield from glaciated catchments.

The variable and sometimes very crude quality of the sediment yield data from this sample of glaciated drainage basins cannot be over-stressed. Not only were different methods employed in data gathering but the length of records varied from 1 to 82 years and the sampling was often very infrequent. In some studies great care had been taken in allowing for the significance of low-frequency high magnitude events, in other studies a simple average of widely spaced temporal observations was included and in a few studies no indication was given of the method of derivation of the sediment yield estimates. The only previous attempt to analyse suspended sediment yield data for a large sample of glaciated catchments was by Guymon (1974), when he considered data from twelve partly glaciated and one glacier-free catchment in Alaska, and so the present study attempts a new perspective by including information from glaciated catchment studies throughout the world.

The data are presented graphically in Figures 12.2 and 12.3 with the sites grouped into the three categories previously employed by Church and Gilbert (1975): glaciated mountain catchments, sub-arctic mountain catchments and arctic catchments. Drainage basins with less than 10% glacier cover are also indicated. Figure 12.2 illustrates the range in magnitude of the mean annual sediment yield and discharge volume for the 43 drainage basins. It is not surprising that there is a strong relationship between total suspended sediment yield and discharge volume since the former is calculated from suspended sediment concentration and discharge and there is inevitable scale dependence in the data, but the enormous range in the data is worthy of note, with sediment yield and discharge varying from less than $700\,t\,yr^{-1}$ and $1 \times 10^6\,m^3\,yr^{-1}$, respectively (Hilda glacier, Hammer and Smith, 1983) to over $63 \times 10^6\,t\,yr^{-1}$ (Hunza river, Ferguson, 1984) and over $82\,000 \times 10^6\,m^3\,yr^{-1}$ (Yukon river, Guymon, 1974), respectively. It is of much greater interest to investigate the

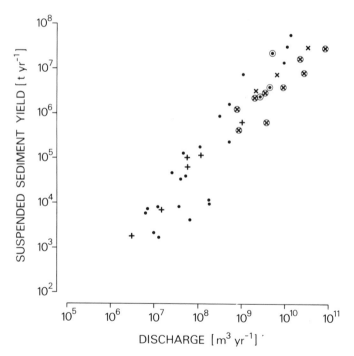

FIGURE 12.2   Relationship between suspended sediment yield and discharge volume for 43 drainage basins (the symbols are explained in Figure 12.3)

processes underlying the scale dependence of the data and the degree to which catchment characteristics influence sediment yield from drainage basins of similar size.

Figure 12.3 shows three different plots of suspended sediment yield data against explanatory variables that have been presented in other studies of suspended sediment yield patterns (in the following discussion all the relationships assume $\log_{10}$ transformations of the variables). One of the problems encountered in using suspended sediment yield estimates as an index

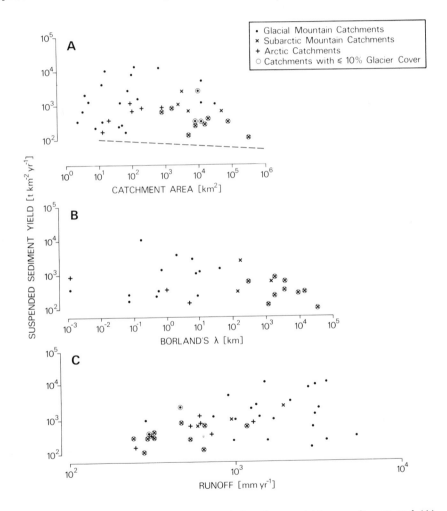

FIGURE 12.3 Relationship between suspended sediment yield per unit area and (A) catchment area [the dashed line represents the world suspended sediment yield relationship as derived by Walling and Kleo (1979)]; (B) Borland's λ index; (C) depth of runoff

of erosion rates is to overcome the effects of the intermediate processes involved in sediment delivery to the monitoring site. Since larger catchments provide more opportunities for sediment to be deposited and re-eroded before removal from the catchment, basin area is often used as a surrogate for intermediate storage opportunities when investigating data from catchments of different magnitudes. Walling (1983) discussed the sediment delivery problem in detail and illustrated the significance of catchment area in controlling the sediment delivery ratio (Roehl, 1962). Walling and Kleo (1979) derived a relationship between suspended sediment yield per unit area and catchment area from their world suspended sediment data base. Their relationship, which is shown in Figure 12.3A, has a slope of −0.065. Although this compares favourably with the slope of −0.058 derived from the current sample of catchments, it must be stressed that the slope of the regression relationship derived from the glaciated catchment data was not significantly different from zero. Nevertheless, it should be noted that all the glaciated catchment observations plot above the Walling and Kleo world relationship. The present data compare favourably with information from humid agricultural basins in the United States, where sediment source areas are plentiful (see Walling, 1983). The larger basins in the present study are also those with low glacier cover and, therefore, any relationship between sediment yield and basin area for glaciated catchments may be partly influenced by changing opportunity for sediment storage but it may also be a function of a change in the mix of erosion processes in catchments of different size.

Borland (1961) derived an index (λ) for estimating sediment yield from glaciated catchments in Alaska:

$$\lambda = (A_T L_G)/(A_G)$$

where A_T is the total catchment area, A_G is the glaciated area of the catchment and L_G is the length of river from the glacier snout to the monitoring site. λ was found by Guymon (1974) to be unreliable when a larger sample of Alaskan catchments was analysed and Figure 12.3B shows the weak relationship between suspended sediment yield and λ in the present sample of catchments with sufficient information on catchment characteristics to evaluate λ. The three variables included in the index represent source areas of high erosion (A_G) and opportunity for sediment storage (A_T, L_G), but Guymon found that the weighting between the variables was inappropriate and that additional data on the size distribution of the particles transported improved the reliability of the relationship. The particle size distribution of suspended sediment is controlled by rock type, weathering, erosion and transport processes and would be a suitable additional variable to the present study if the data were available.

In the absence of particle size information for the sample of 43 catchments, it was impossible to reproduce Guymon's analysis of Alaskan data. Nevertheless,

TABLE 12.3 Regression models estimated from a sample of 43 glacierized catchments

$$SSY_T = 0.0005\ Q_{T(18.13)^{**}}^{1.011} \qquad\qquad R^2 = 0.889$$

$$SSY = 1162\ A_{T(0.99)}^{-0.058} \qquad\qquad R^2 = 0.023$$

$$SSY = 884\ L_{G(2.87)^{**}}^{-0.111} \qquad\qquad R^2 = 0.168$$

$$SSY = 839\ A_{G(0.05)}^{-0.004} \qquad\qquad R^2 = 0.000$$

$$SSY = 8.84\ Q_{(3.34)^{**}}^{0.660} \qquad\qquad R^2 = 0.214$$

$$SSY = 986\ \lambda_{(2.94)^{**}}^{-0.092} \qquad\qquad R^2 = 0.174$$

$$SSY = 805\ A_{T(3.16)^{**}}^{-0.521} A_{G(2.97)^{**}}^{0.667} \qquad\qquad R^2 = 0.200$$

$$SSY = 468\ A_{T(1.08)}^{-0.233} A_{G(1.76)}^{0.440} L_{G(1.98)}^{-0.122} \qquad R^2 = 0.273$$

$$SSY = 387\ A_{G(2.11)^{*}}^{0.188} L_{G(3.66)^{**}}^{-0.168} \qquad\qquad R^2 = 0.251$$

$$SSY = 1.91\ Q_{(3.35)^{**}}^{0.822} A_{T(1.10)}^{0.072} \qquad\qquad R^2 = 0.237$$

$$SSY = 1.57\ Q_{(3.79)^{**}}^{0.823} A_{(1.66)}^{0.133} \qquad\qquad R^2 = 0.264$$

$$SSY = 24.9\ Q_{(1.66)}^{0.513} L_{G(0.62)}^{-0.036} \qquad\qquad R^2 = 0.221$$

$$SSY = 7.52\ A_{T(1.36)}^{-0.233} Q_{(2.33)^{*}}^{0.615} A_{G(1.84)}^{0.440} \qquad R^2 = 0.298$$

SSY_T, suspended sediment yield in 10^6 tonnes yr^{-1}.
Q_T, discharge volume in 10^6 m^3 yr^{-1}.
SSY, suspended sediment yield in tonnes km^{-2} yr^{-1}.
Q, discharge in mm yr^{-1}.
A_T, total catchment area in km^2.
A_G, glacerized area in km^2.
L_G, distance from measurement site to nearest glacier snout in km.
t statistics for the slope parameters of the regression relationships based on log-transformed variables are given in parentheses. Many of the slope parameters are not significantly different from zero but the estimated questions are listed to illustrate the range of relationships that were considered. * and ** represent slope parameters that are significantly different from zero ($P<0.05$ and 0.01, respectively).

several simple and multiple regression models were estimated from the sample of catchments (Table 12.3). In the light of the variable quality of the sample data and the complexity of the processes involved in delivering suspended sediment from glaciated drainage basins of varying size, it is not surprising that none of the regression models explained a large proportion of the variance in suspended sediment yield. Figure 12.3C shows the relationship between suspended sediment yield per unit area and the discharge expressed in millimetres. Table 12.3 shows that discharge gave the best explanation of suspended sediment yield of any of the variables considered. The directions of the single variable relationships were generally as expected with suspended sediment yield per unit area positively linked with discharge (Q) and negatively linked with A_T, L_G and λ. Only A_G provided an unexpectedly negative and non-significant relationship. Three multivariate relationships are worthy of note. When A_T and A_G are included as independent variables in a multiple regression analysis, they both have slopes that are significantly different from zero ($P<0.01$) with A_T showing the expected negative slope and A_G a positive slope, illustrating the significance of glaciated areas as sediment source areas. If L_G is introduced as an additional variable, the directions of slope conform to those suggested by Borland (1961) and confirmed by Guymon (1975), but none of these slopes are significantly different from zero. Finally, it is noteworthy that the multiple regression relationship which includes L_G and A_G as the two independent variables provides a higher coefficient of determination than that which includes A_G and A_T. The distance of the measurement site from the glacier snout appears to have an important effect on the sediment yield of the catchment. The effects of sampling location will be considered in further detail in Section 12.6. Apart from a scale effect, the small sample of information presented in Figures 12.2 and 12.3 gives no indication of differences in the generation of suspended sediment from the three types of glacierized catchment.

12.4   THE RELATIONSHIP BETWEEN SUSPENDED SEDIMENT CONCENTRATION AND DISCHARGE

It is extremely labour intensive to obtain good records of suspended sediment concentration for a proglacial stream. Even if a recording turbidity meter is used to obtain continuous records, the probe needs to be cleaned frequently and the meter recalibrated by hand sampling in order to ensure consistency in the records. Apart from the academic implications of obtaining detailed information on the suspended sediment load of proglacial streams, there are significant economic and design reasons for doing so. The problems created by high concentrations of suspended sediment in the meltwater are considerable, since this sediment may quickly silt up any engineering structures installed to store or divert the water and may cause a great deal of damage to turbines and other components of hydroelectric power generation machinery (see Chapter 17). There is, therefore, a great need to find a way of interpolating estimates of suspended sediment concentration so that the effort involved in collecting field observations may be reduced. A number of techniques have been investigated with this problem in mind and these are based on the relationship between suspended sediment concentration and discharge.

Suspended sediment rating curves have frequently been employed to describe and estimate the level of suspended sediment transport in proglacial streams. These rating curves are derived by applying linear regression analysis to concurrent suspended sediment concentration and discharge observations for the same site. Since the flow of the water in a stream transports the sediment, it seems logical to anticipate a simple relationship between the suspended sediment concentration and the discharge of the stream. The plentiful supply of sediment in the rock debris of glaciated catchments should reduce the regulating effect of sediment supply on the suspended sediment concentration — discharge relationship, so that proglacial streams should be particularly suited to this type of analysis. Mathews (1964), Borland (1961) Bezinge (1978), Østrem (1975a), Collins (1979a) and Bogen (1980) have all established suspended

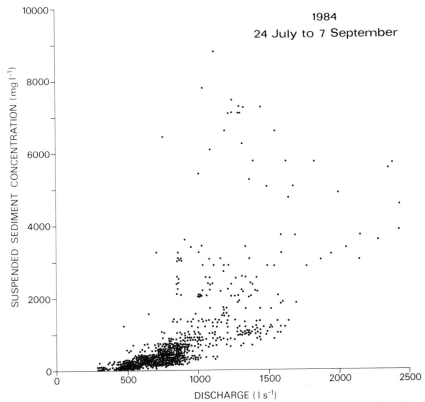

FIGURE 12.4 Scatter plot of concurrent observations of suspended sediment concentration and discharge for a part of the 1984 ablation season for the proglacial stream of Glacier de Tsidjiore Nouve

sediment rating curves for proglacial streams with varying levels of success.

Figure 12.4 displays some observations of suspended sediment concentration and discharge for the proglacial stream of the Tsidjiore Nouve glacier. If linear regression analysis is to be used to estimate sediment rating curves, then the assumptions of the regression model should be adhered to. In particular, the variables should be transformed to provide a scatter in the observations indicating a linear relationship between suspended sediment concentration and discharge with residuals in the dependent variable (suspended sediment concentration) that are random, have zero mean, are normally distributed with constant variance and are serially uncorrelated (Gurnell and Fenn, 1984a). It is clear that such assumptions are not met by the data presented in Figure 12.4. The scatter in the plot can be reduced by transforming the axes to a logarithmic scale (Figure 12.5) and this transformation has frequently been applied prior to estimating suspended sediment rating curves. Although a logarithmic

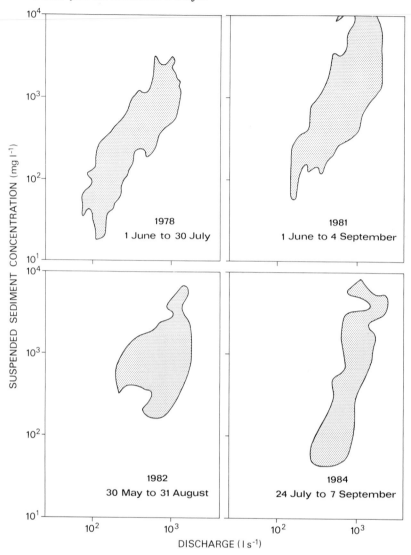

FIGURE 12.5   Logarithmically transformed plots of suspended sediment concentration and discharge for four ablation seasons for the proglacial stream of the Glacier de Tsidjiore Nouve. The data are all enclosed by the shaded area in each of the four diagrams

transformation appears to yield a near-ideal distribution of the observations, there are three major problems which remain when estimating a suspended sediment rating curve from sample data and then using that rating curve to calculate suspended sediment transport patterns and suspended sediment load from discharge observations. First, sediment rating curves derived from different

TABLE 12.4 Suspended sediment rating curves estimated from the data presented in Figure 12.5

Monitoring period	Suspended sediment rating relationships estimated from log-transformed data	Sample size	R^2
1/6 to 30/7/1978	$\log_{10} S = -0.789 + 1.286\log_{10} Q$ (79.62)	1440[†]	0.82
1/6 to 4/9/1981	$\log_{10} S = -0.549 + 1.236\log_{10} Q$ (54.22)	2288[†]	0.56
30/5 to 31/8/1982	$\log_{10} S = 2.022 + 0.307\log_{10} Q$ (7.57)	565[‡]	0.09
24/7 to 7/9/1984	$\log_{10} S = -5.314 + 2.734\log_{10} Q$ (47.96)	1002[†]	0.70

S, suspended sediment concentration in mg l^{-1}; Q, discharge in l s^{-1}; t statistics for the slope parameters of the regression relationship are given in parentheses.
[†]Sampling interval 1 h.
[‡]Sampling interval 4 h.

time periods may be very different in form (Figure 12.5, Table 12.4) so that it would be totally inappropriate to estimate suspended sediment transport from discharge unless a sediment rating curve was available which spanned the period for which the estimate was required. Second, detailed examination of the pattern of the residuals of all sediment rating curves estimated by simple linear regression analysis for the proglacial stream of the Tsidjiore Nouve glacier has shown that they are very strongly serially autocorrelated and that this autocorrelation biases the rating curve and any estimates of suspended sediment concentration derived from it (Fenn *et al.*, 1985). Third, because the rating curve is estimated from log-transformed data, estimates of concentrations and yields of suspended sediment will, on average, be too low (Walling, 1977; Ferguson, 1984; Fenn *et al.*, 1985). It is for these reasons that Fenn *et al.* (1985) concluded that it is impossible to estimate an unbiased simple regression model relating suspended sediment concentration to discharge for the proglacial stream of the Tsidjiore Nouve glacier. There is evidence that identical problems have arisen when data from other proglacial streams have been analysed.

Østrem (1975a) presented suspended sediment rating curves for proglacial streams in Norway. He noted that it was possible to estimate sediment rating curves which fitted the observations of suspended sediment concentration and discharge extremely well for some time periods, but that curves providing a good fit were unusual.

'Separate curves for some subperiods could probably be established, but would be valid for very short periods only; investigations have shown that they would not be valid for the corresponding period in a following year. . . . the relationship between discharge and transport is changing more or less continuously during a runoff season and from year to year.'

(Østrem, 1975a, p. 109)

The attractions of the simplicity of suspended sediment rating curves have led many researchers to attempt to devise reliable and useful curves by splitting observations of suspended sediment concentration and discharge into subgroups. Thus Collins (1979a) estimated suspended sediment rating curves for individual rising and falling limbs of diurnal discharge hydrographs generated by the Gorner glacier, Switzerland; Church and Gilbert (1975) presented rising and falling stage sediment rating curves for the Decade River, Baffin Island, and Hammer and Smith (1983) estimated early and late ablation season suspended sediment rating curves for the proglacial stream of Hilda glacier, Alberta.

In considering the problems of estimating suspended sediment rating curves for proglacial streams, the hysteresis in the relationship between suspended sediment concentration and discharge has often been stressed (Østrem, 1975a; Collins, 1979a). If observations on a sediment rating plot are linked according to their time sequence, they frequently show a looped relationship, which gives rise to the serial autocorrelation in the residuals from suspended sediment rating curves which has already been mentioned. Daily 'rating loops' may be defined (Collins, 1979a) and these may continue through sequences of days (Østrem, 1975a) and may be apparent in data collected over a whole ablation season (Lemmens, 1978). There are two elements in the relationship between suspended sediment concentration and discharge in proglacial streams which give rise to the looped patterns. First, there is usually a lag between the peaks in the time series of the observations of the two variables at a range of temporal scales. The peak of suspended sediment concentration normally occurs in advance of the discharge peak within diurnal discharge cycles, over periods of several days and at the scale of the whole ablation season. This results in higher suspended sediment concentration during periods of rising discharge. Second, suspended sediment concentration falls away more abruptly than discharge from peak levels. This suggests that, in spite of the apparently unlimited sources of suspended sediment, the transport of suspended material exhibits exhaustion effects during periods when the current discharge level has recently been exceeded.

Thus the concentration of suspended sediment in proglacial streams is affected not only by the current discharge of the proglacial stream but also by the recent discharge history of the stream and the degree to which past discharges have exhausted the supply of sediment from the sediment source areas. Two approaches have been employed to overcome the problem of serial autocorrelation in the residuals from simple suspended sediment rating curves resulting from lags and exhaustion effects in the relationship between suspended sediment concentration and discharge. These two approaches can be termed multivariate regression approaches and time series analytical approaches.

Richards (1984) adopted a multivariate regression approach by incorporating information on preceding discharge in estimating suspended sediment rating curves for the proglacial stream of Storbreen, Jotunheimen, Norway. He included the rate of change of discharge per hour over the preceding 2 h as an additional independent variable to current discharge level in his sediment rating equations. He showed that during dry periods this additional variable was far more significant than discharge in explaining levels of suspended sediment concentration and that during rainy periods it was a valuable additional variable to discharge, providing a more reliable sediment rating curve. Richards' multivariate sediment rating curve takes account of the recent discharge levels in the stream in estimating suspended sediment concentration and so accounts for some of the hysteresis in the data. Rainfall- and meltwater-induced floods transported different amounts of suspended sediment and Richards stated that if a longer series of data were available for analysis, it might be possible to incorporate other independent variables in the multivariate sediment rating curve, including temperate and rainfall. Such an analysis would further clarify the controls on suspended sediment concentration in proglacial streams.

Gurnell and Fenn (1984a,b) suggested two approaches to constructing suspended sediment rating curves which are based on the analysis of time series of observations of suspended sediment concentration and discharge and which benefit from the availability of long time series of hourly observations of these two variables for the proglacial stream of the Tsidjiore Nouve glacier. Gurnell and Fenn (1984a) dealt with the problem of strong serial autocorrelation in hourly observations from the proglacial stream of the Tsidjiore Nouve glacier for two separate ablation seasons (1978 and 1981), by estimating a Box and Jenkins (1970) transfer function from the two logarithmically transformed, first-differenced time series after filtering both series by a univariate time series model estimated from the discharge series. The transfer function related the hourly change in suspended sediment concentration to the hourly change in discharge and incorporated a 1 h lag. The cross-correlation function between the two first-differenced series indicated that the change in discharge one and two days before might also be relevant to the estimation of suspended sediment concentration, although these were not found to contribute sufficient information to the transfer function (estimated from 25 days' data) to warrant increasing its complexity. The time series approach to estimating suspended sediment rating curves was extended by Gurnell and Fenn (1984b) by first separating the discharge series into flow components prior to considering the links and the lags between suspended sediment concentration and discharge. A simple two-reservoir mixing model of a type frequently employed in hydrological studies (see, for example, Walling, 1974) and previously used in the context of glacial meltwater separation by Collins (1979b) provided a means of separating the total discharge into 'subglacial' and 'englacial' components. In the context of proglacial streams, the separation is achieved by identifying

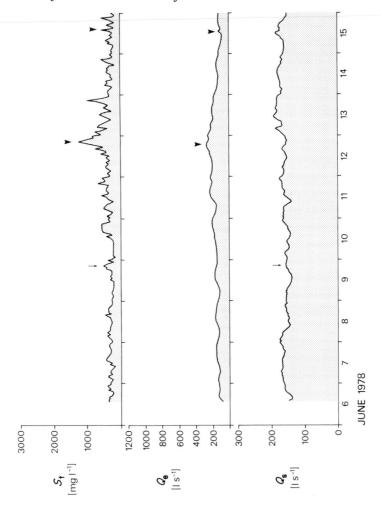

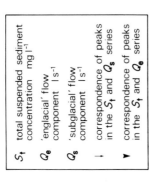

S_t total suspended sediment concentration   mg l^{-1}

Q_e 'englacial' flow component   l s^{-1}

Q_s 'subglacial' flow component   l s^{-1}

→   correspondence of peaks in the S_t and Q_s series

▶   correspondence of peaks in the S_t and Q_e series

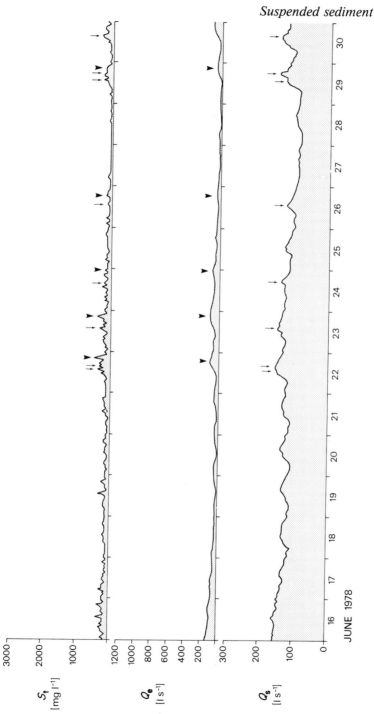

FIGURE 12.6 Hourly total suspended sediment concentration, Q_e, and Q_s for 6th to 30th June 1978. *Reproduced by permission of Catena Verlag from Catena, Supplement 5, 1984, 112*

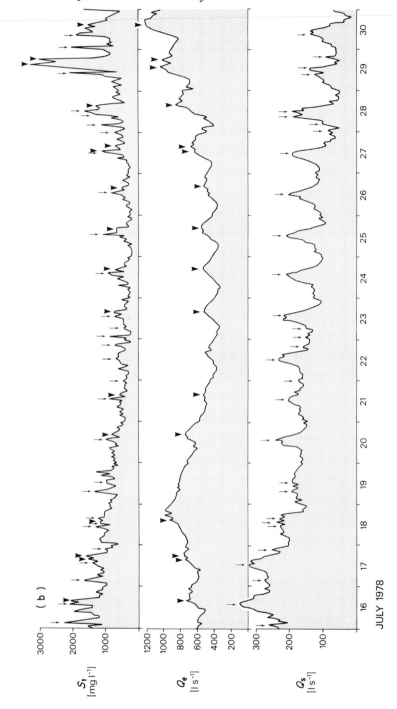

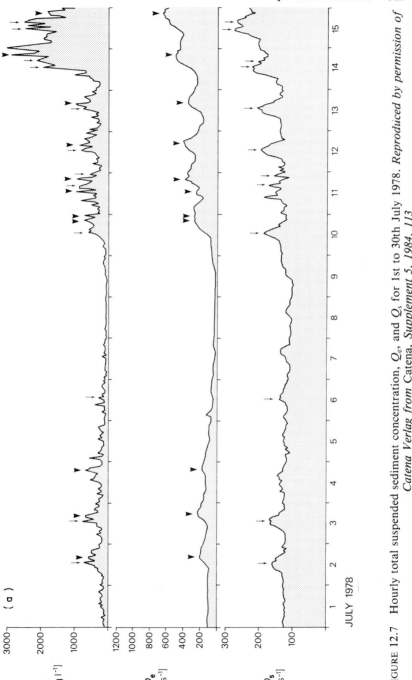

FIGURE 12.7 Hourly total suspended sediment concentration, Q_e, and Q_s for 1st to 30th July 1978. *Reproduced by permission of Catena Verlag from* Catena, *Supplement 5, 1984, 113*

the electrical conductivity of the long residence time 'subglacial' flow component and the shorter residence time 'englacial' flow component (chosen to be 50 and 18 μS cm^{-1}) and then substituting these values into the mixing model:

$$Q_S = [(C_t - C_e)/(C_s - C_e)] Q_t$$

where Q represents discharge, C represents electrical conductivity and the subscripts t, e and s refer to the total, 'englacial' and 'subglacial' flow components, respectively. Figures 12.6 and 12.7 show the results of this separation of data for June and July 1978. Peaks of suspended sediment concentration match peaks of discharge in both the 'englacial' and 'subglacial' discharge component series and suggest that a multivariate transfer function employing the two flow components as input variables to predict suspended sediment concentration would be a possible method of improving the prediction of suspended sediment concentration in this proglacial stream. Figures 12.6 and 12.7 also provide information on the possible locations and roles of different sediment source areas which will be discussed in Section 12.6.

Although the methods of estimating suspended sediment rating curves have become increasingly complex, incorporating transformations, leads, lags and rates of change in variables, all of these techniques still assume a linear relationship between the included variables and none of them have progressed sufficiently far to cope, even in a linear manner, with the problems of sediment availability and exhaustion. Any detailed time series plots of discharge and suspended sediment concentration for a proglacial stream show that although there are clear links between the two variables in their diurnal and seasonal rhythms, the suspended sediment concentration series nearly always shows marked and rapid exhaustion after a major discharge event. This implies that even in the environment of alpine glacierized catchments, where sediment appears to be freely available and where there is little vegetation cover to stabilize the sediment, there is still the queuing problem of sediment supply in relation to frequency of flood events to transport the sediment, which has been identified in other types of drainage basin. Thus, time between events and flow magnitude are both important in explaining suspended sediment transport. In addition, sediment availability may vary markedly between ablation seasons so that the level of response of suspended sediment transport to discharge variations must be calibrated seasonally. It would appear that the best current approaches to estimating suspended sediment transport involve either the estimation of a transfer function between the suspended sediment and discharge series combined with the use of occasional (weekly?) hand samples to re-target the link between the two variables, or a multivariate regression model which incorporates information on discharge history as well as current discharge levels.

12.5 ANOMALIES FROM THE SUSPENDED SEDIMENT-DISCHARGE RELATIONSHIP

The individual time series of suspended sediment concentration and discharge observed by sampling at one location on a proglacial stream exhibit some similarities. Figure 12.8A illustrates some results of hourly monitoring of suspended sediment concentration and discharge on the proglacial stream of the Glacier de Tsidjiore Nouve and shows the clear diurnal cycles in both time series, the lag of discharge behind suspended sediment concentration and the exhaustion effect in the suspended sediment concentration series over periods of similar diurnal discharge pattern. These properties are typical of proglacial suspended sediment and discharge series but a further property is also typical: the apparently random sudden changes of suspended sediment concentration, often without any noticeable change in discharge. These anomalies in suspended sediment concentration may be generated by sediment inputs within the glaciated zone of the catchment or they may reflect or be moderated by proglacial sediment sources and sinks. Whatever their source, they are significant components of the transport of suspended sediment load in proglacial streams which may not be estimated by employing suspended sediment rating curves.

Small anomalies in suspended sediment concentration have been observed in many studies. Østrem (1975a) and Bogen (1980) both commented on the sudden and short-lived 'clouds' of suspended sediment that passed their sampling sites. Østrem emphasized the need for continuous turbidity records to ensure that water samples were not unrepresentative as a result of sampling during such events. Figure 12.8B shows the continuous turbidity record for a part of the period presented in Figure 12.8A and illustrates the microfluctuations in suspended sediment which occur in proglacial streams often with no accompanying discharge variation. Studies of suspended sediment transport from the Tsidjiore Nouve glacier indicate that proglacial suspended sediment sources only cause minor variations in transport rates. The major anomalies in suspended sediment concentration are generated in the glaciated part of the catchment and in particular during outbursts of meltwater which release large discharges to proglacial streams over relatively short periods of time.

Glacial meltwater outburst events are accompanied by marked variations in discharge but the sediment transport is well in excess of the levels that might be predicted by any relationship between discharge and suspended sediment concentration. There are very few suspended sediment records for meltwater outburst events. Beecroft (1983) observed an outburst from the Tsidjiore Nouve glacier during which approximately 19% of the 1981 ablation season's (June to August inclusive) suspended sediment load was transported by less than 5% of the discharge volume. Other suspended sediment transport anomalies of varying magnitude have been reported during the 1981 ablation season from the Tsidjiore Nouve glacier (Gurnell, 1982; Gurnell and Fenn, 1985) and a second

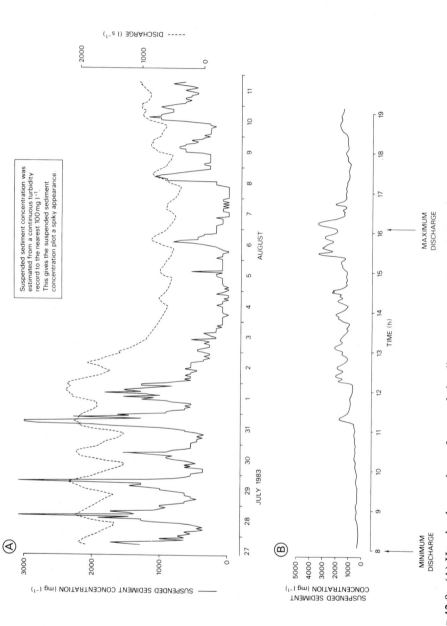

FIGURE 12.8   (A) Hourly observations of suspended sediment concentration and discharge for the proglacial stream of the Glacier de Tsidjiore Nouve, 27th July to 11th August 1983. (B) Continuous pattern of turbidity variations, 28th July 1983, showing numerous short-term flushes of sediment on the rising limb of the diurnal discharge hydrograph

major meltwater outburst occurred in August where a further 35% of the season's suspended sediment load was transported by 15% of the discharge volume. Thus, the two main outburst events accounted for over 50% of the suspended sediment transport and 20% of the discharge volume in less than 15% of the 3-month 1981 monitoring period. The first outburst was relatively more effective than the second in transporting suspended sediment in relation to the discharge volume. This was because the outburst was the first major discharge event of the ablation season and so it was able to tap the large amounts of sediment that are usually available to the first major floods, as well as eroding material from an old lateral moraine along the outburst water's unusual route from the side of the glacier.

Meltwater outbursts may result from the storage and sudden release of large amounts of water from within, under or on a glacier (this is the cause of the annual flood from the Gorner glacier, Collins, 1979a), from accumulation of water behind moraines on or to the side of glacier tongues which may then be released by failure of the moraine dams (Lliboutry *et al.*, 1977; Yesenov and Degovets, 1979) and which are frequently accompanied by mudflows (Nikitin and Shchetinnikov, 1980; Jackson, 1979), from the damming of streams by glacier tongues (Young, 1980) or by avalanche debris. Haeberli (1983) provided an excellent review of the causes and effects of major historical meltwater outburst floods in the Swiss Alps. He noted that 60–70% of 'unusual' or non-periodic glacier floods observed in the Swiss Alps since the beginning of the Little Ice Age were caused by outbursts from marginal glacier lakes or sudden breaks in ice dams, whereas the remaining 30–40% resulted from ruptures of water pockets located within the glacier. The meltwater involved in such outbursts may result from ablation of the ice under the effects of solar radiation or warm air masses but may accumulate more rapidly and catastrophically in association with volcanic activity. The jökulhlaups of Iceland often result from meltwater generated by volcanic activity. Tómasson *et al.* (1980) described the Skeidará jökulhlaups which are frequently associated with volcanic erruptions in Grímsvöten. These jökulhlaups transport large amounts of suspended sediment which are petrographically different from the non-jökulhlaup suspended load, indicating their different source area.

There exists a great range of magnitudes of suspended sediment transport anomalies which may be encountered on proglacial streams. Such anomalies may or may not be accompanied by discharge variations and may be generated by many different processes and sources. Meltwater outbursts are the most significant events in transporting large suspended sediment volumes but there are many other processes which may affect the suspended sediment load and which are significant in developing understanding of the glacial sediment system. This section has considered suspended sediment anomalies as they would be monitored at a single site on a stream receiving the total meltwater discharge from a glacial drainage basin. However, anomalies of suspended sediment transport can be more clearly identified and attributed to specific source areas within and around the glacier when both a spatial and a temporal sampling scheme are adopted. Since single suspended sediment monitoring sites are usually located at a distance from the glacier snout, the observed suspended sediment load will also be affected by the influence of proglacial sources and sinks of sediment in addition to inputs of sediment from the glaciated part of the catchment. Table 12.2 summarizes the temporal and spatial sampling strategies adopted in studies of suspended sediment transport in glaciated catchments. There are very few studies where detailed temporal sampling has been undertaken at more than one location within the drainage basin, but the significance of sediment sources associated with or near the glacier and in the proglacial zone will be considered in the next two sections in the light of the systematic studies presented in Table 12.2 and spot samples taken by other researchers.

12.6   SPATIAL ASPECTS OF SUSPENDED SEDIMENT DELIVERY TO THE PROGLACIAL STREAM NETWORK

Suspended sediment may be delivered to the main proglacial stream by transport in meltwater along supraglacial, englacial, subglacial or ice-marginal routes. It may also be transported in streams from non-glacial environments, including groundwater and snowmelt sources. Figure 12.9 illustrates a variety of routways for suspended sediment transport in the catchment of the Mont Miné glacier, Valais, Switzerland. Figure 12.9A provides a general view of the glacier. There are inputs of meltwater to the ablation zone of the Mont Miné glacier from three streams shown enlarged in Figure 12.9B, and these appear to be draining from snowmelt (stream 1), icemelt from a separate and higher glacier (stream 2) and meltwater from an overhanging section of the upper part of the Mont Miné glacier itself (stream 3). The highly crevassed surface of the Mont Miné glacier tongue precludes the development of major supraglacial streams. Meltwater and sediment are delivered to the proglacial zone of the Mont Miné glacier from a major stream which appears subglacially from the left side of the Mont Miné glacier snout, where it joins the main proglacial stream from the adjacent Ferpècle glacier (site 4). Another major stream (enlarged in Figure 12.9C) appears from the base of the dead ice core of an old lateral moraine. A series of minor streams and moraine mass movements deliver sediment from the glacier snout into the proglacial stream network of distributaries which feed a proglacial lake. This section will consider observations of suspended sediment transport and delivery from the glaciated part of a drainage basin, including transport in supraglacial, englacial, subglacial and ice marginal streams, and also in streams fed by snowmelt and groundwater seepage.

Supraglacial streams transport very variable concentrations of suspended sediment and the main control on this variability seems to be the distance of the monitoring site from the glacier snout. Lister (1981), studying sediment

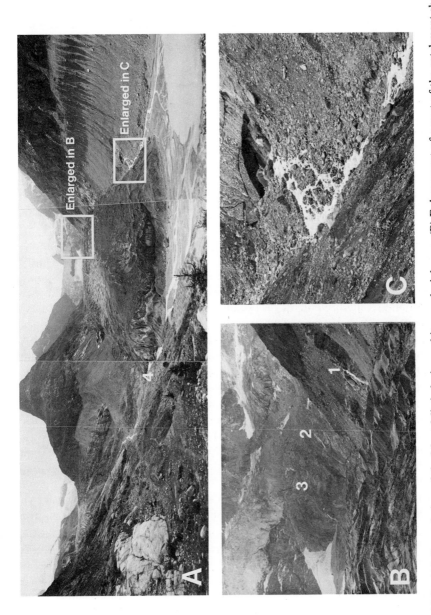

FIGURE 12.9 (A) General view of the Mont Miné glacier and its proglacial zone. (B) Enlargement of a part of the catchment showing three inputs of meltwater to the glacier tongue. (C) Enlargement of a stream issuing from the base of an old ice-cored lateral moraine

particle size, shape and load in a cold glacier in east Greenland, noted that the mean concentration of suspended sediment in a supraglacial and a subglacial stream and in a stream draining an area of ice-cored moraines were in proportions 100:900:300 mg l^{-1}. Dowdeswell (1982) noted supraglacial suspended sediment concentrations in the range 61–430 mg l^{-1} on Sylgjujökull, Iceland. However, neither of these studies concerned alpine glaciers, for which there appear to be few field measurements. Hammer and Smith (1983) monitored suspended sediment concentrations in two supraglacial streams on the Hilda glacier, Canadian Rocky Mountains. They considered that the suspended sediment in these streams was contributed by 'sloughing of surface mantle, melting out of high englacial debris along channel walls and redistribution of fine sediment by surface runoff.' In comparison with the suspended sediment concentrations in the proglacial streams, they considered that concentrations were low in the supraglacial channels, probably because of the paucity of fine material available for these streams to transport. Hammer and Smith observed concentrations below 200 and 400 mg l^{-1} in the two supraglacial streams monitored. The availability of sediment for transport by supraglacial streams increases towards the snouts of alpine glaciers and as the meltwater drains down the snout of the glacier it is able to tap englacial debris. The Tsidjiore Nouve glacier, for example, contains rich englacial debris bands which are exposed at the glacier snout (Figure 12.10). Fenn (1983) sampled water from two sites on the main stream draining the face of the Tsidjiore Nouve glacier snout during June and July 1978. He noted that the daily samples from this stream contained suspended sediment concentrations of at least an order of magnitude higher than any of the tributary and main stream sampling sites on the proglacial stream network. Maximum sediment concentrations exceeded 60 000 mg l^{-1} in the supraglacial stream compared with less than 3000 mg l^{-1} in samples derived from proglacial sampling sites at the same time. The suspended sediment concentration of supraglacial streams on alpine glaciers has received little attention. Although the majority of such streams appear to transport very little sediment, this is probably largely a result of the limited availability of fine material. Near and at the snout of alpine glaciers, sufficient fine material may be available for the suspended sediment concentration to be very high. Streams flowing down the steep snouts of glaciers will also have sufficient energy to transport coarser material than upstream supraglacial streams, flowing on the more gently sloping glacier surface.

There are even fewer observations of suspended sediment concentrations in englacial streams than there are for supraglacial streams because of the enormous difficulties in accessing and maintaining monitoring sites. The results of such observations are very variable but, as with supraglacial streams, distance from the glacier snout or margins may explain some of the variability. Since discharge influences a stream's ability to transport sediment, the magnitude and nature of the flow in englacial streams will have a major impact on the amount of

FIGURE 12.10 Bands of debris exposed at the snout of the Glacier de Tsidjiore Nouve

sediment transported and englacial streams will tend to increase in size down-glacier. In addition, sediment supply will regulate the amount of sediment transported and englacial streams will have increasing opportunity to tap englacial debris towards the snout, margins and bed of a glacier. Gustavson and Boothroyd (1982) noted the existence of a phreatic zone within the Malaspina glacier and the near permanent and stable tunnels occupied by many englacial and subglacial streams. The presence of a large fan of debris at the outlet of one englacial stream confirmed that significant amounts of material were being transported by this stream. They observed apparently high suspended sediment loads in the turbulent flows of englacial streams exposed by collapse of the overlying ice. Suspended sediment concentrations observed where englacial streams issued as fountains at the glacier margin ranged from 1700 to 4700 mg l^{-1} and ice-cored eskers revealed around the glacier margins also indicated high suspended sediment and bedload in former englacial streams. The suspended sediment was thought to be derived from sources near or at the bed of the glacier, since the water was flowing under hydrostatic pressure. The Malaspina glacier is a piedmont rather than an alpine glacier but no equivalent observations exist for alpine glaciers. Streams sometimes appear from englacial tunnels at alpine glacier snouts but there have been few attempts to sample the suspended sediment in such streams.

The suspended sediment concentration of the subglacial streams of alpine glaciers has received more attention than supra- and englacial streams. Vivian (1970) and Vivian and Zumstein (1973) reported observations from the Argentière glacier, France. Suspended sediment concentrations were observed at a location well above the snout of the glacier where water is abstracted through a subglacial water intake forming part of a hydroelectric power scheme. Two water samples were abstracted from the subglacial stream each month over a 2-year period (51 samples in total). These samples indicated very low suspended sediment concentration in the low winter flows (November to March, inclusive) of less than 50 mg l^{-1} but an increase to over 200 mg l^{-1} from June to September. Hagen *et al.* (1983) and Kjeldsen (1981) described suspended sediment transport in a subglacial stream feeding a water intake below the Bondhusbreen glacier, Norway. This stream transported an estimated suspended sediment load of 9300 and 7400 tonnes during the 1979 and 1980 ablation seasons with suspended sediment concentrations in the range 15–70 mg l^{-1}, rising to a maximum of 785 mg l^{-1} during flash floods. Hagen *et al.* (1983) noted that the greatest concentration of sediment incorporated in the ice occurred in a 1–2 m thick layer at the sole of the glacier. Ice near the 'normal' course of the main

---

FIGURE 12.11 *(opposite)* Spatial survey observations of suspended sediment concentration in relation to total stream discharge on 26th, 22nd, 6th and 2nd July 1981, Glacier de Tsidjiore Nouve. *Reproduced by permission of the International Association of Hydrological Sciences from* IAHS Publ. 138, p. 326 (1982)

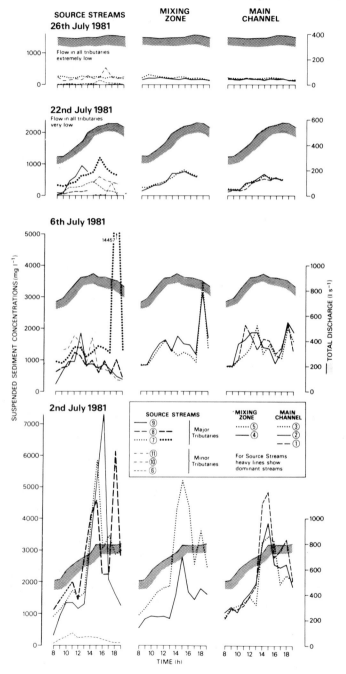

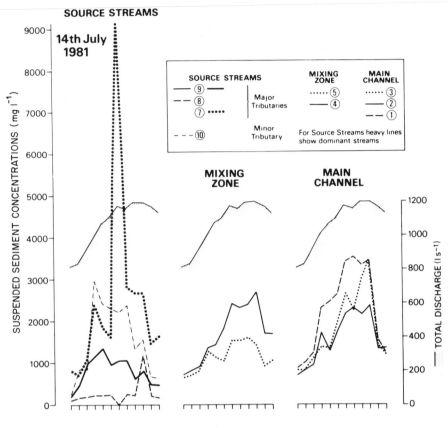

FIGURE 12.12

subglacial stream was relatively clean, probably as a result of melting of 'dirty' ice by the stream and its replacement by the inflow of 'clean' ice from above. During flash floods, sediment was probably derived from lateral melting of the debris-laden ice beyond the margins of the 'normal' channel. These sediment loads, in proportion to the catchment area, compare favourably with suspended sediment transport rates monitored in proglacial streams of other Norwegian glaciers.

Non-glacial sources and transport rates of suspended sediment within glaciated catchments have rarely been quantified, although research in non-glacierized catchments (e.g. McCann and Cogley, 1973) suggests that the early season snow melt floods can transport large loads of suspended sediment but thereafter the flow is insufficient to erode or transport significant amounts of sediment. Gurnell (1983) reported suspended sediment concentrations from a single spatial sample of streams in the Val d'Arolla, Switzerland, and showed that streams fed entirely from snowmelt contained less than $20\,mg\,l^{-1}$ of suspended sediment in

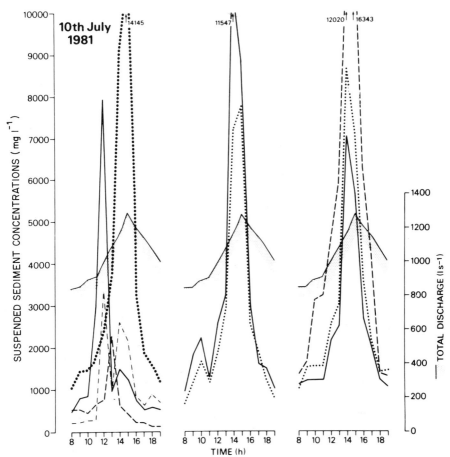

FIGURE 12.12 Spatial survey observations of suspended sediment concentration in relation to total stream discharge on 14th and 10th July 1982, Glacier de Tsidjiore Nouve. *Reproduced by permission of the International Association of Hydrological Sciences from IAHS Publ. 138, p. 327 (1982)*

comparison with concentrations of over 300 mg l^{-1} in the proglacial streams of the Tsidjiore Nouve and Lower Arolla glaciers, and concentrations in excess of 50 mg l^{-1} in proglacial streams below water intakes (where sedimentation occurs). Gurnell (1982) showed that a small stream fed entirely by groundwater seepage near the snout of the Tsidjiore Nouve glacier contained negligible quantities of suspended sediment. A more extensive study of groundwater seepage on the Skeidarársandur by Hjulström (1955) suggested that the groundwater seepage was fed by glacial meltwater and that its turbidity was related to the distance that the water had travelled within the sandur plain material before appearing as a stream on the surface. The water with lowest suspended sediment load was found in streams draining from the valley side and from snowmelt sources. Gustavson (1975) reported concentrations of suspended sediment in the range 135–540 mg l^{-1} for a stream draining from an area of dead ice near the Malaspina glacier.

Observations of suspended sediment transport from different types of stream in alpine glaciated catchments are extremely sparse. One reason for this is the difficulty of identifying and sampling streams which represent only one type or source of flow. In general, flow sources will become increasingly complex and sediment will become more freely available down-glacier. The sampling of tributaries to the main proglacial stream, regardless of the source area or the complexity of the routing of flow from that source area, should provide valuable additional information on the yield and transport of suspended sediment in glaciated alpine catchments. Such spatial sampling of suspended sediment concentration has been undertaken on the tributaries to the main proglacial stream of the Tsidjiore Nouve glacier. Figures 12.11 and 12.12 illustrate six separate periods of 12 h when sampling was undertaken at a series of sampling sites throughout the proglacial stream network. The data for 22nd July illustrate clear diurnal cycles in suspended sediment concentration in the tributary streams draining the glacier snout which are similar in form to the pattern of suspended sediment concentration in the main stream to which the tributaries contribute flow, but which reach peak concentration at different times. The difference in the timing of peak concentration in the tributaries may reflect differences in the timing of discharge and delivery of sediment from the source areas contributing to each tributary stream. The data for 14th July show flushes of sediment of different magnitude and timing superimposed on the diurnal cycles of sediment concentration in each tributary, but there is little evidence of these flushes of sediment once the tributary flow is mixed in the main stream. Data for 10th July illustrate the degree to which a major input of sediment from one tributary can influence the sediment concentration pattern in the main stream. The data for 14th and 10th July also provide examples of instances when flushes of suspended sediment are delivered to the main stream simultaneously from more than one tributary. Gurnell (1982) described further details of such spatial patterns in suspended sediment concentration in the proglacial stream network of the Tsidjiore Nouve glacier and explained the flushes as follows:

'Single-tributary flushes must be attributed to the tapping of new local sediment sources within the glacier system, perhaps as a result of ice collapse or the local shifting of drainage routes. When flushes occur in more than one tributary at the same time they may be the result of similar processes to those responsible for a single tributary flush, but they may also be caused by sudden movements of the glacier releasing sediment to isolated drainage lines within the glacier at the same time.'

(Gurnell, 1982, p. 329)

Figures 12.6 and 12.7 illustrate the links between peaks of suspended sediment concentration and the 'englacial' and 'subglacial' flow components of the flow of the main stream. These data suggest some links between suspended sediment discharge source areas and the routing of the two flow components (Gurnell and Fenn, 1984b). Hourly sampling of suspended sediment concentration and electrical conductivity of the tributary streams suggest that the source areas and routing of meltwater to the tributary streams may help to explain some of the larger flushes of suspended sediment which appear in the tributaries (Gurnell and Fenn, 1985). Daily sampling of suspended sediment concentration in the proglacial tributary streams of the Tsidjiore Nouve glacier during July 1983 revealed two other properties of sediment delivery to the proglacial zone. Firstly, the tributary which received the majority of the supraglacial drainage of the snout zone of the glacier and which was also the largest tributary, transported the highest concentrations of suspended sediment. Secondly, the collapse of a large boulder which was leaning against the snout of the glacier yielded a notable flush of suspended sediment. Both of these properties suggest that supraglacial meltwater activity near the snout of an alpine glacier may have a significant effect on suspended sediment transport in proglacial streams, although the largest quantities of sediment are undoubtedly delivered by streams which issue from beneath the ice.

12.7 PROGLACIAL SEDIMENT SOURCES AND SINKS

The observed suspended sediment yield of a glaciated drainage basin will depend on the yield of sediment from glacial and non-glacial sources and the proportion of that sediment which is deposited within the basin above the sediment monitoring site. Analysis of the suspended sediment yield records of 43 glaciated drainage basins earlier in this chapter illustrated that sediment yield decreased with increasing distance of the monitoring site from the glacier snout and this section will consider the processes which contribute to that general trend.

Proglacial inputs of sediment have been identified which will counteract the effects of sediment deposition in the proglacial environment. Bogen (1980) suggested that the 'clouds' of suspended sediment that were occasionally observed in the meltwater streams from Austerdalsbreen and Tunsbergdalsbreen were probably caused by the temporary deposition and subsequent erosion of sediment associated with discharge fluctuations in the proglacial stream network, whereas Hammer and Smith (1983) reported clear inputs of sediment from the erosion of proglacial channel banks. Richards (1984) noted the accumulation of sediment in the troughs of fluted moraine in front of the Storbreen glacier during periods of dry weather and the augmentation of the suspended sediment load of the proglacial stream by sediment transported from mountain slopes and the till plain during wet weather. Thus the proglacial zone provided a 'buffer' storage area between the snout of the glacier and the point at which the suspended sediment concentration was monitored. The microfluctuations in suspended sediment concentration in the proglacial stream of the Tsidjiore Nouve glacier illustrated in Figure 12.8B resulted from the progressive inundation of the proglacial flood plain as the low-discharge, single-thread flow pattern was transformed by overbank flows into a high-discharge, multi-thread flow pattern. All of these proglacial, fluvially controlled sediment inputs are small and the turbidity record for the Tsidjiore Nouve glacier indicates that these inputs provide a far smoother and more attenuated time pattern than the short-lived and spiky inputs of sediment that may be traced back to the flows that issue from the glacier snout. Nevertheless, there are more catastrophic proglacial processes which can yield sudden and large volumes of sediment to the proglacial stream. Catastrophic drainage of lakes has already been mentioned as a source of sediment and, although the release of many lakes may be glacially controlled, proglacial lakes dammed behind old terminal moraines (Lliboutry *et al.*, 1977) can also release large volumes of sediment. Avalanches, major mudflows and local collapse of saturated morainic deposits can also deliver large volumes of sediment to proglacial streams and may also cause ponding of the streams (Lliboutry *et al.*, 1977; Nikitin and Shchetinnikov, 1980; Theakstone, 1970; Yesenov and Degovets, 1979). There is no information on the form of the

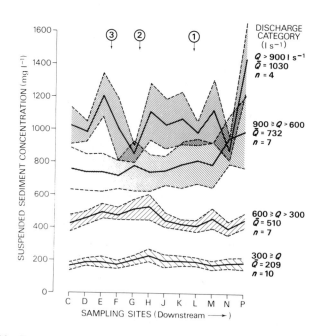

FIGURE 12.13   Trends in mean suspended sediment concentration down the main proglacial stream of the Glacier de Tsidjiore Nouve during different ranges of discharge (one standard deviation bands are shown)

discharge and suspended sediment concentration patterns that would be induced by such events.

The potential for proglacial inputs of suspended sediment would appear to contradict the previously proposed trend of decreasing suspended sediment concentration with increasing distance from the glacier snout in proglacial streams. Multiple site observation of suspended sediment concentration along the main channel of the proglacial stream network of the Tsidjiore Nouve glacier also suggest that sediment concentration increases rather than decreases away from the glacier snout (Gurnell, 1982). Figure 12.13 presents suspended sediment concentration data from twelve sampling sites located along the main proglacial stream of the Tsidjiore Nouve glacier downstream of the point where all the major tributaries have joined the stream and, therefore, do not influence the downstream trend. The raw data (Fenn, 1983) exhibited enormous scatter, but Figure 12.13 shows the average concentration of suspended sediment for each of four ranges of discharge. There is no indication of a downstream decrease in suspended sediment concentration in Figure 12.13, but the trend in the average values and the one standard deviation bands suggests that concentration may increase downstream during discharges in excess of $600 \, \mathrm{l \, s^{-1}}$. Other information on sediment concentration in this stream suggests that flood plain inundation and the accompanying development of increasingly complex multi-thread channel patterns are initiated at a discharge of approximately $600 \, \mathrm{l \, s^{-1}}$ and so the role of flood plain sediment source areas in proglacial suspended sediment transport may be indicated by Figure 12.13. However, these data should be viewed with some caution for a number of reasons. First, the proglacial stream of the Tsidjiore Nouve glacier has a high angle of slope (over $10°$ minimum slope, with a mean channel slope over $15°$), which will affect the stream's ability to transport sediment, and the stream's long profile provides few areas where sediment deposition can easily take place. Second, all the data presented in Figure 12.13 were derived from sampling on the rising limb of the diurnal discharge hydrograph and so the falling limb might present a very different picture.

Increases in suspended sediment concentration in a downstream direction on alpine proglacial streams have been noted by Mills (1979), Fahnestock (1963), Ferguson (1984) and Maizels (1978), but there have also been studies which have revealed a downstream decrease. Details of the discharge regime and hydraulic characteristics of the proglacial river and the availability of sediment are essential in explaining why such differences occur. Boothroyd and Ashley (1975), working on the outwash deposits from the Malaspina glacier, illustrated the initial increase and subsequent decrease in suspended sediment concentration across an outwash during rising discharge. This pattern in suspended sediment concentration was mirrored in the pattern of flow velocity. The zone of increasing velocity and sediment concentration was an unvegetated, steep, coarse gravel area of the fan. Downstream decreases in sediment concentration and velocity occurred across an area which graded from fine gravel to sand with well established vegetation on the inter-stream areas, a decrease in slope and a change in channel pattern from incised to braided. Thus adjustments in suspended sediment concentration are a response to a complex of factors.

There are two properties of the proglacial river network which are particularly important in affecting downstream change in suspended sediment concentration. First there is the degree to which major proglacial sediment sinks are present. Proglacial lakes have an enormous effect on the suspended sediment transport pattern. The size and shape of these lakes will affect their trap efficiency and there is a wealth of field data on sedimentation processes in proglacial lakes. Mathews (1956, 1964) presented an early indication of the importance of proglacial lakes in trapping sediment and subsequent studies have confirmed the significance of proglacial lakes in this respect (e.g. Gilbert, 1975; Gilbert and Shaw, 1981; Gustavson, 1975; Gustavson *et al.*, 1975; Østrem, 1975b; Smith *et al.*, 1982). Second, there is the concept of paraglacial sedimentation (Church and Ryder, 1972). Church and Ryder (1972) suggested the term 'paraglacial' to 'define non-glacial processes that are directly conditioned by glaciation. It refers both to proglacial processes, and to those occurring around and within the margins of a former glacier that are the direct result of the earlier presence of ice.' The increase in suspended sediment concentration for short distances in a downstream direction on proglacial streams reflects the availability of sediment for fluvial sediment transport. This sediment availability is a paraglacial effect, since the sediment was originally produced by the glacier even if it has subsequently been eroded, sorted, transported and redeposited by fluvial processes. The degree to which such sediment can have an impact on fluvial sediment transport will be governed by the geology and climate of the area as they controlled the original rate of glacial erosion; the time since the sediment was eroded, since this will affect the amount of sediment still available for transport; and the distance from the glacier snout, since this combined with time will affect the degree to which the sediment has been removed by fluvial processes or stabilised by the development of a vegetation cover. The complex effects of these factors together with the discharge of the proglacial stream, the slope of the valley train and any changes in base level which may have occurred, will explain why there is an increase or decrease in suspended sediment transport along a proglacial stream and also why an increase in suspended sediment concentration for short distances from the glacier snout will change into a decrease as sediment supply becomes more limited downstream. Thus the global pattern of decreasing sediment transport with increasing distance from the glacier snout will frequently be reversed near to the snout of the glacier as a result of these paraglacial effects.

12.8 PARTICLE SIZE CHARACTERISTICS OF PROGLACIAL SUSPENDED SEDIMENT

The particle size characteristics of the suspended sediment in proglacial streams must be affected by the rock from which the sediment was derived and its weathering, erosion and transport history. Guymon (1974) showed how the inclusion of a parameter of the particle size distribution (P_s, the dimensionless slope of the particle size curve at the d_{65} divided by the particle size at 65% finer than) with the variables of Borland's λ index improved the estimates of sediment yield for glaciated catchments in Alaska. Guymon considered that P_s was a measure of the soil characteristics of the basin and suggested that glaciated catchments produced a noticeably different particle size distribution from non-glaciated catchments although he had insufficient data to attempt to prove this. Other studies have emphasized the dynamic nature of the proglacial suspended sediment distribution in time and space and this will be illustrated in the remainder of this section.

The variations in the grain size of material deposited in lakes fed by glacial meltwater is well known. Østrem (1975a) described varved sediments from lakes in Norway, showing that summer layers were generally coarser than winter layers and that there was a trend towards increasingly coarse sediment in the bottom layers. These variations in particle size reflected two timescales of processes; the seasonal variations in the discharge of the proglacial stream and supply of sediment and the effect on sediment supply of the retreat of the glacier after the initial development of the proglacial lake. The seasonal variations in the particle size of deposited sediments allowed Østrem to estimate annual volumes of sediment deposited in a lake over a 29-year period and to match these with the difference in monitored suspended sediment inputs and outputs from the lake.

Rainwater and Guy (1961) presented observations of the diurnal pattern of the concentration of clay, silt and sand sized fractions of the suspended sediment transported from Chamberlain glacier, Alaska. They found that each of the particle size ranges produced diurnal cycles which were in phase with one another but which had a very different range or amplitude in concentration. Finer material showed less variation in concentration during the diurnal discharge cycle than coarser material. The diurnal trends in the proportions of different grain sizes transported reflect the variations in discharge of the proglacial stream and thus its ability to transport the coarser particles. As a result of the extremely variable discharge and associated flow velocities in proglacial streams, it is possible for fairly coarse particles to form a component of the suspended load at certain times. Church and Gilbert (1975) noted that:

> 'Particle size of suspended materials in glacial streams appears to vary over a range larger than between 1 mm and 0.001 mm. The size

of smallest material will probably depend upon local geology, and upon weathering and glacial comminution. Amongst coarser materials, the distinction between bed and suspended load is arbitrary, since material may move on the bed for some distance, be swept into suspension where turbulence is very intense, and be deposited again on the bed downstream. In very exceptional floods, materials larger than 8 mm may move in long saltation jumps.'

(Church and Gilbert, 1975, p. 41)

Fenn (1983) collected daily water samples at 10.00 h throughout July 1978 from the proglacial stream of the Glacier de Tsidjiore Nouve. He found that although there was a clear link between discharge and total suspended sediment concentration greater than 8 μm, the concentrations of suspended sediment coarser than 63, 250 and 500 μm showed little relationship with discharge or with the concentration greater than 8 μm. Correlation between the 63, 250 and 500 μm series of observations was high but the concentrations of these coarser fractions appeared to be largely independent of flow magnitude although there was some evidence of flushing of coarser sediments with rising discharge. Fenn concluded that most of the suspended sediments in the proglacial stream of the Glacier de Tsidjiore Nouve are in the 8–63 μm size range and may be interpreted as the product of 'glacier grinding' and of subglacial origin. The transport of coarser particles is in the form of short-lived flushes, usually in response to high flows but obviously subject to complex interactions between magnitude and routing of stream flow and sediment supply.

12.9 CONCLUSION

This chapter has attempted to identify trends in suspended sediment concentration and yield in proglacial streams at a wide range of scales as revealed by different temporal and spatial sampling schemes. As with many environmental processes, increasing frequency of temporal and spatial sampling reveals more and more variability in the transport of suspended sediment in proglacial streams and so this summary will attempt to identify trends and patterns noted by the many studies which have been discussed. Table 12.5 lists some of the properties of suspended sediment concentration dynamics in proglacial streams which have been noted and classifies them according to their spatial and temporal context. Table 12.5 emphasizes the shortage of information on patterns of suspended sediment in tributary streams and this shortage of information becomes even more severe in the glaciated part of the catchment. Although it is possible to apply increasingly complex analyses to very detailed, single-site, proglacial suspended sediment and discharge data, there is a great need to expand and confirm the theories produced by such analyses by attempting detailed, shorter term, nested sampling of suspended sediment

TABLE 12.5   Some general trends in suspended sediment concentration noted in studies of glacierized catchments

Location	Trend
Supraglacial, englacial, subglacial, ice-marginal streams	Suspended sediment concentration generally lower in supraglacial than in englacial, subglacial or ice-marginal streams.
	There is a possible trend towards increasing suspended sediment concentration towards the glacier snout, which has been demonstrated for supraglacial streams.
Tributary streams from the glacier snout	Diurnal hysteresis in the suspended sediment concentration discharge relationship.
	Major and minor single and multiple tributary suspended sediment flushes; minor to major meltwater outbursts.
	Rationalization of glacial meltwater drainage system during ablation season leading to diversion of majority of flow and suspended sediment transport to one (or more) increasingly dominant stream(s).
Main proglacial river	Diurnal, sub-seasonal and seasonal hysteresis in the suspended sediment concentration—discharge relationship.
	Major differences in the level of suspended sediment transport between years (presumably as a result of differences in the sediment supply).
	Great impact of meltwater outbursts on suspended sediment transport.
	Flushes of sediment from glacial and proglacial sediment sources. Reduction of suspended sediment transport by proglacial sediment sinks, particularly lakes.
	Downstream initial increase followed by decrease in suspended sediment concentration, in the absence of major lateral inputs of sediment.

transport nearer to its source areas. This discussion has stressed the dearth of information on sediment transport in supraglacial, englacial, subglacial and ice-marginal streams and the proposals concerning spatial patterns, notably down-valley increases, in suspended sediment concentration have yet to be confirmed by rigorously designed and analysed programmes of field sampling. Observations of suspended sediment concentration in streams on or near the snout of alpine glaciers have indicated the importance of this near snout zone in contributing and moderating suspended sediment transport, but even here there is a need for data on a larger sample of alpine, glaciated basins. Most information on suspended sediment transport is available for the proglacial zone of alpine, glaciated catchments and here the significance of proglacial sources and sinks, particularly the role of proglacial lakes, has been widely studied. Two trends in research are required: first the more effective design and implementation of spatial and temporal sampling of suspended sediment concentration, and second the wider application of other laboratory analyses of particle size, texture, shape and mineralogy of collected samples so that these two threads of research can combine to reveal the locations of sediment source areas and the methods and routes of supply of sediments to the proglacial fluvial system. More intensive and carefully structured sampling of a large number of proglacial streams and over longer periods of time would allow an evaluation not only of sediment sources, transport and routing but also of the impact of glacier activity on these processes. Kamb *et al.* (1985), in their fascinating study of the effect on the meltwater discharge of surging of the Variegated glacier, Alaska, in 1982–83, noted that the turbidity of the outflow stream was much higher during the surge than before or after it. Work on suspended sediment concentration at a fixed point on the proglacial stream of the Tsidjiore Nouve glacier during a period of advance in recent years has also indicated a possible positive link between rate of advance and suspended sediment yield but as yet there is insufficient information to link suspended sediment yield to glacier activity with any degree of confidence.

REFERENCES

AMERICAN SOCIETY OF CIVIL ENGINEERS (1969a), Sediment measurement techniques: A. Fluvial sediment. Task Committee on Preparation of Sedimentation Manual, *Proceedings of the American Society of Civil Engineers, Journal of the Hydraulics Division*, HY5, 1477–1514.

AMERICAN SOCIETY OF CIVIL ENGINEERS (1969b), Sediment measurement techniques: B. Laboratory procedures. Task Committee on Preparation of Sedimentation Manual, *Proceedings of the American Society of Civil Engineers, Journal of the Hydraulics Division*, HY5, 1515–1543.

BEECROFT, I. (1981), Variations, over a 24 h period, in suspended sediment concentration and size distribution in meltwater from the Tsidjiore Nouve glacier, Arolla, Valais, Switzerland, *Wessex Geographer*, **16**, 12–18.

BEECROFT, I. (1983), Sediment transport during an outburst from Glacier de Tsidjiore Nouve, Switzerland, 16–19 June 1981, *Journal of Glaciology*, **29**, 185–190.

BEZINGE, A. (1978), Torrents glaciaires, hydrologie et charriages d'alluvions, *Sonderdruck aus dem Jahrbuch der Schweizerischen Naturforschenden Gesellschaft, Wissenschaftlicher*, Teil 1978, 152–169 (translated and modified as Chapter 17 in this book).

BOGEN, J. (1980), The hysteresis effect of sediment transport systems, *Norsk Geografisk Tidsskrift*, **34**, 45–54.

BOOTHROYD, J. C., and ASHLEY, G. M. (1975), Processes, bar morphology and sedimentary structures on braided outwash fans, northeastern gulf of Alaska, in JOPLING, A. V., and MACDONALD, B. C. (Eds.), *Glaciofluvial and Glaciolacustrine Sedimentation*, Society of Economic Palaeontologists and Mineralogists Special Publication **23**, 193–222.

BORLAND, W. M. (1961), Sediment transport of glacier-fed streams in Alaska, *Journal of Geophysical Research*, **66**, 3347–3350.

BOX, G. E. P., and JENKINS, G. M. (1970), *Time Series Analysis, Forecasting and Control*, Holden-Day, San Francisco, 553pp.

BRYAN, M. L. (1972), Variations in quantity and quality of Slims river water, Yukon Territory, *Canadian Journal of Earth Sciences*, **9**, 1469–1478.

BURROWS, R. L., EMMETT, W. W., and PARKS, B. (1981), Sediment transport in the Tanana River near Fairbanks, Alaska, 1977/79. *United States Geological Survey Water Resources Investigation*, 81–20.

CHERNOVA, L. P. (1981), Influence of mass balance and run-off on relief-forming activity of mountain glaciers, *Annals of Glaciology*, **2**, 69–70.

CHURCH, M. A. (1974), On the quality of some waters on Baffin Island, North West Territories, *Canadian Journal of Earth Sciences*, **11**, 1676–1688.

CHURCH, M., and GILBERT, R. (1975), Proglacial fluvial and lacustrine environments, in JOPLING, A. V., and MACDONALD, B. C. (Eds.), *Glaciofluvial and Glaciolacustrine Sedimentation*, Society of Economic Palaeontologists and Mineralogists Special Publication 23, pp. 22–100.

CHURCH, M., and RYDER, J. M. (1972), Paraglacial sedimentation: a consideration of fluvial processes conditioned by glaciation, *Geological Society of America Bulletin*, **83**, 3059–3072.

COLLINS, D. N. (1979a), Sediment concentration in melt waters as an indicator of erosion processes beneath an alpine glacier, *Journal of Glaciology*, **23**, 247–257.

COLLINS, D. N. (1979b), Quantitative determination of the subglacial hydrology of two alpine glaciers, *Journal of Glaciology*, **23**, 347–361.

DOWDESWELL, J. A. (1982), Supraglacial re-sedimentation from melt-water streams on to snow overlying glacier ice, Sylgjujökull, west Vatnajökull, Iceland, *Journal of Glaciology*, **28**, 365–375.

ELVERHOI, A., LIESTØL, O., and NAGY, J. (1980), Glacial erosion, sedimentation and microfauna in the inner part of Kongsfjorden, Spitsbergen, *Norsk Polarinstitutt Skrifter*, **172**, 33–61.

FAHNESTOCK, R. K. (1963), Morphology and hydrology of a glacial stream, White River, Mount Rainier, Washington, *United States Geological Survey Professional Paper*, 422-A, 70pp.

FENN, C. R. (1983), Proglacial streamflow series: measurement, analysis and interpretation, unpublished PhD Thesis, University of Southampton, 425pp.

FENN, C. R., GURNELL, A. M., and BEECROFT, I. (1985), An evaluation of the use of suspended sediment rating curves for the prediction of suspended sediment concentration in a proglacial stream, *Geografiska Annaler*, **67A**, 71–82.

FERGUSON, R. I. (1984), Sediment load of the Hunza river, in MILLER, K. J. (Ed.), *International Karakoram Project 2*, Cambridge University Press, pp. 581–598.

GILBERT, R. (1975), Sedimentation in Lillooet Lake, British Columbia, *Canadian Journal of Earth Sciences*, **12**, 1697–1711.

GILBERT, R., and SHAW, J. (1981), Sedimentation in proglacial Sunwapta Lake, Alberta, *Canadian Journal of Earth Sciences*, **18**, 81–93.

GREGORY, K. J., and WALLING, D. E. (1973), *Drainage Basin Form and Process: A Geomorphological Approach*, Edward Arnold, London, 456pp.

GURNELL, A. M. (1982), The dynamics of suspended sediment concentration in a proglacial stream, in GLEN, J. W. (Ed.), *Hydrological Aspects of Alpine and High Mountain Areas (Proceedings of the Exeter Symposium, July 1982)*, International Association of Hydrological Sciences Publication 138, pp. 319–330.

GURNELL, A. M. (1983), Downstream channel adjustments in response to water abstraction for hydro-electric power generation from alpine glacial melt-water streams, *The Geographical Journal*, **149**, 342–354.

GURNELL, A. M., and FENN, C. R. (1984a), Box–Jenkins transfer function models applied to suspended sediment concentration–discharge relationships in a proglacial stream, *Arctic and Alpine Research*, **16**, 93–106.

GURNELL, A. M., and FENN, C. R. (1984b), Flow separation, sediment source areas and suspended sediment transport in a pro-glacial stream, in SCHICK, A. P., (Ed.), *Channel Processes: Water, Sediment, Catchment Controls, Catena*, Suppl. 5, pp. 109–119.

GURNELL, A. M., and FENN, C. R. (1985), Spatial and temporal variations in electrical conductivity in a proglacial stream system, *Journal of Glaciology*, **31**, 108–114.

GUSTAVSON, T. C. (1975), Sedimentation and physical limnology in proglacial Malaspina lake, southeastern Alaska, in JOPLING, A. V., and MACDONALD, B. C. (Eds.), *Glaciofluvial and Glaciolacustrine Sedimentation*, Society of Economic Palaeontologists and Mineralogists Special Publication 23, pp. 249–263.

GUSTAVSON, T. C., ASHLEY, G. M., and BOOTHROYD, J. C. (1975), Depositional sequences in glaciolacustrine deltas, in JOPLING, A. V., and MACDONALD, B. C. (Eds.), *Glaciofluvial and Glaciolacustrine Sedimentation*, Society of Economic Palaeontologists and Mineralogists Special Publication 23, pp. 264–280.

GUSTAVSON, T. C., and BOOTHROYD, J. C. (1982), Subglacial fluvial erosion: a major source of stratified drift, Malaspina glacier, Alaska, in DAVIDSON-ARNOTT, R., NICKLING, W., and FAHEY, B. D. (Eds.), *Glacial, Glaciofluvial and Glaciolacustrine Systems*, Proceedings 6th Guelph Symposium on Geomorphology, 1980, pp. 93–116.

GUYMON, G. L. (1974), Regional sediment yield analysis of Alaska streams, *American Society of Civil Engineers, Journal of the Hydraulics Division*, **100**, 41–51.

HADLEY, R. F., and WALLING, D. E. (Eds.), (1984), *Erosion and Sediment Yield*, Geo Books, Norwich.

HAEBERLI, W. (1983), Frequency and characteristics of glacier floods in the Swiss Alps, *Annals of Glaciology*, **4**, 85–90.

HAGEN, J. O., WOLD, B., LIESTOL, O., ØSTREM, G., and SOLLID, J. L. (1983), Subglacial processes at Bondhusbreen, Norway: preliminary results, *Annals of Glaciology*, **4**, 91–98.

HAMMER, K. M., and SMITH, N. D. (1983), Sediment production and transport in a proglacial stream: Hilda glacier, Alberta, Canada, *Boreas*, **12**, 91–106.

HASHOLT, B. (1976), Hydrology and transport of material in the Sermilik area 1972, *Geografisk Tidsskrift*, **75**, 30–39.

HJULSTRÖM, F. (1955), The ground water, *Geografiska Annaler*, **37**, 234–245.

JACKSON, L. E. (1979), A catastrophic glacial outburst flood (jokulhlaup) mechanism for debris flow generation at the Spiral Tunnels, Kicking Horse River basin, British Columbia, *Canadian Geotechnical Journal*, **16**, 806–813.

KAMB, B., RAYMOND, C. F., HARRISON, W. D., ENGELHARDT, H. ECHELMEYER, K. A., HUMPHREY, N., BRUGMAN, M. M., annd PFEFFER, T. (1985), Glacier surge mechanism: 1982–1983 surge of Variegated glacier, Alaska, *Science*, **227**, 469–479.

KJELDSEN, O. (1981), *Materialtransportundersøkelser i Norske Bre-elver 1980*, Vassdragsdirektoratet Hydrologisk Avdeling Rapport 4–81, 41pp.

KJELDSEN, O., and ØSTREM, G. (1980), *Materialtransportundersøkelser i Norske Bre-elver 1979*, Vassdragsdirektoratet Hydrologisk Avdeling Rapport 1–80, 43pp.

LANSER, O. (1958), Reflexions sur les débits solides en suspension des cours d'eau glaciaires, *International Association of Scientific Hydrology Bulletin*, **4**, 37–43.

LEMMENS, M. (1978), Relations entre concentration en cations dissous et débit de l'émissaire du glacier de Tsidjiore Nouve (Valais), *Catena*, **5**, 227–236.

LISTER, H. (1981), Particle size, shape, and load in a cold and a temperate valley glacier, *Annals of Glaciology*, **2**, 39–44.

LLIBOUTRY, L., ARNAO, B. M., PAUTRE, A., and SCHNEIDER, B. (1977), Glaciological problems set by the control of dangerous lakes in Cordillera Blanca, Peru. 1. Historical failures of morainic dams, their causes and prevention, *Journal of Glaciology*, **18**, 239–254.

McCANN, S. B., and COGLEY, J. G. (1973), The geomorphic significance of fluvial activity at high latitudes, in FAHEY, B. D., and THOMPSON, R. D. (Eds.), *Research in polar and alpine geomorphology*, 3rd Guelph Symposium on Geomorphology, 1973, Geo Abstracts, Norwich, pp. 118–135.

MAIZELS, J. K. (1978), Débit des eaux de fonte, charges sedimentaires et taux d'erosion dans le massif du Mont Blanc, *Revue de Geographie Alpine*, **66**, 65–91.

MATHEWS, W. H. (1956), Physical limnology and sedimentation in a glacial lake, *Bulletin of the Geological society of American*, **67**, 537–552.

MATHEWS, W. H. (1964), Sediment transport from Athabasca glacier, Alberta, *International Association of Scientific Hydrology Publication* 65, 155–165.

METCALF, R. C. (1979), Energy dissipation during subglacial abrasion of Nisqually glaciers, Washington, *Journal of Glaciology*, **23**, 233–246.

MILLS, H. H. (1979), Some implications of sediment studies for glacial erosion on Mt. Rainier, *Northwest Science*, **53**, 190–199.

NIKITIN, A. M., and SHCHETINNIKOV, A. S. (1980), Mudflows in the Isfairamsai river basin in 1977, in *International Symposium on the Computation and Prediction of Runoff from Glaciers and Glacierised Areas, Tbilisi, USSR, 3–11 September 1978*, Akademiya Nauk SSR, Institut Geografii, Materialy Gliatsiologicheskikh Issledovanii Khronika, Obsuzhdeniya 39, pp. 175–179.

ØSTREM, G. (1975a), Sediment transport in glacial meltwater streams, in JOPLING, A. V., and MACDONALD, (Eds.), *Glaciofluvial and Glaciolacustrine Sedimentation*, Society of Economic Palaeontologists and Mineralogists Special Publication 23, pp. 101–122.

ØSTREM, G. (1975b), Sediment transport studies at selected glacier streams in Norway 1969, in JOHNSON, P. G. (Ed.), *Fluvial Processes: Cueillette et Analyse des Donnees*, Occasional Papers, Department of Geography and Regional Planning, University of Ottawa, 3, University of Ottawa Press, pp. 5–23,

ØSTREM, G., BRIDGE, C. W., and RANNIE, W. F. (1967), Glacio-hydrology, discharge and sediment transport in the Decade glacier area, Baffin Island, N.W.T., *Geografiska Annaler*, **49A**, 268–282.

ØSTREM, G., ZIEGLER, T., EKMAN, S. R., OLSEN, H. ANDERSSON, J. E., and LUNDÉN, B. (1971), *Slamtransportstudier i Norska Glaciärälvar 1970*, Stockholms Universitet Naturgeografiska Institutionen, 133pp.

RAINWATER, F. H., and GUY, H. P. (1961), Some observations on the hydrochemistry and sedimentation of the Chamberlain glacier area, Alaska, *United States Geological Survey Professional Paper*, 414-C.

RICHARDS, K. S. (1984), Some observations on suspended sediment dynamics in Storbregrova, Jotunheim, *Earth Surface Processes and Landforms*, **9**, 101–112.

ROEHL, J. E. (1962), Sediment source areas, delivery ratios and influencing morphological factors, *International Association of Scientific Hydrology Publication*, **59**, 202–213.

SHAW, J., GILBERT, R., and ARCHER, J. J. (1978), Proglacial lacustrine sedimentation during winter, *Arctic and Alpine Research*, **10**, 689–699.

SMITH, N. D. (1978), Sedimentation processes and patterns in a glacier-fed lake with low sediment input, *Canadian Journal of Earth Sciences*, **15**, 741–756.

SMITH, N. D., VENOL, M. A., and KENNEDY, S. K. (1982), Comparison of sedimentation regimes in four glacier-fed lakes of western Alberta, in DAVIDSON-ARNOTT, R., NICKLING, W., and FAHEY, B. D. (Eds.), *Glacial, Glaciofluvial and Glaciolacustrine Systems*, Proceedings 6th Guelph Symposium on Geomorphology, 1980, pp. 203–238.

THEAKSTONE, W. (1970), Sediments, structures and processes. Studies at the Østerdalsisen glacier-dammed lake, 1970, *Aarhus Universitet. Skrifteri Fysisk Geografi*, **2**, 11pp.

TKACHEVA, L. G. (1974), Suspended sediment discharge in the rivers of Soviet Central Asia, *Soviety Hydrology: Selected Papers*, **3**, 161–167.

TÓMASSON, H., PÁLSSON, S., and INGÓLFSSON, P. (1980), Comparison of sediment load transport in the Skeidará Jökulhlaups in 1972 and 1976, *Jökull*, **30**, 21–33.

VIVIAN, R. (1970), Hydrologie et erosion sous-glaciaire, *Revue de Geographie Alpine*, **58**, 241–265.

VIVIAN, R., and ZUMSTEIN, J. (1973), Hydrologie sous-glaciaire au glacier d'Argentière (Mont Blanc, France), in *Symposium on the Hydrology of Glaciers, (Proceedings of the Cambridge Symposium, 7–13 September 1969)*, International Association of Scientific Hydrology Publication 95, pp. 53–64.

WALLING, D. E. (1974), Suspended sediment and solute yields from a small catchment prior to urbanisation, in GREGORY, K. J., and D. E. WALLING, (Eds.), *Fluvial Processes in Instrumented Watersheds*, Institute of British Geographers Special Publication 6, pp. 169–192.

WALLING, D. E. (1977), Assessing the accuracy of suspended sediment rating curves for a small basin, *Water Resources Research*, **13**, 531–538.

WALLING, D. E. (1983), The sediment delivery problem, *Journal of Hydrology*, **65**, 209–237.

WALLING, D. E., and KLEO, A. H. A. (1979), Sediment yield of rivers in areas of low precipitation: a global view, in *The Hydrology of Areas of Low Precipitation (Proceedings of the Canberra Symposium, December 1979)*, International Association of Hydrological Sciences Publication 128, pp. 479–493.

WEIRICH, F. H. (1982), Site location and instrumentation aspects of a study of sedimentation processes in a proglacial lake in southeastern British Columbia, Canada, in DAVIDSON-ARNOTT, R., NICKLING, W., and FAHEY, B. D. (Eds.), *Glacial, Glaciofluvial and Glaciolacustrine Systems*, 6th Guelph Symposium on Geomorphology, 1980, pp. 239–260.

YESENOV, U. YE., and DEGOVETS, A. S. (1979), Catastrophic mudflow on the Bol'shaya Almatinka river in 1977, *Soviet Hydrology: Selected Papers*, **18**, 158–160.

YOUNG, G. (1980), The impact of floods from glacier dammed lakes, Yukon, Canada, in *International Symposium on the Computation and Prediction of Runoff from Glaciers and Glacierised Areas, Tbilisi, USSR, 3–11 September 1978*, Akademiya Nauk SSSR, Institut Geografii, Materialy Gliatsiologicheskikh Issledovanii Khronika, Obsuzhdeniya 39, 170–175.

Glacio-fluvial Sediment Transfer
Edited by A. M. Gurnell and M. J. Clark
©1987 John Wiley & Sons Ltd.

Chapter 13

Bedload

B. Gomez

ABSTRACT

This review initially considers the development of sampling and measuring devices which are appropriate for use in gravel bed channels, the findings of recent investigations into bedload transport dynamics in gravel-bed rivers as a whole and their bearing on attempts which have been made to predict bedload discharge. Finally, specific attention is paid to mountain rivers and the degree to which the bedload transport in this environment conforms with observations on sediment movement in gravel-bed rivers in general is assessed.

13.1 INTRODUCTION

Bedload may be defined as that part of the sediment load which is supported by frequent solid contact with the bed and which therefore moves, by rolling or saltation, in close proximity to it. Schoklitsch (1950), when commenting on the (then) current state of research into the bedload transport process, noted that '. . . there is not too much known about it.' However, in the intervening three and a half decades this situation has changed dramatically.

Much of the information that has recently become available has been obtained as a consequence of advances in the design of sampling and measuring devices during the past 15 years, which have at last made it practicable to attempt to obtain comprehensive field data. This has, in turn, led to an increased understanding of the nature of the bedload transport process as it occurs in nature, and thereby to a better appreciation of the difficulties that are involved in the derivation of relationships which seek to predict bedload transport rates in natural alluvial channels.

A distinction may be made between the bedload transport process in sand-bed channels, where the particle size distribution of the bed material is predominantly finer than 2.0 mm and the mobile bed deforms into ripples and dunes, and the bedload transport process in gravel-bed channels, where the bed material is predominantly coarser than 2.0 mm and the mobile material moves in discontinuous, planar sheets or streamers, in the absence of (small-scale) bedforms. Most mountain rivers are characterized by coarse bed material and this review, therefore, concentrates on the bedload transport process in gravel-bed channels. Sections 13.2, 13.3 and 13.4 are concerned with sampling and measurement techniques, the nature of the bedload transport process and the prediction of bedload transport rates in gravel-bed channels as a whole. Finally, section 13.5 offers a perspective on these elements as they apply specifically to mountain rivers.

13.2   TECHNIQUES FOR SAMPLING AND MEASURING BEDLOAD DISCHARGE IN GRAVEL-BED CHANNELS

The bedload is confined to a layer a few grain diameters thick immediately above the stream bed (Abbott and Francis, 1977), and is usually transported intermittently, at a velocity which is less than that of the flow (Leopold and Emmett, 1981; Emmett *et al.*, 1983). Thus, unlike suspended sediment discharge, which may be readily determined by measuring the sediment concentration and the water discharge over a given period of time (since suspended sediment is transported for sustained periods at approximately the same velocity as the flow), bedload discharge is usually determined by measuring the weight of the sediment that passes through a given cross-section. The reliable measurement of bedload discharge rates in natural channels therefore hinges on the efficient collection of all, or a representative portion, of the bedload which passes through a given cross-section in a given period of time.

Apparatus and techniques for measuring bedload discharge fall into two broad categories: indirect and direct methods. Indirect methods, which include the utilization of acoustic collision meters (e.g. Johnson and Muir, 1969; Tywonick and Warnock, 1973; Jonys, 1976; Richards and Milne, 1979) have proved largely unsuccessful, owing to the difficulties that have been encountered in the calibration of these devices, and are not considered further. Direct methods of measuring bedload discharge may be further sub-divided into three categories, which involve the utilization of sampling devices, pit traps and tracers. Comprehensive details of much of the equipment that has been developed to sample or measure bedload discharge in natural alluvial channels have previously been provided by the Federal Inter-Agency River Basin Committee (1940) and Hubbell (1964). Only equipment that has been developed since the publication of Hubbell's report and/or has proved to be particularly appropriate for use in gravel-bed channels is considered in this review.

13.2.1   Sampling devices

The fact that the presence of a sampling device resting on the channel bed distorts the pattern of flow, and hence the bedload discharge in the vicinity of the sampler, has long been recognized (e.g. Davis and Wilson, 1919). The degree to which a particular device will modify the patterns of water and sediment movement is dependent on factors such as its configuration, the flow velocity and depth, the amount and size distribution of the material which is in motion and the duration of the sampling period. Thus, in order to function effectively the sampler must be designed so as to disturb the flow pattern as little as possible, thereby ensuring that its sampling efficiency (the ratio of the amount of bedload collected to that which would have passed through the space occupied by the sampler in the same time, in its absence) is as near to 100% as possible. At the same time, the sampler must also be able to accommodate the entire range of particle sizes that are in motion, as well as to accumulate and retain them whilst remaining in a stable position on the bed surface. Unfortunately, these criteria have proved almost impossible to satisfy collectively. Many bedload sampling devices thus represent a poorly balanced compromise between the most efficient configuration and the design that is deemed to be most appropriate for a particular set of operating conditions. As a consequence, sampling devices exhibit sampling efficiencies which approach 100%.

13.2.2.1 *Basket samplers*

The earliest bedload samplers consisted merely of a box which was open at the top and at the upstream end (e.g. Davis and Wilson, 1919), or a rectangular tube with a mesh screen covering the downstream end (e.g. Ehrenberger, 1932). Later designs evolved into 'baskets,' as the solid sides and top, and eventually the base, were replaced with mesh screens (Ehrenberger, 1932). The most intensive development of basket samplers occurred during the 1930s in Germany and Switzerland, where they were used extensively (e.g. Nesper, 1937; Swiss Federal Authority, 1939). During this period a succession of basket samplers were developed, each representing an improvement in design over its predecessors. The earliest in the series was the Mulhofer sampler (Mulhofer, 1933) and subsequent variations included the Ehrenberger (Ehrenberger, 1931), Nesper (Nesper, 1937) and Swiss Federal Authority (SFA) (Swiss Federal Authority, 1939) samplers.

Sampling efficiencies of 62% (Einstein, 1937) and 60% (Novak, 1957) have been obtained for the Ehrenberger sampler. Tests which were conducted on the Nesper (Einstein, 1937) and the SFA (Swiss Federal Authority, 1939) samplers provided no clear results, since efficiencies were observed to vary in response to changes which occurred in the amount and particle size distribution of the bedload, the bottom velocity and the volume of sediment that was permitted to accumulate in the sampler. However, Einstein (1937) considered that, on average, a sampling efficiency of 45% was representative of the performance of basket-type samplers in general.

Basket samplers have a large capacity and this factor, when taken in combination with their ability to retain a wide range of particle sizes, makes them a particularly attractive device for deployment in gravel-bed rivers. Thus, in spite of their comparatively poor sampling efficiency, they have continued to be utilized (e.g. Hollingshead, 1971; Nanson, 1974). The widespread use of basket samplers in Canadian gravel-bed rivers prompted the Water Survey of Canada (WSC) to undertake an exhaustive series of tests which were designed to evaluate the performance of the WSC Basket Sampler (Gibbs and Neill, 1972; Engel and Lau, 1980, 1981; Engel, 1982). The results of these tests, which reinforced Einstein's (1937) and Novak's (1957) conclusion that basket-type samplers do not have a constant sampling efficiency, suggested that for all practical purposes the sampling efficiency of the WSC Basket Sampler was only 30%.

13.2.2.2   *Pressure-difference samplers*

The introduction of a basket sampler into the flow necessarily imparts increased resistance to the flow and a consequent reduction in its velocity. This, in turn, reduces the velocity at the sampler entrance and causes material to be deposited before it enters the sampler. A pressure-difference sampler is designed to overcome the problem of decreased velocity at the sampler entrance by creating a pressure drop at the exist of the device, which is just sufficient to balance the energy loss and thereby induce an entrance velocity which is the same as that in an undisturbed stream. This is usually achieved by constructing the sampler walls so that they diverge towards the sampler exit.

Some of the earliest pressure-difference samplers were developed in the USSR during the 1920s (Shamov, 1935). Later variations on this theme include the Arnhem (Schaank, 1937), Károlyi (Károlyi, 1947) and VUV (Novak, 1957) samplers. Sampling efficiencies of 70% (Meyer-Peter, 1937), 45% and 70% (Novak, 1957) have been obtained for the Arnhem, Károlyi and VUV samplers, respectively. More recently Gibbs and Neill (1973) have recommended that an average (constant) efficiency of 58% is appropriate for the VUV sampler. However, Engel (1983) found that its sampling efficiency depended on the prevailing hydraulic conditions, and varied between 60% and 30%. Although these efficiencies remain far from satisfactory, the potential of pressure-difference samplers has continued to be explored. For example, Helley and Smith (1971) have developed a pressure-difference sampler which is a modified version of the Arnhem sampler. The Helley–Smith sampler represents a considerable advance in sampler technology and, although lacking formal certification, several versions of the sampler are currently in world-wide use in a variety of rivers which range in size from less than 4 m to more than 6000 m wide.

Laboratory tests have established that the hydraulic efficiency (defined as the ratio of the mean velocity of the flow through the sampler entrance to the mean velocity of flow through the area occupied by the sampler in its absence) of the standard (76.2 mm) Helley–Smith sampler is approximately 1.54 (Druffel *et al.*, 1976), provided that the sample bag is not filled to more than 40% of its capacity. Sediment or organic matter which is close to the mesh size has a tendency to clog the mesh and cause an unpredictable decrease in the hydraulic efficiency (Druffel *et al.*, 1976; Johnson *et al.*, 1977; Emmett, 1980; Beschta, 1981a). A field calibration of the standard Helley–Smith sampler was undertaken by Emmett (1980). This study indicated that, for sediment in the 0.25–0.5 mm size range, the sampler is over-efficient, as a result of the addition of material moving in suspension to the bedload retained by the sampler. For particles in the 0.5–16 mm size range the efficiency of the standard Helley–Smith sampler ranges from 90% to 110%. Insufficient data were available to assess accurately the performance of the standard sampler in relation to material which was coarser than 16 mm. Bagnold (1977) suggested that its efficiency may be impaired if considerable numbers of larger particles are present in the bedload. However, a scaled-up (152.4 mm^2) version was apparently able to cope successfully with coarse gravel particles (Bagnold, 1977). Detailed laboratory calibration of the sampler for particles in the 2–75 mm size range is currently being undertaken by the United States Geological Survey and the Federal Inter-Agency Sedimentation Project (Hubbell *et al.*, 1981).

13.2.2 Pit traps

Pit traps are sunk into the streambed with their upstream lip flush with its surface. They present an obstruction to the bedload which falls into and is thereby retained in the trap. Pit traps have a distinct advantage over bedload sampling devices, in that it is not only possible to catch the total amount of bedload that passes through a given section but also, with more sophisticated traps, to undertake continuous programmes of sampling and measurement. Experiments have also shown (e.g. Poreh *et al.*, 1970; Gribben, 1974) that it is possible to construct pit traps so that they possess an efficiency of approximately 100%. Pit traps consequently represent an extremely attractive proposition if accurate long-term measurements of bedload discharge are required. However, their use has in general been limited to relatively small channels because of their impracticability and the not inconsiderable expense that is associated with their construction on larger rivers.

Several approaches to pit trap design have emerged. The basic design is that of a concrete-lined pit or slot which generally spans the entire width of the channel and is normally emptied after each flood event or at the end of a flood season. The bedload discharge for the period in question is then either determined volumetrically by survey (e.g. Hansen, 1973) or it is weighed (e.g. Newson, 1980). Alternatively, the pit or slot may either be designed so as to generate a vortex which ejects the sediment from the channel (e.g. Parshall, 1952; Robinson, 1962; Klingeman and Milhous, 1970; Hayward and Sutherland, 1974), or the sediment may be pumped from the trap (Dobson and Johnson,

1940; Einstein, 1944), and in this manner a virtually continuous series of measurements may be obtained.

Truly continuously recording pit traps have been developed by the United States Geological Survey (Emmett, 1980) and Reid *et al.* (1980). In the former case the material which falls into the trap lands on a laterally moving conveyor belt which transports it to the bank-side, where it is weighed, whereas in the latter case the changing weight of the trapped material is monitored *in situ* by means of a pressure pad.

13.2.3 Tracers

In channels with very coarse gravel beds, monitoring the movement of individual particles may provide an alternative or a useful adjunct to the use of samplers and traps (Nelson and Coakley, 1974). The simplest way of tagging individual clasts is to paint or number them (e.g. Thorne and Lewin, 1979; Leopold and Emmett, 1984). However, this technique is capable of providing only comparatively limited information about the distance which the clasts have travelled and the proportion and size of the particles which have moved on, for example, a flood-by-flood basis. Rates of recovery of the particles are also low. The tagging of particles (e.g. Butler, 1977) or the enhancement of a clast's natural magnetic remanence (Rummery *et al.*, 1979; Arkell *et al.*, 1983) may facilitate recovery. However, the insertion of magnets into natural clasts (e.g. Ergenzinger and Conrady 1982; Hassan *et al.*, 1984), the fabrication of artificially, magnetized clasts (e.g. Reid *et al.*, 1984) and the utilization of the natural magnetic properties of magnetite- and pyrrhotite-bearing clasts (Ergenzinger and Custer, 1983), coupled with the installation of electromagnetic sensors in the stream bed, have recently provided more detailed and reliable data about the movement of coarse gravel during flood events.

13.3 BEDLOAD TRANSPORT DYNAMICS IN GRAVEL-BED CHANNELS

It is clear from the preceding discussion that it is now possible to make accurate measurements of bedload discharge in natural gravel-bed rivers, and there is a growing body of reliable field data which has been obtained through the application of such devices as the Helley–Smith bedload sampler (e.g. Jones and Seitz, 1980), pit traps (e.g. Leopold and Emmett, 1976, and 1977) and electromagnetic sensors (e.g. Reid *et al.*, 1984). In particular, data collection has been orientated towards the clearer elucidation of those factors which influence particle motion and determine the relationship between the flow conditions and the prevailing bedload transport rate.

13.3.1 Controls upon the initiation of particle motion

It has long been recognized that, in spite of the underlying veracity of fundamental relationships such as those proposed by Hjulström (1935) and Shields (1936) which seek to describe the circumstances which lead to the entrainment of sediment particles, a considerable amount of uncertainty remains as to the exact nature of the hydraulic conditions which are required to entrain a given particle size (Novak, 1973; Miller *et al.*, 1977). Thus, Baker and Ritter (1975), Church (1978) and Williams (1983), who have collated much of the available field data on the entrainment of gravel particles, have observed that the threshold condition for a given particle size may vary by an order of magnitude. Several factors contribute to this uncertainty, the principal of which relate to the fact that neither the influence of hydrodynamic lift nor the effect of non-uniform bed materials on particle entrainment are explicitly considered in such relationships. In addition, as Neill (1968) has noted, in nature it is only possible to observe some finite transport rate and not the exact point at which initiation of particle motion occurs.

Both Helley (1969) and Baker (1973) have emphasized the importance of considering the lift forces, whose influence may cause particles to be entrained at lower threshold values than would be predicted by the existing relationships. Further, Brayshaw *et al.* (1983) and Novak and Nalluri (1984) have observed that the susceptibility of grouped particles resting on a co-planar bed to initiation of motion increases with decreasing spacing, as a consequence of the increased lift (and hence decreased stability) which arises from the acceleration of the flow around the flanks of juxtaposed particles. The effect of proximity apparently becomes negligible when adjacent particles are separated by 8 diameters or more (Leopold *et al.*, 1966).

The influence which the presence of non-uniform bed materals exerts on the threshold of particle motion is manifest with respect to variations in the condition of the bed surface which have been observed to occur in natural gravel-bed channels. Church (1978) noted that particles in non-cohesive beds commonly exist in either a normal, an overloose or an underloose state. In the case of a normal boundary, the particles rest in a non-dispersed state and exhibit a random arrangement without evidence of imbrication. An overloose boundary is composed of particles which are arranged in a dispersed or dilated state and an underloose boundary consists of imbricated or closely packed particles. Most of the relationships which seek to describe the threshold of particle motion were derived with reference to normal boundary conditions, and thus they may not necessarily be directly applicable to the bed materal in natural gravel-bed channels, which commonly exists in an overloose or an underloose state (cf. Reid and Frostick, 1984). This is because the entrainment of particles in an overloose state requires less than the indicated tractive force, whereas particles in an underloose state require a greater tractive force to initiate motion, as a consequence of the variation that is experienced in the magnitude of the inter-granular frictional force.

Lane and Carlson (1954) and Laronne and Carson (1976) have commented on the effect which the imbrication and packing of particles were implicitly observed to exert on the threshold of particle motion in gravel-bed channels. Brayshaw (1984, 1985) and Reid and Frostick (1984) have presented direct evidence obtained from field experiments which clearly indicates that the enhanced resistance to motion which is occasioned both by the interlocking of clustered particles and the infiltration of fine particles into the interstices of the gravel framework retards the entrainment process and thereby delays the point at which bedload transport is observed to commence in some gravel-bed channels. Conversely, Ramette and Heuzel (1962), Fenton and Abbott (1977) and Novak and Nalluri (1984), who considered the case of isolated particles which protrude into the flow, observed that the tractive force which was required to entrain these particles was considerably less than that which was indicated by existing relationships.

Brayshaw (1984) suggested that cluster bedforms occupy approximately 10% of the bed surface and, therefore, their presence exerts a relatively minor influence on the movement of bedload in most gravel-bed channels, where the effect of variations in the exposure of the particles which comprise a co-planar bed, occasioned by the juxtaposition of grains of different sizes, is likely to be of more importance. Both Gessler (1971, 1973) and Günter (1971) have demonstrated that higher critical shear stresses are required to entrain sediment mixtures than uniform materials, since the finer particles are protected from the flow by their coarser neighbours which protrude further into the flow. Einstein (1950) attempted to account for the influence that a sediment mixture exerts on particle motion in sand-bed channels through the medium of a hiding factor, which purported to describe the shielding effect of large particles on their smaller counterparts. Andrews (1983) has demonstrated that in natural gravel-bed rivers the effects of the relative protrusion of larger particles into the flow nearly compensates for differences in particle weight. Thus particles between 0.3 and 4.2 times the median diameter of the (subsurface) bed material are capable of being entrained by approximately the same tractive force. This implies that the process of differential entrainment, whereby progressively larger particles are set in motion as the magnitude of the flow increases, is unlikely to be of importance in gravel-bed channels.

13.3.2   Controls upon the bedload transport–flow relationship

Direct observations in natural gravel-bed channels have indicated that once in motion the bedload exhibits considerable spatial and temporal variability, and that a non-linear relationship exists between the prevailing flow conditions and the instantaneous bedload transport rate. This variability arises from a number of sources which, broadly speaking, are related to the factors which determine sediment availability. At the most fundamental level, Hayward and Sutherland (1974), Takei *et al.* (1975), Griffiths (1980) and Arkell *et al.* (1983) have suggested that variations in bedload transport rates may be influenced by the stability of a restricted area of riparian land, and so will ultimately be governed by the complex relationships which control the supply of sediment to the stream channel during individual storm events (cf. Harvey, 1977).

Nanson (1974) and Leopold and Emmett (1976) have observed that bedload transport rates, for given flow conditions, are higher at the beginning of a runoff season than at the end. Several workers (e.g. Skibinski, 1968, and Károlyi, 1957) have also noted a general tendency for bedload transport rates, for given flow conditions, to be higher on the rising limb than on the falling limb of the flood hydrograph. Such variations may reflect the (temporary) exhaustion of the supply of transportable material and will, as such, impose a hysteretic effect on the observed relationship between flow conditions and bedload transport rates. However, since an alluvial channel is capable of regulating its own sediment supply by making adjustments to the bed over which it flows, changes which occur in the bed material which, for example, are initiated through the mechanisms of armouring and scour and fill, are likely to exert an equal if not greater impact on the instantaneous bedload transport rate.

Armouring has been observed to influence bedload transport rates in natural gravel-bed channels in a number of ways. For example, Gomez (1983) has demonstrated that the development of an armoured surface causes a progressive decline in bedload transport rates. Conversely, Milhous and Klingeman (1973), Emmett (1976) and Jackson and Beschta (1982) have shown that the disruption of an armoured surface leads to a sudden increase in the amount of sediment that is available for transport and the consequent enhancement of bedload transport rates. Differences in the resistance of the bed material to movement, which are promoted by the infiltration of fine particles into the interstices of the gravel framework, exert an effect which is analogous to that of armouring. Thus Reid *et al.* (1985) have demonstrated that bedload transport rates may be enhanced once the fine particles have been removed from the bed surface and/or the surficial bed material is disturbed.

The rapid changes in cross-sectional area which are occasioned by scour and fill have also been observed to be responsible for pronounced fluctuations in bedload transport rates (cf. Andrews, 1979; Lisle, 1979). This mechanism has been employed by several workers (e.g. Milligan *et al.*, 1980; Leopold and Emmett, 1984; Emmett *et al.*, 1983) to account for the occurrence of non-uniform sediment concentrations in a downstream direction. The influence of organic debris, as expressed through the presence of log jams, has also been observed to exert a similar effect on bedload transport rates in forest streams (e.g. Mosley, 1981).

The downstream variations in sediment storage that occur between pools and riffles have also been employed to account for the wave-length nature of bedload discharge in many gravel-bed channels (e.g. Griffiths, 1979; Mead *et al.*, 1981). Two types of wave-like translation of gravel have been observed to occur in natural channels (Weir, 1983). The first occurs in response to an isolated event in which a large body of material enters the stream channel and is subsequently moved downstream (e.g. Gilbert, 1917; Mosley, 1978; Lekach and Shick, 1983), while the second relates to the annual transfer of material which, for example, occurs in the East Fork River (Emmett *et al.*, 1983). An analagous phenomenon to the (long-term) wave-like translation of gravel are the (short-term) pulses which have been observed in many channels. This well documented phenomenon (e.g. Emmett, 1975; Reid *et al.*, 1985), which in sand-bed and laboratory gravel-bed channels is associated with the passage of ripples and dunes (e.g. Ehrenberger, 1931; Zanen, 1967; Ikeda, 1983), is as yet poorly explained in relation to the transport of coarse particles which move as sheets or streamers (Hayward and Sutherland, 1974; Leopold and Emmett, 1984). Several workers, both in the laboratory (e.g. Gessler, 1965) and in the field (e.g. Leopold and Emmett, 1976; Reid *et al.* 1985), have noted that bedload sheets and streamers are discontinuous in both the longitudinal and transverse directions, and Beschta (1981b) has suggested that the lateral migration of these features may provide one mechanism whereby pulsing is induced. Alternatively, Reid *et al.* (1985) have suggested that the movement of groups of coarse particles, possibly in the form of a kinematic wave (cf. Langbein and Leopold, 1968), may contribute to the generation of bedload pulses.

13.4 PREDICTION OF BEDLOAD DISCHARGE IN GRAVEL-BED CHANNELS

The theoretical approach to the estimation of bedload discharge was initiated by du Boys (1879), and centres on the derivation of a relationship between the flow conditions and the bedload transport rate. Theoretical approaches fall into four categories, which involve consideration of: a discharge relationship (e.g. Schoklitsch equation; Shulits, 1935); a shear stress relationship (e.g. Meyer-Peter and Müller, 1948); a stream power relationship (e.g. Bagnold, 1980); or a stochastic relationship (e.g. Einstein, 1950), respectively. Most equations, in spite of the restricted range of data which were usually employed in their development, were ostensibly designed for non-specific applications. However, a number of equations have been specifically applied or modified to enable them to be utilized for the purpose of estimating bedload transport rates in gravel-bed channels (Table 13.1).

There is considerable uncertainty associated with the application of bedload transport equations to the prediction of bedload transport rates in natural channels, and this has inspired a daunting number of comparative studies (e.g. Vanoni *et al.*, 1960; Shulits and Hill, 1968; Vanoni, 1975; White *et al.*, 1975).

Broadly speaking, Emmett and Leopold's (1977) conclusion that existing bedload equations are inadequate to predict transport rates for many rivers echoes the findings of all of these studies. The obvious question is, therefore, why in the face of the plethora of bedload transport equations that are currently available is it not possible to predict accurately bedload transport rates in natural streams?

TABLE 13.1　Bedload transport equations for gravel-bed rivers

Bedload transport equation	Applicable particle size range (mm)
Gravel equations:	
Meyer-Peter (Meyer-Peter *et al.*, 1934)	3.0– 28.7
Meyer-Peter and Müller (1948)	0.4– 28.7
Parker (Parker *et al.*, 1982)	18.0– 28.0
Schoklitsch 1934 (Shulits, 1935)	0.3– 5.0
Schoklitsch 1943 (Schoklitsch, 1950)	0.3– 5.0
Yang (1984)	2.5– 7.0
Sand/gravel equations:	
Ackers and White (1973)	>0.04
Day modification (White and Day, 1982)	0.4– 14.0
Sutherland modification (Proffitt and Sutherland, 1983)	1.7– 4.4
Bagnold (1980)	0.3–300.0
du Boys–Straub (Straub, 1935)	0.1– 4.0
Einstein (1950)	0.7– 28.7
Yalin (1963)	0.7– 28.7

Leaving aside the question of the inherent veracity of any given equation, the performance of an equation depends essentially on the extent to which the bedload transport process in natural channels may be effectively characterized by functional models which assume that bedload transport rates are a unique function of hydraulic parameters, the maximum possible amount of bedload is being transported and the flow and sediment properties (and hence the bedload transport rate) for the period in question are invariant and may be described with reference to a steady state. However, it is clear from the preceding discussion that these assumptions are likely to be fulfilled on only a limited number of occasions in natural channels, where variations in sediment availability profoundly affect bedload transport rates. For example, during those periods when a stream may not be competent to transport the entire range of particle sizes that are available on the stream bed, predicted bedload transport rates will inevitably be higher than observed transport rates (e.g. Emmett, 1976; Gomez, 1983). Bedload transport rates, at a constant discharge, also vary in response to sequential variations in sediment availability which occur both on a seasonal basis (e.g. Mead *et al.*, 1981), as well as in association with the passage of bedforms (e.g. Hubbell *et al.* 1981; Ikeda, 1983), or groups of particles (Reid *et al.*, 1984, 1985). Many equations (e.g. Schoklitsch equation; Shulits, 1935; Bagnold, 1980) also involve the identification of a threshold condition, above which bedload transport is initiated and below which no movement occurs. However, studies in gravel-bed streams (e.g. Reid and Frostick, 1984; Reid *et al.*, 1985) have clearly demonstrated that the threshold at which motion is initiated is both variable and is often greater than that at which motion is observed to cease. This occurs because the difference between static and dynamic friction means that more force is required to initiate than to maintain motion (cf. Hjulström, 1935), and in response to variations in magnitude of the intergranular frictional force which arise from differences in the packing of bed materials (e.g. Church, 1978; Reid and Frostick, 1984).

Bennett and Nordin (1977), Jackson and Beschta (1982) and Borah *et al.* (1982a, b) have attempted to model the routing of bed material more effectively by providing for the changes in sediment storage which occur in response to the disruption of an armoured surface and/or localized episodes of scour and fill. Conversely, Griffiths (1979), Weir (1983) and Pickup *et al.* (1983) have attempted to rationalize the wave-like translation of bedload in terms of kinematic wave models. However, since all these models require access to detailed field measurements for their development and calibration, it is likely that their application will be limited, at least in the immediate future, to the characterization of the bedload transport process in those channels for which such data are already available, rather than for the estimation of bedload discharge in ungauged streams. In this respect, the utilization of regression-based models (e.g. Yang, 1973, 1984) might prove to be a less time-consuming and equally effective method of characterizing non-equilibrium bedload transport rates in natural gravel-bed channels.

Bedload transport rates in natural alluvial channels clearly cannot always be regarded as a direct function of the prevailing flow conditions. Thus, in circumstances where non-equilibrium transport conditions are likely to prevail and the assumptions associated with the application of bedload transport equations are unlikely to be fulfilled, bedload transport equations should be applied with care. In the presence of equilibrium transport conditions bedload transport equations should, in principle, be capable of being applied with reservation. However, the very fact that the development of most of the existing equations was undertaken with reference to limited data bases suggests that their partiality cannot be unequivocally guaranteed, and that careful attention should be paid to the range of conditions for which they were derived before attempting to apply them to natural channels.

13.5 PERSPECTIVE ON BEDLOAD TRANSPORT IN MOUNTAIN RIVERS

Mountain rivers are characterized by steep gradients (usually of the order of a few percent), bed materials which are composed of an extremely wide range of particle sizes and are usually dominated by the coarser fractions (typically 0.5 mm–0.5 m) and roughness elements which are of the same order of magnitude as the flow depth. Generally, because of the large and often extremely variable particle sizes and sediment yields that are encountered in these rivers, the most comprehensive data sets that are available for mountain rivers (in both upland and glacierized basins) have been obtained with the aid of pit traps (e.g. Milhous and Klingeman, 1973; Hayward, 1979; Laufer and Sommer, 1982).

Bedload discharge is commonly reported to constitute less than 25% of the total sediment load in lowland streams (e.g. Johnson and Smith, 1977; Jones and Seitz, 1980). However, in upland rivers the situation appears to be far more variable. Several workers (e.g. McPherson, 1971; Hollingshead, 1971; Nanson, 1974; Carling, 1983a) have reported values which are in agreement with those obtained from lowland rivers. However, other investigators (e.g. Milhous and Klingeman, 1973; Hayward, 1979) have noted that, on occasion, bedload has been observed to constitute in excess of 75% of the total sediment discharge. Data which have been obtained from glacierized basins (e.g. Church, 1972; Østrem, 1975; Laufer and Sommer, 1982) suggest that the bedload discharge is equally variable in glacier-fed rivers. This variability appears to arise from the fact that in mountain rivers differences in the proportion of the total sediment load that is attributable to bedload discharge, between events of a similar magnitude and/or between low-frequency, high-magnitude and high-frequency, low-magnitude events, are much more closely linked to variations in the rate at which sediment is supplied to the stream channel than they are in lowland rivers (Hayward, 1979).

The pronounced differences in the flow structure between low- and high-gradient channels, which result from the presence of roughness elements that protrude significantly into the flow, have been shown to exert a profound influence on the entrainment of sediment particles in steep channels (Bathurst *et al.*, 1982). For example, Ashida and Bayazit (1973) and Bayazit (1978, 1982) have demonstrated that the susceptibility of particles to motion in steep channels increases significantly with increasing relative particle roughness (cf. Fenton and Abbott, 1977). Carling (1983b) has also shown that the threshold shear stress required to initiate motion of a given particle size in narrow upland streams is considerably higher than that which is required to initiate motion in relatively wide (but otherwise hydraulically similar) channels, because of the differences in compound boundary roughness which are experienced between broad and narrow channels.

Hayward (1979) has demonstrated that pool and riffle sequences in upland rivers are also responsible for the wave-like translation of gravel, which occurs in a similar manner to that observed in lowland rivers, and that the magnitude of the wave-form is dependent on the degree to which the pools are filled with sediment. In many upland rivers the high gradients accentuate the pool and riffle sequences into a step-pool morphology, so that the water cascades between relatively deep pools (Kellerhals, 1977). Ashida *et al.* (1981) and Whittaker and Davies (1982) have shown that the wave-like translation of bed material which results from the presence of step-pool sequences in steep channels exhibits similar characteristics to that which is experienced in lower gradient rivers. Solov'ev (1969) has also commented on the presence of bedload pulses in mountain rivers, but offers no clear explanation for their existence.

The inherent unsteadiness which is characteristic of the bedload transport process in lowland stream channels appears to be accentuated by factors such as the presence of step-pool sequences in steep channels and the dependence of bedload yields on the supply of sediment to the stream channel. These and other factors, such as the enhanced form resistance which is characteristic of these channels, pose additional problems when attempts are made to calculate sediment transport rates in mountain rivers. Consequently, existing bedload transport equations appear to be even more unreliable when they are applied to mountain rivers (Haddock, 1978; Smart, 1984). This led Griffiths (1980) to suggest that supply-based models might provide a more effective means of predicting bedload yields in mountain rivers.

Differences in energy distribution and sediment supply cause mountain streams to behave differently from their lowland counterparts. A knowledge of bedload transport dynamics in mountain streams is, therefore, crucial to the understanding of the bedload transport as a whole. Of particular interest, for example, is the extent to which the transport of bedload is dependent on sediment supply in mountain rivers. However, in spite of the increased interest that has been generated in the bedload transport process in recent years, as a result of the advances which have been made in the design of sampling and measuring devices, comparatively little attention has been paid to mountain rivers. There is, therefore, considerable potential for further research in this field.

REFERENCES

ABBOTT, J. E., and FRANCIS, J. R. D. (1977), Saltation and suspension trajectories of solid grains in a water stream, *Philosophical Transactions of the Royal Society of London*, **284A**, 225–254.

ACKERS, P., and WHITE, W. R. (1973), Sediment transport: new approach and analysis. *American Society of Civil Engineers, Journal of the Hydraulics Division*, **99**, 2041–2060.

ANDREWS, E. D. (1979), Scour and fill in a stream channel, East Fork River, Western Wyoming, *United States Geological Survey, Professional Paper*, No. 1117.

ANDREWS, E. D. (1983), Entrainment of gravel from naturally sorted riverbed material, *Geological Society of America Bulletin*, **94**, 1225–1231.

ARKELL, B., LEEKS, G., NEWSON, M., and OLDFIELD, F. (1983), Trapping and tracing: some recent observations of supply and transport of coarse sediment from upland Wales, *International Association of Sedimentologists, Special Publication*, No. 6, 107–119.

ASHIDA, K., and BAYAZIT, M. (1973), Initiation of motion and roughness of flows in steep channels, *Proceedings of 15th Congress of International Association of Hydraulics Research*, 475–484.

ASHIDA, K., TAKAHASHI, T., and SAWADLA, T. (1981), Processes of sediment transport in mountain stream channels, in *Erosion and Sediment Transport in Pacific Rim Steeplands (Proceedings of the Christchurch Symposium, 25–31 January 1981)*, International Association of Hydrological Sciences Publication 132, pp. 166–178.

BAGNOLD, R. A. (1977), Bedload transport by natural rivers, *Water Resources Research*, **13**, 303–312.

BAGNOLD, R. A. (1980), An empirical correlation of bedload transport rates in flumes and natural rivers, *Proceedings of the Royal Society of London*, **372A**, 453–473.

BAKER, V. R. (1973), Palaeohydrology and sedimentology of lake Missoula flooding in eastern Washington, *Geological Society of America Special Paper*, No. 144.

BAKER, V. R., and RITTER, D. F. (1975), Competence of rivers to transport coarse bedload materials., *Geological Society of America Bulletin*, **86**, 975–978.

BATHURST, J. C., GRAF, W. H., and CAO, H. H. (1982), Initiation of sediment transport in steep channels with coarse bed material, in SUMER, B. M., and MULLER, A. (Eds.), *Mechanics of Sediment Transport*, Balkema, pp. 207–213.

BAYAZIT, M. (1978), Scour of bed material in very rough channels, *American Society of Civil Engineers, Journal of the Hydraulics Division*, **104**, 1345–1349.

BAYAZIT, M. (1982), Flow structure and sediment transport mechanics in steep channels, in SUMER, B. M., and MULLER, A. (Eds.), *Mechanics of Sediment Transport*, Balkema, pp. 197–206.

BENNETT, J. P., and NORDIN, C. F. (1977), Simulation of sediment transport and armouring, *Hydrological Sciences Bulletin*, **22**, 555–569.

BESCHTA, R. L. (1981a), Increased bag size improves Helley–Smith bedload sampler for use in streams with high sand and organic matter transport, in *Erosion and Sediment Transport Measurement (Proceedings of the Florence Symposium, 22–26 June 1981)*, International Association of Hydrological Sciences Publication 133, pp. 17–25.

BESCHTA, R. L. (1981b), Patterns of sediment and organic matter transport in Oregon Coast Range streams, in *Erosion and Sediment Transport in Pacific Rim Steeplands (Proceedings of the Christchurch Symposium, 25–31 January 1981)*, International Association of Hydrological Sciences 132, pp. 179–188.

BORAH, D. K., ALONSO, C. V., and PRASAD, S. N. (1982a), Routing graded sediments in streams: formulations, *American Society of Civil Engineers, Journal of Hydraulics Engineering*, **108**, 1486–1503.

BORAH, D. K., ALONSO, C. V., and PRASAD, S. N. (1982b), Routing graded sediments in streams: applications, *American Society of Civil Engineers, Journal of Hydraulics Engineering*, **108**, 1504–1517.

BRAYSHAW, A. C. (1984), Characteristics and origin of clustered bedforms in coarse-grained alluvial channels, *Canadian Society of Petroleum Geologists Memoir*, No. 10, 77–85.

BRAYSHAW, A. C. (1985), Bed microtopography and entrainment thresholds in gravel-bed rivers, *Geological Society of America Bulletin*, **96**, 218–223.

BRAYSHAW, A. C., FROSTICK, L. E., and REID, I. (1983), The hydrodynamics of particle clusters and sediment entrainment in coarse alluvial channels, *Sedimentology*, **30**, 137–143.

BUTLER, P. R. (1977), Movement of cobbles in a gravel-bed stream during a flood season, *Geological Society of America Bulletin*, **88**, 1072–1074.

CARLING, P. A. (1983a), Particulate dynamics, dissolved and tota load in two small basins, northern Pennines, U.K., *Hydrological Sciences Journal*, **28**, 355–375.

CARLING, P. A. (1983b), Threshold of coarse sediment transport in broad and narrow natural streams, *Earth Surface Processes and Landforms*, **8**, 1–18.

CHURCH, M. (1972), Baffin Island sandurs: a study of Arctic fluvial processes, *Geological Survey of Canada Bulletin*, 216.

CHURCH, M. (1978), Palaeohydrological reconstructions from a Holocene valley fill, *Canadian Society of Petroleum Geologists Memoir*, No. 5, 743–772.

DAVIS, A. P., and WILSON, H. M. (1919), *Irrigation Engineering*, Wiley.

DOBSON, G. C., and JOHNSON, J. W. (1940), Studying sediment loads in natural streams, *Civil Engineering*, **10**, 93–96.

DRUFFEL, L., EMMETT, W. W., SCHNEIDER, V. R., and SKINNER, J. V. (1976), Laboratory hydraulic calibration of the Helley-Smith bedload sediment sampler, *United States Geological Survey Open-File Report*, No. 76-752.

DU BOYS, M. P. (1879), Études du regime et l'action exercee par les eaux sur un lit a fond de graviers indefininent affouiable, *Annales de Ponts et Chausses*, Ser 5, **18**, 141–195.

EHRENBERGER, R. (1931), Direkte geschiebemessungen an der Donau bei Wien und deren bisherige Ergebnisse, *Die Wasserwirtschaft*, **34**, 1–9.

EHRENBERGER, R. (1932), Geschiebemessungen an flussen mittels Auffanggeraten und Modellversuche mit Letzteren, *Die Wasserwirtschaft*, **36**, 1–8.

EINSTEIN, H. A. (1937), Die Eichung des im Rhein verwendeten Geschiebefungers, *Schwizerische Bauzeitung*, **110**, 29–32.

EINSTEIN, H. A. (1944), Bedload transportation in Mountain Creek, *United States Department of Agriculture, Soil Conservation Service Technical Paper*, No. 55.

EINSTEIN, H. A. (1950), The bedload function for sediment transportation in open channels, *United States Department of Agriculture, Soil Conservation Service Technical Bulletin*, No. 1026.

EMMETT, W. W. (1975), The channels and waters of the Upper Salmon River area, Idaho, *United States Geological Survey Professional Paper*, No. 870A.

EMMETT, W. W. (1976), Bedload transport in two large gravel-bed rivers, Idaho and Washington, *Proceedings of 3rd Federal Inter-Agency Sedimentation Conference*, 4-101–4-113.

EMMETT, W. W. (1980), A field calibration of the sediment-trapping characteristics of the Helley–Smith bedload sampler, *United States Geological Survey Professional Paper*, No. 1139.

EMMETT, W. W., and LEOPOLD, L. B. (1977), A comparison of observed sediment transport rates with rates computed using existing formulae, in DOEHRING, D. D. (Ed.), *Geomorphology in Arid Regions*, Allen and Unwin, pp. 187–188.

EMMETT, W. W., LEOPOLD, L. B., and MYRICK, R. M. (1983), Some characteristics of fluvial processes in rivers, *Proceedings of 2nd International Symposium on River Sedimentation, Nanjing*, 730–754.

ENGEL, P. (1982), Characteristics of the WSC basket type bedload sampler, *National Water Research Institute, Canada, Report*, No. H81-3345.

ENGEL, P. (1983), Sampling efficiency of the VUV bedload sampler, *National Water Research Institute, Canada, Report*, No. H82-377.

ENGEL, P., and LAU, L. (1980), Calibration of bedload samplers. *American Society of Civil Engineers, Journal of the Hydraulics Division*, **106**, 1677–1685.

ENGEL, P., and LAU, L. (1981), The efficiency of basket type bedload samplers, in *Erosion and Sediment Transport Measurement (Proceedings of the Florence Symposium, 22–26 June 1981)*, International Association of Hydrological Sciences Publication 133, pp. 27–34.

ERGENZINGER, P. J., and CONRADY, J. (1982), A new tracer technique for measuring bedload in natural channels, *Catena*, **9**, 77–80.

ERGENZINGER, P. J., and CUSTER, S. G. (1983), Determination of bedload transport using natural magnetic tracers: first experience at Squaw Creek, Gallatin County, Montana, *Water Resources Research*, **19**, 187–193.

FEDERAL INTER-AGENCY RIVER BASIN COMMITTEE (1940), Equipment used for sampling bedload and bed material, *Federal Inter-Agency River Basin Committee, Report*, No. 2.

FENTON, J. D., and ABBOTT, J. E. (1977), Initial movement of grains on a stream bed: the effect of relative protrusion, *Proceedings of the Royal Society of London*, **352A**, 523–537.

GESSLER, J. (1965), Der Geschiebetriebbeginn bei mischungen Untersucht an naturlichen Abpflasterungserscheinungen in Kanalen, *Mitteilungen VA WE, ETH, Zurich*, No. 69.

GESSLER, J. (1971), Critical shear stress for sediment mixtures, *Proceedings of 14th Congress of the International Association of Hydraulics Research*, C1-1–C-8.

GESSLER, J. (1973), Behaviour of sediment mixtures in rivers. *Symposium on River Mechanics, International Association of Hydraulics Research*, A35-1–A35-11.

GIBBS, C. J., and NEILL, C. R. (1972), Interim report on laboratory study of basket-type bedload samplers, *Research Council of Alberta, Report*, No. REH/72/2.

GIBBS, C. J., and NEILL, C. R. (1973), Laboratory testing of model VUV bedload samplers, *Research Council of Alberta, Report*, No. REH/73/2.

GILBERT, G. K. (1917), Hydraulic-mining debris in the Sierra Nevada, *United States Geological Survey Professional Paper*, No. 105.

GOMEZ, B. (1983), Temporal variations in bedload transport rates: the effect of progressive bed armouring, *Earth Surface Processes and Landforms*, **8**, 41–54.

GRIBBEN, K. (1974), Bedload transport: a literature survey and some laboratory work on the development of pit traps, Unpublished MSc Dissertation, University of Newcastle-upon-Tyne, Department of Civil Engineering.

GRIFFITHS, G. A. (1979), Recent sedimentary history of the Waimakiriri River, New Zealand, *Journal of Hydrology, N.Z.*, **18**, 6–28.

GRIFFITHS, G. A. (1980), Stochastic estimation of bedload yield in pool and riffle mountain streams, *Water Resources Research*, **16**, 931–937.

GÜNTER, A. (1971), Die kritische mittlere Schlenschubspannung bei Geschiebemisschungen, unter Berucksichtigung der Desckschlichtbildung und der turbulentbedingten Sohlenschubspannungsschwankungen. *Mitteilungen der Versuchsanstalt für Wasserbau, Hydrologie und Glaziologie, ETH, Zurich*, No. 3.

HADDOCK, D. R. (1978), Modelling bedload transport in mountain streams of the Colorado Front Range, Unpublished MSc Dissertion, Colorado State University, Earth Resources Department.

HANSEN, E. A. (1973), In-channel sedimentation basins, a possible tool in trout habitat management, *Progressive Fish Culturist*, **35**, 138–142.

HARVEY, A. M. (1977), Event frequency in sediment production and channel change, in GREGORY, K. J. (Ed.), *River Channel Changes*, Wiley, pp. 301–315.

HASSAN, M. A., SCHICK, A. P., and LARONNE, J. B. (1984), The recovery of flood-dispersed coarse sediment particles, *Catena Supplement*, **5**, 153–162.

HAYWARD, J. A. (1979), Mountain stream sediments, in MURRAY, D. L., and ACKROYD, P. (Eds.) *Physical Hydrology*, New Zealand Hydrological Society, pp. 193–212.

HAYWARD, J. A., and SUTHERLAND, A. J. (1974), The Torless Stream vortex-tube sediment trap, *Journal of Hydrology N.Z.*, **13**, 41–53.

HELLEY, E. J. (1969), Field measurements of the initiation of large bed particle motion in Blue Creek, near Klamath, California, *United States Geological Survey Professional Paper*, No. 562G.

HELLEY, E. J., and SMITH, W. (1971), Development and calibration of a pressure-difference bedload sampler, *United States Geological Survey Open-File Report*.

HJULSTRÖM, F. (1935), Studies of the morphological activity of rivers as illustrated by the Rover Fyris, *Bulletin of the Geological Institute, University of Uppsala*, **25**, 221–527.

HOLLINGSHEAD, A. B. (1971), Sediment transport measurements in gravel rivers, *American Society of Civil Engineers, Journal of the Hydraulics Division*, **97**, 1817–1834.

HUBBELL, D. W. (1964), Apparatus and techniques for measuring bedload, *United States Geological Survey Water Supply Paper*, No. 1748.

HUBBELL, D. W., STEVENS, H. H., SKINNER, J. V., and BEVERAGE, J. P. (1981), Recent refinements in calibrating bedload samplers, *Water Forum '81*, **1**, 128–140.

IKEDA, H. (1983), Experiments on bedload transport, bedforms and sedimentary structures using fine gravel in the 4-metre wide flume, *Environmental Research Centre, University of Tsukuba, Paper*, No. 2.

JACKSON, W. L., and BESCHTA, R. L. (1982), A model of two-phase bedload transport in an Oregon Coast Range stream, *Earth Surface Proceses and Landforms*, **7**, 517–527.

JOHNSON, C. W., and SMITH, J. P. (1977), Sediment characteristics and transport from northwest rangeland watershed, *American Society of Agricultural Engineers Paper*, No. 77-2509.

JOHNSON, C. W., ENGLEMAN, R. L., SMITH, J. P., and HANSON, C. L. (1977), Helley -Smith bedload samplers, *American Society of Civil Engineers, Journal of the Hydraulics Division*, **103**, 1217–1221.

JOHNSON, P., and MUIR, T. C. (1969), Acoustic detection of sediment movement, *Journal of Hydraulics Research*, **7**, 519–540.

JONES, M. L., and SEITZ, H. R. (1980), Sediment transport in the Snake and Clearwater Rivers in the vicinity of Lewiston, Idaho, *United States Geological Survey Open-File Report*, No. 80-690.

JONYS, C. K. (1976), Acoustic measurement of sediment transport, *Inland Waterways Directorate, Canada, Scientific Series Report*, No. 66.

KÁROLYI, Z. (1947), Kisérletek a hordalékfogóval, *Különnyomat a Vizügyi Közlemények, Budapest*, 1–12.

KÁROLYI, Z. (1957), A study into inconsistencies in bedload transport on the basis of measurement in Hungary, *International Association of Scientific Hydrology, Proceedings General Assembly, Toronto*, **1**, 286–299.

KELLARHALS, R. (1977), Hydraulic performance of steep natural channels, in SLAYMAKER, H. O., and McPHERSON, H. J. (Eds.), *Mountain Geomorphology*, Tantalus Research, pp. 131–139.

KLINGEMAN, P. C., and MILHOUS, R. T. (1970), Oak Creek vortex bedload sampler, unpublished paper presented at the 17th Annual Pacific North-West Meeting of the American Geophysical Union, Tacoma.

LANE, E. W., and CARLSON, E. J. (1954), Some observations on the effect of particle shape on the movement of coarse sediment, *Transactions of the American Geophysical Union*, **35**, 453–462.

LANGBEIN, W. B., and LEOPOLD, L. B. (1968), River channel bars and dunes—theory of kinematic waves, *United States Geological Survey Professional Paper*, No. 422L.

LARONNE, J. B., and CARSON, M. A. (1976), Interrelationships between bed morphology and bed material transport for a small gravel-bed channel, *Sedimentology*, **23**, 67–85.

LAUFER, H., and SOMMER, N. (1982), Studies on sediment transport in mountain streams of the eastern Alps, in *Commision Internationale des Grandes Barrages, Proceedings of 14th Congress on Large Dams*, pp. 431–453.

LEKACH, J., and SCHICK, A. P. (1983), Evidence for transport of bedload in waves: analysis of fluvial sediment samples in a small upland stream channel, *Catena*, **10**, 267-279.

LEOPOLD, L. B., and EMMETT, W. W. (1976), Bedload measurements, East Fork River, Wyoming, *Proceedings of the National Academy of Sciences, USA*, **73**, 1000-1004.

LEOPOLD, L. B., and EMMETT, W. W. (1977), 1976 bedload measurements, East Fork River, Wyoming, *Proceedings of the National Academy of Sciences, USA*, **74**, 2644-2648.

LEOPOLD, L. B., and EMMETT, W. W. (1981), Some observations on the movements of cobbles on a streambed, in *Erosion and Sediment Transport Measurement (Proceedings of the Florence Symposium, 22-26 June 1981)*, International Association of Hydrological Sciences Publication 133, pp. 49-59.

LEOPOLD, L. B., and EMMETT, W. W. (1984), Bedload movement and its relation to scour in *River Meandering*, American Society of Civil Engineers, pp. 640-649.

LEOPOLD, L. B., EMMETT, W. W., and MYRICK, R. M. (1966), Channels and hillslope processes in a semi-arid area, New Mexico, *United States Geological Survey Professional Paper*, No. 352G.

LISLE, T. E. (1979), A sorting mechanism for a riffle-pool sequence. *Geological Society of America Bulletin*, **90**, 1152-1157.

MCPHERSON, H. J. (1971), Dissolved suspended and bedload movement patterns in Two O'Clock Creek, Rocky Mountains, Canada, Summer 1969, *Journal of Hydrology*, **12**, 84-96.

MEAD, R. H., EMMETT, W. W., and MYRICK, R. M. (1981), Movement and storage of bed material during 1979 in East Fork River, in *Erosion and Sediment Transport in Pacific Rim Steeplands (Proceedings of the Christchurch Symposium, 25-31 January 1981)*, International Association of Hydrological Sciences Publication 132, pp. 225-235.

MEYER-PETER, E., and MÜLLER, R. (1948), Formulas for bedload transport, *Proceedings of 2nd Meeting of the International Association of Hydraulics Research, Stockholm*, **3**, 39-64.

MEYER-PETER, R., FAVRE, H., and EINSTEIN, H. I. (1934), Neuere Versuchsresultate uber den Geschiebetriel, *Schweizerische Bauzeitung*, **103**, 147-150.

MILHOUS, R. T., and KLINGEMAN, P. C. (1973), Sediment transport system in a gravel-bottomed stream, *Proceedings of 21st Annual Speciality Conference of the Hydraulics Division, American Society of Civil Engineers*, 293-303.

MILLER, M. C., MCCAVE, I. N., and KOMAR, P. D. (1977), Threshold of sediment motion under unidirectional current, *Sedimentology*, **24**, 507-527.

MILLIGAN, J., CHACHO, E., and MOLNAU, M. (1980), Sediment transport in a pool-riffle stream, *Symposium on Watershed Management, American Society of Civil Engineers*, 430-441.

MOSLEY, M. P. (1978), Bed material transport in the Tamaki River, near Dannevirke, North Island, New Zealand, *N.Z. Journal of Science*, **21**, 619-626.

MOSELY, M. P. (1981), The influence of organic debris on channel morphology and bedload transport in a New Zealand forest stream, *Earth Surface Processes and Landforms*, **6**, 571-579.

MULHOFER, L. (1933), Untersucheingen über die Schwebstaff und Geschiebeführung des Inn nächst Kirchbichl (Tirol), *Die Wasserwirtschaft*, **1-6**, 1-23.

NANSON, G. C. (1974), Bedload and suspended load transport in a small steep mountain stream, *American Journal of Science*, **274**, 471-486.

NEILL, C. R. (1968), A re-examination of the beginning of movement for coarse granular bed materials, *Hydraulic Research Station Report*, No. INT68.

NELSON, D. E., and COAKLEY, J. P. (1974), Techniques for tracing sediment movement. *Inland Waterways Directorate, Canada, Scientific Series Report*, No. 32.

NESPER. F. (1937), Ergebnisse der Messungen uber die Geschiebe- und Schlammfuhrung des Rheins an der Brugger Rheinbrucke, *Schweizerische Bauzeitung*, **110**, 143–148 and 161–164.

NEWSON, M. (1980), The geomorphological effectiveness of floods—a contribution stimulated by two recent events in mid-Wales, *Earth Surface Processes* **5**, 1–16.

NOVAK, I. D. (1973), Predicting coarse sediment transport: the Hjulström curve revisited, in MORISAWA, M. (Ed.), *Fluvial Geomorphology*, Allen and Unwin, pp. 13–25.

NOVAK, P. (1957), Bedload meters—development of a new type and determination of their efficiency with the aid of scale models, *International Association of Hydraulics Research, Proceedings 7th Meeting, Lisbon*, **1**, A9-1–A9-11.

NOVAK, P., and NALLURI, C. (1984), Incipient motion of sediment particles over fixed beds, *Journal of Hydraulics Research*, **22**, 181–197.

ØSTREM, G. (1975), Sediment transport in glacial meltwater streams, in JOPLING, A. V., and MACDONALD, B. C. (Eds.), *Glaciofluvial and Glaciolacustrine Sedimentation*, Society of Economic Paleontologists and Mineralogists Special Publication No. 23, pp. 101–122.

PARKER, G., KLINGEMAN, P. C., and MCLEAN, D. G. (1982), Bedload and size distribution in paved gravel-bed streams, *American Society of Civil Engineers, Journal of the Hydraulics Division*, **108**, 544–571.

PARSHALL, R. L. (1952), Model and prototype studies of sand traps, *Transactions of the American Society of Civil Engineers*, **117**, 204–212.

PICKUP, G., HIGGINS, R. J., and GRANT, I. (1983), Modelling sediment transport as a moving wave—the transfer and deposition of mining waste, *Journal of Hydrology*, **60**, 281–301.

POREH, M., SAGIV, A., and SEGINER, I. (1970), Sampling efficiency of slots, *American Society of Civil Engineers, Journal of the Hydraulics Division*, **96**, 2065–2078.

PROFFITT, G. T., and SUTHERLAND, A. J. (1983), Transport of non-uniform sediments, *Journal of Hydraulics Research*, **21**, 33–43.

RAMETTE, M., and HEUZEL, M. (1962), Le Rhone à Lyon: étude de l'entraînement des galets à l'aide de traceurs radioactifs, *La Houille Blanche*, **17A**, 389–399.

REID, I. R., and FROSTICK, L. E. (1984), Particle interaction and its effect on the threshold of initial and final bedload motion on coarse alluvial channels, *Canadian Society of Petroleum Geologists, Memoir*, No. 10, 61–68.

REID, I., LAYMAN, J. T., and FROSTICK, L. E. (1980), The continuous measurement of bedload discharge, *Journal of Hydraulics Research*, **18**, 243–249.

REID, I., BRAYSHAW, A. C., and FROSTICK, L. E. (1984), An electromagnetic device for automatic detection of bedload motion and its field applications, *Sedimentology*, **31**, 269–276.

REID, I., FROSTICK, L. E., and LAYMAN, J. T. (1985), The incidence and nature of bedload transport during flood flows in coarse-grained alluvial channels, *Earth Surface Processes and Landforms*, **10**, 33–44.

RICHARDS, K. S., and MILNE, L. M. (1979), Problems in the calibration of an acoustic device for the observation of bedload transport, *Earth Surface Processes*, **4**, 335–346.

ROBINSON, A. R. (1962), Vortex tube sand trap, *Transactions of the American Society of Civil Engineers*, **127**, 391–425.

RUMMERY, T. A., OLDFIELD, F., THOMPSON, R., and NEWSON, M. (1979), Magnetic tracing of stream bedload, *Geophysical Journal of the Royal Astronomical Society*, **57**, 278–279.

SCHAANK, E. M. H. (1937), Diskussion: 'Eine Vorrichtung zur Messung, des Geschiebes an der Flusschles,' *Bericht über Die Erste Tagung Internationaler Verband für Wasserbauliches Versuchswese, Berlin*, 113–116.

SCHOKLITSCH, A. (1950), *Handbuch des Wasserbaues*, Springer.

SHAMOV, G. I. (1935), Kratkoye soobschchgeniye o rezultatakh laoratornikh isslyedovahii batomyetrov, *Scientific Research Institute Hydrotechnics Transactions*, **16**, 218–220.

SHIELDS, A. (1936), Awendung des Achnlichkeitsmechanik und der turbulenzforschung auf die Geschiebebewegung, *Mitteilungen der Preussischen Versuchsanstalt für Wasserbau und Schiffbau*, No. 87.

SHULITS, S. (1935), The Schoklitsch bedload formula, *Engineering*, **139**, 644–646.

SHULITS, S., and HILL, R. D. (1968), Bedload formulas, *United States Department of Agriculture Report*, No. ARS-SCW-1.

SKIBINISKI, J. (1968), Bedload transport at flood time, *International Association of Scientific Hydrology Publication*, No. 75, 41–47.

SMART, G. M. (1984), Sediment transport formula for steep channels *American Society of Civil Engineers, Journal of Hydraulic Engineering*, **110**, 267–276.

SOLOV'EV, N. YA. (1969), Pulsation in the movement of bedload in mountain streams, *Transactions of the State Hydrological Institute (TRUDY-GGI)*, **175**, 119–123.

STRAUB, L. G. (1935), Missouri River Report, Appendix 15, *United States Serial No. 9829, House Document No. 238, 73rd Congress, 2nd Session*.

SWISS FEDERAL AUTHORITY (1939), Untersuchungen in der Natur uber Bettbildung Geschiebe- und Schwebestffuhrung, *Mitteilungen Nr. 33, des Amtes für Wasserwirtschaft*.

TAKEI, A., KOBASHI, S., and FUKUSHIMA, Y. (1975), The analysis of runoff from two small catchments in granitic mountains, *International Association of Scientific Hydrology Publication*, No. 117, 29–34.

THORNE, C. R., and LEWIN, J. (1979), Bank processes, bed material movement and planform developments in a meandering river, in RHODES, D. D. , and WILLIAMS, G. P. (Eds.), *Adjustments of the Fluvial System*, Allen and Unwin, pp. 117–137.

TYWONICK, N., and WARNOCK, R. G. (1973), Acoustic detection of bedload transport, *9th National Research Council, Canada, Hydrology Symposium*, 728–743.

VANONI, V. A. (1975), *Sedimentation Engineering*, American Society of Civil Engineers — Manuals and Reports on Engineering Practice, No. 54.

VANONI, V. A., BROOKS, N. H., and KENNEDY, J. F. (1960), Lecture notes on sediment transportation and channel stability, *California Institute of Technology Report*, No. KHOR-1.

WEIR, G. J. (1983), One-dimensional bed wave movement in lowland rivers, *Water Resources Research*, **19**, 627–631.

WHITE, W. R., and DAY, T. J. (1982), Transport of graded bed material, in HEY, R. D., BATHURST, J. C., and THORNE, C. R. (Eds.), *Gravel-Bed Rivers*, Wiley, pp. 181–213.

WHITE, W. R., MILLI, H., and CRABBE, A. D. (1975), Sediment transport theories: a review, *Proceedings of the Institution of Civil Enginers*, **59**, 265–292.

WHITTAKER, J. G., and DAVIES R. H. (1982), Erosion and sediment transport processes in step-pool torrents, in *Recent Developments in the Explanation and Prediction of Erosion and Sediment Yield (Proceedings of the Exeter Symposium, 19–30 July 1982)*, International Association of Hydrological Sciences Publication 137, pp. 99–104.

WILLIAMS, G. P. (1983), Palaeohydrological methods and some examples from Swedish fluvial environments, I — cobble and boulder deposits, *Geografiska Annaler*, **65A**, 227–243.

YALIN, M. S. (1963), An expression for bedload transportation, *American Society of Civil Engineers, Journal of the Hydraulics Division*, **89**, 221–250.

YANG, C. T. (1973), Incipient motion and sediment transport, *American Society of Civil Engineers, Journal of the Hydraulics Division*, **99**, 1805–1826.

YANG, C. T. (1984), Unit stream power equation for gravel, *American Society of Civil Engineers, Journal of Hydraulic Engineering*, **110**, 1783–1797.

ZANEN, I. A. (1967), The results of continuous bedload measurements related to fluctuations of the river bed, *International Association of Scientific Hydrology Publication*, No. 75, 255–275.

Glacio-fluvial Sediment Transfer
Edited by A. M. Gurnell and M. J. Clark
©1987 John Wiley & Sons Ltd.

Chapter 14

Electrical Conductivity

C. R. FENN

ABSTRACT

This chapter examines the ways in which electrical conductivity (EC) has been employed in hydroglaciological studies. The first part (Section 14.2) considers the uses and limitations of EC as a parameter of meltwater quality. The second part (Section 14.3) reviews results obtained from studies of meltwater conductivity in supraglacial environments. The third part (Section 14.4) considers the EC characteristics of proglacial streams, focusing on the nature and significance of spatial and temporal variability in reported results, and on the characteristics of conductivity–discharge relationships. The final part (Section 14.5) highlights the ways in which EC studies have contributed in the investigation of problems in glacier hydrochemistry and glacier hydrology.

14.1 INTRODUCTION

The use of electrical conductivity (EC) in drainage basin research is essentially predicated on two grounds: (1) its ability to act as a water quality determinand representing solute concentration and (2) the ease and economy with which it may be measured. The uses and limitations of EC as a parameter of water quality have been examined with respect to a number of research applications, including studies of natural waters and soil solutions (e.g. Tanji and Biggar, 1972; Foster *et al.*, 1981) and glacial meltwaters (e.g. Behrens *et al.*, 1971; Collins, 1977). Notwithstanding the recognition of important qualifications and limitations (Table 14.1), such studies have invariably recommended EC as a useful surrogate measure for total solute concentration. Such an assessment, coupled with the comparative ease with which EC may be measured electrometrically with relatively cheap instrumentation, has encouraged many researchers working in many areas to employ EC as a useful hydrochemical parameter.

The purpose of this chapter is to examine the bases, nature and significance of studies of EC undertaken in glacial drainage basins. Although a global

TABLE 14.1   The bases and limitations of EC as a water quality parameter (Sources: Hem, 1970; Tanji and Biggar, 1972; Collins, 1977; Golterman *et al.*, 1978; Foster *et al.*, 1981)

Theoretical Bases

Electrochemical Base: EC a measure of the ability of a solution to conduct an electrical current. Electrical current transferred by ions (electrically charged particles) in solution. EC related to the concentration and nature of the ions present. Total conductivity the summation of individual ion conductivity ($EC = \Sigma c_i$).

Hydrogeochemical Application: EC proportional to the concentration of ions in solution. TDS (total dissolved solids) concentration proportional to EC: empirically verified. Individual ion concentration proportional to EC: empirically queried.

Limitations: The relationship between EC and ionic concentration in natural waters is affected by:

(i) Ionic composition of the solution
'Natural waters are not simple solutions. They contain a variety of ionic and undissociated species, and the amounts and proportions of each may range widely' (Hem, 1970, p. 99). Different ions contribute unequally to EC acording to their valency, size, mass and ionic mobility. Any change in the relative ionic composition of a solution consequently introduces complexities in the EC–ion concentration relationship, e.g. solutions of differing TDS concentration and differing ionic composition may yield the same EC. Strictly, variations in EC therefore provide a reliable indication of variations in ionic concentration only when the proportion of the individual ions present remains appreciably unchanged. Only charged ion species impart conductance. Ion association or ion pairing may produce neutrally charged compounds, and/or species which impart less conductance in their paired state than they would if their component ions were completely dissociated. EC may therefore underestimate total solute concentration if uncharged or undercharged species are present. Ion pairing typically develops as ionic concentration increases: the error is accordingly likely to be of significance in solute rich soil and ground waters, but of limited importance in dilute surface waters.

(ii) Ionic concentration of the solution
Ionic mobility (i.e. the potential velocity of an ion through an electrolyte) is decreased by particle interaction and interference at high levels of ionic concentration. The relationship between EC and ionic concentration is accordingly non-linear.

(iii) Water temperature
EC is dependent on the degree of ionization of the electrolyte and the mobility of its constituent ions as well as on the composition and concentration of the ions present. Both the ionic mobility and the degree of ionization (i.e. the fraction of an electrolyte which dissociates into ions) are temperature dependent: the higher the temperature, the higher the degree of ionization and the ionic mobility. EC readings are accordingly often standardized to a reference temperature (20 °C, 25 °C) to control for the temperature effect. A correction factor of 2% per °C is the accepted norm. However, the rate of increase is higher and more variable at low water temperatures (Østrem, 1964; Collins, 1977).

(iv) pH
The ionic specific conductance of the hydrogen ion varies substantially with pH: it adds less than $0.2\,\mu S\,cm^{-1}$ to total conductivity at a pH of 6.5, but over $100\,\mu S\,cm^{-1}$ at a

(continued)

TABLE 14.1 *(continued)*

pH of 3.5 (Foster *et al.*, 1981). (The change reflects the greater dissociation of the water molecule in more acidic solutions.) Further, metal ions may be preferentially mobilized in acidic waters. In low pH waters (<5.5), a change in EC may thus result from a change in hydrogen activity as well as from a change in total ionic concentration. pH data may be valuable in interpreting EC in such circumstances.

(v) CO_2
Conductance may be achieved by bicarbonate and hydrogen ions acquired by dissolution of carbon dioxide from the air. CO_2 is preferentially dissolved in cold, dilute, turbulent waters.

(vi) Sediment load
Sediments in river water (or water samples) may exert an influence on its chemical quality via: (1) sorption phenomena; ions may adsorb on to wash, suspended or bed sediments, and desorb from them (during transport, during storage or on filtration); (2) dissolution; fine particles may dissolve (wholly or partially) during transport, or during storage as an unfiltered water sample. Errors may arise when measurements are taken from water samples which have reacted differently or have been treated differently.

(vii) Dissolved organic material (DOM)
The dissolution of organic materials in natural waters may produce polyelectrolytes which impart significant conductance. Most surface waters probably contain insufficient amounts of DOM to influence EC, but organics may be sufficiently plentiful in soil solutions or polluted waters to produce significant effects on EC.

perspective is taken, data from the author's studies at the Glacier de Tsidjiore Nouve, Switzerland, are used to highlight particular issues, and to draw comparisons with the comprehensive results obtained by Collins (1977, 1979a, b, 1981, 1983) from the catchment of the Gornergletscher, which is also located in Switzerland.

14.2 THE BASES OF EC INVESTIGATIONS IN GLACIER BASINS

14.2.1 The uses of EC in glacier basin research

EC measurements may be made electrometrically (i.e. by immersing a conductivity cell within the fluid to be measured) with apparent ease and accuracy. Continuous records may be obtained by transferring the signal from a cell fixed within the flow to a chart recorder or data logger (Figure 14.1). Meltwater conductivities, whether derived from continuous in-stream monitoring programmes or from discrete water sampling schemes, have found application in a variety of contexts within the overall research field of glacier hydrochemistry and hydrology and have been directed to the following specific objectives:

FIGURE 14.1   Recording EC in the Tsidjiore Nouve proglacial stream. The signal from a robust shielded cell attached to a WPA CM-25 meter is recorded by a Squirrel data logger

1. To indicate the nature and extent of temporal variations in meltwater quality, whether during the ablation season (e.g. Collins, 1979a; Gurnell and Fenn, 1985) or over the annual runoff cycle (e.g. Collins 1981; Oerter *et al.*, 1980).
2. To indicate the nature and extent of spatial variations in meltwater quality, whether within a single catchment (e.g. Gurnell and Fenn, 1985) or between different basins (e.g. Collins and Young, 1981).
3. To provide quantitative estimates of solute load and yield (e.g. Collins, 1979a, 1981, 1983).
4. To facilitate separation of hydrochemically distinct contributions to total meltwater runoff in order to examine the source, storage and routing characteristics of the glacial runoff system (e.g. Collins, 1977, 1979b; Oerter *et al.*, 1980; Gurnell and Fenn, 1984).
5. To identify solute sources and pathways (e.g. Collins, 1981).
6. To provide a check on measured specific ion concentrations (e.g. Reynolds and Johnson, 1972).

(Tracing experiments using EC to detect artificially introduced salt waves are not considered in this discussion.)

   In practice, most investigations have embraced a number of these objectives. Many studies have additionally included consideration of the characteristics of specific ion concentrations (e.g. Collins, 1979b). Indeed, many studies of glacier basin hydrochemistry have concentrated exclusively upon the concentration dynamics of specific ions (e.g. Slatt, 1972; Lemmens and Roger, 1978; Raiswell and Thomas, 1984). The greater ease with which EC may be determined in comparison with the determination of concentrations of specific ions makes the former particularly attractive in remote alpine basins, however, especially as EC may be monitored continuously whereas individual ion concentrations have necessarily to be determined from spot samples. This holds even when ion-selective electrode procedures of the type advocated by Reynolds (1971) are employable. The advantages of the continuous record over the discrete sample set constitute a compelling justification for the monitoring of EC.

14.2.2 The limitations of EC studies in glacier basin research

Ease of measurement is evidently a necessary but not a sufficient condition for the suitability of monitoring EC as a meltwater quality parameter: its electrochemical bases must also be acceptable. The general limitations of EC as a quality parameter are outlined in Table 14.1. Applying such considerations to the alpine environment, the following comments may be made:

14.2.2.1 *Ionic composition of meltwaters*

Church (1974) considered that the ionic balance of meltwaters on Baffin Island was sufficiently stable to enable EC to act as a reliable surrogate for total dissolved solids (TDS) concentration. Collins (1979a), however, determined that the proportions of different ions present in the waters of the Gornera do not remain constant, to the extent that he considered EC unable to properly represent fluctuations in the concentration of individual ions. Nevertheless, he did consider that EC was able to act as a useful overall indicator of TDS concentration. In a subsequent analysis of waters from the Gornera, Collins (1983) showed that relationships between EC and individual cation concentration were poor ($R^2 \leqslant 0.33$), that the relationship between EC and the sum of cationic equivalents (Σc_i) was significantly better ($R^2 = 0.88$), but that the estimation of Σc_i from EC may still entail errors of up to $\pm 25\%$. Taking meltwater environments overall, it would seem reasonable to expect that the proportion of individual ions present in outflow from a glacierized basin (in which regolith development is invariably restricted and biological and pedological activity is minimal) would be at least as stable as that in outflow from a non-glacierized basin. It would seem prudent, however, to regard EC as a useful indicator of hydrochemical behaviour, but not as a reliable means of estimating individual (or even total) ionic concentrations.

14.2.2.2   Ionic concentration of meltwaters

Ion pairing is unlikely to exert a significant influence in waters containing such low solute concentrations as have been reported for glacial meltstreams.

14.2.2.3   Temperature effects

Glacial meltwaters are close to 0 °C on emergence from the glacier, and remain at very low temperatures for some distance downstream. Collins (1977) showed that standard compensation factors are unreliable in such circumstances, and warned against the use of automatic temperature adjusting EC meters in glacial meltwaters. Most EC values taken from in-stream surveys reported subsequent to Collins' (1977) study have accordingly been given as measured (i.e. uncorrected) at ambient water temperature. Given a restricted range of temperature fluctuation, such a practice would seem justifiable. [It should, perhaps, be noted that the report on conductivity records in the Vernagtbach by Oerter *et al.* (1980) stands as an exception: water temperatures of up to 8.0 °C led them to standardize EC readings (to 20 °C) using correction factors determined from multiple water samples.] The error due to not standardizing for temperature will manifest itself as a reduced EC range (i.e. the application of temperature correction factors accentuates the range of the original EC measurements). Care is accordingly needed in comparing EC data reported at different temperature levels.

14.2.2.4   pH effects

Raiswell (1984), surveying results from 9 separate hydrochemical studies embracing 40 different melt systems, noted that meltwater pH varies from 4.0 to 9.5 but that most values lie between 6.5 and 8.0. Hydrogen ion activity is consequently unlikely to substantially influence EC determinations in the majority of glacial catchments, although particular cases to the contrary have been identified (e.g. Reynolds and Johnson, 1972).

14.2.2.5 CO_2 effects

Metcalf (1979) contended that meltwaters emerging from intraglacial conduits will be undersaturated with respect to atmospheric CO_2, and may rapidly dissolve CO_2 in the proglacial zone. Reynolds and Johnson (1972) also commented that the turbulent, well aerated flow conditions characteristic of meltwater environments may promote chemical weathering by maintaining the acid potential of the water. However, the effect on EC is unlikely to be great.

14.2.2.6 Sediment load effects

Glacier-fed streams invariably transport high, although very variable, quantities of sediment in suspension (Østrem, 1975; Collins, 1979c; Fenn *et al.*, 1985). Sorption and dissolution effects on meltwater quality, therefore, need to be considered. Collins (1979b) contended that both dissolution of particles and cation exchange between meltwater and suspended particles play only a minor role in modifying the chemical composition of meltwaters once they reach large channels. The importance of sorption in solute transport in meltwater streams has, however, been highlighted by the work of Lorrain and Souchez (1972) in the Moiry glacier basin and by Lemmens (1975) in the Tsanfleuron basin. Slatt (1972) noted that changes in the chemical characteristics of water samples may occur as a result of partial dissolution during storage of (formerly) suspended particles. Meltwater samples for ion concentration determinations are often filtered immediately on collection in order to avoid the influence of subsequent desorption and/or dissolution on their chemical composition (Slatt, 1972; Lorrain and Souchez, 1972; Lemmens, 1975; Collins, 1979a; Eyles *et al.*, 1982). Quite the opposite approach seems to be appropriate when dealing with EC, however. Collins (1977) noted that desorption of ions from sediments during filtration leads to an increase in the EC of filtered waters relative to unfiltered waters. Since continuous in-stream EC measurements are made in the presence of whatever sediment is being transported, EC should, for comparative purposes, be taken from unfiltered water samples. Given differences in sample treatment, care should be taken when comparing EC data obtained from continuous and discrete sampling frameworks.

14.2.2.7 Organic material effects

The source-route characteristics of glacial waters ensures that they invariably contain minimal, if any, amounts of dissolved organic matter (DOM) (Raiswell, 1984).

Overall, the potential drawbacks associated with the use of EC as a water quality parameter for meltwater studies appear to remain well within acceptable bounds. The basic qualifications on the use of EC in meltwater environments may be identified as follows:

(a) EC may be used as a useful indicator of hydrochemical response in meltwater systems, but should not be uncritically used as an indicator or predictor of individual ion concentration.

(b) The effects of water temperature and suspended sediment load should be taken into consideration in treating water samples, and in comparing results from continuous in-stream and discrete water sample studies.

TABLE 14.2 Summary details from studies of the electrical conductivity of supraglacial waters

Glacier	Site characteristics	Dates	Sampling details			Electrical conductivity (μS cm^{-1})			Source
			No. of sites	No. of samples per site	Temperature range (°C)	Min.	Max.	Mean	
Gornergletscher	Small icemelt streams	1974–77		12 samples	0.1	0.1	1.6		Collins (1977, 1979b)
	Large icemelt streams	13–15 Aug. 1976	1	Cont.	0.1	2.7	5.4		Collins (1979b)
Findelengletscher	Snowmelt stream	1974–77		6 samples	0.1	1.3	1.6		Collins (1979b)
Theodulgletscher	Snowmelt stream	1974–77		2 samples	0.1	1.3	1.9		Collins (1979b)
Haut Arolla	Streams draining clean ice	Sept. 1983	23			0.4	5.1	2.4	Southampton Univ. (1983) (personal communication)
	Streams draining moraine		7			1.2	8.5	4.7	

Bas Arolla	Streams draining clean ice	1–3 Aug. 1984	5	18		3.0	7.0		Stevens (1985) (personal communication)
	Streams draining moraine		6	18		4.0	7.0		
Tsidjiore Nouve	Streams draining clean ice	22–27 July 1984	3	42	0.0–4.0	1.0	6.0	2.7	Stevens (1985) (personal communication)
	Streams draining moraine		5	36	0.0–5.0	1.0	7.0	3.7	
	Pools in moraine areas		2	36	0.1–5.1	3.0	10.0	5.6	
Vernagtferner	Snow and icemelt streams	1976	?		S @ 20 °C*	1.0	20.0		Oerter. et al. (1980)
Hintereisferner	Snow and icemelt water	July 1970	?		S @ 20 °C*	2.5	10.0		Behrens et al. (1971)
Chamberlin	Surface ice	20 Aug. 1958						9.0	Rainwater and Guy 1961

*S @ denotes standardization to the stated reference temperature.

HAUT AROLLA: SEPTEMBER 1984

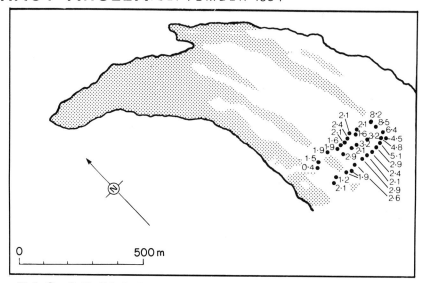

BAS AROLLA: AUGUST 1984

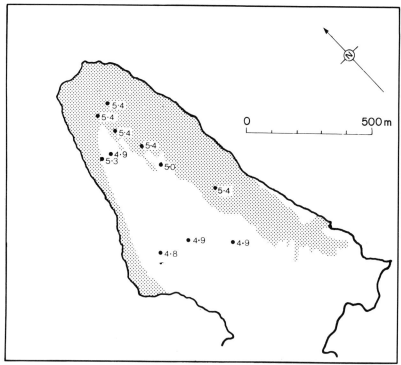

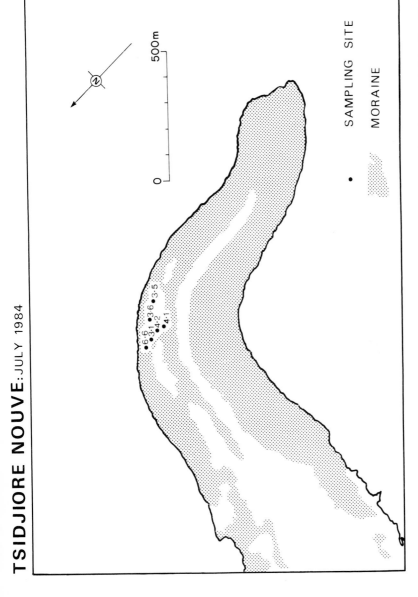

FIGURE 14.2 Temporal mean ECs (μS cm^{-1}) measured at supraglacial sites on three glaciers in the Val d'Arolla. The Haut Arolla data were determined by geography students of the University of Southampton. The Bas Arolla and Tsidjiore Nouve data are due to Stevens (personal communication)

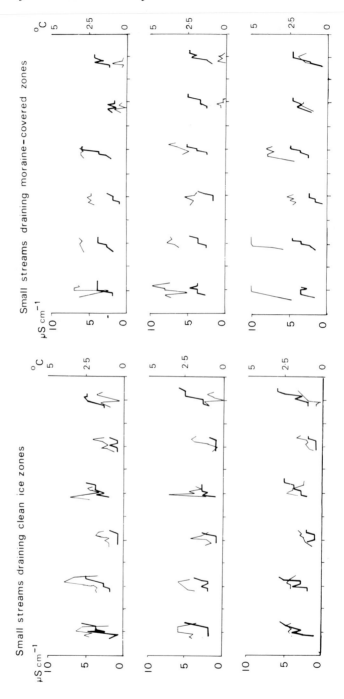

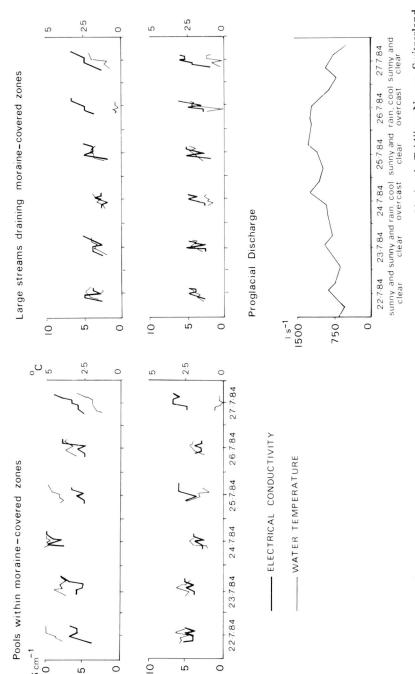

FIGURE 14.3   Spatial and temporal variations in the EC (μS cm⁻¹) of supraglacial waters, Glacier de Tsidjiore Nouve, Switzerland, August 1984. (Source: Stevens, personal communication.)

14.3   SUPRAGLACIAL EC STUDIES

A knowledge of levels of EC in supraglacial meltwater is of significance in relation to establishing the relative importance of the atmosphere, the supraglacial environment (including snow-covered, bare ice and moraine-covered zones) and the subglacial environment as solute sources. Supraglacial waters are invariably taken to have a relatively low solute concentration. The few empirical studies of supraglacial EC which have been undertaken verify that solute levels are indeed low (Table 14.2). The results of Oerter *et al.* (1980) from the Vernagtferner, where values as high as 20.0 μS cm^{-1} were obtained, stand out as an exception in comparison with the predominantly single-figure EC values obtained elsewhere.

Some degree of spatial and temporal variability in supraglacial EC levels has been detected. Collins (1977) reported that meltwaters flowing in large supraglacial streams on the Gornergletscher exhibited significantly higher EC (2.7–5.4 μS cm^{-1}) than did waters in small surface streams (0.1–1.6 μS cm^{-1}); no significant difference was found in the EC of snowmelt and icemelt waters, however. Multiple site measurements on three glaciers in the Val d'Arolla (Figure 14.2) show that supraglacial conductivities tend to be greater in waters draining from zones of surface moraines, particularly if water is held in store in surface depressions or fissures. Lemmens and Roger's (1978) experimental studies are of relevance here: they showed that the rate of reaction between dilute meltwaters and fresh supraglacial sediment is exceptionally rapid, with solute levels approaching equilibrium within 10 min. Ricq de Bouard (1973) reported similarly.

Temporal variations in supraglacial waters have also been noted. Collins (1977) showed that EC levels in a large meltstream on the Gornergletscher tended to exhibit transient temporal peaks on an otherwise constant background level. The peaks were explained in terms of the arrival of waters whose chemical quality is concentrated during periods of ice crystal formation. Measurements of waters on the ablation tongue of the Glacier de Tsidjiore Nouve (Figure 14.3) reveal EC variations on a number of spatial and temporal scales. Complementary variations in water temperature clearly account for some of the fluctuations in EC, but by no means all of them, as can be seen by comparing the EC and water temperature profiles in Figure 14.3. At a given site, EC varies from hour to hour and from day to day, according to climatic and flow influences. The effect of rainfall on EC levels is especially clear. At a given time, EC varies from site to site, probably in relation to the degree and duration of meltwater–moraine contact. The data presented in Figure 14.3 indicate that the degree of at-a-site temporal variation is as great as between-site variation.

Available evidence on supraglacial EC levels suggests that surface meltwaters acquire their limited solute load from direct atmospheric input (Collins, 1977) and (preferentially) through contact with surface moraines. Wind-blown

TABLE 14.3   Summary details from studies employing continuous monitoring designs to measure electrical conductivity in proglacial streams

Glacier	Survey dates	Length of record (days)	Temperature range (°C)	Distance from glacier (km)	Electrical conductivity (μS cm^{-1})				Diurnal range in summer		Source
					Autumn range	Winter range	Spring range	Summer range	Smallest	Largest	
Gornergletscher, Switzerland	15 July–2 Sept. 1975	50	0.1–1.0	0.25				6.5–44.0	9.5–11.0	10.0–44.0	Collins (1979b)
	28 Dec. 1978–17 Oct. 1979	197	0.1–1.2		40.0–51.0	85.0–115.0	30.0–89.0	8.0–58.5	22.0–28.0	14.0–50.0	Collins (1981, 1983)
Findelengletscher, Switzerland	1–24 Aug. 1977	24	0.1–4.0	0.40				18.6–68.2			Collins (1979b)
Tsidjiore Nouve, Switzerland	6 June–30 July 1978	55	0.1–0.9	0.36				18.5–50.0	25.0–25.5	38.0–46.5	Fenn (1983)
	21 July–9 Sept. 1984	23	0.1–1.2	0.36				18.1–45.9	24.3–27.8	20.9–34.8	
Vernagtferner, Austria	Jan–Dec. 1976		0.0–8.0	1.20							Oerter et al. (1980)
Peyto, Canada	21 May–2 Sept. 1979	104	0.7–1.5	1.00		145.0–170.0	180.0–100.0	22.0–42.0			Collins and Young (1981)

TABLE 14.4   Summary details from studies employing discrete sampling designs to measure electrical conductivity in proglacial streams (partly after Collins, 1979a)

Glacier	Survey dates	Sampling details*				Temperature range (°C)	Distance from glacier (m)	Electrical conductivity (μS cm⁻¹)		Source
		Δt(h)	n	U/F	Notes			Min.	Max.	
Tsidjiore Nouve, Switzerland	12–14 June 1977	1	30	U	Samples taken	0.7–3.9	280	21.0	74.0	Fenn (1983)
	3–7 July 1977	1	50	U	from 0800–1700 h	1.3–3.6	280	10.5	16.2	
	24–28 July 1977	1	50	U		0.7–2.7	280	6.5	15.5	
	14–18 Aug. 1977	1	44	U		0.9–2.5	280	5.0	18.5	
	3–5 Sept. 1977	1	27	U		1.0–2.0	280	9.0	20.0	
Bas Arolla, Switzerland	23–27 June 1977	1	45	U	Samples taken	0.21–1.3	250	7.0	11.0	Fenn (1983)
	25–29 Aug. 1977	1	45	U	from 0900–1700 h	0.11–0.6	250	5.0	10.0	
	13 June 1977	1	9	U		0.5–1.0	250	6.6	8.1	
	20 June 1977	1	9	U		0.7–1.1	250	6.0	9.0	
	2 Aug. 1977	1	9	U		0.8–1.2	250	4.0	8.5	
	9 Aug. 1977	1	9	U		0.7–1.1	250	6.0	8.0	
Hintereisferner, Austria	22–23 July 1970	3	8	U		S @ 20 °C†	—	30.0	45.0	Behrens *et al.* (1971)
S.Cascade, USA	11 Aug. 1970	—	5	U		S @ 25 °C	—	6.3	12.9	Reynolds and Johnson (1972)
Chamberlin, Alaska	5–6 Aug. 1958 7 July–	2	12	U		S @ 25 °C	244	9.0	16.0	Rainwater and Guy (1961)
	28 August 1958	24	9	U		S @ 25 °C	244	10.0	60.0	

Location	Period	Δt	n	U/F	Notes	S @ t		EC		Reference
Castner, Alaska	Aug. 1968	—	13	F		S @ 25 °C	—	144.0	316.0	Slatt (1970)
Norris, Alaska	July 1969	—	7	F		S @ 25 °C	—	38.0	86.0	Slatt (1970)
North R., Baffin Is.	10 July–19 Aug. 1967	—	10	U	Samples taken at irregular intervals and stored for long periods	S @ 20 °C	—	6.8	12.4	Church (1974)
Middle R., Baffin Is.	10 July–19 Aug. 1967	—	10	U		S @ 20 °C	—	7.0	15.3	
South R., Baffin Is.	19 Aug. 1967	—	10	U		S @ 20 °C	—	12.8	49.1	
Lewis R., Baffin Is.	21 June–9 Aug. 1963	—	5	U		S @ 20 °C	—	11.2	620.0	
Maktak R., Baffin Is.	20 June–12 Aug. 1964	—	14	U		S @ 20 °C	—	9.1	195.0	
	19–28 July 1972	—	9	U		S @ 20 °C	—	6.4	31.2	
Nugarssik, E. Greenland	10–16 Aug. 1982	2	84	U		0.5–2.0	200	7.1	14.0	Fenn (unpublished)
Unnamed 1, E. Greenland	10–16 Aug. 1982	2	84	U		1. 0–4.0	500	6.5	19.0	
Unnamed 2, E. Greenland	10–16 Aug. 1982	2	84	U		1. 0–4.0	500	3.0	18.0	

* Δt = sampling interval; n = total number of samples; U/F denotes whether the sample was unfiltered (U) or filtered (F) prior to EC determination.

† S @ denotes standardization to the stated reference temperature.

sediments may also contribute to the solute load of supraglacial streams (Collins, 1979b). Ions contributed to the supraglacial environment during precipitation events appear to be rapidly and differentially leached from firn by percolating meltwaters (Davies *et al.*, 1982). It has also been noted that the ionic content of supraglacial waters may be enhanced by a number of 'concentrating effects' such as firnification (Slatt, 1972), 'squeezing out' during regelation (Slatt, 1972) and evaporation (Stednick, 1981).

14.4 PROGLACIAL EC STUDIES

Proglacial waters are significantly ionically enriched relative to the supraglacial waters which contribute to them; the difference amounts to an order of magnitude in most cases (cf. Tables 14.2, 14.3 and 14.4).

Many more studies have been undertaken of the EC of proglacial waters than of either supraglacial or subglacial waters. Few studies, however, have monitored EC with a detail sufficient to enable results to be used to characterize meltwater solute dynamics comprehensively. The continuous records obtained from the Gorner (1975, 1979), Findelen (1977), Tsidjiore Nouve (1978, 1984), Vernagtferner (1976) and Peyto (1979) glacier basins are notable in this respect (Table 14.3). Many other studies have used fairly simple sampling programmes, either to fix the level of meltwater EC or to demonstrate the nature and extent of EC variations with flow on a diurnal or a seasonal timescale (Table 14.4). Many of these studies have employed EC in an adjunctive role, with emphasis being variously placed on isotopic and ionic determinations (e.g. Behrens *et al.*, 1971). Unfortunately, variation in sampling design and standardization procedures mitigate against an analytical comparison of the results produced. Tables 14.3 and 14.4 therefore differentiate between studies according to their sampling design, and also (in the light of the issues discussed in Section 14.2.2) according to the measurement practices adopted.

The variation in EC values illustrated in Tables 14.3 and 14.4 embodies both spatial and temporal aspects. The following discussion attempts to highlight the nature and significance of the available data through consideration of spatial and temporal variations in the EC of proglacial streams (Section 14.4.1), proglacial hydrograph–chemograph patterns (Section 14.4.2) and models of the proglacial EC–discharge (Q) relationship (Section 14.4.3).

14.4.1 Spatial and temporal variations in the EC of proglacial streams

14.4.1.1 Between-basin spatial variations

A consideration of, amongst others, the papers of Keller and Reesman (1963), Reynolds and Johnson (1972), Slatt (1972), Church (1974) and Collins (1983) provides evidence of variability in solute activity between glacier basins. Slatt (1972) noted that differences in the ionic concentration of meltwaters of catchments of the same rock type were no smaller than those of catchments of varying rock type, and concluded that rock type is probably not the most important variable controlling the chemical quality of meltwaters.

The conflation of varied hydrometeorological conditions with whatever 'static' environmental differences may exist precludes a quantified expression of between basin differences in EC on the basis of the data in Tables 14.3 and 14.4. However, Collins (1979a) noted that the conductivities of summer season meltwaters issuing from alpine glacier basins floored by igneous and metamorphic rocks lie within the range 6.0–60.0 μS cm^{-1}, but commented that EC typically varies considerably from catchment to catchment according to glaciological, lithospherical, sedimentological and hydrological characteristics. For example, the relatively restricted ablation season range of conductivities in Peyto Creek, Canada (22–42 μS cm^{-1}) has been explained in terms of a comparatively high groundwater flow contribution (Collins and Young, 1981). Data produced from point sampling of late winter/early spring meltwaters issuing from glaciers in the French and Swiss Alps (Table 14.5) are probably sufficiently comparable with respect to climatic influences to indicate basin to basin variability in EC resulting from catchment rather than climatic characteristics; it would appear that such differences exist regardless of season.

TABLE 14.5 Electrical conductivities of meltwaters issuing from selected glaciers in the French and Swiss Alps during a spring snowmelt period. Measurements are single-point determinations for each stream taken during a period of fine settled weather in April 1982

Glacier	Date and time of measurement	Water temperature (°C)	Electrical conductivity (μS cm^{-1})	
			As measured	Standardized to 25 °C
Vallée de Chamonix, France:				
Argentière	16.4.82, 1800 h	2.2	174.0	283.6
Mer de Glace	1300 h	1.9	136.0	223.0
Bossons	1900 h	1.9	81.0	132.8
Taconnaz	1200 h	6.5	141.0	208.7
Val d'Hérens, Switzerland:				
Ferpècle/Mont Miné	17.4.82, 1030 h	11.8	140.0	183.4
Tsidjiore Nouve	1730 h	5.5	181.0	275.1
Bas Arolla	1345 h	10.8	180.0	241.2
Mattertal, Switzerland:				
Z'mutt	18.4.82, 1200 h	12.3	222.0	286.4
Gorner	1400 h	12.2	183.0	236.1
Findelen	1615 h	11.8	247.0	323.6

14.4.1.2   *Within-basin spatial variations*

Most ionic enrichment is believed to occur at subglacial sites, where the disintegration of mineral lattices by glacier grinding provides a potentially rich reservoir of rock debris capable of reacting rapidly with dilute waters (Collins, 1977, 1981). However, Metcalf (1979) pointed out that CO_2 may be preferentially dissolved as subglacial waters emerge from the glacier; Lemmens (1975) contended that desorption is particularly favoured if a large spreading of water prevails in the outwash zone and Collins [in discussion of Metcalf (1979)] commented that meltwater conductivity may increase through the proglacial zone as groundwater contributions are released. Moreover, many proglacial streams are fed by multiple outflow channels draining contributing areas which may be physically separated and hydrochemically dissimilar; Collins (1979a) suggested that the Gornera may be fed by waters derived from hydrochemically dissimilar subglacial routes (an hypothesis reached through a consideration of changes in the relative ionic composition of outflow waters). Spatial variations in the EC of proglacial waters may, therefore, occur.

Little empirical information on site to site variations in EC within proglacial stream systems is available, since most studies have focused on temporal dynamics observed at single sites. Some data from the Tsidjiore Nouve basin are available, however. Lemmens and Roger (1978) detected no downstream change in the concentration of dissolved cations in the proglacial stream of the Glacier de Tsidjiore Nouve. However, Gurnell and Fenn (1985) detected significant site to site variations in the conductivity characteristics of the main outflow streams issuing from the snout of the Glacier de Tsidjiore Nouve, and concluded that the characteristics of bulk meltwaters may vary temporally according to the quantity–quality characteristics of partial area contributions supplied to the stream. It would appear that the incorporation of the spatial dimension into meltwater hydrochemical response models may be as fruitful as has been shown to be the case in non-glacial basins (e.g. Walling and Webb, 1980).

14.4.1.3   *Season-to-season temporal variations*

Vivian and Zumstein (1973) noted enhanced solutional activity under the Glacier d'Argentière in the winter period, and Collins (1981) reported that conductivities in the Gornera are two to ten times higher in winter than in summer. Oerter *et al.* (1980) have also reported significant seasonal differences in the conductivity of the meltwaters of the Vernagtbach ranging from a spring season maximum

---

FIGURE 14.4   (*opposite*) Annual EC records of (a) the Vernagtferner, Austria (1976) and (b) the Gornera, Switzerland (1979). (b) *Reproduced by permission of the International Glaciological Society and Dr D. N. Collins from* Annals of Glaciology, *1981, 2, 13.*

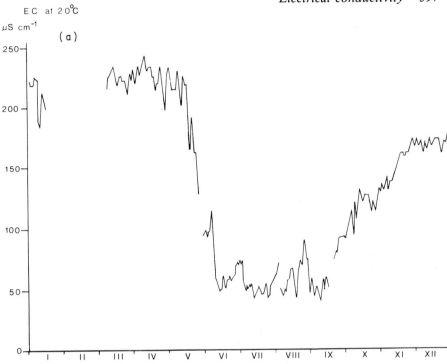

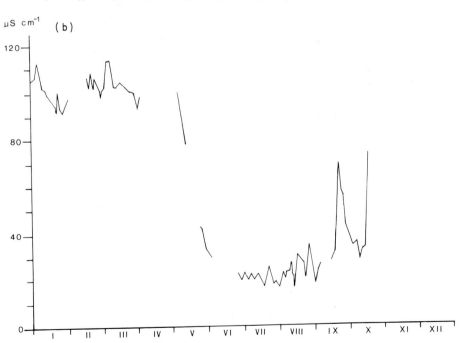

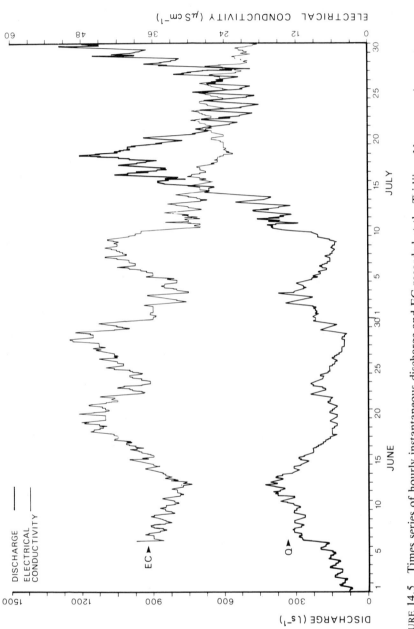

FIGURE 14.5   Times series of hourly instantaneous discharge and EC recorded at the Tsidjiore Nouve gauging station during June and July 1978

of (approximately) 245 μS cm^{-1} to an ablation season minimum of (approximately) 45 μS cm^{-1}. The EC of the Spring season meltwaters of glaciers in the French and Swiss Alps (Table 14.5) are also well above their summer season equivalents.

The annual EC records obtained from the Vernagtferner and Gornergletscher catchments [by, respectively, Oerter *et al.* (1980) and Collins (1981)] demonstrate a high degree of similarity in their serial behaviour (Figure 14.4). Winter and summer conditions define hydrochemically distinct responses, with low-amplitude fluctuations around a high average level maintained by groundwater contributions in winter, and high-amplitude diurnal fluctuations around a low average level maintained by subglacially enriched meltwaters in summer. Spring snowmelt and autumn snowfall conditions generate transitional periods, with the fall in EC attendant on the activation of the ablation season flow cycle in spring being the more abrupt. The long-period record from the Peyto glacier basin (Collins and Young, 1981) conforms to the same pattern. While there is a demonstrably consistent match in the pattern of annual EC fluctuations, the actual level of EC differs from basin to basin and, for a given basin, may differ from year to year. Collins (1981), for example, observed conductivities in the Gornera ranging from 95 to 115 μS cm^{-1} in the period from February to March 1979, whereas in the same period the following year the EC ranged from 66 to 84 μS cm^{-1}.

14.4.1.4 *Temporal variations within the summer ablation season*

Summer season records of EC and stream discharge (Q) in glacial outflow streams typically exhibit a pattern of EC cycles occurring approximately out of phase with the quasi-regular repeating hydrographs generated by snow and ice ablation (Figure 14.5).

The range of conductivities experienced varies from basin to basin, and from year to year for a given basin (Tables 14.3 and 14.4). The minimum conductivity of the summer season flow from some glaciers (e.g. Gornergletscher, Lower Arolla glacier) is apparently contributed almost wholly by atmospheric inputs of solutes (i.e. minimum proglacial EC is equivalent to supraglacial EC values); in other basins (e.g. Findelengletscher, Glacier de Tsidjiore Nouve, Peyto glacier), however, subglacial and/or groundwater flows presumably contribute solutes even during periods of the melt season when flows have a very low EC (i.e. the minimum proglacial EC lies well above the maximum supraglacial EC). The peak conductivity of the summer season in most basins is achieved during streamflow recession periods, when the diurnal cycle is suppressed and chemically enriched waters released from subglacial or groundwater stores constitute most of the total flow. The EC of proglacial streams may approach the EC of subglacial waters at such times (Collins, 1979a).

The ablation season range of EC of proglacial waters subsumes marked diurnal fluctuations, whose form, amplitude and timing vary over the season and differ from basin to basin. Section 14.4.2 examines these properties in some detail. For the present, we can note that variations in EC, whether on a seasonal or diurnal timescale, are, in general, inversely related to seasonal and diurnal fluctuations in stream discharge.

The inverse relationship between variations in Q and EC in proglacial streams has been explained in terms of temporal changes in the relative contribution made to total flow by those waters which remain hydrochemically dilute in passing through the glacier and those which become solute-enriched. Put simply, the former appear as a diurnal flow peak which dilutes the chemical quality of the baseflow maintained by the latter (in much the same manner as stormflow inputs to non-glacierized river basins dilute antecedent baseflow solute levels). Collins (1977) formalized a simple two-component mixing model based on a distinction between those flows routed rapidly without chemical enrichment through the 'englacial' drainage system (Q_e) and those routed more slowly, with attendant chemical enrichment, through the 'subglacial' drainage system (Q_s) to explain temporal variations in the quantity and quality of bulk meltwaters (Q or Q_t). Raiswell (1984) has presented chemical models capable of explaining the hydrochemical evolution of the two components in terms of their open-system and closed-system characteristics. A different explanation for the inverse relationship between proglacial discharge and solute concentration has been advanced by Lemmens and Roger (1978). Their model is based on variations in the degree of enrichment of meltwaters according to discharge-related changes in hydraulic radius and velocity in the subglacial drainage system: efficient enrichment is possible during periods of low flow, when intimate water–boundary contact is favoured, while during periods of high flow, enrichment is limited by the spatial extent and temporal duration of water–boundary contact.

Whichever of the two models is taken (i.e. the mixing model or the variable enrichment model), it is clear that solute concentrations fall as flow level rises, and that the fundamental control on solute uptake is the duration of contact between meltwater and rock material. (It is noteworthy that it is theoretically feasible under the mixing model for solute concentrations to vary positively with diurnal flow fluctuations, were Q_s the dominant component of Q_t, and were fluctuations in Q_t and Q_s in-phase. Empirical evidence, however, demonstrates Q_e to be the dominant component, and EC to be inversely related to diurnal flow changes.)

Sections 14.4.2 and 14.4.3 examine the nature of observed EC–Q relationships in relation to these theoretical models.

14.4.2 Diurnal hydrograph-chemograph patterns

The detailed Q and EC records of the Gorner, Findelen, Tsidjiore Nouve, Vernagtferner and Peyto glacier basins permit a reasonably detailed comparative

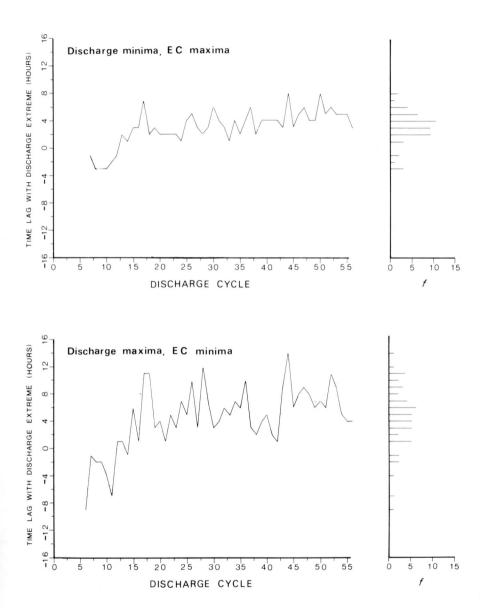

FIGURE 14.6 Lag times between diurnal discharge and EC extremes in the 1978 Tsidjiore Nouve record. Lag times, measured in hours, between diurnal discharge minima and EC maxima (upper plot), and diurnal discharge maxima and EC minima (lower plot) are shown

analysis of the form and timing of diurnal hydrographs and chemographs in proglacial streams. Some notable between-basin differences emerge. The EC record of the Gornergletscher basin, for example, is characterized by diurnal EC cycles with steep falling limbs and high amplitudes (the ratio of diurnal maximum to minimum EC ranges from 1.16 to 4.40), while diurnal fluctuations of EC in the proglacial stream of the Tsidjiore Nouve glacier have steep rising limbs and damped amplitudes (the maximum:minimum EC ratio ranges from 1.02 to 1.67) (see Figure 14.5). Further, while all glacier basins reportedly release a series of approximately out-of-phase diurnal Q and EC cycles, the average lag time between diurnal Q and EC extrema shows some degree of difference both between basins and over time. The Q and EC events of the Gorner, Findelen and Peyto basins have been described as being in inverse phase (Collins, 1979b; Collins and Young, 1981). Collins (1977), in presenting details of the Gornera data, reported that diurnal EC minima precede the Q maxima by 1–2 h, and that diurnal EC maxima occur 2–3 h after diurnal Q minima. A similar situation of EC peaking on the rising limb of the diurnal flow hydrograph has also been reported for the Tsidjiore Nouve basin, where diurnal EC maxima occur around 4 h after Q minima (but see Figure 14.6), and (by assumption in Gurnell and Fenn, 1984) for the Vernagtferner basin. Such a circumstance implies that the onset of the diurnal flow cycle is accompanied by a period of solute flushing, whether resulting from the release of waters from subglacial cavities (Collins, 1977, 1979b) or from the forcing out of waters held in the conduit system via a translatory flow mechanism (Gurnell and Fenn, 1984). The presence or absence, and the duration, of such an event apparently varies from basin to basin; as lag times between diurnal Q and EC extrema fluctuate over the season (Figure 14.6), the form of the flushing event in a particular basin may also be subject to variability over time.

Cross-correlation analysis has potential alongside the scrutiny of diurnal events in the present context. The form of the cross-correlation function (ccf) of the 1978 Tsidjiore Nouve Q and EC data (Figure 14.7a) shows a strong inverse match with only damped cyclicity (the latter indicating a relatively limited diurnal dilution effect) and a short lag effect. The form of the cyclical and lag effects is shown more clearly in the ccf of the data after the removal of the trend by first differencing (Figure 14.7b). The best (inverse) match position indicates the presence, overall, of a $+2$ h lag between the Q and EC series (Q leading). The irregularities superimposed upon the 24 h cycle suggest the operation of solute supply/exhaustion effects other than those related simply to the general diurnal discharge cycle.

14.4.3   EC–Q relationships

The EC–Q scatter plot for proglacial streams invariably indicates an inverse relationship between the two variables which is often non-linear and shows a

(a) ORIGINAL DATA

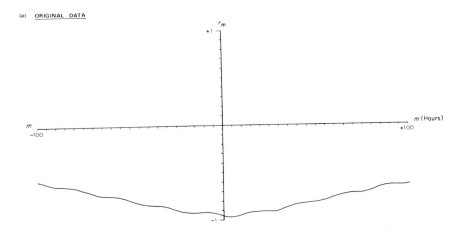

(b) FIRST DIFFERENCES

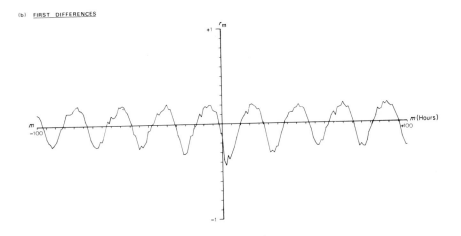

FIGURE 14.7 Cross-correlograms of the 1978 Tsidjiore Nouve discharge (*Q*) and EC series. The upper diagram (a) is the plot of the ccf of the EC and log *Q* series. The lower diagram (b) is the plot of the ccf of the EC and log *Q* series after first differencing

widely varying degree of scatter. The degree of scatter differs from basin to basin (and, potentially, from time to time for a given basin) according to the extent of variability in the quantity–quality characteristics of runoff components, and also according to the extent of hysteresis effects generated by diurnal and longer period lags between the *Q* and EC series.

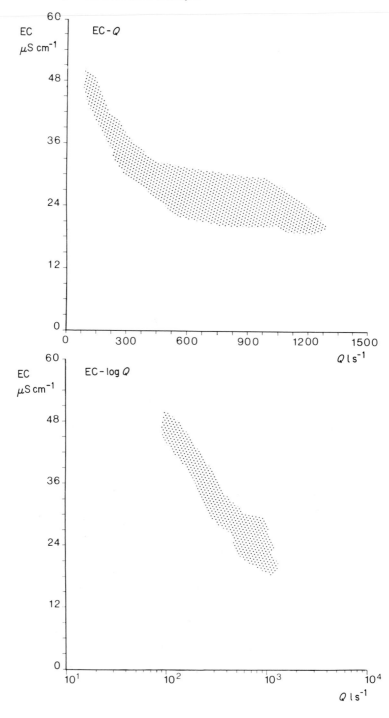

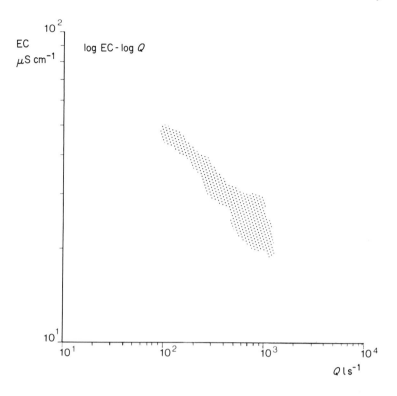

FIGURE 14.8   Scatter plots of the 1978 Tsidjiore Nouve EC–Q data (6th June to 30th July) (1307 values). The EC–log Q format gives a distribution which is more linear and monotonic than either the raw format or the log–log format

Collins' (1981) full-year results from the Gornera are better described by the margins of 'a triangle with spike' than by a straight line or a curve, such is the diurnal, intra-seasonal and inter-seasonal variability in the proportion of englacial and subglacial flow contributions. The ablation season scatter plot for the Gornera (Collins, 1979a) reduces to a trapezium, but retains considerable scatter. The EC–Q scatter plot for the considerably smaller Tsidjiore Nouve proglacial stream (Figure 14.8) is, in contrast, relatively well defined, reflecting a more conservative relationship between variations in streamflow and EC than that reported for the Gornera. However, the relationship between EC and Q indicated by the Tsidjiore Nouve data is markedly non-linear, suggesting the operation of solute flushing effects at high flow levels (possibly in association with the extension and/or reorganisation of subglacial drainage networks across fresh zones of reactive rock debris) and/or enhanced ion acquisition at low flows (possibly in association with extended meltwater–moraine contact durations) (Gurnell and Fenn, 1985).

The traditional EC-Q rating curve may provide an unsatisfactory representation of the bivariate relationship in the presence of diurnally and seasonally variable mixing ratios and hysteresis effects. Such a contention holds good for even the relatively simple Tsidjiore Nouve data. Standard ordinary least-squares rating relationships derived from the Tsidjiore Nouve data (Table 14.6) yielded highly autocorrelated residuals, indicating the presence of patterns other than trends in the data. In these circumstances, a transfer function fitted according to the methods proposed by Box and Jenkins (1970) proved to be a considerably more efficient and appropriate way of expressing the EC-Q relationship (Gurnell and Fenn, 1985).

TABLE 14.6   Rating equations and regression statistics fitted to the 1978 Tsidjiore Nouve EC-Q data. The low Durbin–Watson values (d) indicate the presence of high positive serial correlation in the residuals from the rating models

Data grouping	n	Rating equation	Regression statistics		
			F	R^2	d
All data	1307	EC = $95.2 - 24.4 \log Q$	13682.9	0.91	0.0825
Low flow data	906	EC = $104.1 - 28.2 \log Q$	7740.0	0.90	0.1438
High flow data	401	EC = $55.1 - 10.4 \log Q$	96.5	0.19	0.0499
Rising stage data	499	EC = $95.6 - 24.2 \log Q$	4567.5	0.90	0.1867
Falling stage data	808	EC = $95.3 - 24.7 \log Q$	11057.9	0.93	0.1079
Rising ablation periods	796	EC = $95.1 - 24.4 \log Q$	5524.2	0.87	0.0906
Falling ablation periods	511	EC = $94.6 - 24.1 \log Q$	9524.4	0.95	0.1215

14.5   THE CONTRIBUTION OF EC STUDIES TO HYDRO-GLACIOLOGICAL RESEARCH

14.5.1   Glacier hydrochemistry

14.5.1.1   Solute sources
The enrichment of meltwaters during their passage from the supraglacial to the proglacial domain is usually explained in terms of the acquisition (whether by ion exchange and/or direct solution) and transport (whether in fluid or attached to sediments) of solutes during passage across subglacial surfaces (and/or through subglacial moraines) (Rainwater and Guy, 1961; Collins, 1977, 1981; Lemmens and Roger, 1978). While specific ion analyses have yielded information of direct value in the investigation of subglacial chemical processes (e.g. Souchez and Lorrain, 1978; Lemmens and Roger, 1978; Eyles *et al.*, 1982), EC studies have contributed little of direct relevance to the detailed identification of solute sources. The absence of any sign of exhaustion in summer season solute loads estimated from EC records did, however, lead Collins (1983) to suggest that summer meltwaters probably acquire ionic material directly from subglacial

surfaces, rather than simply by evacuating a solute reservoir established over the winter season.

14.5.1.2 Solute loads and yields

Collins (1979a, 1981, 1983) quantified the solute load of the Gornera using an index S, obtained as $S = EC \cdot Q_t$. On a diurnal timescale, the index S decreases as Q_t increases, with fluctuations in S being similar to and in-phase with those in EC. The peak diurnal S follows the minimum total flow by a lag time of 1–3 h, indicating that the supply of solutes increases during the early stages of the diurnal flow cycle (as Q_s flows peak), but then decreases sharply (as Q_e flows rise). The relationship between Q_t and S over the ablation season exhibits a positive trend with considerable scatter, indicating that estimates of load from flow discharge would involve large errors. Peak levels of solute transport in the Gornera are apparently achieved by flows of intermediate magnitude.

Collins (1983) has also produced estimates of total monthly and annual solute load for the Gornera by estimating the sum of cationic equivalents from EC data. Computation of solute yield from these data (after correction for atmospheric solute input) indicates a chemical yield of 454 mequiv.m^{-2}yr^{-1} ($\pm 25\%$) for the Gornergletscher catchment. This figure may be compared with estimates of solute yield produced from cation determinations in other glacierized catchments:

S. Cascade, USA ... 960 mequiv.m^{-2}yr^{-1}(Reynolds and Johnson, 1972)
Berendon, Canada... 947 mequiv.m^{-2}yr^{-1}(Eyles *et al.*, 1982)
Tsidjiore Nouve, ... 508 mequiv.m^{-2}yr^{-1}(Souchez and Lemmens; see
    Switzerland   Chapter 11)

The North American basins have higher cationic denudation rates than the Swiss basins, but it is evident that the rates of both sets are internally comparable, and are above the global mean rate of 390 mequiv.m^{-2}yr^{-1}. It would seem that glacier basins deliver relatively high solute yields in comparison with their non-glacierized counterparts. Most of the load is probably evacuated during the summer season. Collins (1983), for example, determined that 95.5% of the annual solute yield of the Gornergletscher catchment is exported during the period May to September.

14.5.2 Glacier hydrology

14.5.2.1 Flow separation

Hydrochemical characteristics have been used to distinguish between melt-waters according to both their sources and their pathways. Source-specific discriminations have been based on the varying isotopic and ionic characteristics of snowmelt, icemelt and ground waters (e.g. Behrens *et al.*, 1971; Zeman and Slaymaker, 1975; Eyles *et al.*, 1982). Route-specific discriminations have been based on the differing EC of quick-flow and delayed flow waters (e.g. Collins, 1977, 1979b; Oerter *et al.*, 1980; Gurnell and Fenn, 1984). This section summarizes the principles of EC-based flow separation. The following section outlines the significance of an ability to deconvolute the flow record in such a manner, notably in relation to the interpretation of the form and functioning a glacier drainage systems.

Collins (1977, 1979b) showed that a simple mixing model of the form

$$Q_s = [(EC_t - EC_e)/(EC_s - EC_e)] Q_t$$

(where Q and EC are defined as before and the subscripts t, e and s represent, total, englacial and subglacial components, respectively) may be used to separate englacially routed (dilute) waters from subglacially routed (chemically enriched) waters. A concomitant of the spatial and temporal variation in meltwater conductivities described in Sections 14.3 and 14.4 is that appropriate values for the operative terms EC_e and EC_s are likely to vary from glacier to glacier. Collins (1977) determined 2 and 44 μS cm^{-1} to be appropriate values for EC_e and EC_s for the Gornergletscher; Gurnell and Fenn (1984) used 18 and 50 μS cm^{-1} for the Tsidjiore Nouve glacier; and Oerter *et al.* (1980), using a variant of the above mixing model, took 18 and 350 μS cm^{-1} to represent lumped meltwaters and groundwaters, respectively.

Values for EC_e and EC_s may best be determined through direct measurement of supraglacial and subglacial conductivities. It may not, however, always be possible to determine values for EC_e and EC_s from direct measurements. In such cases, estimation is required. On the basis of available measurements of supraglacial EC (see Table 14.2), a value of 3–5 μS cm^{-1} would seem to be a reasonable one to take for EC_e. If the glacier is predominately free of debris, EC_e may be lower, whereas if the glacier surface carries a large morainic cover, EC_e may be higher. The EC of waters emerging from any groundwater springs in the proglacial zone may provide an indication of EC_s. Alternatively, EC_s may be approximated directly from the proglacial EC record (the availability of which is, of course, a prerequisite for flow separation). The value of EC during recession flow periods, when most of the runoff is contributed by waters held in storage within the system, may be assumed to represent EC_s reasonably well. Available proglacial EC data (see Tables 14.3 and 14.4) suggest that 60 μS cm^{-1} may be a reasonable value to expect for EC_s during the summer ablation period.

The need to use suitable values for EC_e and EC_s lies in their effects on the Q_e and Q_s series generated. Gurnell and Fenn (1984) have shown that the actual values taken for EC_e and EC_s substantially affect the absolute and relative magnitude of the resultant Q_e and Q_s components. It is notable, however, that the serial pattern in Q_e and Q_s series remains effectively stable whatever values of EC_e and EC_s are used (Gurnell and Fenn, 1984).

14.5.2.2 *Volumes and patterns of englacial and subglacial flow components*

Concurrent Q_t, Q_e, Q_s and EC series have been used to investigate the hydrological systems of three Swiss glaciers (Gorner, Findelen and Tsidjiore Nouve) and one Austrian glacier (Vernagtferner). In all cases, both Q_e and Q_s series exhibit diurnal cycles. The relative magnitude, shape and timing of englacial and subglacial flow cycles differs from basin to basin, however. Table 14.7 presents details relating to each of these properties, as they are described in the original studies. It is clear that (a) the relative importance of Q_e and Q_s components of total flow differs from glacier to glacier and (b) the shape and relative timing of diurnal Q_e and Q_s fluctuations differ dramatically from glacier to glacier, varying from an almost perfectly out-of-phase pattern in one direction [Gornergletscher, with a lag (Q_e leading) of 12–14 h], through a perfectly in-phase pattern (Findelengletscher, with a lag of 0 h), to an out-of-phase pattern in the opposite direction [Tsidjiore Nouve, with a lag (Q_s leading) of 1–13 h].

Such observations suggest that different alpine glaciers possess different flow routing system characteristics. Collins' (1977) englacial and subglacial model has proved to be a useful structural basis for interpreting observations from a particular glacier. Collins' (1979b) model for the Findelengletscher involves widespread interconnection between Q_e and Q_s pathways, with diurnal water pressure fluctuations generating in-phase responses in Q_t, and Q_e and Q_s. Collins' (1977, 1979b) model for the Gornergletscher, in contrast, involves a greater degree of buffering between Q_e and Q_s flows, with diurnal water pressure peaks forcing some basal waters into cavity storage, resulting in a lag between the Q_e peak occurring in-phase with water pressure fluctuations and the Q_s peak which is activated as water pressure falls and ice pressure rises. Gurnell and Fenn's (1984) model for the Glacier de Tsidjiore Nouve involves a translatory flow effect with waters delayed in the basal conduit system overnight being forced out to form a Q_s peak ahead of water queued in the englacial system. Each of the above models invokes a different hydraulic effect operating within an interconnected englacial–subglacial drainage system to explain flow routing phenomena particular to the glacier in question. It can be seen that EC-based flow separation studies enable characteristics of the quick-flow and delayed-flow contributions from a given glacier to be identified, compared with those of other basins and interpreted. Such a process may provide valuable insights into the nature of the structure and performance of glacier drainage systems.

The above discussion has focused on the contribution that EC-based flow separation studies have made to studies of glacier hydrology. The ability to

TABLE 14.7   Magnitude, form and timing properties of the Q_e, Q_s and Q_t series of the proglacial streams of the Gornergletscher, Findelengletscher, Vernagtferner and Glacier de Tsidjiore Nouve

	Gornergletscher, 68.9 km² (Collins, 1977, 1979b)	Findelengletscher, 19.1 km² (Collins, 1979b)	Vernagtferner, 9.6 km² (Oerter et al., 1980)	Tsidjiore Nouve 3.1 km² (Gurnell and Fenn, 1984)
Relative proportions of Q_e and Q_s in Q_t	Q_e 10–80% of Q_t, 60–80% during normal ablation. 75% of ablation season drainage Q_s: 30–70% for a few hours each day. <30% of Q_t during high flows. Ca.40% of Q_t during intermediate flows. Up to 89% of Q_t during low flows. No secular change over ablation season	Q_e: maximum around 60% of Q_t in early August. Q_s: Increases in importance in late season, and during low ablation periods. Large diurnal variation in relative proportions of Q_e and Q_s	Q_e $(Q_1 + Q_2)$: 76%. Q_s $(Q_3 + Q_4)$: 24% (data determined for one day)	
Shape of Q_e	Asymmetric diurnal cycles with steep rising limb (steeper than those of Q_t)			Smooth cycles
Shape of Q_s	High-amplitude, asymmetric diurnal cycles with slow rise to peak and short falling limbs. Mirror image of cycles in Q_t and Q_e	Low amplitude diurnal cycles		Cycles with a steep rise to peak (4–7 h), and a sharp fall from peak, followed by a period of fairly stable flow preceding the next peak

Timing $Q_e - Q_s$	Out-of-phase: Max. Q_s 0–2 h after min. Q_e Min. Q_s within 3 h of max. Q_e	In-phase	Out-of-phase: Max. Q_s 1–13 h before max. Q_e (mean = 5 h) Min. Q_s 0–12 h after max. Q_e (mean = 4 h)
Timing $Q_t - Q_e$	In-phase: Similar in shape and timing	Perfectly in-phase	Out-of-phase Q_s $(Q_3 + Q_4)$ leads Q_e $(Q_1 + Q_2)$ (by inference, Gurnell and Fenn, 1984)
Timing $Q_t - Q_s$	Out-of-phase: Max. Q_s 0–2 h after min. Q_t	In-phase	

separate proglacial discharge into component Q_e and Q_s series may also have potential in relation to the modelling of suspended sediment concentration dynamics. Gurnell and Fenn's (1984) study revealed a far better correspondence between fluctuations of subglacial flow and suspended sediment concentration (S_t) that between total flow and S_t. Further, many of the secondary daily peaks present in the S_t series matched the peaks of the Q_e series, suggesting that the occurrence of multiple daily peaks in suspended sediment transport may be related to the stepped arrival of Q_s and Q_e flows, each of which may integrate sediments from different source areas.

REFERENCES

BEHRENS, H., BERGMAN, N., MOSER, H., RAUERT, W., STICHLER, W., AMBACH, W., EISNER, H., and PESSL, K. (1971), Study of the discharge of alpine glaciers by means of environmental isotopes and dye-tracers, *Zeitschrift für Gletcherkunde und Glazialgeologie* 7, 79–102.

BOX, G. E. P., and JENKINS, G. M. (1970), *Time Series Analysis, Forecasting and Control*, Holden-Day, San Francisco, 553pp.

CHURCH, M. (1974), On the quality of some waters on Baffin Island, Northwest Territories, *Canadian Journal of Earth Sciences*, 11, 1676–1688.

COLLINS, D. N. (1977), Hydrology of an alpine glacier as indicated by the chemical composition of meltwater, *Zeitschrift für Gletscherkunde und Glazialgeologie*, 13, 219–238.

COLLINS, D. N. (1979a), Hydrochemistry of meltwaters draining from an alpine glacier, *Arctic and Alpine Research*, 11, 307–324.

COLLINS, D. N. (1979b), Quantitative determination of the subglacial hydrology of two Alpine glaciers, *Journal of Glaciology*, 23, 347–362.

COLLINS, D. N. (1979c), Sediment concentration in meltwaters as an indicator of erosion processes beneath an alpine glacier, *Journal of Glaciology*, 23, 247–257.

COLLINS, D. N. (1981), Seasonal variation of solute concentration in meltwaters draining from an alpine glacier, *Annals of Glaciology*, 1, 11–16.

COLLINS, D. N. (1983), Solute yield from a glacierised high mountain basin, in WEBB, B. W. (Ed.) *Symposium on Dissolved Loads of Rivers and Surface Water Quantity/Quality Relationship (Proceedings of the Hamburg Symposium, 15–17 August 1983)*, International Association of Hydrological Sciences Publication 141, pp. 41–50.

COLLINS, D. N., and YOUNG, G. J. (1981), Meltwater hydrology and hydrochemistry in snow and ice-covered mountain catchments, *Nordic Hydrology*, 12, 319–334.

DAVIES, T. D., VINCENT, C. E., and BRIMBLECOMBE, B. (1982), Preferential elution of strong acids from a Norwegian ice cap, *Nature (London)*, 300, 161–163.

EYLES, N., SASSEVILLE, D. R., SLATT, R. M., and ROGERSON, R. J. (1982), Geochemical denudation rates and solute transport mechanisms in a maritime temperate glacier basin, *Canadian Journal of Earth Sciences*, 9, 1570–1581.

FENN, C. R. (1983), *Proglacial streamflow series: measurement, analysis and interpretation*, unpublished PhD Thesis, Southampton University, 425pp.

FENN, C. R., GURNELL, A. M., and BEECROFT, I. (1985), An evaluation of the use of suspended sediment rating curves for the prediction of suspended sediment concentration in a proglacial stream, *Geografiska Annaler*, 67A, 71–82.

FOSTER, I. D. L., GRIEVE, I. C., and CHRISTMAS, A. D. (1981), The use of specific conductance in studies of natural waters and soil solutions, *Hydrological Sciences Bulletin*, **26**, 257–269.

GOLTERMAN, H. L., CLYMO, R. S., and OHNSTAD, M. A. M. (1978), *Methods for Physical and Chemical Analysis of Fresh Waters*, IBP Handbook No. 8, 2nd ed., Blackwell, Oxford, 214pp.

GURNELL, A. M., and FENN, C. R. (1984), Flow separation, sediment source areas and suspended sediment transport in a pro-glacial stream, in SCHICK, A. P. (Ed.), *Channel Processes—water, sediment, Catchment Controls*, Catena Supplement 5, pp. 110–119.

GURNELL, A. M., and FENN, C. R. (1985), Spatial and temporal variations in electrical conductivity in a proglacial stream system, *Journal of Glaciology*, **31**, 108–114.

HEM, J. D. (1970), Study and interpretation of the chemical characteristics of natural waters, *United States Geological Survey Water Supply Paper*, 1473, 362pp.

KELLAR, W. D., and REESMAN, A. L. (1963), Glacial milks and their laboratory-simulated counterparts, *Geological Society of America Bulletin*, **74**, 61–76.

LEMMENS, M. (1975), De l'exportation du sodium et du potassium par les eaux de fusion d'un glacier Alpin en région calcaire, *Catena*, **2**, 215–262.

LEMMENS, M., and ROGER, M. (1978), Influence of ion exchange on dissolved load of Alpine meltwaters, *Earth Surface Processes*, **3**, 179–188.

LORRAIN, R. D., and SOUCHEZ, R. A. (1972), Sorption as a factor in the transport of major cations by meltwaters from an Alpine glacier, *Quaternary Research*, **2**, 253–256.

METCALF, R. C. (1979), Energy dissipation during subglacial abrasion of Nisqually glacier, Washington, *Journal of Glaciology*, **23**, 233–246.

OERTER, H., BEHRENS, H., HIBSCH, G., RAVERT, W., and STICHLER, W. (1980), Combined environmental isotope and electrical conductivity investigations of the runoff of Vernagtferner (Oetztal Alps), in *International Symposium on the Computation and Prediction of Runoff from Glaciers and Glacierised Areas, bilisi, 3–11 September 1978*, Akademiya Nauk SSSR, Institut Geografii, Materialy Gliatsiologicheskikh Issledovanii Khronika, Obsuzhdeniya 39, pp. 86–91 and 157–161.

ØSTREM, G. (1964), A method of measuring water discharge in turbulent streams, *Geographical Bulletin*, **21**, 21–43.

ØSTREM, G. (1975) Sediment transport in glacial meltwater stream in JOPLING, A. V., and McDONALD, B. C. (Eds.), *Glaciofluvial and Glaciolacustrine Sedimentation*, Society of Economic Palaeontologist and Mineralogists Special Publication 23, pp. 101–122.

RAINWATER, F. H., and GUY, H. P. (1961), Some observations on the hydrochemistry and sedimentation of the Chamberlain Glacier Area, Alaska, *United States Geological Survey Professional Paper*, 414-C, 14pp.

RAISWELL, R. (1984), Chemical models of solute acquisition in glacial meltwaters, *Journal of Glaciology*, **30**, 49–57.

RAISWELL, R., and THOMAS, A. G. (1984), Solute acquisition in glacial meltwaters.I. Fjallsjokull (South East Iceland): bulk meltwaters with closed system characteristics, *Journal of Glaciology*, **30**, 35–43.

REYNOLDS, R. C., JR. (1971), Analysis of alpine waters by ion electrode methods, *Water Resources Research*, **7**, 1333–1337.

REYNOLDS, R. C., JR., and JOHNSON, N. M. (1972), Chemical weathering in the temperate glacial environment of the Northern Cascade Mts., *Geochimica and Cosmochimica Acta*, **36**, 537–554.

RICQ DE BOUARD, M. (1973), Interpretations de mesures chimiques et physicochimiques sur des eaux de fusion de neige et de glace, *Zeitschrift für Gletscherkunde und Glacialgeologie*, **9**, 169–180.

SLATT, R. M. (1970), *Sedimentological and geochemical aspects of sediment and water from ten Alaskan valley glaciers*, unpublished PhD Thesis, University of Alaska, 125pp.

SLATT, R. M. (1972), Geochemistry of meltwater streams from nine Alaskan glaciers, *Geological Society of America Bulletin*, **83**, 1125–1132.

SOUCHEZ, R. A., and LORRAIN, R. D. (1978), Origin of the basal ice layer from Alpine glaciers indicated by its chemistry, *Journal of Glaciology*, **20**, 319–328.

STEDNICK, J. D. (1981), Hydrochemical balance of an alpine watershed in Southeast Alaska, *Arctic and Alpine Research*, **13**, 431–438.

TANJI, K. K., and BIGGAR, J. W. (1972), Specific conductance model for natural waters and soil solutions of limited salinity levels, *Water Resources Research*, **8**, 145–153.

VIVIAN, R. A., and ZUMSTEIN, J. (1973), Hydrologie sous-glaciare au glacier d'Argentière (Mont-Blanc, France), in *Symposium on the Hydrology of Glaciers (Proceedings of the Cambridge Symposium, 7–13 September 1969)*, International Association of Scientific Hydrology Publication 95, pp. 53–64.

WALLING, D. E., and WEBB, B. W. (1980), The spatial dimension in the interpretation of stream solute behaviour, *Journal of Hydrology*, **47**, 129–149.

ZEMAN, L. J., and SLAYMAKER, H. (1975), Hydrochemical analysis to discriminate variable source areas in an Alpine basin, *Arctic and Alpine Research*, **7**, 341–351.

Glacio-fluvial Sediment Transfer
Edited by A. M. Gurnell and M. J. Clark
©1987 John Wiley & Sons Ltd.

Chapter 15

Fluvial Sediment Yield from Alpine, Glacierized Catchments

A. M. GURNELL

ABSTRACT

This chapter draws together the components of total sediment yield from alpine glacierized catchments and briefly considers the problems of quantifying these components, their relative importance as components of total sediment load and fluvial sediment transfer in alpine glacierized drainage basins.

15.1 INTRODUCTION

Chapters 6–9 emphasized the processes of production, transport and deposition of sediment by glaciers in alpine drainage basins. Chapters 10–14 considered the hydrology of glacierized catchments and the individual components of the total sediment load transported by the fluvial system from these catchments. This brief chapter will draw together discussion of the components of fluvial sediment transport from glacierized catchments and will consider three main themes: the problems of specifying and quantifying components of the total fluvial sediment load in alpine proglacial streams (Section 15.2); the relative importance of each of the components of the sediment load in the alpine context (Section 15.3); and the generation of sediment yield from alpine glacierized drainage basins (Section 15.4).

15.2 SPECIFICATION OF THE COMPONENTS OF THE FLUVIAL SEDIMENT LOAD IN GLACIERIZED CATCHMENTS

The fluvial sediment yield from a glacierized catchment depends, as in any catchment, on two groups of processes: the processes controlling the supply of sediment to the fluvial system and the ability of the fluvial system to transport the material supplied to it. Transport of dissolved and very fine suspended sediment (washload—generally taken to consist of particles less than 0.064 mm in diameter) is controlled more by supply than by transport processes. This explains the typical inverse relationship between solute concentration and discharge (Chapters 11 and 14) and it also explains the peaking of electrical conductivity on the rising limb of the diurnal discharge hydrograph (Chapter 12), which reflects the build-up of material within the englacial and subglacial drainage system during low flows and its removal ahead of the diurnal discharge peak. The concentration of the coarser bed-material load (generally taken to be the component of the undissolved load which consists of particles larger than 0.064 mm) becomes increasingly dependent on transport-related processes the coarser the particles being transported. This link between bed-material concentration and fluvial transport is likely to be particularly strong in glacial and proglacial environments where material of a wide range of particle sizes is usually freely available and it explains the typical positive relationship between the concentration of transported bed material and discharge. Bed-material load is usually subdivided into suspended sediment load and bedload. Suspended sediment load is considered to be the portion of the bed-material load which is maintained in the flow by turbulent mixing processes, whereas bedload is transported by saltation, rolling or sliding along the channel bed. This distinction between the two components of the bed-material load is somewhat arbitrary in any fluvial environment but is particularly problematic in the context of proglacial, subglacial and englacial streams. Steep alpine proglacial streams frequently exhibit a step-pool morphology (Section 13.5) so that water cascades rapidly and turbulently between fairly deep pools within which flow velocities may be relatively low. The very variable flow velocities and levels of turbulence associated with this type of stream morphology lead to enormous overlap in the particle size ranges of material moving in suspension or as bedload in different segments of the stream system. The overlap in particle size range of suspended material and bedload is further complicated in the subglacial and englacial environments, where water can be flowing under strong hydrostatic pressures over significant distances (Chapter 10). Difficulties in accurately monitoring bed-load transport in the high-energy, predominantly braided channel environments of proglacial streams and the very coarse particles which have been trapped using some techniques for sampling suspended sediment (Chapter 12), indicate that the distinction between the two types of load is not only difficult on theoretical grounds in glacierized catchments, but estimates of sediment transport are likely to be very strongly influenced by the monitoring techniques employed.

15.3   COMPONENTS OF THE FLUVIAL SEDIMENT YIELD OF ALPINE GLACIERIZED CATCHMENTS

In spite of the difficulties in monitoring the components of the fluvial sediment load in alpine glacierized catchments, some attempts have been made to

TABLE 15.1 Ablation season yield of suspended sediment and bedload from some alpine glacierized catchments. Data sources: Beecroft (personal communication; Hammer and Smith (1983); Kjeldsen (1981); Kjeldsen and Østrem (1980)

Glacier	Catchment area (km²)	Glacier cover (%)	Year	Bedload t	%*	Suspended t	%*
Tsidjiore Nouve, Switzerland	4.8	70	1981	6400	40	9500	60
			1982	4300	33	8700	67
Bondhusbreen, Norway	12.6	87	1979	5200	56	4100	44
			1980	3060	42	4300	58
Engabreen, Norway	50	76	1979	7300	37	12200	63
			1980	8600	36	15500	64
Nigardsbreen, Norway	65	72	1979	14000	43	18400	57
			1980	5200	30	11900	70
Hilda, Canada	2.24	?	1977	784	59	548	41
			1978	981	55	808	45

*Percentage of total bed material load.

quantify the relative magnitude of bedload and suspended sediment yield. Table 15.1 presents estimates of the bedload yield and suspended sediment yield over complete ablation seasons from four proglacial streams and one subglacial stream (Bondhusbreen). Suspended sediment concentration was estimated by filtration of water samples in every case but three different methods were employed to estimate bedload transport. In the case of the Hilda glacier study (Hammer and Smith, 1983) bedload was estimated by the resurvey of the surface of bed material as it accumulated behind a wire mesh fence which traversed the proglacial stream bed. Bedload transport rates at Bondhusbreen (Kjeldsen, 1981; Kjeldsen and Østrem, 1980) and at Tsidjiore Nouve were estimated from the amount of material that accumulated in a gravel sedimentation chamber, whereas at Engabreen and Nigardsbreen (Kjeldsen, 1981; Kjeldsen and Østrem, 1980) the rate of accumulation of deltas where the proglacial streams entered lakes formed the basis of bedload yield estimation. Table 15.1 covers only a very small sample of drainage basins but it suggests that the yields of suspended sediment and bedload from alpine glacierized catchments may be of a similar order of magnitude, with suspended sediment yield representing a slightly larger component of the total sediment yield in the small sample of basins considered. It is impossible to estimate how representative of alpine glacierized drainage basins these data may be, but the figures in Table 15.1 show a marked contrast with estimates from the much colder environment of Baffin Island, where Church and Gilbert (1975) suggested bedload: suspended sediment load ratios of 9:1 (Ekalugad Fiord, South River) and 8:2 (Lewis River, 1963 and 1964). However, these estimates from Baffin Island were produced using very different methods from those employed in the studies in Table 15.1. They were based on more restricted field monitoring and on the application of bedload transport equations, which may partly explain the contrast with the estimates presented in Table 15.1.

It has been suggested that the solute yield (Chapter 11) and the suspended sediment yield (Chapter 12) of alpine glacierized catchments are well in excess of the global average. However, the relative magnitude of solute, suspended sediment and bedload yield from the same alpine glacierized drainage basin and for the same ablation season has only been quantified in detail for the Hilda glacier basin (Hammer and Smith, 1983). Hammer and Smith (1983) estimated loads of 784; 548; 42 t for the 1977 ablation season and 981; 808; 28 t for 1978 (bedload; suspended sediment; solute yield). These loads can be translated into percentage of total sediment yield for the three components as 57; 40; 3% and 54; 45; 2% for 1977 and 1978, respectively. This illustrates that the yield of suspended sediment and bedload is greatly in excess of solute yield for this catchment and a similar result would be highly probable for other alpine glacierized catchments.

15.4   GENERATION OF SEDIMENT YIELD FROM ALPINE GLACIERIZED DRAINAGE BASINS

Processes of sediment production and transport in alpine glacierized catchments and their variability in time and space have been reviewed in Section 4.2. Chapters 10–14 have provided detailed accounts concerning many of the fluvial sediment transfer processes in alpine glacierized catchments and it is the aim of this section merely to summarize and contrast these detailed accounts in order to provide a brief perspective on the transfer of the total sediment load to and through the fluvial system. The primary sediment sources are the atmosphere and the bedrock of the basin and sediment is released from these sources into the glacio-fluvial system by precipitation, wind transport, weathering and erosion processes. It is improbable that sediment will move directly from its primary source to form a part of the sediment yield of the catchment. Instead, it will accumulate and move between intermediate areas of storage until it is finally transported from the drainage basin by the proglacial stream.

In Chapter 11 many processes were described which indicate potential intermediate source areas of solutes. Thus solutes may be leached from snow and firn areas by meltwater (Section 11.3), whereas dust in summer firn layers may trap solutes by membrane filtration (Section 11.4). Ice may reject solutes into surrounding meltwater during recrystallization and refreezing processes (Sections 11.3 and 11.5). Membrane filtration may also be significant in association with mud layers at the glacier bed (Section 11.4) and ion exchange in association with clay minerals in meltwater can affect the concentration of particular solutes (Section 11.6). Thus the solute load of the proglacial stream may be affected by a wide range of supra-, en- and subglacial processes so that intermediate solute sources exist widely within the glacierized portion of an alpine catchment.

The suspended sediment load and bedload portions of the sediment yield of alpine glacierized catchments are fed from supra-, en- and subglacial and ice marginal debris sources. The yield of sediment depends on the tapping of these sources by flowing water. Although large amounts of debris may be present on the glacier surface in the ablation zone of the glacier, the transport of material from this area is relatively small because of the generally coarse calibre of the rock debris, because of the fairly low power of all but the largest supraglacial streams and because of the location of the main medial moraines which form the interfluves of supraglacial catchment areas and so are remote from the main supraglacial streams draining the central portions of these supraglacial drainage basins.

Very little is known about the transport of suspended sediment and bedload in englacial streams, but it seems that sizeable loads of sediment are only likely to be transported by such streams where the flow is under hydrostatic pressure and where the stream has been able to entrain debris upstream from the debris-rich lower layers of the ice or from the glacier bed. Where rock debris exists at the base of a glacier, a wide range of particle sizes is likely to be present to feed both the suspended sediment load and bedload of subglacial and in some cases englacial streams. However, the continued supply of sediment from subglacial stores and from debris-rich layers in the lower sections of the glacier can only be maintained if different areas of these sediment stores are continually exposed to meltwater flow, and this can only be achieved by movements in the course of subglacial streams or movements of material by the ice body. It might, therefore, be expected that a high yield of sediment from an alpine glacierized catchment would be promoted, at least in the short term, by high meltwater discharges, very active stream systems with frequent changes of course and expansion of the stream network in the headwater areas and very active ice movements. These conditions are most likely in drainage basins where the glacier is advancing and in particular where the glacier is surging.

It would appear that sources of solutes are generally more widely distributed in the glacierized portions of alpine catchments than are sources of suspended sediment load or bedload. However, in the immediate proglacial zone of alpine drainage basins, adjustments in the solute load appear to be negligible whereas suspended sediment transport and bedload transport can vary widely. Chapter 16 will consider fluvial processes in the proglacial environment and their relationship with the suspended sediment load and bedload provided by meltwaters draining the glacierized portion of alpine drainage basins. The glacio-fluvial sediment system will then be considered within the applied context of the Grande Dixence hydroelectricity scheme in Chapter 17.

REFERENCES

CHURCH, M., and GILBERT, R. (1975), Proglacial fluvial and lacustrine environments, in JOPLING, A. V., and MACDONALD, B. C. (Eds.), *Glaciofluvial and Glaciolacustrine*

Sedimentation, Society of Economic Palaeontologists and Mineralogists Special Publication 23, pp. 22–100.

HAMMER, K. M., and SMITH, N. D. (1983), Sediment production and transport in a proglacial stream: Hilda glacier, Alberta, Canada, *Boreas*, **12**, 91–106.

KJELDSEN, O. (1981), *Materialtransportundersøkelser i Norske Bre-elver 1980*, Vassdragsdirektoratet Hydrologisk Avdeling Rapport 4–81, 41pp.

KJELDSEN, O., and ØSTREM, G. (1980), *Materialtransportundersøkelser i Norske Bre-elver 1979*, Vassdragsdirektoratet Hydrologisk Avdeling Rapport 1–80, 43pp.

SECTION IV

Implications

Glacio-fluvial Sediment Transfer
Edited by A. M. Gurnell and M. J. Clark
©1987 John Wiley & Sons Ltd.

Chapter 16

Proglacial Channel Processes

C. R. FENN and A. M. GURNELL

ABSTRACT

This chapter emphasizes proglacial channel processes within alpine proglacial (valley train) environments, although studies of these zones are set within the broader context of studies of the continuum of sandur environments from broad sandur plains to very confined valley trains. The discussion emphasizes the rates and modes of river channel adjustment in planform and in cross-section in proglacial valley train environments rather than the sedimentary consequences of such adjustments. Four main themes are considered: the nature of the constraints which differentiate valley train from sandur plain processes; channel braiding processes; adjustments in proglacial channel pattern at a range of spatial and temporal scales; and the interactions between proglacial braided channel form and size and the transmitted discharge and sediment regime.

16.1 INTRODUCTION

Following the discussion of sediment transport processes in alpine, glacierized drainage basins (Chapters 11–15), this chapter will consider the ways in which the alpine proglacial fluvial system adjusts to variations in discharge and sediment load delivered from the glacierized part of the basin. The proglacial area is defined as 'the area which is about to be or has been covered by glacial ice, along with the narrow zone (1–2 km) beyond the maximum extent of an ice mass in which meltwater activity can result in significant landform modification' (Price, 1980, p. 79). Such areas are characterized by great aggradation and are collectively termed sandurs. There are essentially two types of sandur: those formed where meltwater from a glacier flows out on to an extensive plain and those where the meltwater is confined within a valley. The former type may be called an 'outwash plain', 'sandur' or 'plain sandur' and the latter type may be called a 'valley sandur' or 'valley train.' Krigström (1962) described how the development of a plain sandur occurs when a 'number of large meltwater streams separately build up extensive aggradational plains, upon

FIGURE 16.1   (A) An arctic plain sandur and stream: Kverkjökull, Iceland. (B) An
alpine valley train and stream: Tsidjiore Nouve, Switzerland

which they branch out more and more to produce an increasingly complicated network of braided rivers. The separate networks widen and often join further downstream and a vast composite sandur plain is formed.' A valley train, in contrast, is usually characterized by having 'only one main water channel, which, however, branches out and creates a large network of smaller, secondary channels. This network does not fill up the whole valley continuously (Krigström, 1962, p. 329). Although this chapter is concerned with processes associated with valley trains, some reference is necessary to studies undertaken on plain sandurs for two main reasons. Firstly, because the pioneering studies of the hydrology and morphology of proglacial environments were undertaken on arctic and subarctic sandur plains (e.g. Thorarinsson, 1939; Hjulström, 1952, 1954; Arnborg, 1955). Secondly, sandur plains and valley trains are essentially subject to the same glacio-hydrological influences: while lateral constriction of the outwash zone induces definable differences in the fluvial geomorphology of the two environments (Section 16.2), such differences are invariably of degree rather than kind. The fluvial processes associated with a valley train can be viewed as an environmental variant of those associated with a plain sandur, and the plain sandur and valley train may profitably be regarded as end members of a continuum of proglacial outwash environments, the morphological characteristics of any particular stream system appearing from the current literature to be prescribed more by local conditions than by latitudinal location (Figure 16.1), although future research in high latitude areas may suggest otherwise.

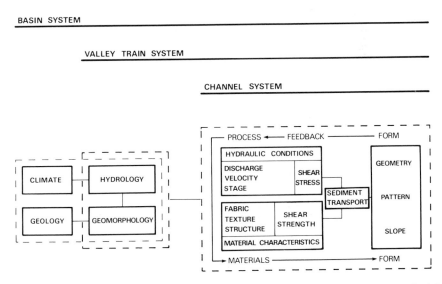

FIGURE 16.2 Factors influencing form, process and material relationships in proglacial channels

As in any alluvial stream system, the size and shape of proglacial channels (in plan, in profile and in section) are related to the character of channel bed and bank materials and to the quantity and quality of water and sediment discharged into and through the stream system. Proglacial channel form is accordingly influenced by a variety of factors which operate at a variety of levels from the basin scale, where general glaciological, geological, climatic, hydrological and geomorphological characteristics are important, to the channel scale, where flow–particle interactions are important. Figure 16.2 summarizes the nature of the process system involved. Given the instability of outwash zone deposits and the considerable variability of outflow water and sediment discharges, proglacial stream channels are morphodynamically unstable. A stage of dynamic equilibrium is characteristic, with channels adjusting their dimensions and their geometry to accommodate rapidly and widely varying water and sediment inputs.

The focus of proglacial studies, irrespective of the spatial extent and morphological type (plain sandur or valley train) of the particular site, has almost invariably been with the whole outwash plain and river channel environment, including considerations of the river hydrology, hydraulics and mechanics as well as sedimentary processes and structures in general (e.g. Boothroyd and Ashley, 1975; Church, 1972; Churski, 1973; Fahnestock, 1963, 1969). This chapter will not attempt to be so comprehensive but will emphasize the rates and modes of channel adjustments in both plan and cross-section in proglacial valley train environments rather than the sedimentary consequences of such adjustments. Four main themes will be considered: the nature of the constraints which differentiate valley train from sandur plain processes (Section 16.2); the nature of the braiding process in proglacial stream systems (Section 16.3); the adjustments in proglacial channel pattern at a range of temporal and spatial scales (Section 16.4); and the interactions between proglacial channel form and size and the discharge and sediment transport regime (Section 16.5).

16.2   DIFFERENTIATION OF VALLEY TRAIN FROM SANDUR PLAIN PROCESSES

Sandur plain deposits are built up by the accumulation of glacial and fluvioglacial sediments.

> 'In these areas eskers, kames, delta-moraines and large outwash accumulations may be found in juxtaposition, occasionally partly or completely burying the morainic topography. . . . Generally these expanses of fluvio-glacial landforms are related to mid-latitude ice caps and ice sheets which have disintegrated rapidly by downwasting. They are less common in glaciated troughs in the uplands. This is partly because valley glaciers often remain

active during retreat, inhibiting the survival of subglacial and englacial meltwater forms . . . Also, meltwater features built on valley floors are often destroyed or buried by pro-glacial fluvial processes, which have to operate on a relatively narrow valley floor.'

(Sugden and John, 1976, p. 133)

This description of outwash accumulations shows how the constriction of deposits into a valley causes them to be more likely to reflect the influence of recent processes, so that valley train deposits would be expected to reflect proglacial fluvial processes more than processes associated with glacial activity or subglacial and ice-marginal fluvial activity during glaciation. The valley location has very specific implications for fluvial activity in the proglacial zone, notably the degree of significance of low-frequency, high-magnitude discharge events; the mode of delivery of water and sediment to the proglacial zone; the extent of the influence of local base levels and sediment sinks; the role of vegetation in stabilizing proglacial sediments. Each of these influences will be examined in turn.

Low-frequency, high-magnitude meltwater discharge events would be expected to influence an increasing proportion of the valley train deposit the more restricted the valley bottom. Major meltwater outburst events could well modify the entire river channel system because they provide a 'source of energy which is expended in the transport of huge quantities of sediments and in locally severe erosion of bedrock surfaces and previously deposited proglacial sediment'

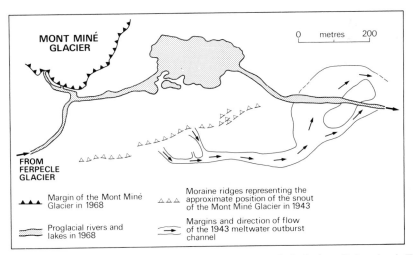

FIGURE 16.3   Outwash zone of the Ferpècle and Mont Miné glaciers, Switzerland. The map was produced in 1968 by members of the Department of Geography, University of Southampton

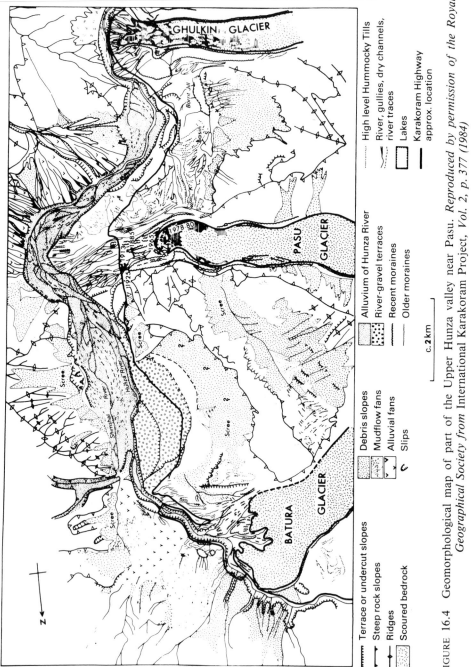

FIGURE 16.4   Geomorphological map of part of the Upper Hunza valley near Pasu. *Reproduced by permission of the Royal Geographical Society from International Karakoram Project, Vol. 2, p. 377 (1984)*

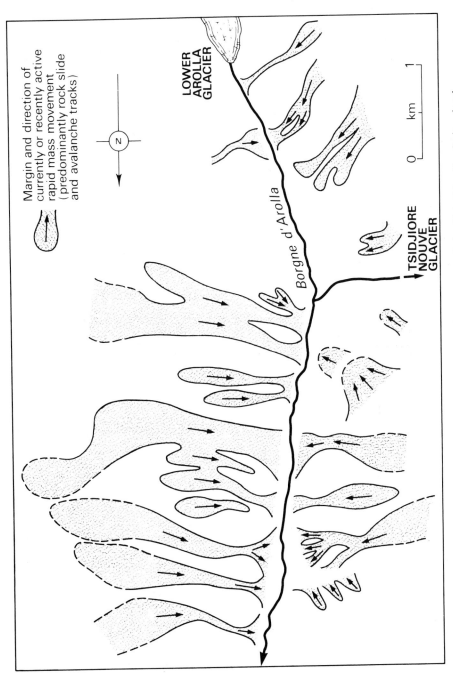

FIGURE 16.5   Avalanche and debris slide tracks along a part of the Val d'Arolla, Switzerland

(Derbyshire *et al.*, 1979, p. 270). Haeberli (1983) described the frequency and characteristics of glacier floods in the Swiss Alps. He identified two frequent types of meltwater outburst hydrograph: the 'sudden break' type yields an almost immediate hydrograph peak followed by a rapid recession, where the peak discharge is very high in comparison with the volume of water delivered during the outburst; the 'progressive enlargement of channels' type results from the progressive growth of intra- and subglacial channels during the outburst and so exhibits a gradual rising limb followed by an abrupt falling limb, where the peak discharge is much smaller in comparison with the discharge volume than in the sudden break type. A sudden break outburst flows down valley behind a frontal wave and is by far the more dangerous of the two outburst types. It will also have the greater impact on the valley train. Figure 16.3 shows the morphological evidence of a relatively small sudden break outburst which occurred in 1943 in the Ferpècle valley, Switzerland. In spite of its relatively small peak discharge and volume in comparison with many of the meltwater outbursts that have been recorded in Switzerland (peak discharge was estimated at $230 \, \text{m}^3 \, \text{s}^{-1}$ and discharge volume was estimated at $255\,000 \, \text{m}^3$; Haeberli, 1983), this outburst caused channel enlargement along the Borgne de Ferpècle and reworking of the majority of the surface of the valley train as far as the junction of the Borgne de Ferpècle with the Borgne d'Arolla (almost 6 km downstream). Flooding occurred at many sites for at least a further 20 km downstream along the Borgne to its junction with the river Rhône at Sion. Figure 16.3 shows the position of the channel cut by the outburst near to the snout of the Mont Miné glacier on a map produced in 1968. The channel is still clearly visible today over 40 years after the outburst.

A valley location permits the valley side slopes as well as the glacierized head of the valley to provide sediment and discharge to the valley train. Figure 16.4 illustrates a whole range of sediment inputs to a part of the Upper Hunza valley, Karakoram mountains, Pakistan. In this area, the relief is rarely less than 4000 m in the main valleys and the river flows along a constricted floodplain which is subject to inputs of water and sediment from many different sources. 'Nearly everywhere the higher peaks and ridges descend precipitously, in snow- and ice-covered avalanche slopes to glacier basins below, in jagged aiguilles and bare rock faces scarred with avalanche chutes, gullies and mudflow channels, or in massive debris slopes (screes) toward the fans, moraines, low terraces, valley fills and channel gravels in the floors of the main valleys' (Goudie *et al.*, 1984, p. 378). Goudie *et al.* (1984) assessed the magnitude and frequency of delivery of sediment by mass movements on the valley sides. Large deep-seated failures were estimated to have a frequency of 1 in 50 years and an associated recovery time of 1000 years. Large mudflows had a return period of 5–10 years with a recovery period of 100–1000 years. Small debris flows had a 1–10 year return period with a 100 year recovery time and small rock falls had a 1 year return period with a recovery period of 10–100 years. Clearly, this landscape is exceptionally active and the mass movements and flows of sediment from glaciated side valleys are having a marked effect on the form of the valley bottom and the route and pattern of the Hunza river. This example shows a suite of extremely active fluvial, glacial and mass movement processes operating alongside one another. Similar processes can also be observed in less active mountain environments. Figure 16.5, for example, shows the extent of inputs of debris along the Val d'Arolla, Valais, Switzerland, below the Lower Arolla and Tsidjiore Nouve glaciers. Lateral inputs of sediment from such processes have a notable impact on the down-valley trend in particle size of the valley train deposits as well as their volume and surface form (e.g. Bradley *et al.*, 1972; Fahnestock and Bradley, 1973). Mass movements from the valley sides not only deliver water and sediment to the valley train, but they may also block the valley, causing a change in the local base level for the fluvial system.

Local interruptions to fluvial processes can have major effects on the fluvial system of valley trains because of the confined, valley-bottom, location. On a plain sandur, an individual small lake, for example, may have very little influence on the overall pattern, capacity or form of the drainage network, but in a narrow valley a small lake could extend across the major part of the valley bottom and so could intercept all of the valley bottom drainage. The larger the lake, the greater its effect would be on the sediment load and the flow routing of the river system. Valley train deposits often change significantly where the river system reaches a major lake. Fahnestock (1969) described how the Kluane

FIGURE 16.6 Channel responses to a snow-cored debris slide: initial diversion, subsequent dissection — proglacial zone of the Lower Arolla glacier, Switzerland

Lake trapped all of the sediments apart from the silt and clay delivered to it by the Slims river from the Kaskawulsh glacier. The Slims river system was adjusted to a base level controlled by the extension of delta deposits into the lake and the fluctuations of the lake water level. Fans of debris derived from mass movement on the valley sides can also create significant local base levels. Small (1973), investigating 'braiding terraces' in the Val d'Hérens, Switzerland, concluded that adjustments of channel pattern during the development of the terrace system resulted from phases of spasmodic channel incision related to the growth and breakthrough of large detrital fans downvalley from the basins in which terrace development was concentrated. Figure 16.6 illustrates a situation following partial blockage of the valley train of the Lower Arolla glacier by a snow and debris avalanche. The response of the proglacial stream involved channel diversion and regrading prior to and following dissection of the impeding feature. Even more significant than lateral mass movements in creating local base level adjustments are the blocking of main-valley river systems by glacier advance across the valley bottom from side valleys. Hewitt (1982), reviewing the occurrence of glacier dams and associated meltwater outburst activity in the Karakoram Himalaya, found that glacier dams were more common than landslide dams created by mass movement processes. Such dams caused the ponding back of rivers to create local (lake) base levels which provided a control for upstream fluvial activity. Associated outburst floods had enormous effects on the valley train downstream. Tufnell (1984) has recently catalogued the occurrence and impact of similar events in the European Alps.

A final aspect of the valley train environment which has a significant effect on the development and maintenance of the drainage pattern in comparison with plain sandur locations is the development of vegetation cover in the valley bottoms. Whereas plain sandurs occur largely in areas where glaciation is maintained as a result of the latitude, valley trains occur largely in areas where glaciation is maintained as a result of the altitude. The greater the altitudinal effect, the more rapidly vegetation is likely to develop on the floodplain downstream with decreasing altitude. Thus in many valley train deposits, vegetation can stabilize the floodplain and increase the cohesion and resistance to modification of the valley train deposits in comparison with a plain sandur environment. The significance of vegetation in stabilizing the flood plain was discussed by Smith (1976). Nevertheless, with increasingly high latitude, the influence of permafrost may also become very significant in affecting the strength and cohesion of sediments for prolonged periods of the year and this may have a stabilizing influence on some plain sandur environments.

16.3   BRAIDED CHANNEL PROCESSES

Threshold conditions for single-thread meandering, wandering and classic braided channel patterns have been considered in Section 5.3.1. This section will concentrate on the processes associated with the characteristically braided channel patterns of river systems developed on sandur deposits. A range of conditions have been suggested as being conducive to the development of a braided channel pattern (see, for example, Church and Gilbert, 1975; Cheetham, 1979; Knighton, 1984). It appears that the presence of an abundant sediment load, particularly bedload; readily erodible channel banks; very variable discharge; and steep longitudinal channel slopes are often associated with braided channel patterns. The first three of these factors are characteristic of proglacial rivers in general and, additionally, alpine proglacial environments also possess steep longitudinal slopes.

Studies of braided river systems in proglacial environments have often been directed towards detailed consideration of sedimentary structures in the fluvial deposits and notably towards the interpretation of structures associated with the deposition, erosion and movement of both small and large bed forms (e.g. Williams and Rust, 1969; McDonald and Banarjee, 1971; Boothroyd and Nummedal, 1978). Sedimentary structures will not be considered here but information will be presented on the form and adjustment of proglacial braided streams as a background to the discussion of river planform changes (Section 16.4) and river cross-section changes (Section 16.5) at different spatial and temporal scales in proglacial valley train environments. Some consideration of sedimentary evidence in relation to palaeohydrology is included in Chapter 5.

Many studies have described the processes by which the proglacial braiding system is gradually utilized during the ablation season (e.g. Church, 1972; Church and Gilbert, 1975; Krigström, 1962; Fahnestock, 1963, 1969; Maizels, 1979). The most marked changes occur during periods of high discharge and/or sediment load. Channels may not be able to transmit high water and sediment loads and so may adjust to convey the load. Church and Gilbert (1975) considered that such adjustments were likely to take three main forms: general scour of the channel bed, leading to an overall increase in depth; backwater development, leading to overbank flooding; and selective scouring at certain points in the channel. The coarser the valley train deposits, the more likely it is that adjustment will take place through backwater development and overbank flooding rather than channel modification. Thus Church and Gilbert described the development of braiding through hydraulic and sedimentary processes within the channel as 'primary anastomosis,' and the reoccupation of dry channels as a result of overflow from actively flowing channels as 'secondary anastomosis.' Similarly, Cheetham (1979) differentiated between adjustments in the braiding pattern resulting from essentially sedimentological processes and channel wandering resulting from flow instability under high-energy flow conditions, and Carson (1984) differentiated between 'braiding' as a truely depositional pattern and 'wandering' type 1 as an erosional pattern formed by the dissection of the active floodplain. Further discussion of terminology is

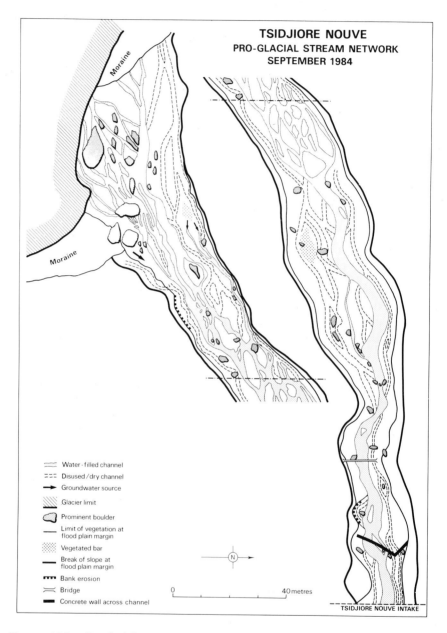

FIGURE 16.7   Proglacial stream network of the Glacier de Tsidjiore Nouve showing active and dry channels as surveyed in September 1984

presented in Section 5.3.1 but in the present context Church and Gilbert's terminology will be employed.

The coarse valley train of the glacier de Tsidjiore Nouve provides an example of a location where the development of a braided channel pattern through secondary anastomosis processes is favoured. The presence of relict channels can be clearly seen on the floodplain (Figure 16.7) and many of these channels are reoccupied during periods of high discharge. On the Tsidjiore Nouve proglacial stream, the degree to which each of the dry channels is reoccupied depends mainly on the magnitude of the discharge and this varies both within diurnal discharge cycles and also through the ablation season. Although the active and dry channel patterns appear to remain remarkably stable in their location, they do change to some degree between years and can be substantially altered as a result of a major meltwater outburst event. Sufficient movement of coarse bed material can occur during very high flows to cause these more permanent changes in the location of the main water-bearing channels. In addition, the individual channels experience continual capacity and form modification (Section 16.5), illustrating that both primary and secondary anastomosis processes occur together within the same proglacial stream system.

Changes in the pattern of braiding of proglacial streams through 'primary anastomosis' processes has been widely studied and has revealed great variability in the form of braiding between different proglacial streams. This is well illustrated by Fahnestock and Bradley's (1973) comparison of braiding patterns on the adjacent Knik and Matanuska rivers. The Knik river system was subject for more than 40 years to annual breakout floods from a lake dammed by the Knik glacier, whereas the Matanuska river experienced lower peak flows and a more gradually varying meltwater flow regime. The Knik river valley train was completely submerged by the breakout floods and its braiding pattern remained remarkably stable with very large bars up to several kilometres in length, corresponding to the magnitude of the large outbreak flows. In contrast, the Matanuska valley train was rarely completely inundated with water and the braiding pattern was much more active than that of the Knik, with areas of active change moving freely back and forth across the valley train. The Matanuska bars were only of the order of a few hundred metres in length and with approximately half of the relief of the Knik bars. Sections of the Knik channel system which had undergone change since the cessation of the annual breakout floods were found to have bars of a similar size to those of the Matanuska although they exhibited lower relief and a generally finer calibre of sediment. Fahnestock and Bradley concluded that the Knik's response to repeated catastrophic floods was stability of pattern and channel form on a relatively low gradient in comparison with the Matanuska. In contrast, a much wider range of flows was capable of modifying the Matanuska channel pattern and this appeared to be a response to the large amount of relatively coarse bed material and the steeper gradient in the Matanuska river in comparison with that of the Knik.

The particle sizes of valley train deposits can be extremely variable, as can the proglacial discharge and sediment regime, with the result that different size fractions of proglacial channel bed and bank materials may be selectively scoured and deposited by different flows and under different hydraulic conditions across the proglacial zone as the braided channel system evolves. Rust (1972), for example, noted a range from -7 to $+8 \phi$ in the particle sizes of bedload in the Donjek river, Canada. Any consideration of variations in proglacial braided channel patterns must be set against information on channel bar formation and bar types, since these are the fundamental, large scale depositional features of these channel patterns. Krigström (1962) observed changes in the braided channel pattern on Icelandic sandur plains. Here, the accumulation of bedload in the form of mid-channel bars was seen to be accompanied by bank erosion. These processes can be explained in the light of the characteristically non-cohesive banks of proglacial streams, their high bedload transport and their very variable discharge regime which encourages frequent deposition and entrainment of bed material and bank erosion and thus high channel width–depth ratios. The wide and shallow stream cross-sections foster complex secondary flow patterns which, coupled with the high bedload transport and variable discharge regime, lead to deposition of coarse material in the channel and subsequent bar development. Thus bank erosion and bar development frequently occur together, although the direction of cause and effect is difficult to specify. Nevertheless, once a bar starts to form, it will cause diversion of flow which would be expected to further enhance bank erosion. Transport of material would continue over and around the bar and finer material would be deposited in the lee of the bar. Krigström (1962) described how in relatively straight sections of channel, bar development caused the flow to divide, resulting in clearly defined channels on either side of the bar, whereas in more sinuous sections of channel and at channel junctions, bar formation often resulted in diversion of flow to one side of the bar. Although the bars were essentially depositional features, flow diversion was accompanied by lateral and vertical channel modification. This modification occurs because the distributary channels are less hydraulically efficient than the parent channel and so an overall increase in slope and a mutual adjustment in other hydraulic variables occur to maintain the stream power necessary to transmit the same amount of sediment (Richards, 1982). Krigström (1962) observed the process of bar formation and channel adjustment continuing in the newly formed flanking channels. Cheetham's (1979) detailed field study of a braided meltwater stream in the Okstindan region, Norway, provides further information on the braiding process. He analysed the at-a-station hydraulic geometry (Leopold and Maddock, 1953) of parent channel (single-thread) and distributary channel (multiple-thread) reaches of the stream. There was little difference in the depth exponents but the distributaries had higher width exponents and lower width constants than parent channels, indicating a contrast in shape. They also had lower velocity exponents and higher velocity constants than the parent channels. The differences in the velocity–discharge relationships indicated that distributary channels would become sediment bottlenecks at high discharges, resulting in continued bar growth and possibly the production of second-order distributaries and bars. The differences in the width–discharge relationships suggested that once a braid bar had begun to form, it was likely to continue to develop until the distributary channels attained similar dimensions to the parent channel.

Many different forms of bar can develop in proglacial braided channel systems. Smith (1974) undertook a detailed study of bar development along a reach of the Upper Kicking Horse river, British Columbia, Canada. He noted the enormous variety in the forms of exposed braid bars and concluded that this was a result of their complex depositional and erosional histories. He stressed the impact of the daily discharge variations during summer on sediment erosion, transport and deposition in proglacial streams and its implications for the variability in the form, location and persistence of the braid bars. He identified four simple (unit) bar types which appeared to form the basis for the development of more complex bar forms: longitudinal, transverse, point and diagonal bars. Each of these bar types was initiated when a local hydraulic change reduced the stream's transporting ability and gravel was deposited. The unit bars could develop over one or a few discharge cycles. Longitudinal, transverse and point bars all had long axes parallel to the flow, tended to fine downstream and formed under slightly different hydraulic conditions. Longitudinal bars were elongated in form with long axes parallel to the channel banks and they persisted in channels where the flow was straight and equally distributed on either side of the axis of the bar. Channel sections with straight parallel banks and with flow that was parallel to the banks and symmetrically distributed in the cross section were conducive to this type of bar development. Transverse bars were fan shaped in plan with broad flat surfaces bounded by slip faces at the curved downstream margin. They tended to develop where there was flow divergence as a result of channel widening, abrupt channel deepening or at the confluence of two channels of unequal flow strength. Point bars occurred in curved channel sections with their long axes parallel to the channel banks. Smith noted that all of these three unit bar types with long axes parallel to the flow grew predominantly in a downstream or lateral direction, although some upstream growth could occur in association with the vertical development of the bar. Diagonal bars developed with their long axes transverse to the direction of the channel banks and occurred where channel cross-sectional flow was asymmetrically distributed as, for example, occurs at channel junctions, channel bends and downstream of exposed bars.

Longitudinal, transverse and point bars all tended to fine downstream, whereas particle size grading in diagonal bars was found to be more complex since it was controlled by flow variations and thus variations in the transporting ability of the flow over the bar margins during the development of the bar. Fining upwards was also characteristic of longitudinal, transverse and point bars, possibly as a result of lateral migration of the bar margins; vertical deposition of sediment during falling discharges; and sorting as a result of avalanching at foreset margins. Although this simple model of the fining upwards and downstream of the internal grain size distribution of unit bars with flow-parallel long axes is particularly applicable to bars formed during a single discharge cycle, Smith suggested that if a unit bar persisted through several discharge cycles a similar grain size pattern might be expected to that produced over a single cycle. Nevertheless, simple unit bars would be rapidly destroyed by erosion or by changing depositional patterns and 'complex braid bars' would result. Although this four-fold classification of unit bars envisaged simple features which developed over one or a few discharge cycles, the classification can be applied descriptively to the morphology of bars which have developed and persisted over long periods of time and which may have complex histories of development. The scale of the bars can vary enormously [see the discussion of Fahnestock and Bradley (1973) earlier in this section] and Williams and Rust (1969) described a hierarchy of three order categories of longitudinal, transverse and point bars, according to the size and character of the channels which delineated them.

The pattern of channels and bars in proglacial braided streams varies both in time and in space. The size and form of the bars may be linked to grain size changes (Knighton, 1984). In the absence of major lateral inputs of sediment valley train deposits tend to exhibit decreasing particle size in a down-valley direction (e.g. Bradley *et al.*, 1972; Smith, 1974; Ballantyne, 1978). Boothroyd and Nummedal (1978) indicated how this downstream trend in clast size is accompanied by a downstream adjustment from predominantly longitudinal bars in the 'coarse-braided' proximal zone to predominantly linguoid bars (which they liken to Smith's transverse bars) in the 'fine-braided' distal zone of some Alaskan and Icelandic outwash fans. Further, both Smith (1974) and Ballantyne (1978) noted that the material associated with the bars is finer than material associated with the adjacent channel bed, although Ballantyne pointed out that the magnitude of the difference decreases downstream.

Temporal variations in the pattern and size of bars and channels can also be large. If the braiding system is in equilibrium with the discharge and sediment regime over a timescale of several years, bars and channels will continue to change, but over a period of years the total sinuosity and complexity of the channel network will remain the same. However, adjustments in the flow and/or sediment regime will cause adjustments in the channel system and these can be achieved through processes of both primary and secondary anastomosis within the same channel system, although one of these will usually predominate. Section 16.4 will consider observations of changes in proglacial channel pattern. Thereafter, changes in proglacial channel cross-sections will be considered (Section 16.5).

16.4 CHANGES IN PROGLACIAL CHANNEL PATTERNS AT VARYING TIME AND SPACE SCALES

Section 16.3 considered how proglacial streams adopt a predominantly braided channel pattern. This section will examine changes in proglacial channel patterns at the following three scales:

(i) The interactions of the proglacial channel pattern with the entire valley train over prolonged periods of time (Section 16.4.1). This section will provide a brief introduction to timescales relevant to palaeohydrology.
(ii) Inter-annual changes (Section 16.4.2).
(iii) Intra-annual changes (Section 16.4.3).

16.4.1 Long-term process and form adjustments

Krigström (1962) and Fahnestock (1969) have both identified a gradual spatial adjustment in river channel pattern across valley train deposits. Krigström noted the existence of three distinct zones in active outwash environments: firstly, a proximal zone, where the river channels are 'comparatively deeply incised . . . with well defined, on the whole parallel sides. The channel usually oscillates gently. Often there is a distinct main channel' (Krigström, 1962, p. 329); secondly, a zone further downstream 'which occupies the larger part of the sandur . . . (where) the channel gradually widens and becomes shallower. The banks become less well defined and the channel may merge into a plain on which the streams can no longer keep together . . . the bedload is the dominant formative element . . . it becomes increasingly more difficult to determine the boundaries between the individual streams' (Krigström, 1962, p. 329); thirdly, a zone furthest downstream where the 'intertwined streams merge into a shallow bay with streaming water' (Krigström, 1962, p. 331). Although these three zones may not be apparent on all valley train deposits, Fahnestock (1969) was able to identify three zones on the valley train of the Slims river. The upper zone contained one to three channels which were incised into coarse gravel deposits; the intermediate zone was multi-channelled with considerable bedload transport which led, at all flow stages, to the channel pattern being dominated by the development and movement of channel bars; and the downstream zone was developed on sand-sized material which graded into a sheet of flowing water. Fahnestock linked the three zones to changes in the size of valley train deposits (which were the result of selective fluvial transport rather than of abrasion or weathering), to degradation in the upstream zone, high sediment transport and gradual degradation or aggradation in the intermediate zone and aggradation in the downstream zone. These zones of degradation, transport and aggradation reflected the gradual retreat of the Kaskawulsh glacier since the late 1800s and the associated decrease in discharge and sediment yield from the glacier. The boundaries between the three zones would be expected to shift up and downstream with fluctuations in the discharge and sediment regime.

Maizels' (1979) very detailed study of the proglacial zone of the advancing Bossons Glacier, France, provides an interesting contrast to that of Fahnestock's (1969) study of the valley train below a retreating glacier. Over a 6-year period, during which the glacier was advancing, the sediment loads delivered to the proglacial zone were in excess of the transporting ability of the stream. The valley train gradients consistently steepened as coarse debris accumulated on the proximal slope of the valley train and localized aggradation caused burial of outwash deposits of a former terminal moraine. The proximal zone of the valley train steepened as a result of rapid aggradation and became more restricted in its down-valley extent during this period. In addition, the number of streams crossing this area and the complexity of their pattern increased. Thus, the valley train maintained a clear but more restricted proximal zone of aggradation, consisting of the coarsest ice-marginal and fluvioglacial debris. The distal zone also aggraded considerably with a significant increase in 'the degree of braiding as defined by the number of channel segments, total channel length, the number of bars and the total number of nodes, bifurcations and links' (Maizels, 1979, p. 99). In interpreting the results presented by Fahnestock (1969), Krigström (1962) and Maizels (1979), it is interesting to speculate on the degree to which their observations are controlled by the changing position of the glacier snout relative to the study site and the sequence of positions of the snout in the past, or by the changing discharge and sediment transport regime associated with the changing size and character of the ice body in response to changes in climate.

Boothroyd and Ashley (1975) provided some details of the spatial variation in processes across the Yana outwash fan of the Malaspina glacier. The fan, like the valley train of the Bossons Glacier, exhibited a decreasing slope and sediment calibre in a downstream direction. The stream channels on the lower sections of the fan were stabilized by vegetation. A downstream transect of observations of flow velocity and suspended sediment concentration during the rising limb of the diurnal discharge hydrograph showed how the different zones of the fan, as defined by decreasing grain size and slope and increasing degree of vegetation cover, were matched by corresponding adjustments in the suspended sediment concentration and flow velocity, both of which decreased downstream.

We can conclude from the above discussion that valley trains, to a greater or lesser degree, contain distinct zones defined by changes in slope, sediment size, degree of vegetation cover and channel pattern. The degree to which these zones are clearly distinguishable will largely depend on the rate of sediment supply to the proglacial area and the meltwater regime available to transport and sort the sediment. Proglacial zones may aggrade, degrade or show areas of both accumulation or removal of debris over time, and although these differences may reflect the level of activity and mass balance of the glacier and, in particular, whether it is in advance or retreat, the proglacial zone actually responds to the competence of the discharge regime to transport the supplied sediment. Thus it is possible to conceive of both general aggradation and degradation occurring under conditions of both advance and retreat. Maizels (1979) has formalized a useful conceptual model capable of (tentatively) establishing the degree to which proglacial aggradation or degradation is likely under given systems conditions (Figure 16.8). In particular, the model indicates how the meltwater discharge will be low during periods of low ice cover and also relatively low during periods of extensive ice cover, with maximum discharge occurring at some intermediate volume or spatial extent of ice. The exact relationship between discharge and ice volume will vary enormously between drainage basins since it represents the response of discharge to climate (through both precipitation and ablation) as they are buffered and transmitted through the intervening mass of ice. The sediment load of the meltwater stream is even

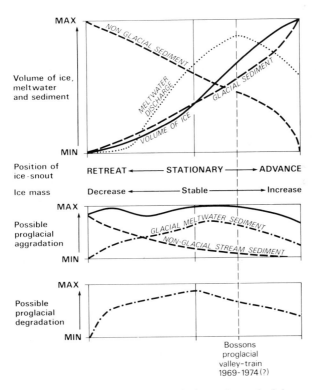

FIGURE 16.8 Tentative model of the association of proglacial aggradation and degradation with the mass balance and water and sediment supply of the source glacier. *Reproduced by permission of the Swedish Society for Anthropology and Geography from* Geografiska Annaler, *1979, 61A, 99*

more difficult to predict because both glacial and non-glacial sediment sources are able to contribute to the sediment load of the proglacial stream system and the significance of both of these sediment source areas will vary enormously between drainage basins. The combined effect of discharge and sediment yield determine whether proglacial aggradation or degradation are likely.

More recently, Maizels (1983) has refined her model and has placed her studies of channel pattern adjustments on the Bossons glacier valley train into the much longer context of prolonged periods of deglaciation as exemplified by a sample of field areas which have experienced different timescales of deglaciation. In general, Maizels indicates that deglaciation is associated with a reduction in both discharge and sediment yield. It is the balance of these two variables which is critical to the mode of change of the proglacial channel system during deglaciation. Maizels (1983) suggests that the ratio between sediment and water yield may initially increase as deglaciation commences but, thereafter, it decreases. This corresponds to Church and Ryder's (1972) model of the adjustment in the sediment yield during 'proglacial' and 'paraglacial' periods. The channel pattern adjusts to these changes in discharge and sediment supply by undergoing a change in channel pattern from a predominantly braided to a predominantly meandering habit via a series of 'transitional' channel systems and the valley train undergoes initial aggradation followed by degradation.

The significance of proglacial stream studies for palaeohydrology are discussed in Section 5.4.

16.4.2   Inter-annual changes in proglacial channel patterns

Annual resurveys of proglacial channel systems have shown that morphological adjustments to changes in sediment and flow regime are achieved very rapidly. Maizels (1979) showed that the proglacial stream system of the Bossons glacier quickly developed an increasingly sinuous and complex multiple channel network on an aggrading valley train during a period of increased sediment delivery associated with glacier advance. The channel system of the proglacial stream of the Lower Arolla also responded rapidly to a sudden input of debris from a series of lateral avalanches occurring (in October 1977) between surveys made in July 1977 and July 1978 (Fenn, 1983): the channel network experienced a general increase in sinuosity and a number of pronounced local changes in pattern at points where avalanches debouched on to the valley train (Figure 16.6 illustrates the effect of one such track), while the long profile of both the main channel and valley train suffered a net reduction in slope involving proximal zone degradation (achieved by the flood flows generated by the same storm which triggered the avalanches) and downstream aggradation (Figures 16.9 and 16.10).

Figure 16.11 shows the channel pattern in the proglacial zone of the Glacier de Tsidjiore Nouve in early July and early September over several years. The stream channels of the proglacial zone can be subdivided into 'tributary

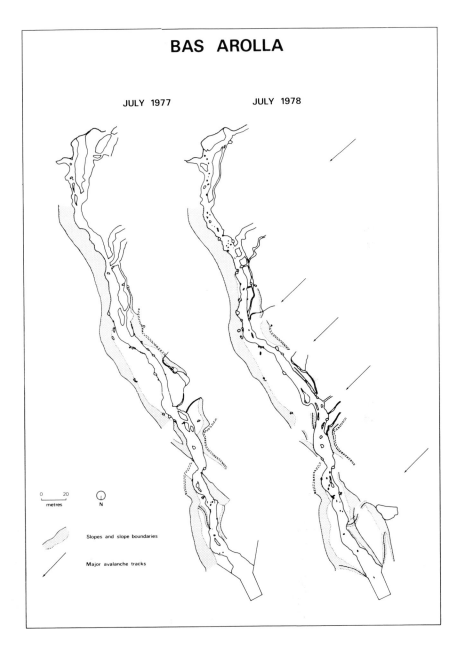

FIGURE 16.9 Planimetric geometry of the Lower Arolla proglacial stream system as surveyed in July 1977 and July 1978

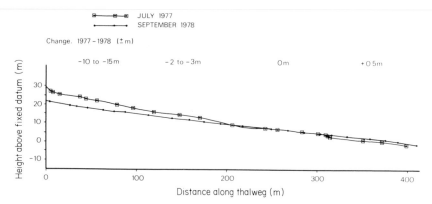

FIGURE 16.10   Long profile of the Lower Arolla proglacial stream as surveyed (along the thalweg) in July 1977 and July 1978

channels' on the proximal zone of the valley train which are isolated channels or source streams which provide inputs of water and sediment to the proglacial zone, and the 'main channel system' which develops across the valley train once the inputs of water and sediment are merged. There is evidence of modification of the channel pattern both between and within years. The tributary zone, in particular, shows clear contrasts between the large number of individual channels in the July maps and the confinement of the flow to one major channel in September (reasons for this are presented in Section 16.4.3). There is some evidence to suggest that the morphological sensitivity of the tributary streams 'buffers' the impact of fluctuating outflow discharges and sediment loads on the main channel network downstream. Very rapid and major changes occur in the relative size and position of these streams both within and between years. The main channel does, however, exhibit some variations in the pattern of braiding from year to year, with the biggest differences in pattern occurring just upstream of the position of still-stands in the glacier's retreat since the mid-1800s. Ribs of debris are present on the north-facing slope of the old lateral moraine bordering the valley train. The bases of these ribs mark the ends of areas of locally shallower valley bottom slope and a marginally broader flood plain area where braiding can occur. Church and Gilbert (1975), when considering large Baffin Island sandurs, identified sudden changes in the channel pattern associated with local areas of very active sedimentation corresponding to zones where the river's competence was exceeded. These areas were characterized by wide, shallow channels and were bounded downstream by steepening of the river's profile such that renewed entrainment of the material could occur. Thus, these areas of local erosion and deposition formed steps on the long profile of the sandur deposits. There would appear to be some parallel in the behaviour of the alpine (Tsidjiore Nouve) and arctic (Baffin Island) streams.

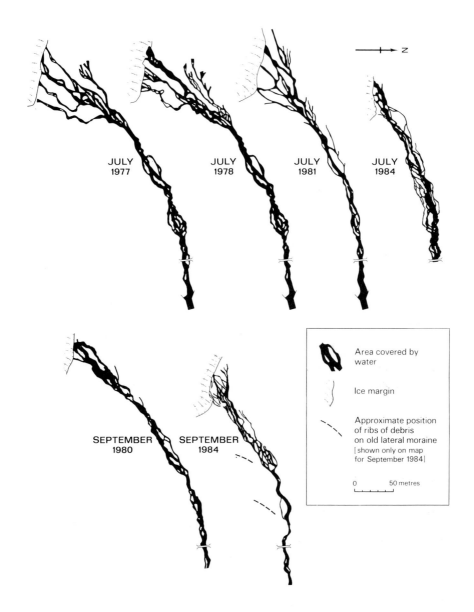

F<small>IGURE</small> 16.11   Channel patterns in the proglacial zone in the Glacier de Tsidjiore Nouve in July 1977, 1978, 1981 and 1984 and September 1980 and 1984

16.4.3 Intra-annual changes in proglacial channel patterns

Adjustments in channel pattern occur more or less continuously through the ablation season. Fahnestock (1963) noted that the analysis of channel pattern at high flows was very difficult because of the great rapidity of change, but that over the ablation season the White river, Mount Rainier, USA, showed a marked change from a meandering pattern to a braided pattern, with the onset of high summer flows, and then back to a meandering pattern with low autumn flows. Figure 16.11 shows how a similar pattern occurs on the Tsidjiore Nouve proglacial stream, which is predominantly confined to a single channel in early July and September, but which develops more complex channel braiding towards the end of July and early August with the highest flows of the ablation season. These changes in pattern are achieved mainly by the reoccupation of channels on different parts of the valley train. Williams and Rust (1969) and Ballantyne (1978) have emphasized the subdivision of the sandur surface into different levels according to the degree to which they are occupied by different flows. Ballantyne (1978) identified a main flow surface (level 1) which carried water even during low discharges, a flood surface (level 2) occupied during fairly regular high discharges and a high flood surface (level 3) which was only occupied under exceptional flood conditions on a small sandur on Ellesmere Island, N. W. T., Canada. Valley trains may not have such distinct zones when they are strongly laterally confined or if they are subject to major meltwater outbursts as in the case of the valley train of the Knik river, which until 1966 was completely washed over by outburst floods from a glacier-dammed lake at the commencement of each ablation season (Bradley *et al.*, 1972; Fahnestock and Bradley, 1973).

At a more detailed scale, it is worth considering seasonal variations in the locations of streams which deliver meltwater and sediment to the proglacial zone. At the Glacier de Tsidjiore Nouve distinct patterns of channel change occur within the tributary zone during the ablation season. Figure 16.11 shows the contrast in the pattern and number of large tributary streams between the beginning and end of the ablation season. The process of change from a multiple tributary to a single main stream draining the glacier snout occurs in a series of stages throughout the summer season. Gurnell (1982) and Gurnell and Fenn (1985) described an event in 1981 when a small outburst of water from one tributary and an accompanying sediment flush occurred. The electrical conductivity of the tributaries draining the glacier snout provided evidence for this flush being the result of temporary storage and rerouting of water from one tributary to another. Further, monitoring of electrical conductivity on a 24 h and 48 h sampling interval throughout July 1978 and July 1981, respectively, showed gradual complementary shifts in tributary conductivity which were interpreted as the result of gradual rerouting of the water during the rationalization of the intraglacial drainage network (Gurnell and Fenn, 1985). Stenborg (1973) showed how separate streams issuing from a glacier front may correspond to completely separate catchment areas within the glacier. In view of the pattern of electrical conductivity, suspended sediment concentration and discharge to proglacial tributaries at the Glacier de Tsidjiore Nouve, it seems likely that independent or semi-independent catchment areas feed these tributaries in the early part of the ablation season, but by mid-August the mixing of waters from these source areas occurs within rather than in front of the glacier system. The valley location of the glacier and the concave form of the valley bottom beneath the ice encourages the gradual movement of water towards the centre of the valley as the englacial and subglacial drainage network develops during the ablation season. This seasonal adjustment in tributary locations might be expected to be characteristic of proglacial tributaries in confined valley locations.

In addition to meltwater drainage from the glacier, another, less important source of tributary flow which seems to occur widely in proglacial areas is 'groundwater' seepage (Hjulström, 1955; Fahnestock, 1969). Such seepage may not be from long-term groundwater storage but (in igneous or metamorphic basins of limited permeability) is more likely to be the result of meltwater penetrating the coarse proximal valley train deposits and then reappearing on the surface downstream where the deposits are finer or have become clogged with a fine matrix of material between the coarser components of the valley train. Flow from such tributaries tends to be more nearly constant, with a markedly lower suspended sediment concentration than the tributaries that respond directly to meltwater inputs from the glacier. The morphological pattern of such drainage channels is likely to be more stable than those fed directly by glacial meltwaters.

16.5 CHANGES IN PROGLACIAL CHANNEL CROSS-SECTIONS AT VARYING TIME AND SPACE SCALES

The resolution with which changes in stream channel form may be identified is at its greatest at the scale of the cross-section, because resurveys can be most easily undertaken with an accuracy and frequency appropriate to the nature of the form-process adjustments occurring, and where observations of change in the data derived can be made with most veracity. Changes in channel shape may be identified through the superimposition of resurveyed cross-profiles and changes in channel capacity may be identified through the measurement of area below a fixed datum.

Very little research has focused on the detailed adjustment of proglacial channel cross-sections to the time-variant streamflow and sediment volumes fed into and through them. Morphological and hydraulic geometry surveys of proglacial channels have been undertaken (e.g. Fahnestock, 1963; Broscoe, 1972; Church, 1972; Rice, 1982) but, by and large, such surveys have occupied an adjunctive role in 'whole environment' studies and have, in consequence,

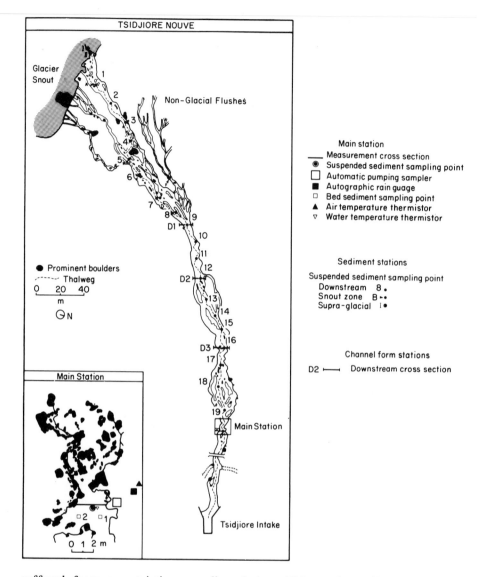

suffered from a restrictive sampling design. This section will, therefore, concentrate on the results of detailed form-process surveys undertaken in the proglacial zones of the Tsidjiore Nouve and Lower Arolla glaciers, Valais, Switzerland, although the results of these surveys will be compared with those of other studies where possible. Three main aspects of cross-sectional adjustment are considered. A discussion of annual, seasonal, daily and hourly changes in the size and shape of cross-profiles of the Lower Arolla and Tsidjiore Nouve

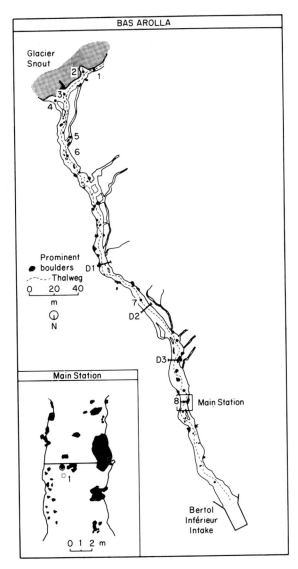

FIGURE 16.12 Locations of channel cross-section study sites employed on the proglacial streams of the Tsidjiore Nouve and Lower (Bas) Arolla glaciers

proglacial stream channels (Section 16.5.1) is followed by a comparison of the characteristics of these channels with other proglacial channels through a consideration of the at-a-station hydraulic geometries of proglacial channels (Section 16.5.2). Finally, the temporal changes in form and capacity of the

Tsidjiore Nouve stream channel are related to temporal variations in streamflow and sediment transport (Section 16.5.3).

Figure 16.12 shows the location of the sites employed in the Tsidjiore Nouve and Lower Arolla surveys. Detailed cross-section surveys were taken at a standard time and at regular intervals at a main station on each stream. The main stations were used for monitoring streamflow and sediment transport as well as channel form changes, and were sited against a fixed boulder bank to provide a stable reference point. Three other survey sites on each stream (D1, D2, D3) were also employed in order to assess the degree of spatial variability in channel response within each stream system. A fixed datum was established at each site to permit measurement of morphological cross-sectional area, hereafter referred to as channel size (note that in all cases channel size relates to the cross-sectional area of the channel below a constant datum, and not to water section area). The datum at the Tsidjiore Nouve main station represents a bankfull level, but those of all other sites are arbitrary; in the discussion that follows, attention should accordingly be placed on temporal changes in channel size and not on the absolute level of channel size (i.e. between-site comparisons of absolute channel size are invalidated). Most emphasis is placed on measurements at the Tsidjiore Nouve main station, where continuous streamflow and sediment transport data were available.

16.5.1   Adjustments in cross-section shape and size

The adjustments observed on the Tsidjiore Nouve and Lower Arolla streams were examined over a range of progressively shortening timescales.

Figure 16.13 shows the cross-sectional geometry of the main sections on the Tsidjiore Nouve and Lower (Bas) Arolla streams as surveyed in July 1977 and July 1978. Superimposition of the 1977 and 1978 profiles shows the location and extent of scour and fill occurring between the annual surveys. Both sections experienced changes in channel shape and size, but to significantly differing extents. Channel change in the Tsidjiore Nouve section involved separate zones of scour and fill, the location of which effected a more regular profile in 1978 in comparison with that of 1977, and the magnitude of which (fill greater than scour) resulted in a marginally smaller channel size in 1978 ($1.75 \, \text{m}^2$) than in 1977 ($1.81 \, \text{m}^2$), constituting a change of -3% in channel size. No major change in channel status occurred. Conversely, the changes evident in the Lower Arolla section represent a significant alteration in channel status, involving left bank fill and considerable right bank scour below and beyond the 1977 right bank survey point (from point B to point C, Figure 16.13). Channel shape was dramatically altered from a deep, single section to a wide, twin channel section (the mid-channel bar was sometimes submerged, according to stage). Channel size was also significantly altered. The size of the 1978 section ($4.26 \, \text{m}^2$) was only slightly larger than the 1977 section adjusted to the same width ($4.04 \, \text{m}^2$,

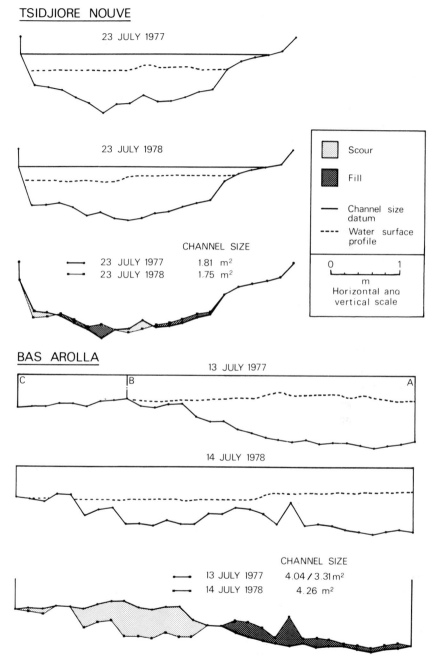

FIGURE 16.13   The Tsidjiore Nouve and Lower (Bas) Arolla main station cross-sections
as surveyed in July 1977 and July 1978

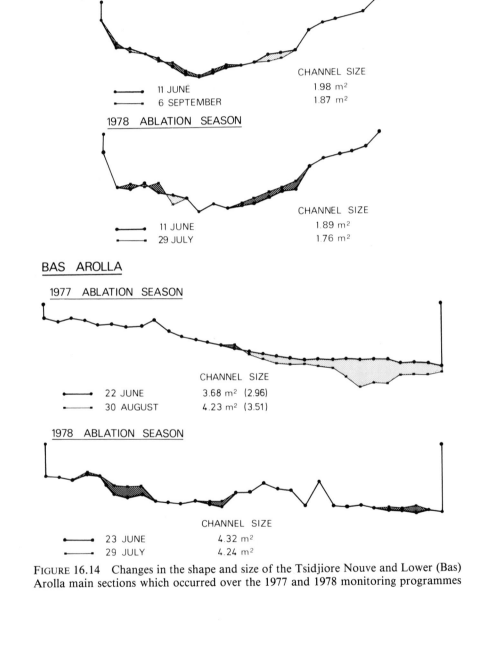

FIGURE 16.14   Changes in the shape and size of the Tsidjiore Nouve and Lower (Bas) Arolla main sections which occurred over the 1977 and 1978 monitoring programmes

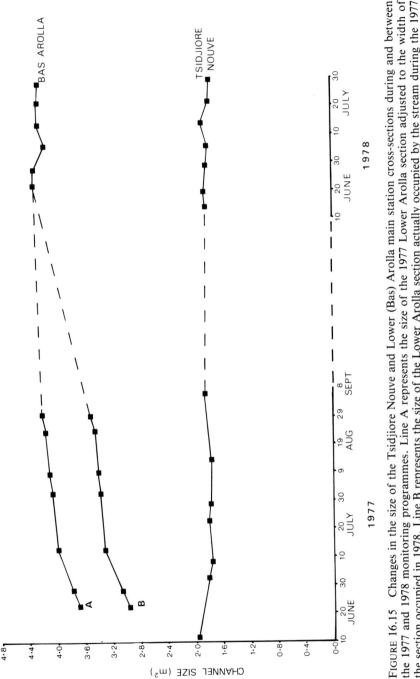

FIGURE 16.15 Changes in the size of the Tsidjiore Nouve and Lower (Bas) Arolla main station cross-sections during and between the 1977 and 1978 monitoring programmes. Line A represents the size of the 1977 Lower Arolla section adjusted to the width of the section occupied in 1978. Line B represents the size of the Lower Arolla section actually occupied by the stream during the 1977 monitoring programme

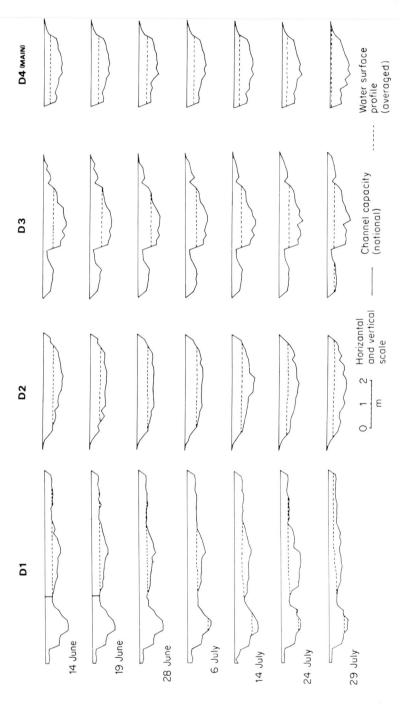

FIGURE 16.16   Channel cross-sections at stations D1, D2, D3 and D4 (main section) of the proglacial stream of the Glacier de Tsidjiore Nouve at various times during the 1978 monitoring programme

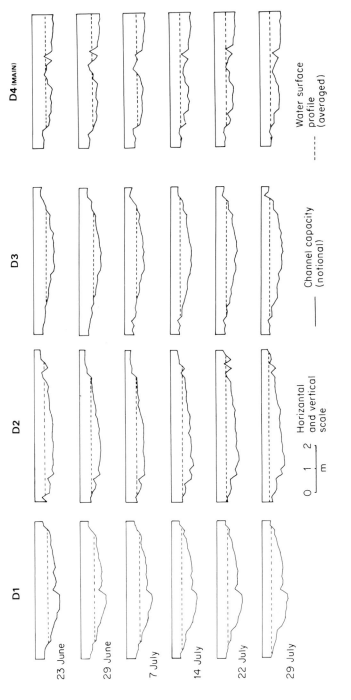

FIGURE 16.17   Channel cross-sections at stations D1, D2, D3 and D4 (main section) of the proglacial stream of the Lower (Bas) Arolla glacier at various times during the 1978 monitoring programme

i.e. a change of + 5%), but was dramatically larger than the size of the section actually occupied by the 1977 flows (i.e. 3.31 m² at 4.2 m width, a change of + 29%). From such evidence, it would appear that the Tsidjiore Nouve section maintained an equilibrium form over the two seasons, with minor adjustments in size and shape occurring as dictated by prevailing in-channel conditions, while the Lower Arolla section experienced a fundamental change in the status of its equilibrium between the two surveys. The pattern of right bank scour and left bank fill in the Lower Arolla section suggests that the left bank avalanche-debris slide event of October 1977 was instrumental in the change in channel form.

Figure 16.14 shows the changes in the shape and size of the Tsidjiore Nouve and Lower Arolla main sections which occurred during the 1977 and 1978 ablation seasons. Figure 16.15 plots the temporal change in size of the Tsidjiore Nouve and Lower Arolla sections which occurred within and between the two measurement seasons. Line A in Figure 16.15 traces the changes in the size of the 1977 Lower Arolla section adjusted to the larger 1978 width; line B shows the changes in the size of the section actually occupied by the flows of 1977. Figures 16.14 and 16.15 show that changes in boundary profile in the Tsidjiore Nouve section occurred via bottom scour and fill over each ablation season, but that no fundamental changes occurred in either the mean status (size = 1.84 m²) or in the variation of the size (0.15 m², i.e. ± 8%) of the section, either from year to year or from week to week. Such observations suggest that the Tsidjiore Nouve section experienced continuous, minor adjustments of form to process which effectively maintained the channel in a state of dynamic equilibrium. In contrast, Figures 16.14 and 16.15 show that the Lower Arolla section experienced a considerable shift in its status over the study programme. The previously noted dramatic change in channel shape (from a deep single to a wide double flow section) evidently occurred between the two ablation seasons, and may be explained directly in terms of the adjustments induced by the input of debris from the October 1977 avalanche and debris slide event. No evidence of readjustment towards the former state is shown in the behaviour of the section during the 1978 season (i.e. the adjustment in channel size observed amounted to only − 2%). However, Figures 16.14 and 16.15 illustrate that the increased size of the 1978 configuration relative to that of 1977 (channel size increased from 3.51 m² at the end of the 1977 season to 4.32 m² at the start of the 1978 season, a change of + 23%) must be seen within the context of a trend towards increasing channel size throughout the 1977 season; changes in channel status were evidently also proceeding through the normal, continuous adjustment process. Given such developments, the acceptance of the new 'imposed' status is perhaps more understandable. Indeed, when the size of the 1977 section is adjusted to the width of the 1978 section, it becomes apparent that little change in size accompanied the considerable change in shape effected by the 'allogenic' event (standardized to constant width, the size of the section increased from

4.23 m² in September 1977 to 4.32 m² in June 1978, a change of only $+2\%$).
Both autogenic and allogenic effects were obviously involved in the changes
which occurred over the study period.

Differing degrees and types of channel form adjustment therefore occurred in
the two proglacial streams examined. A variety of channel form-process
adjustments also occurred in the different sections monitored within each of the
two stream systems. Figures 16.16 and 16.17 show, respectively, the geometries
adopted at the four separate stations monitored over the 1978 ablation season in
the Tsidjiore Nouve and Lower Arolla stream systems. The surveys were taken at
approximately 8-day intervals during the 1978 measurement programme, and the
sections [D1, D2, D3, D4 (i.e. main section)] were located at approximately
50 m intervals along the stream course (see Figure 16.12 for locations). Figure
16.18 shows the intra-seasonal variations in channel size which occurred in each
section. As has been noted, the adjustments displayed by the two main sections
throughout the 1978 season are relatively conservative; both involved only limited
changes around a dynamic mean state. Greater variations in boundary profile,
in channel shape and in channel size apparently occurred at upstream sites in
both stream systems. In the Tsidjiore Nouve stream, the range of adjustments
adopted decreased in the downstream direction; the buffering effect previously
postulated is again supported. In the Lower Arolla stream most change occurred
in sections D1 and D3, sites at which the impact of debris slides was especially
marked. However, while the adjustment dynamics of the Lower Arolla sections
varied in detail, all sections tended towards a similar pattern of adjustment (fill
during June, scour during July). Comparisons between such changes and those
in flow regime suggest that the low flows of June effected fine-material
sedimentation in the Lower Arolla stream, while the higher flows of July flushed
accumulated sediments through and out of the channel system.

Considering day to day changes in channel form, Figure 16.19 shows time
series of channel size, and changes in channel size, compiled from daily resurveys
of the Tsidjiore Nouve main section at 10.00 h from 10th June to 30th July
1978. The range of sizes adopted by the section can now be seen to be from
1.56 to 2.01 m², which is significantly larger than that evident from the quasi-
weekly survey (which gave values of 1.76 to 1.88 m²). Figure 16.19A shows that
the channel size varied from day to day, with most change occurring with the
onset of a high flow period after 14th July. However, it is clear that while
adjustments in channel size in the Tsidjiore Nouve section were of high
frequency, they were of low magnitude. Figure 16.19B shows that daily changes
in channel size were typically of the order of only ± 0.05 m² (i.e. ± 50 cm²),
and that all except two changes involved an area less than 5% of the mean
channel size. The observation that the Tsidjiore Nouve main section had
developed a balanced form is again supported.

Figure 16.20 shows changes in channel size observed by hourly resurveys of
channel form at the Tsidjiore Nouve main station during three differing diurnal

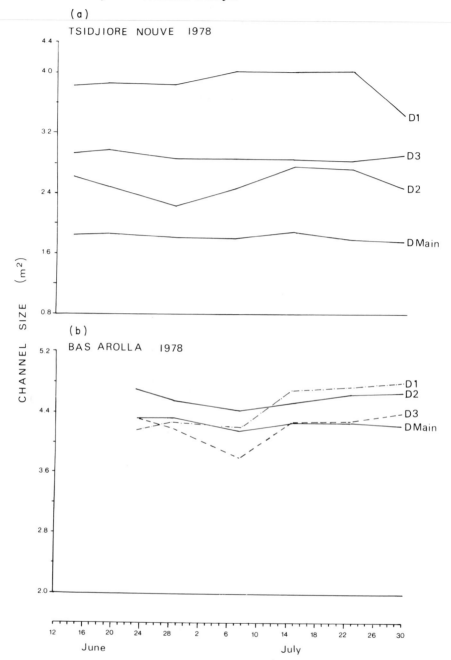

FIGURE 16.18   Changes in the size of the four cross-sections on (A) the Tsidjiore Nouve and (B) the Lower Arolla proglacial streams during the 1978 survey programme

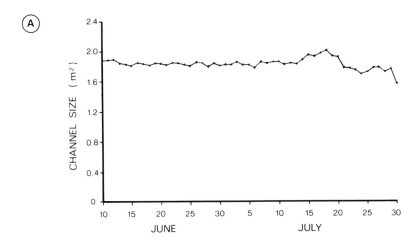

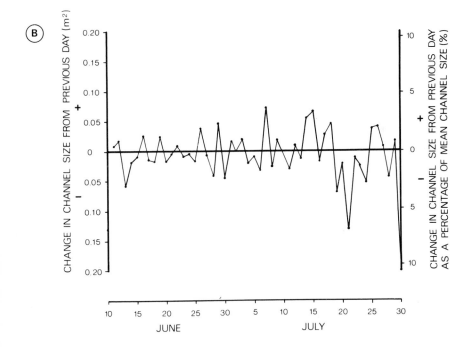

FIGURE 16.19   Changes in the size of the cross-section of the Tsidjiore Nouve main section during the 1978 monitoring programme. (A) Daily channel size series. (B) Day-to-day change in channel size series

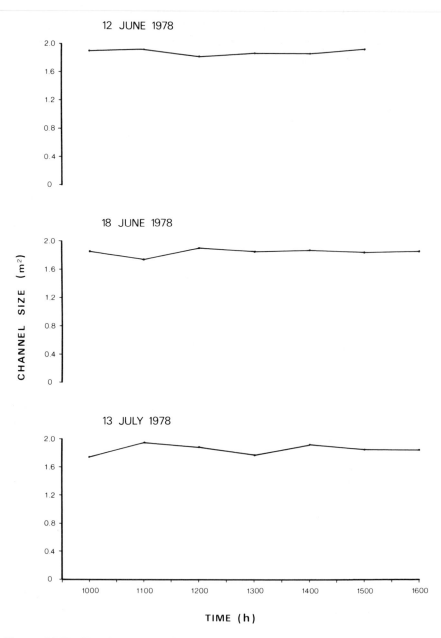

FIGURE 16.20   Hourly variations in the size of the cross-section of the Tsidjiore Nouve main station from 10.00 to 16.00 h during three separate days in the 1978 monitoring programme

streamflow events in 1978 (a high flow, high suspended load, low bedload event (12th June), a low flow, low suspended load, low bedload event (18th June) and a high flow, high suspended load, high bedload event (13th July)). In general, it appears that as much change may occur from hour to hour as may occur over the complete 6 h period, and, on the basis of the data presented in Figure 16.19A, as may even occur over daily and weekly timescales. In detail, most change on an hourly timescale appears to occur during high flow, high bedload transport conditions. In such circumstances, differing rates of transport of the varying particle size ranges present in the channel section may result in material segregation and re-arrangement of the bed surface (Garda and Ranga Raju, 1977).

It would appear from the results presented above that change in the shape and size of proglacial channel cross-sections represent varying degrees of transient shifting bed conditions and more permanent alterations in channel geometry. The most significant conclusion that can be drawn from this multiple scale study is that it is essential for studies at any particular timescale to include some information from finer timescales in order to avoid spurious results. As noted above, changes over an hourly timescale were as large as those identified by sampling at daily and weekly intervals. This implies that within the observation period channel changes were essentially operating at an hourly timescale and that interpretations based on the other scales are spurious. Whatever the relative contributions of such effects, change evidently occurs rapidly and continuously during summer flow periods.

16.5.2 Adjustments in hydraulic geometry

The at-a-station hydraulic geometry (Leopold and Maddock, 1953) represents the way in which a range of hydraulic variables including flow width, mean hydraulic depth and mean flow velocity adjust to changes in discharge. Figure 16.21 shows hydraulic geometry relations for the Tsidjiore Nouve main station during the 1978 measurement programme. While slight non-linearities are detectable in the response of mean depth and mean velocity to discharge variations, the usual linear power function form of the relationships adequately represents the hydraulic geometry:

$$w = 3.119 \, Q^{0.04} \qquad R^2 = 0.91$$
$$d = 0.446 \, Q^{0.23} \qquad R^2 = 0.93$$
$$v = 0.721 \, Q^{0.73} \qquad R^2 = 0.99$$

Since the conditions described by the constants relate to flows equalled or exceeded less than 5% of the time (i.e. when $Q = 1 \, \text{m}^3 \, \text{s}^{-1}$) attention will be focused on the adjustments indicated by the exponents. These exponents show that changes in discharge in the Tsidjiore Nouve section were primarily accommodated through changes in velocity, with changes in depth being less pronounced and channel width being almost invariant.

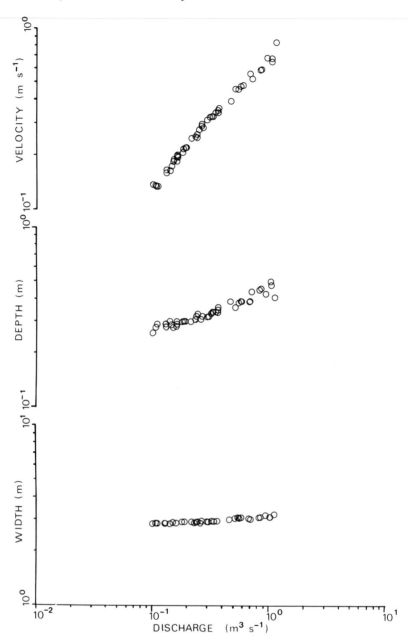

FIGURE 16.21   At-a-station hydraulic geometry relationships for the cross-section at the Tsidjiore Nouve main station from 10th June to 30th July 1978. Data were derived from daily measurements at 10.00 h. Mean velocities were estimated from $v = Q/(wd)$

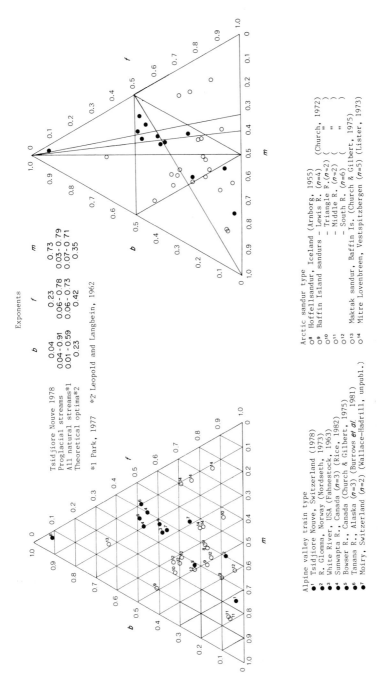

FIGURE 16.22   At-a-station hydraulic geometry exponents of proglacial streams

Following Rhodes (1977) and Park (1977), at-a-station hydraulic geometry exponents may be usefully plotted on a triaxial graph. Figure 16.22 shows the distribution of reported proglacial hydraulic geometry exponents. Proglacial streams evidently adjust to changes in flow in a variety of ways. The exponent sets of Figure 16.22 fall into all but one (i.e. region 1) of the hydraulically distinct zones defined by Rhodes (1977). The channel type frequency distribution of the 33 proglacial exponent sets given in Figure 16.22 differs from that of the 315 sets given in Rhodes (1977), with as many of the exponent sets lying above the $b = f$ line as below it, indicating that individual sections in proglacial streams may experience either decreasing or increasing width–depth ratios during discharge rises. The response displayed by the Tsidjiore Nouve main section tends towards the upper limit of the velocity exponent range and the lower limit of the width exponent range displayed by proglacial streams; this undoubtedly reflects the stable bank characteristics of the selected section, though it should be noted that such sections are by no means unusual in coarse valley train environments.

Rice (1982) suggested that the hydraulic geometry of proglacial streams in arctic (sandur plain) environments differs definably from that of proglacial streams in alpine (valley train) environments, with most adjustment in the former occurring through changes in flow velocity and that in the latter occurring via changes in stream channel width. Rice's contention was, however, based on a consideration of only two arctic and two alpine cases. When the broader range of data plotted in Figure 16.22 is considered, it becomes clear that wide variations in response occur within both arctic and alpine environments, and, indeed, within individual stream systems (note the results from the Sunwapta and Mitre Lovenbreen studies). There would thus appear to be limited grounds for suggesting any environmental distinction in behaviour. Such observations verify Park's (1977) conclusion that considerable variation in at-a-site hydraulic geometry exists both within and between streams of a given type, Knighton's (1984) conclusion that spatial variations in streamflow, sediment transport and channel boundary properties produce spatially variable morphological responses within a stream system and Richards' (1982) suggestion that 'local hydraulic conditions control the exponent set rather than regional climatic or hydrological factors' (Richards, 1982, p. 150). Proglacial streams apparently adjust their morphology and hydraulics to their changeable regime in a variety of ways; a general hydraulic geometry would, therefore, seem to be as untenable in the proglacial environment as it is in any other.

16.5.3   Adjustments of channel size and shape to variations in streamflow and sediment transfer

'Within the constraints imposed by perimeter sediment, the size and shape of a river channel is determined by the whole sequence of sediment-bearing flows which it experiences. Each flow produces its own particular channel response which depends upon the magnitude of the event and the initial state of the channel. The condition of the channel at any time is therefore the product of all previously occurring events.'

(Pickup, 1976, p. 366)

This section attempts to explain the morphological behaviour of the Tsidjiore Nouve main cross-section by relating quantitative descriptions of channel size and shape to streamflow and sediment transport variations. Channel size (morphological cross-sectional area) has already been used as a measure of the capacity of the section. Following Fahnestock (1963), the width–depth ratio and a shape factor (the ratio of maximum to mean hydraulic depth) are used to characterize the shape of the section.

The absolute values of the width–depth and shape factor descriptors of the Tsidjiore Nouve section are of interest in relation to those reported for other proglacial zones. Width–depth ratios (W/D) for the Tsidjiore Nouve section varied around a mean value of 8.83, within the range 6.26–10.96, which is substantially lower than the range of ratios reported by Fahnestock (1963) for proglacial streams in the USA (W/D = 20–30) and by Church (1972) for sandur streams in Baffin Island (W/D = 43–154). The stable bank characteristics of the Tsidjiore Nouve section are undoubtedly partially responsible for the low W/D range obtained. The range of values of the shape factor for the Tsidjiore Nouve section (1.53–1.96) indicates that the cross-section adopted shapes varying between parabolic and triangular. Fahnestock (1963) quoted a mean shape factor of 1.62 for the channels of the White river draining from the Emmons glacier, USA; the Tsidjiore Nouve section (mean shape factor = 1.68) seems to be similar in this respect at least.

Figure 16.23A shows the daily interval time series of the size of the Tsidjiore Nouve section (as surveyed at 10.00 h for the 51-day period from 10th June to 30th July 1978) alongside time series of total daily discharge and suspended sediment load. Figure 16.23B shows channel size plotted with instantaneous values (measured at 10.00 h) of streamflow discharge, mean flow velocity and suspended sediment concentration. Figure 16.23C shows the daily time series of channel size, width–depth ratio and shape factor. Inspection of Figure 16.23 reveals a substantial degree of complexity in morphological response. The first major flow event of the 1978 ablation season [beginning on 14th July (day 35)], for example, scoured bed material from the section, resulting in a significant increase in channel size, while the ensuing second major peak [beginning on 25th July (day 46)] was accompanied by a marked aggradation in the section. The fact that the channel responds to the total character of the discharge and the sediment regime and not simply to the quantity of flow is reflected in a marked inability of simple bivariate statistical models (employing stream

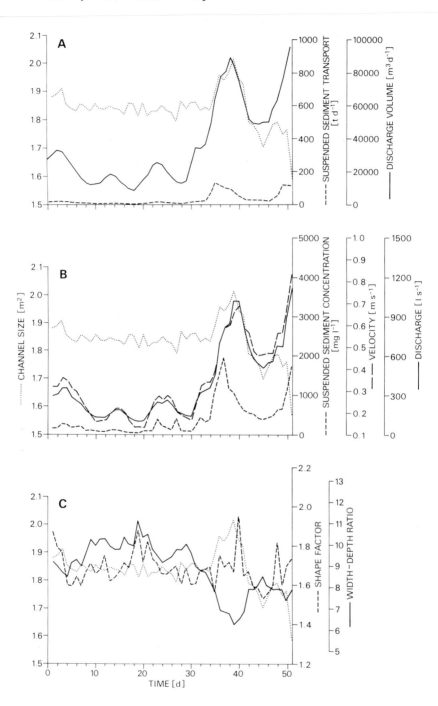

discharge, flow velocity and suspended sediment concentration as controls) to account for any significant degree of the variation in channel size and shape. Further, the morphological behaviour of the channel evidently involves lags and changes in response. Thus, cross-correlograms of each of the form indices (size, width–depth ratio and shape factor) against each of the 'process' variables (discharge, velocity and suspended sediment concentration) demonstrate low levels of association (the high correlations in the width–depth correlograms reflecting the close algebraic link with discharge and velocity) and variable lag times (Figure 16.24). It would appear that the cross-section does not demonstrate a simple process–response behaviour, but exhibits a complex response to more than one control and with varying lags as is suggested for non-proglacial river channels by the above quotation from Pickup (1976).

The measurements, analyses and interpretations presented above represent preliminary contributions to a field in which further research effort is required. The development of form-process models capable of adequately handling the complexities in response identified herein demand detailed, multivariate studies of streamflow–sediment transfer–channel form interactions.

16.6 CONCLUSIONS

This chapter has described the characteristics of proglacial channel systems developed on alpine valley train deposits by focusing on the morphological adjustments which occur over a range of spatial and temporal scales. Emphasis has deliberately been placed on the adjustments in planform and cross-sectional geometry which accompany long- and short-term variations in streamflow and sediment transport. Information derived from studies in plain sandur environments has been included to illustrate the ways in which the processes associated with valley trains exhibit different levels of adjustment to the confinement of their valley location.

A number of general observations may be made from the data examined. In summarizing the nature of morphological adjustments (Section 16.5), it is clear that changes in the form of proglacial channels:

(i) occur in pattern, profile and in section;
(ii) occur over a range of timescales, from the secular to the hourly;
(iii) occur over a range of spatial scales, from the network to the section;
(iv) occur frequently and are achieved rapidly;

FIGURE 16.23 *(opposite)* Daily observations from 10th June to 30th July 1978 at the main section of the Tsidjiore Nouve stream of channel size (at 10.00 h). (A) Total diurnal discharge volume and suspended sediment transport. (B) Suspended sediment concentration, discharge and mean velocity at 10.00 h. (C) Width–depth ratio and channel shape factor at 10.00 h

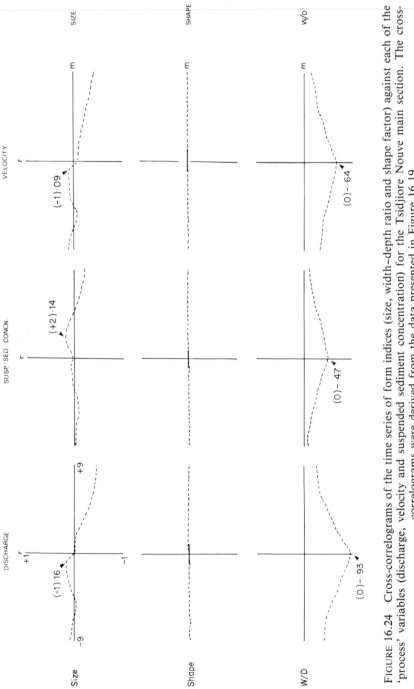

FIGURE 16.24   Cross-correlograms of the time series of form indices (size, width–depth ratio and shape factor) against each of the 'process' variables (discharge, velocity and suspended sediment concentration) for the Tsidjiore Nouve main section. The cross-correlograms were derived from the data presented in Figure 16.19

(v) vary from the large to the small, and from the 'permanent' to the transient;
(vi) vary from local adjustments around an equilibrium form to large-scale alterations in status.

Whilst the nature of morphological adjustment in proglacial channel systems may be conveniently summarized in such general terms, it is clear that it is not as simple to attach specific causes to specific responses, although the discharge and sediment transport history are responsible for the morphology of the proglacial channel system. The locations of the release of water and sediment from the glacier to the proglacial zone influence the channel pattern in the proximal zone of the valley train (Sections 16.4.2 and 16.4.3). Discharge and sediment load variations not only influence processes of erosion and sedimentation within the channel system (Section 16.5), but discharge fluctuations also result in variations in flow depth, avulsion and reoccupation of sections of the flood plain. Thus continual change in the magnitude of the discharge and sediment calibre and load of proglacial streams leads to continual changes in erosion, transport and deposition of sediment, which in turn lead to continual changes in the pattern, profile and cross-sectional size and morphology of proglacial river channels.

Although this state of continuous change of hydrological, sedimentological and morphological conditions is characteristic of proglacial river systems, the end effect of these changing processes is the development of structured proglacial sandur deposits, with clearly distinguishable zones which are essentially a long-term response to the overall competence of the discharge regime to transport the supplied sediment (Section 16.4.1). Thus the character of the valley sandur environment is a product of its glacial and fluvial history. Despite the range of description now possible in relation to a wide variety of proglacial environments, it is very difficult to make generalizations from such description. Perhaps the major contributions of this chapter are to identify deficiencies and to indicate two areas of research priority. First, the status of current research is insufficient to permit the establishment of firm local conclusions about proglacial channel form change, let alone to permit the specification of a global typology. The understanding of the morphological response of proglacial channel systems is a higher order problem than that of the process variations which control the morphological response and so research on the controlling processes must advance concurrently with research on morphological response. Second, multi-scale variability (particularly multi-temporal variability) is a major problem which must be accounted for in research design. Strictly, significant change at any one scale should only be claimed if the observed change exceeds the variability displayed at finer scales, and yet no proglacial study has really attempted to confront this problem in detail. The achievements in research on proglacial channel processes are thus considerable but there are substantial challenges remaining.

REFERENCES

ARNBORG, L. (1955), Hydrology of the glacial river Austerflot, Chapter 7, the Hoffellssandur — a glacial outwash plain, *Geografiska Annaler*, **37**, 185–201.

BALLANTYNE, C. K. (1978), Variations in the size of coarse clastic particles over the surface of a small sandur, Ellesmere Island, N.W.T., Canada, *Sedimentology*, **25**, 141–147.

BOOTHROYD, J. C., and ASHLEY, G. M. (1975), Processes, bar morphology, and sedimentary structures on braided outwash fans, northeastern Gulf of Alaska, in JOPLING, A. V., and MACDONALD, B. C. (Eds.), *Glaciofluvial and Glaciolacustrine Sedimentation*, Society of Economic Palaeontologists and Mineralogists Special Publication 23, pp. 93–222.

BOOTHROYD, J. C., and NUMMEDAL, D. (1978), Proglacial braided outwash: a model for humid alluvial fan deposits, in MIALL, A. D. (Ed.), *Fluvial Sedimentology*, Canadian Society of Petroleum Geologists, Calgary, pp. 641–668.

BRADLEY, W. C., FAHNESTOCK, R. K., and ROWEKAMP, B. T. (1972), Coarse sediment transport by flood flows on Knik river, Alaska, *Geologial Society of America Bulletin*, **83**, 1261–1284.

BROSCOE, A. J. (1972), Some aspects of the geomorphology of meltwater streams, Steele Glacier terminus, in BUSHNELL, V. C. and RAGLE, R. H., (Eds.), *Icefield Ranges Research Project*, Vol. 3, American Geographical Society, New York, and Arctic Institute of North America, Montreal, pp. 47–51.

BURROWS, R. L., EMMETT, W. W., and PARKS, B. (1981), Sediment transport in the Tanana river, near Fairbanks, Alaska, *United States Geological Survey Water Resources Investigation*, 81-20, 56pp.

CARSON, M. A. (1984), The meandering braided river threshold: a reappraisal, *Journal of Hydrology*, **73**, 315–334.

CHEETHAM, G. H. (1979), Flow competence in relation to stream channel form and braiding, *Geological Society of America Bulletin*, **90**, 877–886.

CHURCH, M. (1972), Baffin Island sandurs: a study of arctic fluvial processes, *Geological Survey of Canada Bulletin*, 216, 208pp.

CHURCH, M., and GILBERT, R. (1975), Proglacial fluvial and lacustrine environments, in JOPLING, A. V., and MACDONALD, B. C. (Eds.), *Glaciofluvial and Glaciolacustrine Sedimentation*, Society of Economic Palaeontologists and Mineralogists Special Publication 23, pp. 22–100.

CHURCH, M., and RYDER, J. M. (1972), Paraglacial sedimentation: a consideration of fluvial processes conditioned by glaciation, *Geological Society of America Bulletin*, **83**, 3059–3072.

CHURSKI, Z. (1973), Hydrographic features of the proglacial area of Skeidarajökull, *Geographia Polonica*, **26**, 209–254.

COATES, D. R. (1969), Hydraulic geometry in a glaciated region, *Transactions of the American Geophysical Union*, **50**, 149.

DERBYSHIRE, E., GREGORY, K. J., and HAILS, J. R. (1979), *Geomorphological processes*, Studies in Physical Geography, Dawson, Folkestone.

FAHNESTOCK, R. K. (1963), Morphology and hydrology of a glacial stream — White river, Mount Rainier, Washington, *United States Geological Survey Professional Paper*, No. 422-A, 70pp.

FAHNESTOCK, R. K. (1969), Morphology of the Slims River, in BUSHNELL, V. C., and RAGLE, R. H. (Eds.), *Icefield Ranges Research Project, Scientific Results*, Vol. 1, American Geographical Society, New York, and Arctic Institute of North America, Montreal, pp. 161–172.

FAHNESTOCK, R. K., and BRADLEY, W. C. (1973), Knik and Matanuska rivers, Alaska: a contrast in braiding, in MORISAWA, M. (Ed.) *Fluvial Geomorphology*, George Allen and Unwin, London, pp. 220-250.

FENN, C. R. (1983), Proglacial streamflow series: measurement, analysis and interpretation, unpublished PhD Thesis, University of Southampton, 425pp.

GARDA, R. J., and RANGA RAJU, K. G. (1977), *Mechanics of Sediment Transportation and Alluvial Stream Problems*, Wiley, New Delhi.

GOUDIE, A. S., BRUNSDEN, D., COLLINS, D. N., DERBYSHIRE, E., FERGUSON, R. I., HASHMET, Z., JONES, D. K. C., PERROTT, F. A. SAID, M. WATERS, R. S., and WHALLEY, W. B. (1984), Geomorphology, in MILLER, K. J. (Ed.), *International Karakoram Project*, Vol. 2, Cambridge University Press, pp. 359-410.

GURNELL, A. M. (1982), The dynamics of suspended sediment concentration in a proglacial stream, in GLEN, J. W. (Ed.), *Hydrological Aspects of Alpine and High Mountain Areas (Proceedings of the Exeter Symposium, July (1982)*, International Association of Hydrological Sciences Publication 138, pp. 319-330.

GURNELL, A. M., and FENN, C. R. (1985), Spatial and temporal variations in electrical conductivity in a proglacial stream system, *Journal of Glaciology*, **31**, 108-114.

HAEBERLI, W. (1983), Frequency and characteristics of glacier floods in the Swiss Alps, *Annals of Glaciology*, **4**, 85-90.

HEWITT, K. (1982), Natural dams and outburst floods of the Karakoram Himalaya, in GLEN, J. W. (Ed.), *Hydrological Aspects of Alpine and High Mountain Areas (Proceedings of the Exeter Symposium, July 1982)*, International Association of Hydrological Science Publication 138, pp. 259-269.

HJULSTRÖM, F. (1952), The geomorphology of the alluvial outwash plains (sandurs) of Iceland and the mechanics of braided rivers, *International Geographical Union 17th Congress Proceedings, Washington*, pp. 337-342.

HJULSTRÖM, F. (1954), An account of the expedition and its aims. Chapter 1, The Hoffellssandur, *Geografiska Annaler*, **36**, 135-145.

HJULSTRÖM, F. (1955), Chapter IX. The ground water, *Geografiska Annaler*, **37**, 234-245.

KNIGHTON, D. (1984), *Fluvial Forms and Processes*, Arnold, London.

KRIGSTRÖM, A. (1962), Geomorphological studies of sandur plains and their braided rivers in Iceland, *Geografiska Annaler*, **44**, 328-346.

LEOPOLD, L. B., and MADDOCK, T. (1953), The hydraulic geometry of stream channels and some physiographic implications, *United States Geological Survey Professional Paper*, 252.

LISTER, H. (1973), Glacial origin of proglacial boulders, in *Symposium on the Hydrology of Glaciers (Proceedings of the Cambridge Symposium, 7-13 September 1969)*, International Association of Scientific Hydrology Publication 95, pp. 151-156.

MAIZELS, J. K. (1979), Proglacial aggradation and changes in braided channel patterns during a period of glacier advance: an alpine example, *Geografiska Annaler*, **61A**, 87-101.

MAIZELS, J. K. (1983), Proglacial channel systems: change and thresholds for change over long, intermediate and short time-scales, *Special Publication of the International Association of Sedimentologists*, 6, pp. 251-266.

MCDONALD, B. C., and BANERJEE, I. (1971), Sediments and bedforms on a braided outwash fan, *Canadian Journal of Earth Sciences*, **8**, 1282-1301.

NORDSETH, K. (1973), Fluvial processes and adjustments on a braided river: the islands of Kappongsoyenna on the river Gloma, *Norsk Geografisk Tidsskrift*, **27**, 77-108.

PARK, C. C. (1977), World-wide variations in hydraulic geometry exponents of stream channels: an analysis and some observations, *Journal of Hydrology*, **33**, 133-146.

PICKUP, G. (1976), Adjustments of stream channel shape to hydrologic regime, *Journal of Hydrology*, **30**, 365–373.

PRICE, R. J. (1980), Rates of geomorphological changes in proglacial areas, in CULLINGFORD, R. A., DAVIDSON, D. A., and LEWIN, J. (Eds.), *Timescales in Geomorphology*, Wiley, Chichester, pp. 79–93.

RICE, R. J. (1982), The hydraulic geometry of the lower portion of the Sunwapta River valley train, Jasper National Park, Alberta, in DAVIDSON-ARNOTT, R., NICKLING, W., and FAHEY, B. D. (Eds.) *Research in Glacial, Fluvioglacial and Glaciolacustrine Systems, Proceedings of the 6th Guelph Symposium on Geomorphology*, Geo Abstracts, Norwich, pp. 151–174.

RICHARDS, K. (1982), *Rivers: Form and Process in Alluvial Channels*, Methuen, London.

RHODES, D. D. (1977), The *b–f–m* diagram: graphical representation and interpretation of at-a-station hydraulic geometry, *American Journal of Science*, **277**, 73–96.

RUST, B. R. (1972), Structure and process in a braided outwash river, *Sedimentology*, **18**, 221–245.

SMALL, R. J. (1973), Braiding terraces in the Val d'Hérens, Switzerland, *Geography*, **58**, 129–135.

SMITH, D. G. (1976), Effect of vegetation on lateral migration of anastomosed channels of a glacial meltwater river, *Geological Society of America Bulletin*, **87**, 857–860.

SMITH, N. D. (1974), Sedimentology and bar formation in the upper Kicking Horse river, a braided outwash stream, *Journal of Geology*, **82**, 205–223.

STENBORG, T. (1973), Some viewpoints on the internal drainage of glaciers, in *Symposium on the Hydrology of Glaciers (Proceedings of the Cambridge Symposium, 7–13 September 1969)*, International Association of Scientific Hydrology Publication 95, pp. 117–129.

SUGDEN, D. E. and JOHN, B. S. (1976), *Glaciers and Landscape: A Geomorphological Approach*, Arnold, London.

THORARINSSON, S. (1939), Hofellsjökull, its movements and drainage: Chapter 8 in Vatnajökull—scientific results of the Swedish–Icelandic investigations, 1936–1938, *Geofrafiska Annaler*, **21**, 189–215.

TUFNELL, L. (1984), *Glacier Hazards*, Topics in Applied Geography, Longman, London.

WILLIAMS, P. F., and RUST, B. R. (1969), The sedimentology of a braided river, *Journal of Sedimentary Petrology*, **39**, 649–679.

Glacio-fluvial Sediment Transfer
Edited by A. M. Gurnell and M. J. Clark
© 1987 John Wiley & Sons Ltd.

Chapter 17

Glacial Meltwater Streams, Hydrology and Sediment Transport: The Case of the Grande Dixence Hydroelectricity Scheme*

A. Bezinge

ABSTRACT

Since 1948, the Grande Dixence company has been studying many aspects of the glacial meltwater streams that it uses intensively in the production of hydroelectricity. The Grande Dixence hydroelectric scheme utilizes meltwaters from the alpine valleys of St. Nicolas (Mattertal) and Hérens, Valais, Switzerland. The catchment area of the scheme consists of 35 drainage basins ranging in area from 1 to 80 km². Each basin has its own topographic, microclimatic and glacial characteristics.

Numerous engineering structures with associated process monitoring equipment provide information which allows these characteristics to be evaluated and analysed. Operating the scheme has provided an enormous fund of knowledge, particularly in relation to meltwater characteristics (the raw material from which hydroelectric energy is derived) and sediment transport (the cause of wear of hydraulic machines and works and of unwanted deposits in the sedimentation basins).

The results presented in this chapter, which provide a global picture of meltwater discharges and sediment transport in the Alps, are applicable to other regions of the globe with similar climatic and topographic characteristics, for example the Cordilleras, the Hindukush and the Himalayas.

17.1 INTRODUCTION

Since 1948, the Grande Dixence hydroelectric company has been interested in the discharge and sediment transport characteristics of proglacial rivers in the alpine valleys of St. Nicolas (Mattertal) and Hérens, Valais, Switzerland (Figure 17.1). The meltwaters from 35 drainage basins, ranging in area from 1 to

*This chapter was originally published in French by Birkhauser Verlag, 1981. Copyright Societé Hélvétique des Sciences Naturelles.

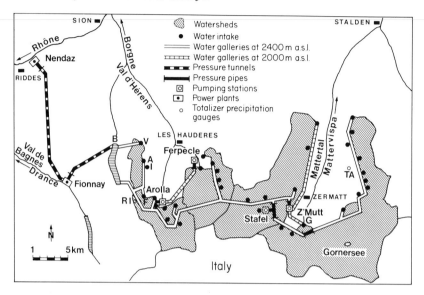

FIGURE 17.1   Grande Dixence hydroelectric scheme. B, Grande Dixence dam (height, 284 m; capacity, 400×10^6 m^3 at the maximum water level of 2364 m a.s.l.); TA, Taschalp (2200 m a.s.l.); RI, Riedmatten (2800 m a.s.l.); V, Vouasson; A, Aiguilles Rouges; I, Ignes; F, Fontanesses; G, Gorner

80 km^2, have been monitored and abstracted since 1953. Whereas the water is the raw material in hydro-power generation, suspended sediment and bedload transported by the water cause wear of hydraulic machinery and unwanted deposits in the accumulation basins. Thus, the operation of the Grande Dixence scheme provides a vast fund of knowledge, particularly as its high mountain, glacier basins are so varied in microclimate, size, altitude, orientation, topography, rock type and degree of glaciation.

17.2   HYDROLOGY

Following Kasser (1959, 1973), mean annual runoff gives a fairly good indication of local mean annual precipitation after two important factors are taken into account: modification of the ice mass and evaporation loss, which is generally accepted to be of the order of 30 cm water equivalent each year.

On the basis of 12 years of ablation season measurements, from 1966 to 1977, average summer flow indices were determined (i.e. $\bar{E}_{5-9}$ expressed as a depth over the catchment in m water equivalent). The summer flow indices represent 96–98% of the value of the annual flow indices.

Figure 17.2 shows flow indices for the Val d'Hérens. In spite of the advance of the Tsidjiore Nouve, Lower Bertol and Ferpècle glaciers, these basins have

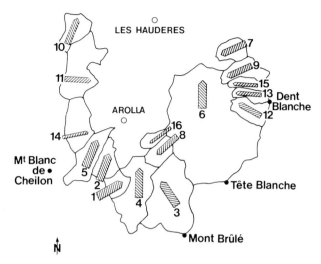

FIGURE 17.2 Average summer flow index, $\bar{E}_{5-9}$ (in m water equivalent) in the Val d'Hérens, 1966–77. The width of the arrows is proportional to the magnitude of $\bar{E}_{5-9}$

Basin		$\bar{E}_{5-9}$	Basin		$\bar{E}_{5-9}$
1	Vuibé	1.65	9	Bricola	1.11
2	Pièce	1.63	10	Vouasson	1.08
3	Haut d'Arolla	1.53	11	Ignes–Aiguilles Rouge	1.02
4	Bertol Inférieur	1.48	12	Manzettes	0.92
5	Tsidjiore Nouve	1.48	13	Rocs Rouges	0.90
6	Ferpècle–Mont Miné	1.32	14	Fontanesses	0.74
7	Les Rosses	1.19	15	Dent Blanches	0.60
8	Bertol Supérieur	1.18	16	Douves Blanches	0.56

relatively high flow indices. The indices are relatively low for the Fontanesses basin (discharge results largely from groundwater seepage, see Section 17.2.1), and for the Douves and Dent Blanche basins which face westwards and have a low glacier cover. The wide range of flow indices illustrates the significance and the variety of microclimates between these basins.

Figure 17.3 shows flow indices for the Zermatt valley. There are very low indices for the west-facing, right bank basins in spite of the high altitude and high glacier cover. This is a semi-arid region, since in the Taschalp valley the mean annual precipitation, measured at an altitude of 220 m a.s.l., is only 60 cm. It should be noted that the Hohberg, Festi and Kin glaciers have been enlarging for the last 5–6 years. The basins of Rotbach and Langfluh experience significant losses as a result of seepage through moraines. The runoff indices are high for the left bank, down-valley basins. The Bis, Trift and Hohwang glaciers have been advancing for several years. These high indices result from high local

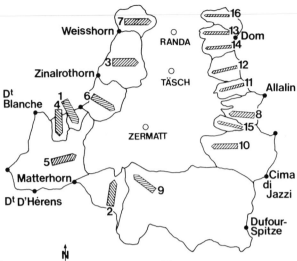

FIGURE 17.3   Average summer flow index, $\bar{E}_{5-9}$ (in m water equivalent) in the Zermatt valley, 1966–77. The width of the arrows is proportional to the magnitude of $\bar{E}_{5-9}$

Basin		$\bar{E}_{5-9}$	Basin		$\bar{E}_{5-9}$
1	Arb	1.54	9	Gorner	1.11
2	Furgg–Obertheodul	1.46	10	Findelen	1.03
3	Schali	1.45	11	Alphubel	0.55
4	Hohwang	1.39	12	Rotbach	0.51
5	Z'Mutt	1.35	13	Festi	0.49
6	Trift	1.27	14	Kin	0.44
7	Bis	1.20	15	Längfluh	0.44
8	Mellichen	1.14	16	Hohberg	0.40

precipitation patterns which result from local winds and from the local extension and overlap of the mediterranean climate.

It is interesting to contrast these data with information from the French Alps. For example, in comparison with the climate of Chamonix, that of the Zermatt area is less favourable for glaciation. If Zermatt experienced the same climate as Chamonix, the glaciers would extend from the 4000 m high summits and combine to form a single glacier which would extend as far as Randa or St. Nicolas. Table 17.1 presents information on the runoff and névés characteristics of these two areas.

TABLE 17.1   Comparison of maximum flow and névé limits at Zermatt and Chamonix

Maximum flow index at Zermatt	~ 150 cm water equivalent
Maximum flow index at Chamonix	~ 270 cm water equivalent
Maximum altitude of névés at Zermatt	3300 m
Maximum altitude of névés at Chamonix	2800 m

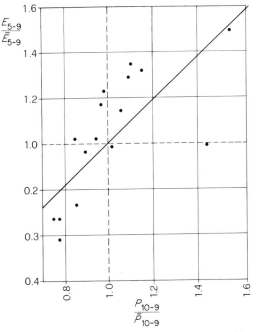

FIGURE 17.4   Correlation between the summer runoff index, $\bar{E}_{5-9}$ of the Fontanesses basin (4.85 km², glacier-free) and annual precipitation, $\bar{P}_{10-9}$, estimates from the totaliser gauge at Riedmatten (2800 m a.s.l.) for the period from 1960/61 to 1976/77. Maximum altitude of the basin, 3393 m a.s.l.; altitude of the water intake (flow gauging station), 2420 m a.s.l. Average values: $\bar{E}_{5-9} = 70$ cm, $\bar{P}_{10-9} = 124$ cm

17.2.1   Annual precipitation and runoff

High-altitude precipitation is difficult to measure because of the effect of snow drifting. Calibrations achieved by probing the depth of snow cover on glaciers has indicated that high-mountain totalizer gauges underestimate precipitation by 10–30%. Since 1960, Grande Dixence has been operating a totalizer gauge at an altitude of 2800 m in a large south-facing depression (Riedmatten). The glacier-free, 4.85 km² Fontanesses basin, in which the gauge is located, permits runoff to be measured with negligible distortion from surface storage effects when the summer is warm and the snow-melt is complete. Characteristics of the precipitation and runoff are presented in Figures 17.4 and 17.5. Although one might initially expect runoff to be proportional to precipitation, this is not confirmed by the precipitation and runoff information presented in Figure 17.5. There are two factors which account for the variability in the annual runoff in comparison with the precipitation: (i) In some years (as for example in 1977), the summer is too cool to achieve complete snow melt. Some of the snow remains in storage in the form of névés to melt in subsequent years when the summer

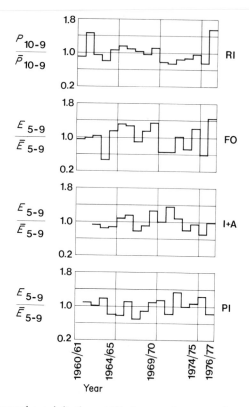

FIGURE 17.5   Annual precipitation at Riedmatten in comparison with summer runoff indices for the Fontanesses, Ignes and Aiguilles Rouges, and Pièce basins (all expressed as proportions of the 1960–61 to 1976–77 mean). Abbrevations as in Figure 17.4

Basin	Area (km²)	Glaciers km²	Glaciers %	Altitude (m a.s.l.) Max.	Altitude (m a.s.l.) Min.	$\bar{E}_{5-9}$ (cm water equivalent)
FO Fontanesses	4.85	0.0	0	3393	2420	70
I + A Ignes and Aiguilles Rouges	7.36	2.03	27	3646	2400	106
PI Pièce	2.7	1.82	67	3796	2450	165

meteorological conditions are favourable. (ii) Water seeps through screes and moraines and is captured by a nearby gallery, so escaping measurement at the gauging station. For this snow basin, years with precipitation of 124 cm water equivalent (50 cm from October to May) result in 74 cm yr^{-1} of runoff (Figure 17.2). The loss of 50 cm consists of 30 cm of evaporation and 20 cm of seepage. There is a correlation of 0.72 between annual runoff volume and annual precipitation (Figure 17.4).

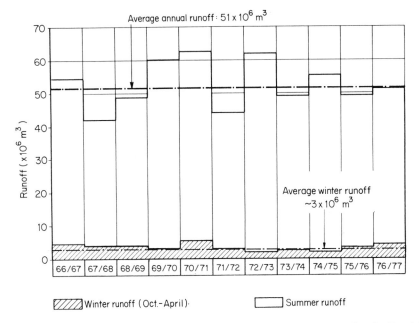

FIGURE 17.6   Winter, summer and annual runoff, 1966–67 to 1976–77, for the Ferpècle basin (catchment area, 36 km²; glacier area, 22 km²; glacier cover, 64%; maximum altitude, 3720 m a.s.l.; altitude of water intake, 1907 m a.s.l.; aspect, north-facing)

When annual runoff volume is compared with precipitation for glacierized basins, it appears that the correlation between the two variables decreases with an increasing glacierized proportion of the catchment. Figure 17.5 illustrates the variations in ablation season runoff from three adjacent basins (Fontanesses, Ignes and Aiguilles Rouges, Pièce) with markedly different percentage glacier cover (0, 27, 67%, respectively), in comparison with the annual precipitation monitored at Riedmatten. Thus, in 1964, which was warm and dry, the snow basin (Fontanesses) exhibited a very low summer runoff in comparison with the average, whereas the volume of runoff in the Pièce basin (67% glacier cover) was well above average. In contrast, the heavy precipitation and cool summer of 1977 yielded the opposite pattern: well above average runoff from Fontanesses and well below average from the Pièce basin.

Figures 17.6 and 17.7 present information on the runoff characteristics of the Ferpècle basin. The annual runoff, monitored at 1907 m, from this 64% glacierized basin, is distributed so that approximately 5% of the runoff occurs in winter (October to April) and about 95% occurs in summer (May to September). It appears that the higher the altitude of the water intake (where discharge is gauged), the lower is the winter percentage of runoff with, for example, about 1% of runoff being generated in winter at an altitude of 2400 m.

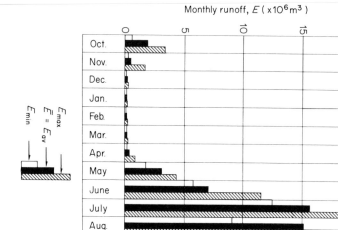

FIGURE 17.7   Mean, minimum and maximum monthly runoff from the Ferpècle basin, 1966–67 to 1976–77 (catchment characteristics as in Figure 17.6). Annual flow index is approximately 1.42 m water equivalent (summer flow index is 1.34 m water equivalent)

Figure 17.7 shows the variability in monthly runoff from year to year for the Ferpècle basin by comparing the maximum (E_{max}), mean (E_{av}) and minimum (E_{min}) runoff in each month over the period of records. Monthly runoff varies with meteorological conditions. From December to March there is little variation in monthly runoff about the mean. However, in October, November and from April to September inter-year variations are very large.

17.2.2   Monthly runoff

The distribution of monthly runoff varies according to the degree of glacier cover in the basin. Figure 17.8 shows the percentage of the ablation season runoff (May to September) generated in each month of the season by three basins (Fontanesses, Ignes and Aiguilles Rouges and Vouasson) with contrasting glacier cover (0, 27 and 61%, respectively). A snow basin, such as the Fontanesses basin, generates relatively high runoff from the onset of the melt season, whereas a glacierized basin yields the major part of its meltwater from mid-summer.

17.2.3   Daily flows

The average flows during each day in summer are a response to variations in the winter snow precipitation and accumulation, climatic conditions, the size

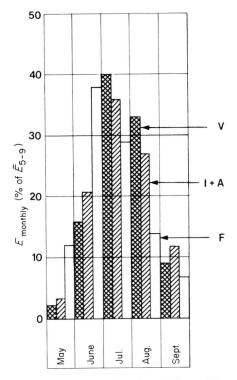

FIGURE 17.8 Average (1963–71) monthly flow indices (E_{monthly}) expressed as a percentage of the average summer flow index ($\bar{E}_{5-9}$) for three adjacent basins with different % glacier cover

Basin		Basin area (km²)	Glaciers		Orientation	$\bar{E}_{5-9}$ (cm water equivalent)
			km²	%		
V	Vouasson	3.14	1.93	61	North-east	105
I + A	Ignes and Aiguilles Rouges	7.36	2.03	27	East	102
F	Fontanesses	4.85	0.0	0	East	71

Note: The winter discharge is approximately 2–3% of the summer discharge for Vouasson and for Ignes and Aiguilles Rouges and 15–20% for Fontanesses.

of the glacier, the time of the season and the névé limits. Figure 17.9, for example, shows the average daily discharge for the large (80 km²) Gorner basin during a warm year (1973). The effects of climatic variations are damped by large glaciers so that flow adjustments are moderated when temperatures rise and fall, particularly early in the ablation season. In 1973, after a winter with light snowfall, the summer was warm and sunny, except at the end of each month when there was snowfall which extended across the zone of ablation on the

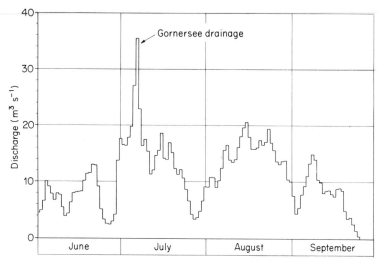

FIGURE 17.9   Average daily flows measured at the Gorner water intake during summer 1973 (catchment area, 80.6 km²; glacier area, 61.6 km²; glacier cover, 76%)

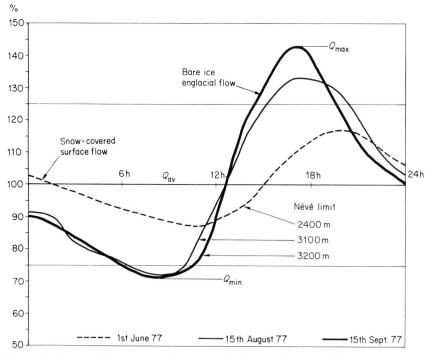

FIGURE 17.10   Discharge from the Z'Mutt basin over 24 h periods, expressed as a percentage of the daily average discharge during 1st June, 15th August and 15th September 1977 (catchment area, 35.2 km²; glacier area, 19.4 km² (in 1955); orientation, east)

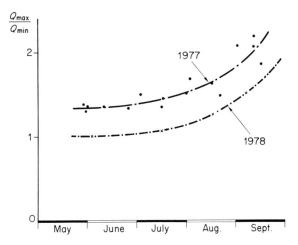

FIGURE 17.11 Changes in the ratio of maximum to minimum discharge (Q_{max}/Q_{min}) for the Z'Mutt catchment during the 1977 and 1978 ablation seasons (maximum névé limit approximately 3200 m a.s.l.)

glacier. The great difference between the maximum and minimum daily mean discharges are particularly noteworthy as a characteristic of flows in a hot summer.

17.2.4 Instantaneous discharges

Discharge exhibits characteristic diurnal variations. At the beginning of the ablation season the diurnal variations in discharge are relatively subdued, following a slightly skewed sinusoidal curve. Towards the end of the summer, however, the ratio of the diurnal maximum (Q_{max}) and minimum (Q_{min}) discharge becomes large. Figure 17.10 shows the change in the distribution of instantaneous discharge (expressed as a percentage of the mean daily discharge) during selected 24 h periods at different times in the ablation season in the Stafel basin (Z'Mutt glacier). Not only does the range of discharge vary within 24 h periods at different times within the ablation season (Figure 17.11 illustrates the gradual adjustment of the ratio of Q_{max} to Q_{min} for the Stafel basin), but also the times of minimum and maximum flow gradually adjust from 11.00 and 20.00 h, respectively, in early summer to 9.00 and 17.00 h towards the end of the ablation season (Figure 17.10). These phenomena result from the snow cover, which retains water, especially early in the season, and which disappears by the end of the season; the altitudinal limit of the névé; and the progressive recreation and enlargement of the subglacial drainage channels through the ablation season to facilitate meltwater drainage. Elliston (1973), in relation to meltwater drainage from the Gorner glacier, has noted that early in the ablation season virtually all the meltwater drainage occurs across the surface of the glacier, but as the névé limit rises the drainage is routed increasingly within and beneath the glacier.

Martinec (1970), as a result of studying glacier runoff curves, has proposed a flow-recession coefficient. This indicates that a proportion of the meltwaters from large glaciers may be retained for several days, in temporary storage, before flowing from the glacier.

Figure 17.11 shows variations in the ratio of Q_{max} to Q_{min} for the Stafel basin (Z'Mutt glacier). Curves representing the adjustment of the ratio through the ablation season are presented for 1977 and 1978. The values for 1978 are approximately 30% smaller than those for 1977 and this difference can be explained by the relatively heavy winter snow fall and cool summer of 1978.

17.2.5   Albedo: an important factor affecting ablation

The albedo is the ratio of reflected to incident shortwave radiation. In 1965, following experiments undertaken in the Antartic, Grande Dixence fitted a helicopter with Morikofer-type solarimeters, one to measure total incoming radiation and the other to measure radiation reflected from the ground. Calibration of these instruments from ground measurements taken at the ice surface by a team from VAW (Versuchsanstalt für Wasserbau, ETH, Zurich, Professor Kasser) showed, firstly, that reflected radiation measurements should not be taken from more than 50 m above the reflecting surface and secondly, that the helicopter blades did not appear to disturb the incident radiation measurements unless their effects were exaccerbated by shadowing from the helicopter cabin.

During August 1965, albedo surveys were rapidly undertaken for the Gorner, Findelen, Z'Mutt and Ferpècle glaciers. The results of the survey of the Findelen glacier are presented in Figure 17.12 and Table 17.2 lists albedo estimates for a sample of locations on the Findelen and Z'Mutt glaciers. The albedo of a glacier is unlikely to vary greatly from year to year, although occasional events

TABLE 17.2   Variations of albedo of different types of glacier surface on the Findelen and Z'Mutt glaciers

Surface type	Altitude (m)	Albedo
Findelen glacier:		
Glacier surface free of snow	2500	0.17
Glacier surface free of snow	2900	0.22
Limit of lower névé	3000	0.35
Limit of upper névé	3150	0.70
Z'Mutt glacier:		
'Black' glacier (i.e. without snow)	2400	0.18
Sérac zone	2900	0.40

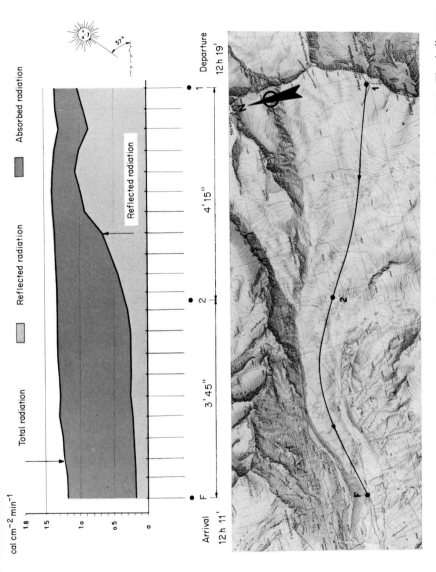

FIGURE 17.12   Albedo measurements along the axis of the Findelen glacier during 21st August 1965. The helicopter was flying horizontally at an altitude of 40–50 m above the glacier surface, with two solarimeters calibrated at Aletsch

resulting from the fallout of particles transported by the atmosphere (such as the deposition of dust blown from the Sahara) can have a significant effect on albedo. In general, a snow-free glacier surface absorbs 2.5 times as much incident radiation as a snow-covered surface, so resulting in 2.5 times more ablation.

17.2.6   Flood discharges generated by processes other than ablation

Certain circumstances may produce floods which interrupt the natural meltwater discharge regime, as follows.

17.2.6.1   Storm precipitation

The effects of storm precipitation vary according to when they occur within the ablation season. At the start of the season the snow mantle can store much of the stormwater so reducing the peak discharge response to the storm. In contrast, at the end of the season the enlarged drainage channels and snow-free surfaces of the glaciers allow rapid and efficient storm drainage.

It is very difficult to estimate the discharge response to extreme storm events, but in the summer of 1971 it was possible to monitor the runoff generated by storm precipitation of 25–35 mm in 24 h. Discharges per unit catchment area were calculated and plotted against catchment area and an envelope curve was estimated. This showed that for a 2 km^2 basin the maximum discharge per unit area was $1 \, m^3 \, s^{-1} \, km^{-2}$, whereas for an 80 km^2 basin the discharge was $0.4 \, m^3 \, s^{-1} \, km^{-2}$. These floods all drained rapidly without significant retention of the water.

Zermatt has experienced two occasions in a period of 100 yr when there was precipitation in excess of 100 mm in 24 h. These storms generated peak discharges 4–5 times greater than would be expected on the basis of the above analysis. EdF (Electricité de France) applied the Gradex method to the Stafel basin (35.2 km^2) and estimated the 100 yr flood at $60 \, m^3 \, s^{-1}$, which is close to the estimate that would be derived from extrapolation of the maximum discharge envelope described above. Nevertheless, this is a great deal less than the $12–15 \, m^3 \, s^{-1} \, km^{-2}$ measured in the Melezza and d'Isorno (Tessin) basins following the catastrophic storms of 7th August 1978.

17.2.6.2   Meltwater outburst floods

The catastrophic outbursts from the Gietroz and l'Allalin glaciers are still clearly remembered. What are less well known are the minor outbursts from small glaciers which release discharges of the order of $1–3 \, m^3 \, s^{-1}$ over periods of a

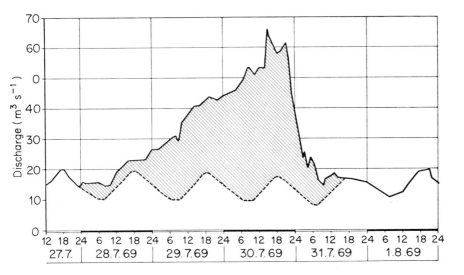

―――― Total discharge

----- Theoretical glacial meltwater discharge

Discharge resulting from the drainage of the Gornersee: ~ 6.2 x 10⁶m³

FIGURE 17.13   Drainage of the Gornersee, July 1969. Discharge measured at the Gorner water intake

few hours. Meltwaters can accumulate over periods of years (en- or subglacially) near the névé limit (e.g. Hohberg glacier, 1966), and are released unpredictably after retention for several years.

   The best known and the most regular meltwater outburst within the Grande Dixence catchment area is the release of waters from the Gornersee, a lake which develops annually at the junction of the Gorner and Grenz glaciers (altitude 2600 m, Figure 17.13). This lake, which was observed by Lambien as early as 1680 (an old engraving), drains every year between the end of June and early September, yielding very high discharges (Figure 17.9) of which the largest known was 200 m³ s⁻¹. Since 1965, Grande Dixence has been observing the level of the lake and the meltwater output. The hydrograph generated by the drainage of the Gornersee in 1969 is shown in Figure 17.13. The discharge can be seen to increase regularly over a period of three days and then to recede rapidly until it reaches the level of the normal glacier meltwater discharge. From the very first day of this event the waters are heavily charged with characteristic yellow sediment. Later pieces of ice, ranging in size from pebble-size to large blocks, are wrenched from the glacier and transported by the intense englacial and subglacial flows. Sometimes, after a fall in summer temperatures, the lake may refill after its initial drainage and a second, smaller, outburst may occur later in the summer, as for example occurred in September 1977. The mechanism by which the meltwater channels are obstructed and reopened is not yet clear, but various phenomena allow the proposal of plausible hypotheses:

(i) the presence of cold ice in the Grenz glacier (temperatures of -2 to $-3\,°C$ have been observed by Professor H. Rothlisberger);
(ii) there is overdeepening of the bedrock to the right of the Gornergrat and a large reserve of water within the glacier.
(iii) the ice flows continuously in the accumulation zone in contrast with differential flow rates in the ablation zone (strong in summer, weak in winter).

The most plausible explanation of the obstruction of the drainage channels is the vertical movement of ice in the absence of subglacial water. It is noteworthy that in 1976 some employees of Grande Dixence were able to penetrate downstream from the Gorner lake along natural galleries 150 m below the surface of the glacier.

17.2.7 Remarks

It is clear from the above discussion that the hydrology of alpine watersheds is very varied and is not regular from year to year. Moreover, long-term hydrology is influenced by changes in glacier area. For example, the Valais glaciers decreased by about 160 km^2 in area between 1915 and 1958. Kasser (1973) has calculated that under similar weather conditions the summer discharge of the Rhône is 700 million m^3 lower as a result of this decrease in glacier area. This phenomenon has also occurred in the runoff of the basins analysed in this study. It is noteworthy that the annual runoff of a glacier basin corresponds to about 1–2% of the volume of ice stored.

17.3 SEDIMENT TRANSPORT

Glaciers and their associated meltwater are powerful erosive agents, displacing material ranging in size from 100 m^3 blocks to the finest particles. Sediment transport is very sensitive to storm events and to the distribution and magnitude

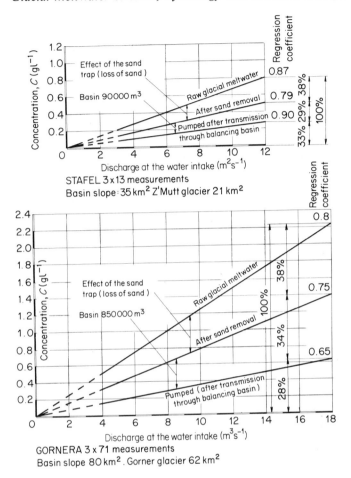

FIGURE 17.14 Suspended sediment concentration as a function of discharge at the water intakes at Stafel and the Gorner. Measured using 50 l cones at about 13.00 h, 1973. Regression equation: $C = f(Q)$

of meltwater discharge, in addition to reflecting the local rock type and the quantity of morainic material available for transport.

17.3.1 Suspended sediment transport in glacial meltwater streams

Fine particles are mainly produced by glacial erosion and by the abrasion caused by the high summer discharges of the glacial meltwater streams. Hydroelectric companies are interested in the concentration and amount of sediment because these data allow the estimation of the potential quantity of sediment deposition. Since 1961, the Grande Dixence company has been monitoring the suspended

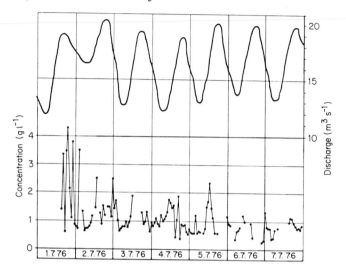

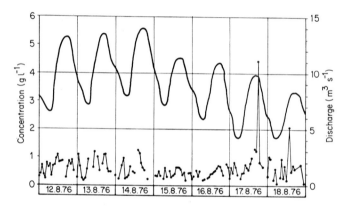

FIGURE 17.15   Discharge and suspended sediment concentration at the Gorner water intake, measured by Dr. D. N. Collins (University of Manchester)

sediment concentration of the Gornera in relation to the performance of water intake structures. The analysis of three samples each day revealed significant variations in sediment concentration which were not correlated with discharge (Figure 17.14).

Since 1972, in response to the results of experiments by Collet (1916) and Bruschin (1971), Grande Dixence has been monitoring sediment concentrations in a number of meltwater streams. In order to avoid using delicate sampling apparatus, a robust 50 l cone with a Plexiglass volumetric tank was developed.

Every day at 13.00 h (the time at which, on average, the discharge is closest to the diurnal mean), 50 l of water were abstracted from the proglacial stream at a point in the cross-section where flow was very fast (to ensure good suspension of the sediment), and the sediment was allowed to settle out in the Plexiglass tank over the next 24 h. The volume of the deposited sediment was then measured and the instrument flushed out so that the next sample could be taken. After calibrating the instrument, the relationship between suspended sediment concentration and discharge was plotted for several proglacial streams. These results are certainly not perfect. Errors result from the precise timing and location of the sampling and from the fact that only one, daily, sample was taken. According to Collins (1978), who has compared the results derived using the Grande Dixence cone sampler with concentrations from his own 1 l hourly samples (Figure 17.15) and continuous records of suspended sediment concentration monitored using a recording turbidity meter, the 50 l cone sampling at 13.00 h provides the best means of estimating the average daily transport of suspended sediment. The sampling technique has two notable advantages: it has excellent reliability and the 50 l sample size is large enough to give representative results; and the time of sampling provides data which have the statistical advantage of avoiding extreme values.

Observations of suspended sediment concentration in the Gornera have yielded maximum concentrations of 5–6 g l^{-1}. Collins (1978) found sporadic concentrations of around 12 g l^{-1} during the drainage of the Gornersee during 31st July 1974. These values are comparable to the concentration of 4.5 g l^{-1} noted by Bruschin (1971) in the Borgne (1966–69). A record concentration of 36 g l^{-1} was measured in 1909 on the Drance and Pardé (1952) measured 150 gl^{-1} on the Isère.

In order to calculate the concentration of suspended sediment from the volume observed using the cone technique, it is necessary to derive a value for the ratio of weight of dry sediment to the volume of sediment deposited without taking long-term compaction into account. Like Collet (1916) and others, calibration

TABLE 17.3 Suspended sediment transport from a sample of glacierized basins

| Glacier | Catchment area (km^2) | Ablation season suspended sediment yield | | | |
| | | (m^3 km^{-2}) | | (t × 10^6 m^3 of discharge) | |
		1973	1974	1973	1974
Gorner	80	1320	835	1470	1070
Stafel	36	400	500*	430	610*
Bertol Inférieur	8.2	110	65	93	83
Tsidjiore Nouve	4.8	500	120	510	180

*Stafel: influence of small lakes upstream.

of the cone technique yielded a ratio of 1.6. In order to assess variability in suspended sediment transport, data were collected from several proglacial streams during two ablation seasons: 1973, a warm summer resulting in high flows, and 1974, a cool summer with relatively low flows. The results of these surveys are presented in Table 17.3. The notable difference in the suspended sediment yield of the Tsidjiore Nouve basin results from the transport of sediment during minor meltwater outbursts with peak discharges of the order of $1.5\,m^3\,s^{-1}$ which carry a great deal of fine sediment.

It is interesting that the drainage of the Gornersee in 1974 transported about $25\,000\,m^3$ of sediment in $4\,d$ in comparison with the total ablation season transport of $67\,000\,m^3$. In 1975 (a cool summer), Collins (1978) measured approximately $30\,000\,m^3$ (around $48\,000\,t$) of suspended sediment transport and $5000\,t$ of dissolved material transport for the Gorner basin.

The estimates of suspended sediment transport presented above compare favourably with previous estimates (supplied by hydroelectricity companies):

$$Massa \ldots\ldots\ldots\ldots 500\,m^3\,yr^{-1}\,km^{-2}$$
$$Dixence \ldots\ldots\ldots 650\,m^3\,yr^{-1}\,km^{-2}$$
$$Drance \ldots\ldots\ldots 850\,m^3\,yr^{-1}\,km^{-2}$$

Pardé (1952) mentioned $4000-5000\,m^3\,yr^{-1}\,km^{-2}$ transport of suspended sediment and bedload for Italy. At Palagnedra (Ticino) the intense storm of 7th August 1978 transported around $15\,000\,m^3\,km^{-2}$ of sediment. Suspended sediment concentration, for the same discharge, is higher in early than in late summer as a result of snowmelt and drainage from saturated moraines.

Figure 17.14 presents information on the effects of the hydroelectric installations on sediment transport. Sand traps are designed to transmit the flow at a nominal $20\,m\,s^{-1}$ with a sediment trap efficiency of 35–50% (according to discharge) and the flow compensation basins of $100\,000-800\,000\,m^3$ capacity have a sediment trap efficiency of 30–40%.

17.3.2 Suspended sediment in lakes

The suspended sediment concentration in the Z'Mutt lake has been studied by two methods: filtration of instantaneous samples abstracted using $1\,l$ (Friedinger) sampling bottles at different depths, and sedimentation measurements over periods of several days using tubes located at various depths in the lake.

For discharges above $11-12\,m^3\,s^{-1}$, a density current develops which appears, on the basis of various estimates, to flow at a speed of $0.4-0.5\,m\,s^{-1}$. Deep in the lake, suspended sediment concentrations derived by the first method of sampling are 3–4 times higher than near the water surface. However, concentrations near the bottom of the lake are only half those of the inflowing water, demonstrating partial dilution and sedimentation (Figure 17.16).

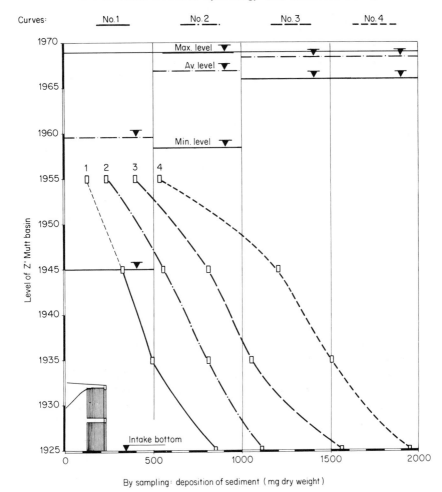

FIGURE 17.16   Suspended sediment concentration in the Z'Mutt lake during summer 1977. The curves represent 10-day averages

Measurements using the second method of sampling give relative rather than absolute values of suspended sediment concentration but the technique integrates time variations well. The results obtained using this technique confirm the results of instantaneous sampling described above and yield lake-bottom concentrations which are 4–7 times greater than those near the surface of the lake. Moreover, analysis of the fine particles trapped by the technique show differences in particle size distribution. Thus, although over 60% of particles are larger than 40 μm near the bottom of the lake, there are no particles as large as this in samples taken 30 m above the bed of the lake (Figure 17.17). Electron microscope

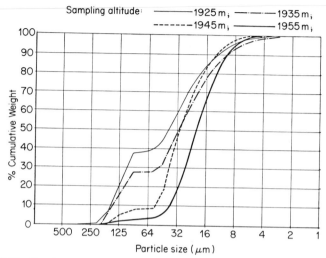

FIGURE 17.17   Particle size distribution of fine suspended sediment sampled at various depths in the Z'Mutt lake (maximum level, 1970 m a.s.l.)

macrography of the sediments allows observation of the details of the individual sediment particles. They are essentially made up of silica grains with very rough, broken or burst surfaces, generally in the form of slivers.

The kinematic viscosity of the lake waters is high because of the low temperatures (1–2 °C). Thus the particles which are subject to significant viscous forces (and comparatively weak gravity forces) settle very slowly. This explains the cloudy appearance of proglacial lakes. The colour of the lakes corresponds to the local rock type.

An examination of the particle size distribution of suspended sediment at different locations illustrates the impact of the local environment on sedimentation. The sand traps are only efficient in trapping particles in the range 30–200 μm, whereas the Z'Mutt lake, which is long and deep, has a trap efficiency of 10–15% for all particle sizes.

The 400×10^6 m³ Grande Dixence lake (located behind the Grande Dixence dam, B in Figure 17.1), which has a maximum depth of 200 m, is a very efficient sediment trap. This essentially results from the fact that the waters captured by the Grande Dixence scheme run into the upstream end of the lake, whereas water is abstracted from the lake at the dam, 5.5 km downstream. In 1976, an average year in terms of hydrology, a mean concentration of 0.4 g l⁻¹ was measured in the waters entering the lake. Large quantities of water are only abstracted from the lake in winter. Suspended sediment concentrations in the range 0.004–0.01 g l⁻¹ have been observed in the water passing through the turbines after leaving the lake. Such concentrations account for only a very small volume of sediment leaving the lake. Since there are 350×10^6 m³ of water in

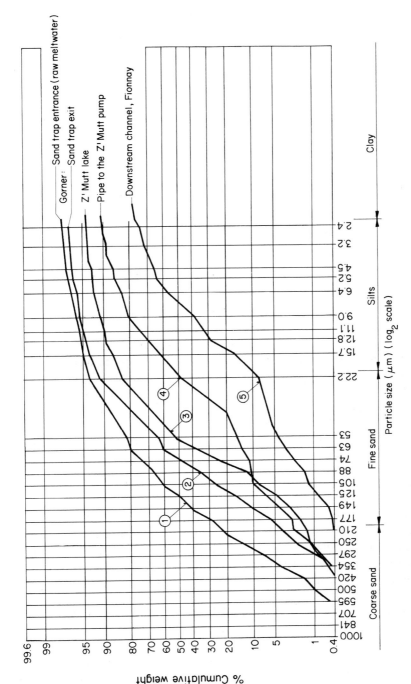

FIGURE 17.18 Particle size distributions for sediment deposited during summer 1972 in the Gorner, Z'Mutt and Fionnay works (analysed by the Laboratory of Geomorphology, Université Libre de Bruxelles)

TABLE 17.4   Estimates of bedload transport from a sample of glacierized basins

Glacier	Catchment area (km²)	Yield of bedload	
		$m^3 yr^{-1}$	$m^3 km^{-2} yr^{-1}$
Gorner	80	17000–35000	200–450
Stafel	36	8000–15000	215–420
Ferpecle	36	5000–12000	140–330
Bertol Inférieur	8.2	1200– 4000	150–500
Tsidjiore Nouve	4.8	400– 2000	80–400

the lake, about 140 000 t or 90 000 m³ of sediment was deposited in the lake during 1976. The yearly extremes of sediment deposition in the Grande Dixence lake vary from 50 000 to 60 000 m³ as a result of a cool summer to 120 000 to 150 000 m³ as a result of a warm summer.

17.3.3   Bedload

There remains the problem of bedload, material transported by saltation or rolling along the bed of the meltwater stream. Bedload can be estimated from the frequency of flushing of gravel traps. The results from such an analysis are presented in Table 17.4. In the Stafel and Tsidjiore Nouve basins the transported volumes of suspended sediment and bedload are very similar, whereas in the Gorner basin the volume of suspended sediment is two to three times greater than the volume of bedload.

17.3.4   Remarks

Some glacierized basins, such as the Arb, Trift, Bis and Festi, yield very little suspended sediment or bedload. Only localized storms, as occur in the Rosses or Vouasson (schists) basins, are able to release debris flows or mudflows of thousands of cubic metres. A similar phenomenon is produced by meltwater outbursts, as has been observed in the Kin basin, where the release of 100 000 m³ of water (peak discharge, 4–5 m³ s^{-1}) transported 70 000–80 000 m³ of debris onto the Visp–Zermatt road and railway line. The discharge and sediment was delivered from source areas at an altitude of between 1400 and 2800 m.

Catastrophic phenomena occur, like the storm in October 1977 over the Ferpècle basin, which transported a sediment load equivalent to 4–5 years of normal sediment yield and which filled up a natural 20 000 m³ lake. It is highly probable that all the large debris cones, which are typical of alpine valleys, originate from extreme phenomena such as violent storms and intense melting of snow and ice.

In view of the large volume of residual moraines and the very variable rate at which they degrade, it is the author's strong opinion that, contrary to current practice, rates of erosion of watersheds should not be estimated on the basis of annual (fluvial) sediment yield.

ACKNOWLEDGEMENT

This chapter is based on a paper which was originally published in French: A Bezinge (1978), Torrents glaciaires, hydrologie et charriages d'alluvions, *Jahrbuch der Schweizerischen Naturforschenden Gesellschaft, Wissenschaftlicher, Teil 1978,* pp. 152–169. Material from this paper is reproduced by permission of the Swiss Academy of Sciences.

REFERENCES AND BIBLIOGRAPHY

BEZINGE, A., and SCHAFER, F. (1968), Pompes d'accumulation et eaux glaciaires, *Bulletin Technique de la Suisse Romande,* **20**, 282–290.

BEZINGE, A. (1971), Déglaciation sur les vals de Zermatt et d'Hérens, paper presented to the glaciological section of the Société Hydrotechnique de France, Grenoble, 4th and 5th March 1971.

BEZINGE, A. (1972), Étude descriptive d'une forte averse sur les Alpes, paper presented to the glaciological section of the Société Hydrotechnique de France, Paris, 2nd and 3rd March 1972.

BEZINGE, A. (1973), Image du climat sur les Alpes, paper presented to the glaciological section of the Société Hydrotechnique de France, Grenoble, 22nd and 23rd February 1973.

BEZINGE, A. (1976), Eaux glaciaires, sédimentations et usures de pompes, Internal study of Grande Dixence S. A., Sion, Valais, Switzerland.

BEZINGE, A., PERRETEN, J.-P., and SCHAFER, F. (1973), Phénomènes du lac glaciaire du Gorner, in *Symposium on the Hydrology of Glaciers (Proceedings of the Cambridge Symposium, 7–13 September 1969),* International Association of Scientific Hydrology Publication 95, pp. 65–78.

BRUSCHIN, J. (1971), Transports solides en suspension dans les rivières Suisses, *Bulletin Technique de la Suisse Romande,* 97th year (14), 10 July 1971, 337–343.

COLLET, L.-W. (1916), Le charriage des alluvions sur certains cours d'eau de Suisse *Annales Suisses d'Hydrographies,* Vol. II, Département Suisse de l'Intérieur, Berne.

COLLET, L.-W. (1925), *Les Lacs. Eléments d'Hydrogéologie,* Gaston Douin, Paris.

COLLINS, D. N. (1978), Hydrology of an Alpine glacier as indicated by the chemical composition of meltwater, *Zeitschrift für Gletscherkunde und Glazialgeologie,* **13**, 219–238.

ELLISTON, G. (1973), Water movement through the Gornergletscher, in Symposium on the Hydrology of Glaciers (Proceedings of the Cambridge Symposium, 7–13 September 1969), International Association of Scientific Hydrology Publication 95, pp. 79–84.

GAUDET, F. (1975), Les cours d'eau alpins de régime glaciaire, *PhD Thesis,* University of Lille.

KASSER, P. (1959), Der Einfluss von Gletscherrückgang und Gletschervorstoss auf den Wasserhaushalt, *Wasser-Energiewirtschaft,* **51**, 155–168.

KASSER, P. (1973), Influences of changes in the glacierized area on summer runoff in the Porte du Scex drainage basin of the Rhône, in *Symposium on the Hydrology of Glaciers (Proceedings of the Cambridge Symposium, 7–13 Septmeber 1969)*, International Association of Scientific Hydrology Publication 95, pp. 219–225.

LUGEON, J. (1928), *Précipitations Atmosphériques. Ecoulement et Hydro-électricité*, Publication de l'Institut Fédéral de la Météorologie, Dunod, Paris.

LUTSCHG, O. (1926), *Über Niederschläge und Abfluss im Hochgebirge*, Sekretariat des Schweizerischen Wasserwirtschaftsverbandes, Zürich.

MARTINEC, J. (1970), Recession coefficient in glacier runoff studies, *Bulletin of the International Association of Scientific Hydrology*, **15**, 87–90.

PARDÉ, M. (1952), *Compte Rendu des Deuxièmes Journées de l'Hydraulique*, Société Hydrotechnique de France, Grenoble, 25–29 June 1952.

REMENIERAS, G. (1960), *L'Hydrologie de l'Ingénieur*, Eyrolles, Paris, pp. 274–413.

SINGH, B., and SHAH, C. (1971), Plunging phenomenon of density currents in reservoirs, *Revue de la Houille Blanche*, **1**, 59–64.

VIVIAN, R. (1973), Les glaciers des Alpes Occidentales, *PhD Thesis*, University of Grenoble, pp. 288–354.

Glacio-fluvial Sediment Transfer
Edited by A. M. Gurnell and M. J. Clark
©1987 John Wiley & Sons Ltd.

Chapter 18

The Glacio-Fluvial Sediment System: Applications and Implications

M. J. CLARK

ABSTRACT

Sediment system studies have wide applicability, but four focuses which typify present concerns are the contemporary processes, the use of sediment budgets to estimate weathering or erosion rates, the renewed emphasis on palaeo-environmental reconstruction and modelling and the practical uses of alpine sediment system studies. Each of these topics represents an important field of application, but each also raises issues which are of relevance to any assessment of the current status of research on the glacio-fluvial sediment system. In particular, the methodological and technical constraints pertinent to such studies are a major influence on their investigative design and validity, and are consequently accorded specific consideration.

18.1 APPROACHES TO THE CURRENT PROCESS SYSTEM

Inevitably, a field as broad as the alpine sediment system spawns such a wide range of component and associated studies that generalization of its research priorities is difficult and potentially misleading. However, if the focus is kept firmly on relevance to the glacio-fluvial sediment system, then it may be permissible to highlight four aspects which combine to reflect the dynamism and constraints of the subject. It is in no way suggested that these represent either a comprehensive illustration of present concerns, or that they would be universally accepted as the research priorities of the subject, but in their diversity they do offer an indication of the range of aspirations and problems to which the contributions in the previous sections of this book might be directed.

As a starting point it is pertinent to note the considerable distinction between the strong theoretical basis of glacial, hydrological and much of glacio-fluvial modelling, and the inductive inferential base of the majority of the periglacial process explanation which dominates consideration of the broader alpine sediment system. Further, periglacial geomorphologists have been slow to examine the detailed physical attributes of materials at low temperatures, and thus have been alerted to phenomena such as permafrost creep and water migration through frozen ground largely as a result of improved field monitoring rather than by prompting from theoreticians. Whilst it is clear that theory can act as a constraint as well as a trigger, serving to deny phenomena that field observation has apparently confirmed, there can be no doubt that an interplay between theoretical, experimental and field inferential approaches offers the strongest basis for robust innovation. Against such standards, geomorphology might be seen to lag behind hydrology and glaciology, and periglaciation to be subservient to glaciation—a supposition that receives some support from the contributions to this volume. However, one of the reasons for such disparities may lie in the greater problems associated with long-term field-based studies, so that existing achievements are to be regarded as creditable even though there is room for further progress.

There is merit in making explicit reference to a premise that was implicit in much of Section 2.2, namely that the strength of, and ultimate justification for, most studies of the alpine sediment system lies in their elucidation of the links between the various subsystems which are normally studied in isolation. For example, in an overall perspective, the major contributions of Rapp's (1960) monumental integrated field study of the Kärkevagge Valley in the mountains of northern Sweden, or Caine's (1974) systems model of alpine sediment transfer, both lie in their synthetic power—indicating the relative importance of components in the former case, and subsystem linkage in the latter—rather than in their detail. In contrast, it is inherent to the instrumental approach of much work on the current process system that the normal investigative design is small scale and process specific, so that integrative studies are both rare and often either deductive or morphologically inferred (a dangerous source of circular argument). Real advance in this area thus awaits either very large instrumental studies which incorporate all the sediment transfer components of an alpine area (an 'instrumental Kärkevagge'), or a deliberate linkage of component studies for a given region or set of comparable regions. These are targets as yet beyond the achievement of glacio-fluvial sediment studies, let alone integrated alpine sediment transfer budgets.

The distinction, seen here, between agglomerative (synthetic) and divisive (analytical) approaches to study is common to much of physical and natural science. Also common at present is the feeling that the power and precision of analytical studies are unlikely in themselves to provide a route towards explanation of the system as a whole. Thus, if such synthetic mastery is regarded as a priority, then it is probably best accessed by synthetic studies. This is the context within which the present volume finds its function, although its aim has been to integrate the elements of the glacio-fluvial subsystem rather than the alpine system as a whole. Many of the contributions are the product of analytical designs, but their agglomeration within a single volume and their evaluation within the framework of an integrated system both express aspirations beyond the description or explanation of isolated components.

In view of the paucity of such cooperative or convergent research programmes, it is worthwhile to recognize the potential of two alternative frameworks. The first capitalizes on the extrapolative power of the large-scale spatial monitoring possible through satellite and aerial remote sensing. Examples relating directly to mountain sediment transfers are the Rocky Mountain alpine tundra stratification from aerial imagery used by Frank and Thorn (1985) as a basis for classifying gross variations in surficial sediment loss, and the evaluation of a digital cover type classification from Landsat imagery of the northern Fennoscandian mountain foothill zone by Clark *et al.* (1985) as a surrogate for snow water equivalent storage. The second alternative to field-based holistic studies of the sediment system derives from the increasing potential for simulation of the system. Except where such models are based on statistical series, they tend to be subject to the constraints of the theoretical base rather than of field calibration. The reliance on theory ensures that effective simulation and prediction are currently limited to atmospheric, glacial and hydrological components, though digital models of ground temperature and some other periglacial elements are becoming available. However, the difficulty of producing verified and representative simulations of systems components is sufficient to indicate that a simulation model of the whole alpine sediment system lies well into the future.

Given the practical, technical and conceptual problems of the integrated investigation of the alpine sediment system, it would be premature to attempt any general statement about the relationship between process and morphology. Even more tantalizing is the prospect of fusing, in rigorous rather than speculative terms, the links between current and past process systems and their associated material and landform assemblages. It thus appears that in the short term the research focus on the current process system is likely to remain multi-focal, with instrumental (field and laboratory) studies, simulation models and extrapolative devices being pursued in parallel. For the moment the holistic ideal of an alpine sediment system incorporating temporal and spatial variants remains an aspiration most closely achieved by a speculative fusion of Church and Ryder's (1972) process phasing, Caine's (1974) process linkage, Rapp's (1960) data and an as-yet unformulated macro-spatial variability model. The challenge is considerable, but the potential rewards are greater.

18.2   EROSION ESTIMATION AND ITS DENUDATIONAL IMPLICATIONS

18.2.1   The aims of the sediment budget approach

One of the most frequent applications of sediment budget studies is the use of sediment output as a first approximation of either erosion (narrowly conceived as bedrock retreat and lowering) or overall denudation (total loss of rock mass from the catchment, including evacuation of surficial sediment). This approach was introduced in Chapter 2 and applied to Swiss alpine data in Chapter 8. The methodological context within which such studies should be viewed is complex, and was cogently introduced by Walling and Webb (1983) in a fluvial context.

By any standards, the sediment budget is a sufficiently vulnerable estimate to render its extrapolation risk-prone, and it can therefore be concluded that there must be compensating advantages to justify the widespread use of such a technique. Two come immediately to mind. First, sediment output integrates the varied yields of the many separate sediment sources in the catchment in a way that can be achieved by no other approach. The end product is a generalization, but if that is what is required then the sediment budget approach is fully warranted. Second, the problems of measuring sediment throughput (usually across a fluvial or glacial channel section) are great, but they may be more amenable to instrumental solution than is the direct field measurement of erosion. Indeed, in the context of upland erosion rate and weathering rate, detailed field studies are both sparse and problematic, thus rendering all the more welcome recent pioneering work such as that of Caine (1976), Thorn and Hall (1980), McGreevy and Whalley (1982), Gardner (1982) and Rapp (1983, 1984). However, in view of the importance that has been attributed to the use of alpine sediment budget as an erosion estimator, it is worth examining briefly the assumptions of the alternative methods available before evaluating the approach.

18.2.2   The methods: sediment budget monitoring and two alternatives

Using general hydrological principles it is possible to infer fluvial sediment budget from hydraulic and geometric attributes (Walling and Webb, 1983), but such estimates are error prone, particularly towards extremes of discharge, roughness and channel gradient. If field monitoring and inclusion of non-fluvial components is preferred, then in principle it would seem that sediment budget might be estimated in three ways—direct measurement at the erosion site, real-time monitoring of the sediment flux from a designated source area (the sediment budget approach), or reconstruction of the rate at which the transported sediment is added to a depositional store (a sediment budget variant extrapolated across a long timescale). The characteristics of these methods are summarized in Table 18.1, from which it can be deduced that each has merits within its own appropriate field of application. Direct measurement appears most rigorous, but only for real-time at-a-site studies. Sediment budget techniques clearly do have considerable potential, but the list of disadvantages is worrying in its length and complexity. A review of these problems must thus precede an evaluation of the merit of the approach.

TABLE 18.1   Characteristics of erosion-rate estimation techniques

Technique	Advantages	Disadvantages
Direct measurement of erosion rate	Precise and unequivocal data	Instrumentation and sampling problems
e.g. micro-erosion meter or erosion pins	Known observation period	Problems extrapolating to long-term trends
	Known observation site	Problems generalising spatially from sample site
	Known environmental conditions	
Rate of sediment transport from source of erosion	Erosion integrated across across catchment	Instrumentation and sampling problems
e.g. real-time river sediment budget	Sub-catchment data indicate crude source areas	Difficult to relate to exact erosion site
	Requires only limited access to site	Problems extrapolating to long-term trend
	Total source area reasonably defined	Problem to distinguish direct erosion from reworking older store
	Amenable to Statistical models and prediction	
Rate of addition of sediment to stores	Erosion integrated across long time period	Instrumentation and sampling problems
e.g. floodplain, fan, moraine or lake aggradation	Source area larger than for direct measurement	Limited data on erosional or depositional environment
	Total source area reasonably defined	Difficult to monitor losses from store
		Difficult to date the period of aggradation and thus to predict
		Problem to distinguish direct erosion from reworking older store

18.2.3   Problems in the extrapolation of sediment budget estimates

Whilst sediment flux is a tantalizingly attractive route to estimating erosion rates, it does suffer a range of potential weaknesses which render it at best tentative and at worst profoundly misleading. The general difficulties of sampling and instrumentation design are shared by all estimation techniques, although there are significant differences in the nature of the errors involved (e.g. Walling and Webb, 1981). For whatever purposes sediment budget monitoring is being undertaken, the design requirements cover criteria for both precision (the fineness and correctness of detail of the spatial, temporal and flux-intensity attributes to be recorded) and accuracy (the extent to which measurements, of whatever detail, represent fully the total, average and range of fluxes involved). Of the two, accuracy is by far the most important, although it is often the case that great effort and expenditure are devoted to instrumentation precision without corresponding respect being paid to the critical significance of sampling design. An attempt has been made in the previous chapters to redress this balance—particularly in the context of glacio-fluvial fluxes. In essence, it is concluded that the variability of sediment flux at all the time and space scales investigated is so great that sampling design is the primary constraint on an investigation's validity, and that this design should be to a major degree site-specific or at least subject to regional zonation.

In part related to the sampling problem is the difficulty of identifying the precise sources of the discharge and sediments that have contributed to budget estimates. Many techniques have been developed to assist in such identification, although the problem is by no means solved. For example, the particulate sediment's mineralogical (Klages and Hsieh, 1975), magnetic (Oldfield, 1983) or palynological (Brown, 1985) characteristics have been regarded as possible indicators of sediment source, whilst solute attributes represented by such indices as electrical conductivity have been incorporated in dilution or mixing models to assess likely glacio-hydrological routes and sources (Sections 12.4 and 14.5). Such inference is a powerful investigatory tool provided that there is a reasonable awareness of its probable confidence limits, and under field conditions it may well be as sensitive as attempts to map sources from surface indicators of erosion. Without some access to sediment source identification, whether directly mapped or inferred from the sediment, the overall budget estimate can only be applied to the whole catchment (slope or fluvial) contributing to the measurement location. This may be entirely satisfactory for some hydrological and geomorphological purposes, particularly at regional scale, but often renders erosion or weathering rate estimations too generalized to be meaningful.

This deficiency becomes of special significance in those many cases where it is imperative to distinguish between actual bedrock erosion and mere evacuation of readily available incoherent material. The distinction between true additions to the sediment system and simple reworking of existing sediment stores is crucial to the interpretation of overall denudation and to the assessment of erosional efficacy of the various processes operative within the catchment. This is, of course, also the fundamental assumption underlying Church and Ryder's (1972) original formulation of the paraglacial model of sediment yield disequilibrium with nonglacial climates, although it has been stressed in Section 2.3.1 that considerably greater attention might also be paid to hydrological relationships in the deglacial system.

Finally, it is clear from all that has already been said about temporal variability, that the data derived from a short (often very short) period of instrumental sediment measurements are most unlikely to have any direct applicability for periods more than a few times that length into the past or future, as was long ago pointed out by Gage (1970). Short-term prediction is becoming an operational possibility, but long-term extrapolation is both impractical and unwarranted. Similarly, palaeoreconstructions based on real-time data may well be no better than speculation, and often do more harm than good since they carry a wholly unjustified image of calibrated respectability. It is not surprising that many authors have preferred to sacrifice the precision of real-time monitoring for the potentially greater temporal representativeness of budget assessments based on accummulation rates in major topographic stores such as lakes (Davis, 1976; Oldfield, 1977; Walling and Webb, 1983), fans (Bauer, 1980), alluvial plains (Church and Ryder, 1972) or moraines (Small *et al.*, 1984)—although in a sense this merely substitutes one set of problems for another. Ultimately the unavoidable conclusion is that the considerable inferential power of sediment budgets should be utilized only with an equally considerable degree of caution.

18.3 PALAEO GLACIO-FLUVIAL SEDIMENT SYSTEMS

18.3.1 The quest for analogues

The long-term extrapolation of sediment budgets, often addressed by way of the glacio-fluvial sediment system, is but one strand in the much broader study of palaeoenvironments—a focus for Quaternary Science as a whole. Such studies have traditionally tended to concentrate on palaeo-forms and stratigraphical sequences, but more recently attention has turned to process reconstruction as such. Thus, the emergence of palaeohydrology as a distinct sub-field in its own right (e.g. Schumm, 1965; Gregory, 1983), represents the inherent desire to achieve a rigorous mastery of the process systems of the past rather than simply describing their morphological or sedimentary end products. Given the extreme difficulty of mastering present-day process systems, it is clear that long-term inference will for some time be more of an aspiration than a practical proposition, but the aim is one that fully justifies the necessary effort, and for individual indices of transport processes considerable progress has been made (e.g. Maizels, 1983). Whilst much work on the alpine and glacio-fluvial sediment system could be regarded as serving an analogue role, only a relatively few authors (e.g. Bryant, 1983) have given specific attention to the constraints of the approach.

In the present context, perhaps the main implication is methodological rather than technical. If modern observations of glacio-fluvial sediment transport processes or of sediment budgets are to be applied to the past, then this application must be handled within a methodological framework that is either explicitly stated or, more often, implicitly assumed. Three possibilities can be suggested as having particular significance:

a. Some form of statistical extrapolation of current rates into the past—an approach discussed in Section 18.2, with some misgivings.

b. Some form of direct analogue—usually a simple and unquestioning assumption that current processes in areas of extreme altitude or latitude could be regarded as representative of conditions pertaining at lower altitudes and latitudes during the Pleistocene. This essentially ergodic assumption is discussed in Section 18.3.2.

c. Some form of synthesis between extrapolation and analogue, which aims to combine the precision of the former with the accuracy of the latter, and substantially to improve upon both. This possibility is briefly considered in Section 18.3.3.

18.3.2   The ergodic fallacy

It has long been assumed that an indication of Pleistocene cold-region environment for a particular site could most easily be achieved by the simple expedient of examining the current conditions in an area colder than the target site by reason of higher altitude or latitude, although otherwise of similar characteristics. Indeed, the birth of modern glacial study with Louis Agassiz in the 1840s, and the increasing pace of mountain and arctic periglacial study in the early decades of the present century, were both motivated powerfully by the belief that an explanation of many present sedimentary and morphological features lies in the study of areas deemed to mirror the conditions of the past. It should be recognized, however, that this assumption, although emotively appealing at a high level of generalization, is as risk-prone as the long-term extrapolation of present-day process data. Without specification and standardization of conditions to a degree that is often in practice impossible to achieve, the present is no more a key to the past than the past is a reliable key to the present.

Nevertheless, the notion that comparisons in space can be substituted for variations in time is deep-rooted in natural science, and is encapsulated in the so-called ergodic theorem (Paine, 1985). The justifications for questioning, and perhaps even rejecting, such a well established principle derive from two main sources. First, the discussion of altitudinal and latitudinal affects on the process system (Section 2.2) reveals sufficient evidence of real contrasts in the operation of the system with gross changes in these two variables to throw doubt on the acceptability of long-distance analogues. It is difficult enough to make meaningful comparisons between two adjacent glaciers in the same region, without expecting to draw enlightened analogies between, for example, present-day Greenland and Pleistocene Switzerland. Interesting and fruitful lines of enquiry could well be prompted by such inter-regional comparisons, but simple interchange of data or explanation is unwarranted.

Second, the temporal basis of the modern analogue approach is equally suspect. There is phasing in the behaviour of the sediment system at every scale (Section 2.3), and unless comparisons are capable of standardizing the phases of the two entities being compared, the analogue is unreliable. Since phasing is poorly understood at present, it follows that rigorous analogue exercises are premature. Both the spatial and temporal difficulties are, in fact, compounded by a strong risk of sedimentary and morphological equifinality. Until unambiguous attributes of form and material can be identified to act as indices of the state of the process system, inter-regional comparisons should be regarded as speculative.

18.3.3 A systems-based synthetic analogue

Given the real concern that both long-term extrapolation and the modern analogue approach currently require considerable refinement before they can be implemented reliably, it is understandable that an alternative should be sought to serve as a route to the better understanding of palaeosystems in general, and the palaeo-sediment system in particular. The greatest potential in this respect appears to lie in the rigorous specification of the present sediment system, such that a functional systems model could then be recalibrated with 'Pleistocene' input variables so as to simulate the likely operation of the palaeosystem. Once again, the approach could well justify the label 'premature,' yet there are signs that the aim is viable even if its expression is, as yet, sub-operational.

Two associated problems are apparent. First, most available data sets of sufficient quality to permit reliable modelling are derived from non-alpine environments, and it is probable that neither their absolute values nor their structure can be safely extended to the mountain zone. This is true, for example, of the Universal Soil Loss Equation (Wischmeier and Smith 1965, 1978) which, although designed for a completely different purpose, does have many attributes that render it potentially suitable for modelling sediment yields under varied and varying conditions. In its present form, its calibrated range excludes the high mountain zone and makes little allowance for nival, glacial or ground freezing inputs—but in principle all could be incorporated. Similarly, the postulated relationship between sediment yield and effective precipitation proposed by Langbein and Schumm (1958) and later developed by Schumm (1965, 1977) may represent latitudinal variation in erosional potential. This relationship derives from non-alpine environments, but the indications are that it may be amenable to extension into the mountain and northern environments, although Walling and Kleo (1979) suggested that factors other than annual precipitation offer better indices of sediment yield, at least for areas of low precipitation. Whether this is true, or whether separate equations are necessary for the alpine zone, the fact remains that such attempts at a general approach may have the potential to model systems which had Pleistocene counterparts and for which it might be possible to estimate at least some of the Pleistocene input variables. One of the most exciting products of this book, and of associated work elsewhere, is the extent to which models of substantial sections of the system begin to appear within reach. The imprecisions remain considerable, but the overall risks are no greater than those which apply to extrapolation or to simplistic analogue arguments.

The second problem of synthetic analogue modelling using a systems approach is that of combining the subsystems which form the focus of most current work. The practical difficulties of model implementation are such that at present all operational models are of subsystems, whilst all the available integrated alpine sediment models are restricted to conceptual structures—fascinating in themselves, but difficult to calibrate to past conditions until they have been rendered functional and then calibrated to a present series of data sets. The remaining problems are legion, but in every sense the goals are rewarding and challenging, and serve to provide a broad conceptual and methodological context for several of the contributions to this book.

18.4   APPLICATIONS FOR STUDIES OF THE ALPINE SEDIMENT SYSTEM

Much of this volume has been derived from and directed towards academic research, but it would be a mistake to assume that the work discussed is in any way lacking in practical application. Indeed, many of the key contributions to the field that have been quoted emanate from authors with professional rather than academic affiliations, for example, the work of Østrem (1975) in Norway, Kuusisto (1984) in Finland and Bezinge *et al.* (1973) in Switzerland. Whilst the range of potential applications is large, it will suffice in the present context to reflect the possibilities through three general categories—mountain hazards, water quantity and water quality. This is not an elegant classification since there is substantial overlap between the classes and they are based on rather different criteria, but it does reflect the focuses of the literature. If a more broadly based introduction to applied science in the alpine zone is required, then an indication of the potential can be obtained from Price (1981) or Messerli and Ives (1984).

Mountain hazards mapping has advanced rapidly during the last decade, particularly in terms of the development of integrated hazard inventories rather than single-hazard (e.g. avalanche) maps. Progress is admirably exemplified by Kienholz (1978) in Switzerland and Dow *et al.* (1981) working in the Colorado Rocky Mountains, and by Kienholz *et al.* (1984) for the Middle Mountains of Nepal. These two study typologies reflect both the physical differences between

TABLE 18.2 Hazard hierarchy for mid-temperate alpine zone. *Reproduced with permission from Dow* et al. *(1981).*

Hazard	Description
Fast movements — Danger to human life and property	
Active avalanche confined path	Rapid downslope movement of dry or wet snow, confined to specific path and having large starting zone, track and well developed runout zone
Active avalanche unconfined slope	Active avalanche area on unconfined slope (steeper than 30°) and having either no vegetation or only sparse conifer vegetation
Potential avalanche slope	Includes timbered slopes steeper than 30° on which avalanches will occur if forest is cut or burned
Debris flow	Rapid mass movement of viscous material saturated by rainfall or snowmelt. Speed increased by added water or steeper slopes. Damage caused by impact pressure or debris inundation
Rockfall	Falling, rolling or bouncing rocks, governed by gravity and commonly initiated by bedrock weathering, erosion of underlying supporting material, seismic activity, etc. Damage by rock impact
Mountain torrent	Creek or river subject to flood if carrying capacity is exceeded. Damage caused by transported detritus, by downcutting or by side-cutting
Flood	Inundation by flowing or standing water overflowing its normal banks, initiated by rainfall, snowmelt or breaking of dams. Damage by water or by deposition
Slow movements — Danger to property	
Landslide	Rotational or translational displacement of an earth mass. Damage by displacement or impact
Subsidence	Downward displacement of surface due to natural or man-made causes (drainage of underground water, mining, etc.)
Soil creep	Downslope displacement due to saturation, mainly in melt season
Expanding soil	Expansion/contraction with water content changes, usually associated with clay content
Potentially unstable soil	Potential occurrence of any of the above hazards

the two types of area (climate, geology, tectonics) and their cultural contrasts (differing land use pressures and conflicts). To an extent, these differences influence the hazard classifications used. Dow *et al.* (1981) presented a structure which divides processes in terms of speed of movement and type of damage (summarized in Table 18.2), and which regards the most likely applications of hazard mapping as lying in two-season tourist management and land use planning. It is clear that the majority of hazards could be interpreted as being elements in the alpine sediment transfer system, and that those with a hydrological input are properly part of the glacio-fluvial system—a duality also apparent in Kienholz (1978).

The tropical mountain hazard inventory proposed by Kienholz *et al.* (1984) has some common physical elements, although its approach to classification is very different (as is indicated by Table 18.3), and the hazards impinge upon considerable amounts of permanent cultivation and residence. Slope instability is manifestly the most significant component, an evaluation already established by Caine and Mool (1982) in the same area. Glacio-fluvial elements are of correspondingly lower ranking, but they remain significant. However, whatever the precise assessment, the unavoidable conclusion to be drawn from hazard inventories such as these is that every aspect of research on the nature, cause, magnitude, frequency and distribution of mountain sediment transfers could be of potential applicability if it referred to a zone within which some human resource exploitation was located.

This type of broad multi-hazard approach displays relationships with two associated groups of application. First, links of both conceptualization and authorship can be traced to the still more general notion of stability and instability of mountain ecosystems, which formed the theme of an important

TABLE 18.3 Type, evidence and corresponding degree of hazard or instability for a tropical mountain area. *Reproduced with permission from Kienholz* et al., *1984)*

Type of hazard	Evidence and degree of intensity
Deep erosion by water (gullies >2 m deep) Shallow water erosion (gullies <2 m deep) Surficial erosion: rills	Hazard or instability is: (a) confirmed/inferred (b) suspected
Deep landslide (>2 m) Shallow landslide (<2 m) Collapsed terraces	Hazard or instability occurs in: (a) more than 50% of unit area (b) less than 50% of unit area
Debris flow Accumulation of debris flow material	Hazard intensity is rated as: (a) degree 2a (highest) (b) degree 2b (c) degree 1 (d) degree 0 (lowest)
Flooding Accumulation of water-transported material	Evidence derived from: (a) Direct examination of unforested areas (b) Indirect mapping of forested areas from air photographs
Major torrent activity Minor torrent activity	
Rockfall source area Accumulation of rockfall material	Full hazard categorization combines Hazard type with the appropriate indices of evidence and degree

international workshop at Berne (Switzerland) in 1981, co-sponsored by the United Nations University, UNESCO, the IGU Commission on Mountain Geoecology and the International Mountain Society. The impetus reflected by this meeting has since spawned a number of studies and papers (e.g. Messerli, 1983). It is not exclusively centred on physical aspects, let alone on the glacio-fluvial sediment system, but does provide a framework of management guidelines within which glacio-fluvial expertise could be applied. A second field associated with the multi-hazard approach is the mountain equivalent of the familiar environmental impact statement. This will incorporate glacio-fluvial elements only to the extent to which they impinge upon a particular problem. Thus Aegerter and Messerli (1983) developed a generalized procedure for the evaluation of the impact of hydroelectric power plants on a mountainous environment which includes non-biotic (radiation, air, water, soil, geology), biotic (soil, vegetation) and cultural indices of which a number relate to fluvial and hydrological elements (Table 18.4). Although the role of sediment system attributes in the total procedure is relatively small, it fully justifies its inclusion and can, at site level, demonstrate significant relationships with the fluvial systems in the impact zone (Gurnell, 1983).

Inevitably, much of the hydrologically based component of the mountain hazard or impact inventory refers to problems of water quantity prediction and management, yet this topic is of sufficient generality to warrant it being treated

TABLE 18.4   Extracts from a heirarchic classification of the impacts of mountain hydroelectric power schemes. *Reproduced with permission from Aegerter and Messerli (1983)*

Atmospheric water	Precipitation, areal distribution
	Density, frequency, duration
Running water	Quantity
	Quality (physical/chemical/biological)
	Self-regenerational potential
	Transportation potential
	Material load
	Danger
Standing water	Quantity/area
	Quality (physical/chemical/biological)
	Temperature
	Self-regenerational potential
	Flow/turbidity
	Sedimentary area
	Compensational basin
	Danger
Aquatic ecosystems	Variations of water level
	Nature of river bed
	Nature of banks
	Flora and fauna

in its own right rather than being regarded simply as one of a list of hazards. However, since much of the present volume has been devoted to the physical background to such applications, it is sufficient in the present context simply to note the extent to which water quantity impinges on channel geometry and stability, sediment transport, aquatic ecosystem, flood frequency and magnitude and water availability for extraction for other purposes. Similarly, water quality is associated with both biotic and non-biotic influences—the latter including channel sedimentation and erosion rates, lake sedimentation (a major problem when lakes perform a flow or sediment regulating function) and water quality for extraction (for industrial, agricultural or domestic use).

18.5   THE STATUS OF THE ALPINE AND GLACIO-FLUVIAL SEDIMENT SYSTEMS

On the basis of the foregoing review of the major attributes of the alpine sediment system and of a representative series of its current research and application focuses, it is appropriate to attempt a provisional assessment of the overall status of research in this field—in effect, a follow-up to Ives (1974). Two decades ago such research was over-concentrated in the mid-temperate zone; it was often damagingly split into sub-disciplines which operated essentially in isolation, it was unacceptably subservient to research in non-alpine areas and it lacked a strong enough theoretical base to provide strongly integrative incentive whether to pure or to applied studies. A very great deal has been achieved during those two decades, on both sides of the Atlantic and in an increasing number of centres elsewhere (Ives, 1985). Both hydrology and periglacial geomorphology have strengthened as disciplines, new journals have appeared, major research programmes have been completed, research institutes and inter-institutional links have been created to focus on the alpine zone, new techniques have been developed, the general level of process understanding has increased enormously and a start has been made on the task of fusing laboratory, field and theoretical thrusts.

Yet despite all this, the research priorities remain disturbingly similar to those of 20 years ago. Then, the challenge was to identify and specify the components and structure of an alpine sediment system; now, the challenge is to do the same for the many different alpine sediment systems that are known to exist at different scales, different latitudes and different altitudes. Then, the priority was to use the systems framework to link the components and subsystems that had hitherto been studied separately; now, the priority is to operationalize systems sufficiently general to permit modelling of different environments, or of one system passing through different states. Then, the future seemed to lie in the development of observational, analytical and modelling techniques; now, perhaps, the key lies in more rigorous theoretical constructs within which to capitalize on the potential that these new techniques has revealed—although contributions such as those of Boulton (1975) heralded a move in this direction.

Not surprisingly, the nature of the subject reflects the aspirations and constraints of the disciplines which contribute to it. At the same time, those disciplines themselves are in a state of flux so that at any one time they display different characteristics. Thus, as has been noted already, the strong theoretical base of glaciology contrasts with the field inferential bias of much glacial geomorphology. For the future scenario postulated above, the theoretical base of glaciology offers it great strength for innovation, and it is most encouraging to see that increasing use of this physical basis is being made by geomorphologists. Both disciplines stand to gain from the combination of theoretical rigour and objective field testing. In parallel with this fusion can be seen a related symbiosis between hydrology and fluvial geomorphology, the former tending towards a greater generality and consequent greater predictive and modelling power at the present time. In comparison, periglacial geomorphology is still under-utilized because it lacks a sufficiently strong theoretical base to merge easily with the other components of the system—although considerable strides have been made in very recent years, and the move towards a more multi-disciplinary geocryology is helpful. Other links which remain inexcusably under-exploited include those with engineering geology and soil mechanics, pedology and the bio-sciences. Perhaps inevitably, each discipline exhibits a certain inertia of ideas, with the result that both concepts and implicit assumptions may vary greatly between them.

Given these variations in status, there are some inevitable contrasts in the preferences and priorities of different groups of scientists, sometimes revealed as national or supra-national stereotypes. In particular, there is a considerable difference in emphasis in those countries where Quaternary Science dominates compared with those which concentrate on the present process system. This is most clearly apparent in the biological, geological and geomorphological fields within which both 'present' and 'palaeo' environments attract strong followings and generate their own research paradigms. The contrasts are less clear in the case of glaciology and hydrology, which have tended to be absolutely dominated by a concern with process. However, with the development of palaeohydrology and with the numerical simulation of past ice sheet behaviour, these subjects too might come to experience a polarization of interests into two groups with past and present labels. Such crystallization of effort can have beneficial repercussions, but there is also a considerable danger of communication lapses, confusingly varied implicit assumptions and markedly variant phasing in the take-up of new ideas. The sociology of science is sufficiently complex to suggest that these tendencies should be openly monitored so as to optimise their beneficial impacts.

That the study of the alpine or glacio-fluvial sediment systems lacks coherence and clearly defined targets may be regarded as unfortunate, but on the other hand it can be seen as reflecting the dynamism of these subjects. It is within this context that benefit can be derived from the foregoing substantive consideration of the glacial and fluvial sediment systems (Chapters 6 to 17) which highlight the components from which any comprehensive integrated model of the system will have to be built. In order to provide a measure of comparability, each of these chapters has incorporated a focus on the currently glaciated southern Swiss Alps. However, every effort has been made to generalize both the results and the conclusions by referring extensively to work elsewhere and by exploring methodological and technical problems, since it is through a strengthening of these aspects that the full potential of the many new techniques can be harnessed in the task of specifying a glacio-fluvial sediment system robust enough to survive application in other places and at other times. However, whilst the challenge of today may seem to lie in data acquisition and analysis, it might be that the greater aspiration for the future rests in theory, concept and synthesis. Only by this route will it be possible to transform sediment transfer processes into a sediment transfer system.

REFERENCES

AEGERTER, S., and MESSERLI, P. (1983), The impact of hydroelectric power plants on a mountainous environment: a technique for assessing environmental impacts, *Mountain Research and Development*, **3**, 157–175.

BAUER, B. (1980), Watershed characteristics in Austria, in DE BOOT, M., and GABRIELS, D. (Eds.), *Assessment of Erosion*, Wiley, Chichester, pp. 67–76.

BEZINGE, A., PERRETEN, J.-P., and SCHAFER, F. (1973), Phénomènes du lac glaciaire du Gorner, in *Symposium on the Hydrology of Glaciers (Proceedings of the Cambridge Symposium, 7–13 September 1969)*, International Association of Scientific Hydrology Publication 95, pp. 65–78.

BOULTON, G. S. (1975), Processes and patterns of subglacial sedimentation: a theoretical approach, in WRIGHT, A. E., and MOSELEY, F. (Eds.), *Ice ages: Ancient and Modern*, Seel House Press, Liverpool, pp. 7–42.

BROWN, A. G. (1985), The potential use of pollen in the identification of suspended sediment sources, *Earth Surface Processes and Landforms*, **10**, 27–32.

BRYANT, I. D. (1983), The utilization of Arctic river analogue studies in the interpretation of periglacial river sediments from southern Britain, in GREGORY, K. J. (Ed.), *Background to Palaeohydrology*, Wiley, Chichester, pp. 413–431.

CAINE, N. (1974), The geomorphic processes of the alpine environment, in IVES, J. D., and BARRY, R. G. (Eds.), *Arctic and Alpine Environments*, Methuen, London, pp. 721–748.

CAINE, N. (1976), A uniform measure of subaerial erosion, *Bulletin of the Geological Society of North America*, **87**, 137–40.

CAINE, N., and MOOL, P. K. (1982), Landslides in the Kolpu Khola Drainage, Middle Mountains, Nepal, *Mountain Research and Development*, **2**, 157–73.

CHURCH, M., and RYDER, J. M. (1972), Paraglacial sedimentation: a consideration of fluvial processes conditioned by glaciation, *Bulletin of the Geological Society of America*, **83**, 3059–3071.

CLARK, M. J., GURNELL, A. M., MILTON, E. J., SEPPÄLÄ, M., and KYÖSTILÄ, M. (1985), Remotely sensed vegetation classification as a snow depth indicator for hydrological analysis in sub-arctic Finland, *Fennia*, **163**, 195–225.

DAVIS, M. B. (1976), Erosion rates and land use history in Southern Michigan, *Environmental Conservation*, **3**, 139–148.

DOW, V., KIENHOLZ, H., PLAM, M., and IVES, J. D. (1981), Mountain hazards mapping: the development of a prototype combined hazards map, Monarch Lake Quadrangle, Colorado, USA, *Mountain Research and Development*, **1**, 55–64.

FRANK, T. D., and THORN, C. E. (1985), Stratifying alpine tundra for geomorphic studies using digitized aerial imagery, *Arctic and Alpine Research*, **17**, 197–188.

GAGE, M. (1970), The tempo of geomorphic change, *Journal of Geology*, **78**, 619–625.

GARDNER, J. S. (1982), Alpine mass-wasting in contemporary time, in THORN, C. E. (Ed.), *Space and Time in Geomorphology*, George Allen and Unwin, London, pp. 171–192.

GREGORY, K. J. (Ed.) (1983), *Background to Palaeohydrology*, Wiley, Chichester.

GURNELL, A. M. (1983), Downstream channel adjustments in response to water abstraction for hydro-electric power generation from alpine glacial melt-water streams, *Geographical Journal*, **149**, 342–354.

IVES, J. D. (1974), Arctic and alpine geomorphology — a review of current outlook and notable gaps in knowledge, in FAHEY, B. D., and THOMPSON, R. D. (Eds.), *Research in Polar and Alpine Geomorphology*, Geo Abstracts, Norwich, pp. 1–10.

IVES, J. D. (1985), Mountain environments, *Progress in Physical Geography*, **9**, 425–433.

KIENHOLZ, H. (1978), Maps of geomorphology and natural hazards of Grindelwald, Switzerland: scale 1:10000, *Arctic and Alpine Research*, **10**, 169–184.

KIENHOLZ, H., SCHNEIDER, G., BISCHEL, M., GRUNDER, M., and MOOL, P. (1984), Mapping of mountain hazards and slope stability, *Mountain Research and Development*, **4**, 247–266.

KLAGES, M. G., and HSIEH, Y. P. (1975), Suspended solids carried by the Gallatin River of southwestern Montana. II. Using mineralogy for inferring sources, *Journal of Environmental Quality*, **4**, 68–73.

KUUSISTO, E. (1984), *Snow Accumulation and Snowmelt in Finland*, Publications of the Water Research Institute, National Board of Waters, Finland, 55, 149pp.

LANGBEIN, W. B., and SCHUMM, S. A. (1958), Yield of sediment in relation to mean annual precipitation, *Transactions of the American Geophysical Union*, **39**, 1076–1084.

MAIZELS, J. K. (1983), Palaeovelocity and palaeodischarge determination for coarse gravel deposits, in GREGORY, K. J. (Ed.), *Background to Palaeohydrology*, Wiley, Chichester, pp. 101–140.

MCGREEVEY, J. P., and WHALLEY, W. B. (1982), The geomorphic significance of rock temperature variations in cold environments: a discussion, *Arctic and Alpine Research*, **14**, 157–162.

MESSERLI, B. (1983), Stability and instability of mountain ecosystems: introduction to the workshop, *Mountain Research and Development*, **3**, 81–94.

MESSERLI, B., and IVES, J. D. (1984), *Mountain Ecosystems: Stability and Instability*, International Mountain Society, Boulder.

OLDFIELD, F. (1977), Lakes and their drainage basins as units of sediment based ecological study, *Progress in Physical Geography*, **1**, 460–504.

OLDFIELD, F. (1983), The role of magnetic studies in palaeohydrology, in GREGORY, K. J. (Ed.), *Background to Palaeohydrology*, Wiley, Chichester, pp 141–166.

ØSTREM, G. (1975), Sediment transport in glacial meltwater streams, in JOPLING, A. V., and MACDONALD, B. C., (Eds.), *Glaciofluvial and Glaciolacustrine Sedimentation*, Society of Economic Palaeontologists and Mineralogists Special Publication 23, pp. 101–122.

PAINE, A. D. M. (1985), 'Ergodic' reasoning in geomorphology: time for a review of the term?, *Progress in Physical Geography*, **9**, 1–15.

PRICE, L. W. (1981), *Mountains and Man*, University of California Press, Berkeley.

RAPP, A. (1960), Recent development of mountain slopes in Kärkevagge and surroundings, northern Scandinavia, *Geografiska Annaler*, **42**, 65–200.

RAPP, A. (1983), Impact of nivation on steep slopes in Lappland and Scania, Sweden, *Nachrichten der Akademieder Wissenschaften im Göttingen*, **35**, 97–115.

RAPP, A. (1984), Nivation hollows and glacial cirques in Soderasen, Scania, south Sweden, *Geografiska Annaler*, **66A**, 11–28.

SCHUMM, S. A. (1965), Quaternary palaeohydrology, in WRIGHT, H. E., and FREY, D. G. (Eds.), *The Quaternary of the United States*, Princeton University Press, Princeton, pp. 783–794.

SCHUMM, S. A. (1977), *The Fluvial System*, Wiley, Chichester.

SMALL, R. J., BEECROFT, I. R., and STIRLING, D. M. (1984), Rates of deposition on lateral moraine embankments, Glacier de Tsidjiore Nouve, Valais, Switzerland, *Journal of Glaciology*, **30**, 275–81.

THORN, C. E., and HALL, K. (1980), Nivation: an arctic-alpine comparison and reappraisal, *Journal of Glaciology*, **25**, 109–124.

WALLING, D. E., and KLEO, A. H. A. (1979), Sediment yields of rivers in areas of low precipitation: a global view, in *The Hydrology of Areas of Low Precipitation (Proceedings of the Canberra Symposium, December 1979)*, International Association of Hydrological Sciences Publication 128, pp. 479–493.

WALLING, D. E., and WEBB, B. W. (1981), The reliability of suspended sediment load data, in *Erosion and Sediment Transport Measurement (Proceedings of the Florence Symposium, June 1981)*. International Association of Hydrological Sciences Publication 133, pp. 177–194.

WALLING, D. E., and WEBB, B. W. (1983), Patterns of sediment yield, in GREGORY, K. J. (Ed.) *Background to Palaeohydrology*, Wiley, Chichester, pp. 177–194.

WISCHMEIER, W. H., and SMITH, D. D. (1965), Predicting rainfall erosion east of the Rocky Mountains, in *United States Department of Agriculture Handbook*, 282.

WISCHMEIER, W. H., and SMITH, D. D. (1978), Predicting rainfall erosion losses — guidebook to conservation planning, in *United States Department of Agriculture Handbook*, 537.

Index